Epigenetics, Nuclear Organization and Gene Function

Epigenetics, Nuclear Organization and Gene Function

with implications of epigenetic regulation and genetic architecture for human development and health

John C. Lucchesi

Epigenetics, Nuclear Organization and Gene Function: with implications of epigenetic regulation and genetic architecture for human development and health. John C. Lucchesi, Oxford University Press (2019).
© John C. Lucchesi 2019. DOI: 10.1093/oso/9780198831204.001.0001

UNIVERSITY PRESS

Great Clarendon Street, Oxford, OX2 6DP,
United Kingdom

Oxford University Press is a department of the University of Oxford.
It furthers the University's objective of excellence in research, scholarship,
and education by publishing worldwide. Oxford is a registered trade mark of
Oxford University Press in the UK and in certain other countries

Published in the United States of America by Oxford University Press
198 Madison Avenue, New York, NY 10016, United States of America

British Library Cataloguing in Publication Data
Data available

Library of Congress Control Number: 2018949227

ISBN 978–0–19–883120–4 (hbk.)
ISBN 978–0–19–883121–1 (pbk.)

DOI: 10.1093/oso/9780198831204.001.0001

Printed and bound by
CPI Group (UK) Ltd, Croydon, CR0 4YY

Links to third party websites are provided by Oxford in good faith and
for information only. Oxford disclaims any responsibility for the materials
contained in any third party website referenced in this work.

Dedication and Acknowledgments

I wish to dedicate this text to the memory of Alan Wolffe whose book entitled *"Chromatin Structure and Function"* provided the heuristic foundation and the framework for modern research on gene function. I would also like to dedicate it to Mitzi Kuroda, David Allis, Jerry Manning, Jacques Coté and Eugene Koonin, five collaborators and friends, to Peter Becker, Tom Cline, Jeff Hansen, Steve Henikoff, John Lis, Brian Oliver, Craig Peterson, Danny Reinberg, Bryan Turner and Jerry Workman, all prized and admired scientific colleagues whose studies and publications have been an inspiration and stimulation for my own work.

I am deeply grateful to Gray Crouse, Arri Eisen, Joel Eissenberg, Peng Jin, David Katz, William Kelly, John McCarrey, Sergio Pimpinelli, Daniel Reines, Max Scott, Kevin Van Bortle and Paula Vertino for agreeing to read individual chapters and for making most useful comments and suggestions.

Finally, I would like to thank Victor Corces for his contribution to the development of the concept of this book.

About the Author

John Lucchesi obtained a PhD in Genetics from the University of California at Berkeley in 1963. Following two years of postdoctoral training in the Institute of Molecular Biology at the University of Oregon, he joined the faculty of the University of North Carolina in Chapel Hill, where he rose to the rank of Cary C. Boshamer Professor of Biology and Genetics. In 1979 he was appointed Adjunct Professor of Genetics, Duke University; in 1983, he was named Senior Fellow of Churchill College, Cambridge University, UK. In 1990, he joined the faculty of the Biology Department at Emory University as Asa G. Candler Professor of Biology and Chair. Dr Lucchesi has served on numerous National Institutes of Health panels including chairing the Genetics Study Section of the Division of Research Grants. He is a Fellow of the AAAS, a former President of the Genetics Society of America and was named Vice-President of the XVII International Congress of Genetics. His research laboratory, continuously supported by funding from the National Institute of General Medical Sciences, has focused on the regulation of transcription, the functional architecture of chromatin and the genetic regulation of development.

Preface

This book is not an attempt to review all of the knowledge accumulated to date in the fields of genetics and epigenetics. Its primary goal is to provide interested students, teachers and researchers with a framework that they can use to understand the basis of epigenetic regulation and to appreciate its derivation from genetics and its interdependence with genetic mechanisms. With rare exceptions, classical transmission genetics, chromosome mechanics and classical cytogenetics are not included. Rather, molecular genetic facts with particular relevance and impact on epigenetic observations and theory were selected for discussion. Another goal of the book is to highlight the role played by the three-dimensional organization of the genetic material, and its distribution within a functionally compartmentalized nucleus. This architectural organization of the genome plays a major role in the retrieval, interpretation and execution of genetic and epigenetic information. As a means of illustrating and emphasizing the evolution of the relationship among genetics, epigenetics and nuclear architecture, from its beginning to the present, wherever possible, references will be made to research articles that first described the new observations.

Approximately three decades ago, most geneticists felt that the mechanism of transcription that underlies all aspects of gene function was largely understood. After all, transcription could be reconstituted *in vitro* with purified factors and enzymes that were highly conserved in eukaryotes and therefore largely interchangeable. Yet, it was also eminently clear that although the fundamental aspects of the transcription process were necessary, they were woefully insufficient to explain the basis of cellular specialization and of organism development; the parameters that defined transcription did not explain the differential expression of genes in time and space during embryogenesis or in the course of adult life. The initial insights on this front had been provided by the observations of Vincent Allfrey and Alfred Mirsky that the chemical nature of the genetic material—the complex of DNA, histones

and non-histone proteins referred to as chromatin—was different in metabolically active cells. The manifestation of this difference, namely the acetylation of histones, indicated that in order to be activated, i.e. to allow the retrieval of genetic information, "a change in the fine structure of the chromatin and in the capacity of the DNA to serve as a template for RNA synthesis" must occur (Pogo et al., 1966). Almost a decade later, Arthur Riggs and Robin Holliday proposed that the methylation of DNA, known to occur in a wide variety of eukaryotic organisms, could be responsible for the repression of gene activity associated with one of the two X chromosomes in female mammals, by altering protein–DNA interactions (Holliday and Pugh, 1975; Riggs, 1975); this suggested the existence of a general mechanism that would be responsible for repressing genes that should remain inactive or that should be made inactive under particular circumstances of cellular differentiation. A difference in the physical organization of active and inactive genes was first demonstrated by Harold Weintraub and Mark Groudine who discovered that the DNA associated with active genes in purified nuclei is more sensitive to enzymatic digestion (Weintraub and Groudine, 1976).

The modern era of research into the chromatin modifications that would allow and regulate gene expression was introduced by separate discoveries, each illustrating a fundamental class of regulatory mechanisms. Craig Peterson and Ira Herskowitz published the purification of a multiprotein complex of yeast that appeared to perform a general function in the transcription of many genes by assisting gene-specific activators (Peterson and Herskowitz, 1992). A few years later, James Brownell and David Allis isolated the first histone acetyl transferase from the highly transcriptionally active macronucleus of *Tetrahymena* and subsequently showed that it was the ortholog of a transcription co-factor long known to be necessary for the activation of numerous genes in yeast (Brownell and Allis, 1995; Brownell et al., 1996). Sealing the dynamic role of histone acetylation in the regulation of gene function was the isolation of the first histone deacetylase by Stewart Schriber (Tauton et al., 1996). These discoveries marked the onset of a major research effort to identify additional covalent modifications of histones and the enzymes responsible for inducing or eliminating these modifications. In parallel, new multiprotein complexes that remodel the conformational state of chromatin were characterized. Such endeavors were aided immeasurably by bioinformatics approaches that revealed the extensive structural and functional conservation of these chromatin-modifying factors. With the advent of microarray technology initially, and subsequently of next-generation DNA sequencing, recent emphasis has been placed on describing global epigenetic patterns across whole chromosomes or entire genomes, with often surprising and illuminating results.

A fundamental insight into the relative position of epigenetic modifications in the overall process of gene function was provided by the discovery of sequence-specific transcription activators—the so-called **pioneer factors**—that are able to access their binding sites in compacted chromatin (Cirillo et al., 2002; Natarajan et al., 1999; Neely et al., 1999). In turn, these factors recruit the modifying complexes responsible for the epigenetic modifications that are associated with gene activation or repression. These discoveries have led to the formulation of some general principles governing the modulation of gene function—opportunistic transcription factors may insert themselves into the genome, attract other factors and initiate covalent modifications of histones that alter inter-nucleosomal interactions, or serve as platforms for transcription activators and repressors; protein complexes change the architectural state of chromatin by displacing nucleosomes or by modifying their association with DNA. The mechanistic basis responsible for these dynamic structural changes in the organization of chromatin is just now beginning to be addressed through a burgeoning effort to redefine them at the biophysical and structural levels.

The eukaryotic nucleus is a highly complex organelle that contains, in addition to the genetic information of the organism, a myriad of distinct subnuclear regions (compartments) and structures. It is not surprising that the orderly reading of the genetic blueprint for cellular maintenance and differentiation engages and affects many aspects of nuclear organization, and that the different nuclear processes involved in gene expression occur in particular nuclear locations. The first suggestion of this spatial dimension was the apparent differential localization of active and inactive regions of the genome, with the former usually more centrally located in nuclei, and the latter usually associated with the nuclear periphery—the so-called lamina. These observations were soon followed by the discovery that the activity of some genes was correlated with their association with nuclear pore complexes or with their presence, together with other genes, in intra-nuclear foci rich in all of the elements of the transcriptional machinery that were named transcriptional factories. Recently, some evidence is accumulating that PML bodies (named after the abnormal distribution of a nuclear protein in promyelocytic leukemia patients) may play a role in transcription and in the response to DNA damage. Nuclear speckles (a.k.a. inter-chromatin granule clusters) are enriched in RNA-splicing factors and seem to associate with active genes.

Clearly, the nuclear address of genes must depend on interactions between specific characteristics of the chromatin wherein they reside and on the properties of the organelles involved. A particularly well-studied case in point is the clustering of insulator sites to organize the genome into cell-specific, large chromatin domains thought to be important in establishing specific programs of gene expression.

It is abundantly evident that absence or malfunction of developmentally regulated transcription factors and errors in histone modifications or remodeling are associated with human diseases, including cancers, and that many cancers exhibit various levels of nuclear disorganization. One of the purposes of this book is to highlight how transcriptional regulation and the concomitant epigenetic modifications are integrated in, and impacted by, the parameters that relate to the architectural organization of nuclei. Beyond contributing to understanding basic cellular differentiation and development, this synthesis should suggest new avenues for therapeutic interventions and lead to more focused approaches to redirect stem cell differentiation for the purpose of regenerative medicine.

To achieve the aim embodied in the title of the book requires a substantial level of selectivity. To this end, fundamental aspects of molecular genetic and epigenetic regulation and of nuclear organization are reviewed in the context of higher metazoans, with an emphasis on mammals in general, and on humans in particular. The same approach is used for a general description of nuclear structures. Examples drawn from unicellular organisms and plants are used when particular regulatory parameters have been studied uniquely in these forms.

References

Brownell, J. E. & Allis, C. D. 1995. An activity gel assay detects a single, catalytically active histone acetyltransferase subunit in *Tetrahymena* macronuclei. *Proc Natl Acad Sci U S A,* 92, 6364–8.

Brownell, J. E., Zhou, J., Ranalli, T., Kobayashi, R., Edmondson, D. G., Roth, S. Y. & Allis, C. D. 1996. *Tetrahymena* histone acetyltransferase A: a homolog to yeast Gcn5p linking histone acetylation to gene activation. *Cell,* 84, 843–51.

Cirillo, L. A., Lin, F. R., Cuesta, I., Friedman, D., Jarnik, M. & Zaret, K. S. 2002. Opening of compacted chromatin by early developmental transcription factors HNF3 (FoxA) and GATA-4. *Mol Cell,* 9, 279–89.

Holliday, R. & Pugh, J. E. 1975. DNA modification mechanisms and gene activity during development. *Science,* 187, 226–32.

Natarajan, K., Jackson, B. M., Zhou, H., Winston, F. & Hinnebusch, A. G. 1999. Transcriptional activation by Gcn4p involves independent interactions with the SWI/SNF complex and the SRB/mediator. *Mol Cell,* 4, 657–64.

Neely, K. E., Hassan, A. H., Wallberg, A. E., Steger, D. J., Cairns, B. R., Wright, A. P. & Workman, J. L. 1999. Activation domain-mediated targeting of the SWI/SNF complex to promoters stimulates transcription from nucleosome arrays. *Mol Cell,* 4, 649–55.

Peterson, C. L. & Herskowitz, I. 1992. Characterization of the yeast *SWI1*, *SWI2*, and *SWI3* genes, which encode a global activator of transcription. *Cell,* 68, 573–83.

Pogo, B. G., Allfrey, V. G. & Mirsky, A. E. 1966. RNA synthesis and histone acetylation during the course of gene activation in lymphocytes. *Proc Natl Acad Sci U S A,* 55, 805–12.

Riggs, A. D. 1975. X inactivation, differentiation, and DNA methylation. *Cytogenet Cell Genet,* 14, 9–25.

Taunton, J., Hassig, C. A. & Schreiber, S. L. 1996. A mammalian histone deacetylase related to the yeast transcriptional regulator Rpd3p. *Science,* 272, 408–11.

Weintraub, H. & Groudine, M. 1976. Chromosomal subunits in active genes have an altered conformation. *Science,* 193, 848–56.

Contents

PART I

An Introduction to Epigenetics and Epigenetic Regulation

CHAPTER 1

Epigenetic phenomena in fungi, plants and animals

The fundamental questions in Biology are how cells differentiate and how organisms develop and adapt to their environment. In all unicellular or multicellular organisms, the information necessary to execute cellular differentiation and the morphological changes associated with development, to perform all physiological reactions and for all behavioral pathways is encoded in the genetic material. The nature, function and transmission of this information have been extensively investigated over the past century, beginning with the inheritance of simple traits, the molecular and physical structure of DNA, the discovery of the genetic code and, in modern times, the genomic and gene product sequencing of organisms. The intellectual euphoria generated by this experimental progression has been occasionally challenged by certain phenomena that could not be explained easily in genetic terms. These instances constitute examples of **epigenetic**, rather than genetic, regulation. The term was first used by the British developmental biologist Conrad Waddington to describe all of the external influences that collaborate with the genetic blueprint to achieve animal development (Waddington, 1942). Waddington further proposed that developmental changes induced by environmental conditions could, in certain circumstances, become heritable.

One of the first examples of epigenetic inheritance was described in corn by Alexander Brink (Brink et al., 1968). The wild type alleles of several genes in the anthocyanin synthesis pathway involved in pigment production were observed to undergo **paramutations**—heritable changes in their level of expression caused by their association with particular mutant alleles said to be **paramutagenic**. In the bread mold *Neurospora*, several instances of mutant alleles affecting the function of the wild type allele in the diploid vegetative phase were noted. Association with a mutant allele affected the wild type allele's expression following several mitotic divisions, although the two alleles had separated long before in meiosis. Many of these cases were explained by the realization that, in order to be properly expressed, an allele had to physically pair, even with a non-functional allele of the same gene, as long as their sequences were sufficiently similar. Failure to activate the wild type allele resulted if the mutant allele was absent or highly deleted (Aramayo and Metzenberg, 1996).

Not surprisingly, Waddington himself generated experimental data that led him to define the concept of **epigenesis**. In a classic letter to *Nature*, he described the results of selecting, over several generations, flies with a particular wing modification induced by heat-shock treatment during development. Following its appearance in the fly cultures, the abnormal phenotype persisted in high frequency for many generations in the absence of any further treatment (Waddington, 1952).

The unexpected "paragenetic" manifestations just described were the harbingers of a universal level of regulation of critical importance to cellular differentiation and organismal development.

Selective gene activity is the basis of cellular and tissue differentiation and of organismal development

All cells in an organism have the same genetic information. Yet, every different cell type is characterized by a specific pattern of gene expression that is transmitted to its daughter cells.

Epigenetics, Nuclear Organization and Gene Function: with implications of epigenetic regulation and genetic architecture for human development and health. John C. Lucchesi, Oxford University Press (2019). © John C. Lucchesi 2019. DOI: 10.1093/oso/9780198831204.001.0001

This differential gene expression consists of activating particular genes and maintaining them in an active state, and repressing particular genes and maintaining them in a repressed state in certain regions of a developing embryo or in certain tissues of the adult. Unicellular organisms, as well, rely on differential gene expression: for example, yeast cells belong to either of two mating types (a and α), each type expressing a different subset of genes.

Differential gene expression is achieved by the selective presence of transcription factors and by a level of regulation interposed between the genotype and the phenotype—**epigenetic regulation**. Epigenetic regulation modulates gene function through modifications that do not involve coding sequence changes. Since it contributes to the selective expression of the genotype, this level of regulation is of fundamental importance to all aspects of cell differentiation and to all aspects of development, physiology and behavior. The existence of epigenetic regulation is made possible by the organization of the genetic material. In cells, DNA is not present in pure form—it is found intimately associated with histones and an extensive variety of other proteins, mostly with regulatory functions, forming a complex referred to as **chromatin**. Some of these regulatory proteins are common to the chromatin of various cell types but may be modified differently in different cells; others are unique to particular cell types.

Another set of considerations that suggest the necessity of a regulatory level for the retrieval of genetic information to achieve the phenotype is the similarity in the size of the genomes of organisms that differ vastly, especially in level of complexity. For example, the genome of the round worm *Caenorhabditis elegans* and our own genome are roughly equivalent— approximately 20,000 protein-coding genes. Yet it is obvious that the complexity of human development, the multiplicity of adult tissue specializations, the vast array of physiological attributes and the complexity of behavior far exceed those of the worm. The immensely more diversified genetic output that we exhibit is largely due to the higher degree of differential gene expression and of differential gene product modifications with which we are endowed (Bulger and Groudine, 2011).

The concept of one genome but multiple epigenomes

As expected, the epigenetic parameters that reflect the activation of a particular set of genes and the inactivation of others in one type of cells are associated with different sets of genes in different cells. The selective expression of the genome accompanied by the concomitant epigenetic modifications that is uniquely characteristic of a cell type is its **epigenome**. Since complex multicellular organisms display numerous cell types, they can be said to have a single genome and multiple epigenomes. Given that the genetic information is identical in all the cells of an embryo that is becoming multicellular by mitotic division, epigenomes must arise through the influence of a wide range of intrinsic and external factors that affect regions of the developing embryo in specific ways.

During the course of oogenesis, as the cytoplasm of the future egg is being synthesized, various maternally expressed morphogens (Turing, 1952) are laid down in a progressive manner (Wolpert, 1969). The undifferentiated cells of the developing embryo respond to differences in concentration along the morphogen's gradient and activate various genes that begin to render them distinct from neighboring cells. A classic example of this differential inception of epigenomes is provided by the Bicoid morphogen responsible for establishing the anterior–posterior embryonic axis in *Drosophila* (Grimm et al., 2010; Porcher et al., 2010). Numerous other morphogens have been identified in a variety of organisms, including, of course, vertebrates (Tabata and Takei, 2004). Following these earliest responses, cellular identities can be further individualized by intercellular signals. These signals consist of molecules produced by certain cells that interact with specific receptors on the surface of other cells. Conformational changes in the receptors initiate a cascade of molecular interactions, leading to the activation of specific genes. The simultaneous presence of various types of receptors with distinct specificities on different groups of cells results in multiple networks of gene regulation. Additional patterns of differential gene activity are generated by intercellular signals that involve the direct interaction of cells with their neighbors as well as with the substrate, and that are responsible for the collective movements of cells during morphogenesis.

Extra-embryonic factors and stimuli can have long-lasting and often transmissible effects on the patterns of gene activation during embryonic development. Experiments performed with rats have shown that stress can alter behavior by inducing the epigenetic repression of specific genes involved in neurotransmission (Youngson and Whitelaw, 2008). It is increasingly clear that the nutritional milieu of developing embryos can impact the metabolism of the resulting adults, later in life—maternal obesity enhances the probability of gestational diabetes and infant obesity (Alfaradhi and Ozanne, 2011).

Epigenetic memory

Regional differentiation during the course of development, and tissue maintenance during adult life require countless somatic cell divisions. In all of these divisions, the genetic material of the parent cell is duplicated and transmitted to the two daughter cells. In the majority of these divisions, the epigenetic information that is responsible for the functional specialization of the parent cell is also transmitted to the daughter cells. A clear example is provided by a particular cell type of the acquired immune system. T lymphocytes that have been exposed to constituents of a pathogen, such as a virus or a bacterium, become activated and are induced to proliferate very rapidly in order to fight the invasion. Once the infection is cleared, the majority of these T lymphocytes die; those few that persist are called memory cells. They become transcriptionally less active and divide at a slower rate. Yet these lymphocytes retain a memory of their exposure to the pathogen— they respond much more rapidly to a re-infection than the first T lymphocytes did after the initial exposure, and they activate all of the genes necessary to restore their identity and rapid division rate of activated T cells.

Clearly, the transmission of epigenetic regulatory elements is required for the maintenance of the differential gene activity specific to a particular cell type. Yet it must overcome the changes that occur during the transition from interphase to mitosis—a substantial abrogation of gene activity as well as a drastic alteration in the structure of chromosomes—that have, as their primary purpose, the faithful transmission of the genome.

Epigenetics and the human condition

Understanding the basis of epigenetic regulation will provide the key to a variety of complex phenomena such as the biology of aging, cancer, acquired immunity, cognition, behavior and psychiatric disorders. Aging is a highly multifaceted process that affects the majority of physiological pathways in single-cell organisms and in multicellular plants and animals. At the basis of age-related changes are the accumulation of somatic mutations (Wallace, 2010) and a very substantial progressive alteration of the epigenetic landscape and nuclear organization (Shin et al., 2011). Epigenetic changes can cause the activation or over-expression of proto-oncogenes or the silencing of tumor suppressor genes,

potentially leading, in both cases, to the onset of cancer (Chi et al., 2010). Cell differentiation and neurogenesis are highly regulated at the level of chromatin modifications (Barber and Rastegar, 2010). Epigenetic regulation is involved in memory formation and storage (Miller et al., 2008), in cognitive diseases such as Rett syndrome or Huntington's disease (Urdinguio et al., 2009) and in psychiatric disorders such as cocaine addiction (Zhou et al., 2014). All aspects of acquired immunity involve chromatin modifications (Green et al., 2006; Jhunjhunwala et al., 2009; Placek et al., 2009).

Understanding epigenetic regulation will provide the tools necessary to implement the therapeutic promise of stem cell differentiation. A number of genes that are required for the differentiation of embryonic stem cells are enriched in histone modifications that are usually associated with active transcription and with repression. The presence of these **bivalent** modifications is thought to poise the genes of stem cells for fast response to extra-cellular morphogens (Landera et al., 2010). A better understanding of the role of these modifications and of epigenetic regulatory mechanisms will have a profound impact on the use of stem cells in regenerative medicine.

In conclusion, epigenetic modifications represent an essential regulatory function for cellular differentiation and organism development by modifying the structure and organization of chromatin and modulating gene expression. The following chapters in Part I describe the basic structure of chromatin and review the basic mechanism of gene transcription.

Chapter summary

The information for all of the physical and physiological characteristics of organisms, from their inception and throughout their life cycle, is encoded in the DNA. DNA does not exist in pure form and is associated with a variety of proteins to form a complex referred to as chromatin. In its initial ground state, the genome is generally repressed. In multicellular organisms, the individualization of cell types during development requires the differential activation of genes, a process that is initiated by the uneven distribution of transcription factors in the fertilized egg. The establishment and maintenance of the transcription of selected genes involve modifications of the proteins that are associated with the DNA and a remodeling of this association. Since these alterations that affect gene activity do not modify the DNA sequence proper, they are referred to as epigenetic modifications. Given their differential gene activity, cell types are characterized by different distributions of epigenetic modifications and, in this sense, can be said to exhibit their own particular epigenome. As cells divide and give rise to different tissues, they must transmit to their daughter cells the information necessary to maintain their particular distribution of active genes. This epigenetic memory may involve the transmission of transcription factors as well as the transmission of epigenetic modifications present in the parent cell's chromatin.

References

Alfaradhi, M. Z. & Ozanne, S. E. 2011. Developmental programming in response to maternal overnutrition. *Front Genet*, 2, 27.

Aramayo, R. & Metzenberg, R. L. 1996. Meiotic transvection in fungi. *Cell*, 86, 103–13.

Barber, B. A. & Rastegar, M. 2010. Epigenetic control of *Hox* genes during neurogenesis, development, and disease. *Ann Anat*, 192, 261–74.

Brink, R. A., Styles, E. D. & Axtell, J. D. 1968. Paramutation: directed genetic change. Paramutation occurs in somatic cells and heritably alters the functional state of a locus. *Science*, 159, 161–70.

Bulger, M. & Groudine, M. 2011. Functional and mechanistic diversity of distal transcription enhancers. *Cell*, 144, 327–39.

Chi, P., Allis, C. D. & Wang, G. G. 2010. Covalent histone modifications—miswritten, misinterpreted and mis-erased in human cancers. *Nat Rev Cancer*, 10, 457–69.

Green, M. R., Yoon, H. & Boss, J. M. 2006. Epigenetic regulation during B cell differentiation controls CIITA promoter accessibility. *J Immunol*, 177, 3865–73.

Grimm, O., Coppey, M. & Wieschaus, E. 2010. Modelling the Bicoid gradient. *Development*, 137, 2253–64.

Jhunjhunwala, S., Van Zelm, M. C., Peak, M. M. & Murre, C. 2009. Chromatin architecture and the generation of antigen receptor diversity. *Cell*, 138, 435–48.

Landeira, D., Sauer, S., Poot, R., Dvorkina, M., Mazzarella, L., Jorgensen, H. F., Pereira, C. F., Leleu, M., Piccolo, F. M., Spivakov, M., Brookes, E., Pombo, A., Fisher, C., Skarnes, W. C., Snoek, T., Bezstarosti, K., Demmers, J., Klose, R. J., Casanova, M., Tavares, L., Brockdorff, N., Merkenschlager, M. & Fisher, A. G. 2010. Jarid2 is a PRC2 component in embryonic stem cells required for multi-lineage differentiation and recruitment of PRC1 and RNA Polymerase II to developmental regulators. *Nat Cell Biol*, 12, 618–24.

Miller, C. A., Campbell, S. L. & Sweatt, J. D. 2008. DNA methylation and histone acetylation work in concert to regulate memory formation and synaptic plasticity. *Neurobiol Learn Mem*, 89, 599–603.

PLacek, K., Coffre, M., Maiella, S., Bianchi, E. & Rogge, L. 2009. Genetic and epigenetic networks controlling T helper 1 cell differentiation. *Immunology*, 127, 155–62.

Porcher, A., Abu-Arish, A., Huart, S., Roelens, B., Fradin, C. & Dostatni, N. 2010. The time to measure positional information: maternal hunchback is required for the synchrony of the Bicoid transcriptional response at the onset of zygotic transcription. *Development*, 137, 2795–804.

Shin, D. M., Kucia, M. & Ratajczak, M. Z. 2011. Nuclear and chromatin reorganization during cell senescence and aging—a mini-review. *Gerontology*, 57, 76–84.

Tabata, T. & Takei, Y. 2004. Morphogens, their identification and regulation. *Development*, 131, 703–12.

Turing, A. M. 1952. The chemical basis of morphogenesis. *Phil Trans R Soc Lond B*, 237, 37–72.

Urdinguio, R. G., Sanchez-Mut, J. V. & Esteller, M. 2009. Epigenetic mechanisms in neurological diseases: genes, syndromes, and therapies. *Lancet Neurol*, 8, 1056–72.

Waddington, C. H. 1942. The epigenotype. *Endeavour*, 1, 18–20.

Waddington, C. H. 1952. Selection of the genetic basis for an acquired character. *Nature*, 169, 278.

Wallace, D. C. 2010. Mitochondrial DNA mutations in disease and aging. *Environ Mol Mutagen*, 51, 440–50.

Wolpert, L. 1969. Positional information and the spatial pattern of cellular differentiation. *J Theor Biol*, 25, 1–47.

Youngson, N. A. & Whitelaw, E. 2008. Transgenerational epigenetic effects. *Annu Rev Genomics Hum Genet*, 9, 233–57.

Zhou, Z., Enoch, M. A. & Goldman, D. 2014. Gene expression in the addicted brain. *Int Rev Neurobiol*, 116, 251–73.

CHAPTER 2

The basic structure of chromatin

Organization of DNA and histones in the nucleosome

In the nuclei of cells, DNA is present in a complex that is called chromatin. Chromatin contains DNA, RNA and two general types of proteins: histones and nonhistones. The basic and universal unit of chromatin is the nucleosome consisting of a segment of DNA wrapped around an octamer of four different histones: H2A, H2B, H3 and H4. These are considered "core" histones (sometimes referred to as "canonical") because they are assembled into nucleosomes that associate with newly replicated DNA. Core histones are highly conserved, and H3 and H4 are among the most conserved proteins in all eukaryotes. Variants that differ in sequence from some of these histones are incorporated into nucleosomes when specific nuclear functions, such as transcription or the repair of DNA damage, are carried out. In multicellular eukaryotes, histone genes can be classified into subtypes on the basis of their expression pattern and genomic organization. Core histone genes are expressed only during the S phase of the cell cycle; these genes do not contain introns and are organized as clusters usually present in multiple copies. These histones form the nucleosomes that associate with the newly replicated DNA and are referred to as replication-coupled (RC) histones. The genes that encode histone variants are present in single or few copies in the genome and are expressed at constant, but low, levels throughout the cell cycle and in differentiated cells. Histone variants are said to be replication-independent (RI). Some histones are expressed in a tissue-specific manner. During spermatogenesis, most histones are replaced by protamines, but a small percentage of nucleosomes remain and some contain testis-specific histone variants. These sperm nucleosomes are thought to have a specific epigenetic function because they are significantly enriched around developmentally important genes, such as the imprinted gene clusters, micro RNA clusters and *HOX* gene clusters discussed in subsequent chapters. All core histones and their variants contain the so-called histone fold at their C-terminal end (three α-helices linked by two loops) and exhibit extended N-terminal tails. Histone H2A also has a C-terminal tail domain (Fig. 2.1).

The histone folds associate with each other to produce H2A–H2B hetero-dimers and H3–H4 hetero-tetramers. Core octamers consist of two molecules of H2A, H2B, H3 and H4 (Fig. 2.2); they can be assembled *in vitro* under conditions of high salt and high protein concentrations by slowly dialyzing away the salt. H3–H4 tetramers can associate with DNA *in vitro*. If H2A and H2B are added, they are recruited to the H3–H4 tetramers.

A fifth type of histone called the linker histone (H1, H5, etc., depending on the particular organism) associates with DNA as it begins and ends its wrapping around the nucleosome (Fig. 2.3). Linker histones show a lot of evolutionary variability, with most organisms expressing several variants in different cell types and developmental stages. Most linker histones possess a short N-terminal sequence, a central globular domain with DNA-binding sites and an extended basic C-terminal tail.

The conclusion that chromatin is organized as a chain of nucleosomes was based on the following observations. When subjected to electrophoresis, cytoplasmic DNA, presumably resulting from nuclear breakdown following cell death, exhibited a series of evenly-spaced bands (Williamson, 1970). Partial digestion of chromatin by endogenous endonucleases or

Epigenetics, Nuclear Organization and Gene Function: with implications of epigenetic regulation and genetic architecture for human development and health. John C. Lucchesi, Oxford University Press (2019). © John C. Lucchesi 2019.
DOI: 10.1093/oso/9780198831204.001.0001

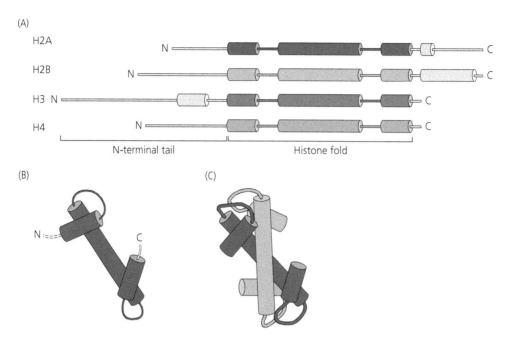

Fig. 2.1 Structure of the four core histones. (A) The regions that undergo the histone fold are indicated, as well as the N-terminal tails where most of the covalent modifications occur. (B) Illustration of the histone fold. (C) Histones H2A and H2B form a dimer; H3 and H4 have the same interaction.

(From *Molecular Biology of the Cell*, 4th Edition)

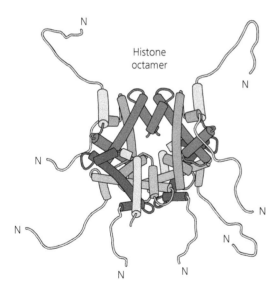

Fig. 2.2 Diagram illustrating the structure of a histone octamer.

(From *Molecular Biology of the Cell*, 4th Edition)

with bacterial enzymes, such as micrococcal nuclease (MNase), yielded an array of DNA fragments representing multiples of a 200 base pair (bp) unit (Hewish and Burgoyne, 1973) (Fig. 2.4). More extensive digestion

with MNase led to DNA fragments of approximately 145 bp. Initially, MNase cleaves the DNA present between nucleosomes and accessible to the enzyme; as digestion proceeds, the remainder of this DNA is degraded, leaving segments of DNA protected from further digestion by their intimate association with histones. Digestion with deoxyribonuclease I (DNase I), an enzyme that induces single-strand cuts, indicated that sensitive sites occurred every 10 bp along the nucleosome-bound DNA (Noll, 1974; Simpson and Whitlock, 1976). Using hydroxyl radical treatment that cleaves the sugar–phosphate backbone, the periodicity of the DNA double helix was found to vary along the nucleosome surface: 10.7 bp per turn in the central region and 10.0 bp per turn in the flanking region (Hayes et al., 1990; Pugl and Behe, 1993). These observations suggested that the DNA is not wrapped uniformly around the histone octamers, a characteristic that may have important implications with respect to the accessibility of nucleosomal DNA to transcriptional factors.

Reconstituted polymers, containing the four core histones and DNA, produced an X-ray diffraction pattern similar to that of native chromatin (Kornberg, 1974; Kornberg and Thomas, 1974). X-ray diffraction studies at 20 Å resolution revealed a disk-shaped structure of

Fig. 2.3 Diagrams illustrating the interaction of a linker histone molecule (green in both illustrations) with a nucleosome and with linker DNA. (From Caterino and Hayes, 2011.)

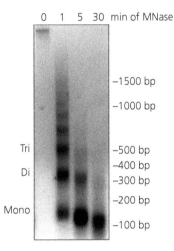

Fig. 2.4 Organization of chromatin. Digestion of chromatin with micrococcal nuclease (MNase) for increasing times; the 1-minute enzyme digest yields a ladder with DNA "rungs" representing mono-, di-, tri-, etc. nucleosomes. The size in base pairs (bp) of the DNA fragments is indicated on the right.

(From *Bio-Protocol*, vol7, Iss 6, 2017.)

approximately 110 Å in diameter and 57 Å in thickness, around which the DNA makes 1.75 turns (Finch et al., 1977). Subsequent studies at 2.8 Å resolution showed that, on average, DNA makes 1.65 turns around the nucleosome, resulting in 147 bp contacting the histones of the octamer at numerous different points in the form of charge–dipole interactions and hydrogen bonds (Luger et al., 1997) (Fig. 2.5). The DNA between two adjacent nucleosomes—linker DNA—has a length that can range from 20 to 90 bp, depending on the particular species or the particular tissue involved within a species.

Assembly of the chromatin fiber

The major cellular processes that involve DNA, replication, damage repair and transcription require chromatin disruption and some form of nucleosome removal or loss. The re-establishment of chromatin structure is mediated by histone **chaperones**. *In vivo*, histones are extensively acetylated before assembly into nucleosomes,

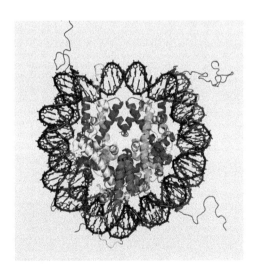

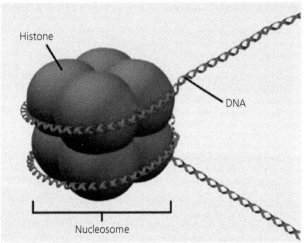

Fig. 2.5 The diagram on the left represents the crystal structure of the nucleosome; the four core histones are represented in color. (From Wikipedia.) The diagrams on the right show the path of the DNA (red ribbon) around the histone octamer.

(From LibreTexts, Biology).

presumably to decrease DNA–protein contacts and increase protein–protein contacts. Nucleosomes connected by linker DNA form a structure referred to as "beads on a string" that is 10 nm in diameter. During DNA replication (S phase), parental histones are segregated to the two nascent DNA molecules in random fashion and new core histones are synthesized in order to produce a full nucleosome complement. Old and new H3–H4 tetramers are deposited first and recruit old or new H2A–H2B dimers. This process requires the action of histone chaperone proteins. Chaperones prevent the highly basic histones from making random interactions with DNA and other negatively charged macromolecules prior to assembly into nucleosomes. In cells, H3–H4 tetramers are found associated with the chromatin assembly factor 1 (CAF-1). This complex is targeted to the newly synthesized DNA duplexes at the replication fork through CAF-1's binding to one of the components of the replication complex—proliferating cell nuclear antigen (PCNA)—that forms a "clamp" around DNA and ensures the processivity of the DNA polymerase complex. Another chaperone, antisilencing factor 1 (Asf1), associates with H3–H4 dimers and is thought to stimulate the activity of CAF-1 by donating its H3–H4 load to allow the deposition of tetramers. The H2A–H2B heterodimers are bound to nucleoplasmin or nucleosome assembly protein 1 (Nap1) chaperones. As will be elaborated in subsequent chapters, histones also play important roles in processes other than replication such as transcription, DNA damage repair, heterochromatin maintenance and control of the cell cycle.

The positioning of nucleosomes along a chromatin fiber is influenced, to some extent, by the DNA sequence but is mostly determined by the action of **remodeling** factors that have generally distinct effects. Some DNA sequences represent high-affinity sites for the deposition of nucleosomes and have been used extensively to generate nucleosomal arrays of particular periodicities *in vitro*. One of these sequences was extracted from the 5S ribosomal RNA gene of chicken erythrocytes (Simpson and Stafford, 1983); another was obtained by multiple rounds of selection from a large pool of randomly synthesized DNA fragments and is called the 601 positioning sequence (Lowary and Widom, 1998). Other sequences, such as poly(dA:dT) tracts, appear to occur frequently in regions of DNA where the nucleosome presence is at low levels.

A number of multiprotein complexes, conserved from yeast to humans, function in cells to position nucleosomes in regular arrays. Complexes such as ACF (assembly of core histones factor), CHRAC (chromatin accessibility complex), NURF (nucleosome remodeling factor) and RSF (remodeling and spacing factor) use the energy of ATP hydrolysis released by a common ATPase subunit—ISWI (imitation switch independent)—to catalyze the formation of regularly spaced nucleosomes that may result in greater compaction. Some of these complexes may also affect gene transcription by varying the distance between nucleosomes (see Chapter 4). Under physiological conditions, the 10-nm chromatin fiber condenses into a secondary structure of 30 nm in diameter (Kruithof et al., 2009) that is widely accepted as the natural state of chromatin in nuclei (Fig. 2.6). This fiber is stabilized by the presence of linker histones. There is approximately one molecule of linker histone per nucleosome in cells. In addition

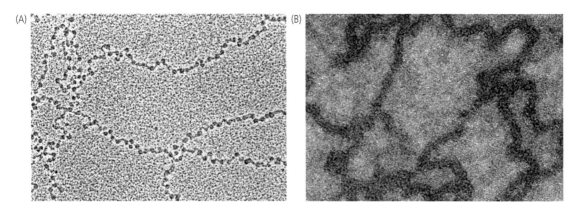

Fig. 2.6 Electron photomicrographs illustrating (A) the uncoiled 10-nm fiber, sometimes referred to as "beads-on-a-string," and (B) the 30-nm chromatin fiber.

(From *Molecular Cell Biology*, 7th Edition.)

to their roles as architectural proteins, linker histones interact with a variety of non-histone proteins.

Transitions of the chromatin fiber into different states of condensation

Several sources of experimental data lead to the conclusion that the condensation from the 10-nm to the 30-nm fiber and beyond depends on charge neutralization, and that condensed chromatin is stabilized by the presence of linker histones. One experimental paradigm that has been extensively used consists of chromatin fibers reconstituted *in vitro* on a DNA molecule that contains a nucleosome-positioning sequence (5S or 601) repeated at regular intervals. The conformation of this fiber under different conditions can be measured by determining its sedimentation coefficient during centrifugation (this parameter is a function of molecular mass and shape; since mass is constant, changes in sedimentation reflect changes in shape). As the salt concentration is increased, the sedimentation coefficient of the chromatin fibers is seen to increase, reflecting a tertiary structure exceeding the level of compaction of 30-nm fibers (Puhl and Behe, 1993). This observation is consistent with the current view of the topological organization of chromatin in the nucleus (see Chapter 10). Similar effects are obtained by increasing the concentration of the divalent cation magnesium (Mg^{++}) (Schwarz and Hansen, 1994). Chromatin fibers reconstituted with histones that lacked their N-terminal tails can be compacted by the addition of Mg^{++}, but not to the extent achieved with normal histones, suggesting that the tails contribute a unique structural input to compaction (Fletcher and Hansen, 1995). Histone tails contain a significant number of basic residues such as lysines (K) or

arginines (R) that confer a negative charge; some of these tails are engaged in interactions between nucleosomes: one of the two N-terminal tails of histone H4 present in one nucleosome associates with a patch formed by the acidic side chains of the globular portion of H2A and H2B on an adjacent nucleosome (Luger et al., 1997). The contribution of this interaction to fiber condensation was predicted by the observation that chromatin fibers assembled with acetylated nucleosomes were unable to undergo the Mg^{++}-induced supercompaction achieved with unmodified nucleosomes (Tse et al., 1998). The various folded forms assumed by chromatin fibers with increasing divalent cation concentrations, including the 30-nm fiber, are not stable. Stability requires not only inorganic cations, but also the presence of linker histones (Carruthers et al., 1998). Linker histones allow chromatin fibers to self-associate at much lower salt concentrations and to form stable 30-nm structures (McBryant et al., 2010). These different observations indicate that the structure of chromatin can be modified by altering the electrostatic status of its components.

A very different approach has been used to study the mechanical properties of 10-nm and 30-nm chromatin fibers. A nucleosomal array is assembled with purified histone octamers on a DNA molecule containing tandem repeats of a nucleosome-positioning element; the DNA is labeled at one end with biotin and at the other with digoxigenin. One end of the array is attached to a coverslip coated with anti-digoxigenin and the other end to an avidin-coated magnetic bead or polystyrene microsphere that binds biotin. Force is exerted by holding the bead with magnetic tweezers or the microsphere with optical tweezers, and moving the coverslip to stretch the nucleosomal DNA. Using this type of assay, the force necessary to disrupt individual nucleosomes (Fig. 2.7)

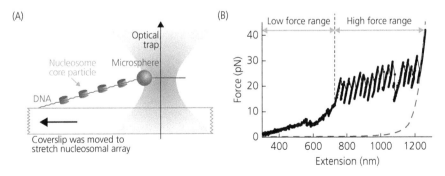

Fig. 2.7 Single-molecule mechanical manipulations of chromatin fibers. (A) A nucleosomal array attached to a microsphere, held by optical tweezers, and to a coverslip can be stretched by moving the coverslip. (B) Force–extension curve illustrating the sequential release of nucleosomes; each peak represents the release of a single nucleosome.

(From Brower Toland et al., 2002.)

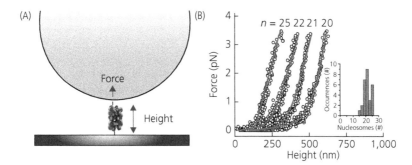

Fig. 2.8   Analysis of chromatin fibers by force spectroscopy. (A) The extension of short 30-nm fibers reconstituted *in vitro* can be measured by tethering them between a glass surface and a paramagnetic bead that can be moved using magnetic tweezers. (B) Force–distance tracings of fibers differing by the number of nucleosomes present. For any given reconstituted fiber, as the bead is raised, increasing the force exerted on the fiber, nucleosomes pop off, unwinding the DNA molecule, and the length of the fiber increases. This approach can be used to determine the effect of various histone modifications on the relative compaction of chromatin fibers.

(From Kruithof et al., 2009.)

and the role of histone tails on the mechanical stability of the 30-nm fiber (Fig. 2.8) were determined (Brower-Toland *et al.*, 2002; Cui and Bustamante, 2000; Kruithof et al., 2009). Histone tail removal decreases the overall affinity of histones for DNA in nucleosomes; acetylation of the lysines converts the amino group to an amide that is not charged, and also decreases the overall affinity of histones for DNA (Brower-Toland et al., 2005). This observation dovetails with the results obtained with the sedimentation analyses of the effect of histone acetylation on fiber compaction discussed earlier.

The different faces of chromatin

One of the major goals accomplished by the unique physical and molecular architecture of chromatin is to fit very long DNA molecules into the cell nucleus during interphase and, following DNA replication, to facilitate their distribution to daughter cells. Another major goal is to protect the genetic material from harmful molecular or physical attacks and to re-establish its integrity following such attacks. A third major goal is to allow the orderly retrieval of genetic information for cellular maintenance, differentiation and function. The compaction necessary to accomplish the first goal is generally not favorable to gene expression, and cells have evolved elaborate mechanisms that dynamically modify and use the features of chromatin compaction to modulate gene function. A discussion of the general characteristics of active and inactive regions of the genome is a useful introduction to this subject. **It also serves as the first illustration of the functional relationship between the genetic material and the architecture of the nucleus.**

Euchromatin vs. heterochromatin

For operational purposes, chromatin can be divided into active and inactive regions, called euchromatin and heterochromatin, respectively, to denote their differential staining characteristics in interphase nuclei (Heitz, 1928). This is a morphological distinction that is still useful today but that has to be re-evaluated as our understanding of its molecular basis increases. Euchromatin is dispersed in appearance and includes the vast majority of active genes or genes that will become activated, in particular groups of cells during tissue differentiation and development. Heterochromatic regions are transcriptionally inactive. Some of these regions, usually present around the centromeres or in the telomeres of chromosomes, remain condensed throughout the cell cycle and are said to represent **constitutive heterochromatin**. These regions are populated by highly repetitive DNA sequences and by transposable elements such as DNA transposons and retroviruses (Box 2.1). One of the characteristics of constitutive heterochromatin is the presence of nucleosomes containing histone H3 tri-methylated at lysine 9 on its N-terminal tail (H3K9me3). Other regions of the genome are condensed and inactive in some cell lineages, but not in others, and are referred to as **facultative heterochromatin**. The nucleosomes in these regions contain histone H3 tri-methylated at lysine 27 (H3K27me3). Chapter 4 provides a comprehensive discussion of these histone modifications. Facultative heterochromatin can involve whole chromosomes, such as the classic inactive mammalian X chromosome (see Chapter 9), or whole sets of chromosomes, as in the case of the paternal genome of some insects (Scarbrough

Box 2.1 Repetitive sequences and transposable elements

The genomes of most eukaryotic organisms are populated by a large number of DNA sequences that are able to insert themselves into new chromosomal locations. These transposable elements (TEs) can be divided into two broad classes: retrotransposons and DNA transposons. Retrotransposons include elements that are flanked at both ends by long repeated sequences (LTR transposons) and elements that lack such terminal repeats (non-LTR transposons). LTR transposons occur in vertebrates. Non-LTR transposons are of two major types: long interspersed nucleotide elements (LINEs) of approximately 5 to 7 kbp in length, and short interspersed nucleotide elements (SINEs) that are often less than 500 bp long.

DNA transposons are flanked by inverted repeats and use a transposase that they encode to generate staggered cuts which free the transposon. The same enzyme induces similar cuts at another site in the genome, allowing the reinsertion of the transposon at a new location. The new insertion site is characterized by the presence of short, direct repeated sequences resulting from the filling in of the staggered cuts by a DNA polymerase (Fig. 2B.1A).

The biogenesis and propagation of retrotransposons involves an RNA intermediate generated by RNA polymerase II. In the case of LTR transposons, this transcript is translated in the cytoplasm into a few proteins that include a reverse transcriptase that copies the transcript into a complementary

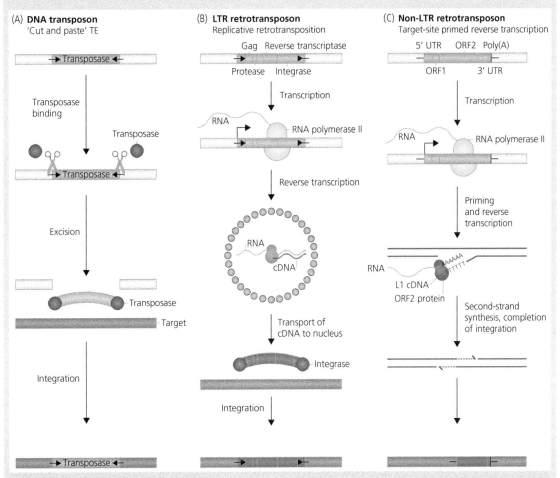

Fig. 2B.1 (A) DNA transposons. (B) LTR retrotransposons. (C) Non-LTR retrotransposons. The proteins that are encoded by autonomous non-LTR retrotransposons can also mobilize non-autonomous retrotransposon RNAs.

(Modified from Levin and Moran, 2011.)

continued

Box 2.1 *Continued*

DNA strand and then into a duplex DNA molecule through its DNA polymerase activity. Another transposon-encoded protein encloses the transposon into a virus-like particle. A third protein is an integrase that makes staggered cuts at random sites in the genome and allows the insertion of the DNA version of the transposon. The original transposon remains in place, and new transposons colonize the genome

(Fig. 2B.1B). Non-LTR transposons encode a protein that has both reverse transcriptase and endonuclease activity. This protein induces a single-strand nick at specific sites in genomic DNA that is used to prime reverse transcription of the RNA as an extension of one of the genomic DNA strands. Second-strand DNA synthesis is directed by the same enzyme or via DNA repair (Fig. 2B1.C).

---

et al., 1984). The facultative heterochromatin concept can also apply to the inactivation of genes, such as homeotic, in those regions of the developing embryo where their expression is inappropriate, or the inactivation of particular alleles depending on their parent of origin (discussed in Chapters 7 and 8).

The molecular biology of heterochromatin

Understanding the molecular biology of heterochromatin formation and maintenance was initiated by the study of position–effect variegation (PEV) in *Drosophila*. The phenomenon derives its name from the observation that genes that are relocated by some chromosomal rearrangement to the neighborhood of constitutive heterochromatin may fail to be transcribed in subsets of those cells where they normally would be active. The inactivation is the effect of the gene's new position along the chromosome leading to the presence in a tissue of sectors (clones) of cells where the gene is inactive and of sectors where it remains active, resulting in a variegated phenotype. An early hypothesis proposed that the special proteins responsible for the distinctive compact structure of constitutive heterochromatin diffuse into euchromatin and influence the expression of juxtaposed genes; however, the process is stochastic and occurs in some embryonic cells and their descendants, but not in others. In the hope of identifying these proteins, genetic screens were carried out in *Drosophila* to discover mutations that would affect a particular example of PEV (Elgin and Reuter, 2013). A number of mutations were found that suppress variegation [Su(var) mutations], leading to the widespread expression of the relocated wild type gene. These mutations identify genes that encode heterochromatin proteins, for example, HP1a (heterochromatin protein 1a), or that encode enzymes that modify histones with covalent modifications found in heterochromatin: Su(var)4-20

that methylates lysine 20 on the N-terminal tail of histone H4 (H4K20me), Su(var)3-64B that removes the acetyl group from lysine 9 on histone H3 (H3K9) and Su(var)3-9 that methylates lysine 9 on histone H3 (H3K9me). Mutations that increased the number of clones where the relocated gene is inactivated, and thereby enhancing the level of variegation [E(var) mutations], often identified genes encoding ubiquitous transcription factors; a reduction in the level of these factors would reduce the frequency of activation of the relocated wild type gene and facilitate its inactivation by encroaching heterochromatin factors. PEV occurs in eukaryotes, ranging from yeast (Grunstein and Gasser, 2013) to humans (Fodor et al., 2010). Many of the key proteins and factors discovered and identified in *Drosophila* are present in these organisms.

A key function of heterochromatin is to prevent the activity of transposable elements that populate the genome of all eukaryotes. Silencing of these elements in the germ line involves a special type of small noncoding RNA that induces the deposition of methylated H3K9 (see Chapter 6). In *Drosophila*, the linker histone H1, which appears to be more prevalent in heterochromatic regions, can specifically recruit the histone H3 lysine 9 methyl transferase Su(var)3-9 (Lu et al., 2013).

Euchromatin and heterochromatin have specific locations in the nucleus

In the 1950s, studies of nuclear structure using the electron microscope revealed that regions of condensed chromatin tended to be associated with the nuclear periphery while more diffuse chromatin had a more central localization. Since those initial observations, a large body of molecular information has been accumulated, describing the nature of this differential association and its role in regulating the transcription process (discussed in Chapters 11 and 12).

Heterochromatic regions lag behind during genome replication

In the genomes of multicellular eukaryotes, DNA replication is initiated at thousands of defined points that exhibit G-rich consensus motifs (Cayrou et al., 2011). Not all of the potential origins of replication are used during the S phase, and some origins are active in certain cell types, but not in others. In addition, the mean replication time of genomic regions in all cells differ with respect to the transcriptional status of the region— transcriptionally active segments, marked by H3K4m3, H3K36me3, H4K20me1, H3K79me2, H3K9ac and H3K27ac, replicate first; repressed regions, marked by H3K27me3 (see Chapters 4 and 5), replicate later; constitutive heterochromatin regions, marked by H3K9me3 and HP1, replicate last (Julienne et al., 2013).

Chapter summary

DNA is not present in pure form in cells. It is associated with histones, non-histone proteins and RNA in a complex referred to as chromatin. Four different types of histones form octamers (nucleosomes), around which DNA is wrapped, yielding a chromatin fiber with the configuration of "beads on a string." Disassembly, followed by reassembly, of this structure occurs during DNA replication, damage repair and transcription. Positioning of nucleosomes on the chromatin fiber is neither uniform nor regular; it is mediated by chromatin remodeling complexes and reflects the functional state of various regions along the fiber. Various biophysical methods have been utilized to study the physical association of nucleosomes and DNA.

Chromatin can be differentiated on the basis of the activity of the genes that are present in a given region. Heterochromatin represents repressed or inactive regions of the genome and exhibits a greater degree of condensation than euchromatin, which refers to more unwound regions where active genes are located. The two types of chromatin are distinguishable by the presence and abundance of repetitive sequences, by their different epigenetic modifications and by their presence in different nuclear locations.

Three basic aspects of DNA metabolism—transcription, replication and damage repair—occur in the context of chromatin. Chapter 3 in Part I and Chapter 5 in Part II discuss the basic mechanism of transcription and its occurrence in the chromatin environment. DNA replication and DNA damage repair are the subjects of Chapters 14 and 15 in Part IV.

References

Brower-Toland, B., Wacker, D. A., Fulbright, R. M., Lis, J. T., Kraus, W. L. & Wang, M. D. 2005. Specific contributions of histone tails and their acetylation to the mechanical stability of nucleosomes. *J Mol Biol*, 346, 135–46.

Brower-Toland, B. D., Smith, C. L., Yeh, R. C., Lis, J. T., Peterson, C. L. & Wang, M. D. 2002. Mechanical disruption of individual nucleosomes reveals a reversible multistage release of DNA. *Proc Natl Acad Sci U S A*, 99, 1960–5.

Carruthers, L. M., Bednar, J., Woodcock, C. L. & Hansen, J. C. 1998. Linker histones stabilize the intrinsic salt-dependent folding of nucleosomal arrays: mechanistic ramifications for higher-order chromatin folding. *Biochemistry*, 37, 14776–87.

Caterino, T. L., Fang, H. & Hayes, J. J. 2011. Nucleosome linker DNA contacts and induces specific folding of the intrinsically disordered H1 carboxyl-terminal domain. *Mol Cell Biol*, 31, 2341–8.

Cayrou, C., Coulombe, P., Vigneron, A., Stanojcic, S., Ganier, O., Peiffer, I., Rivals, E., Puy, A., Laurent-Chabalier, S., Desprat, R. & Mechali, M. 2011. Genome-scale analysis of metazoan replication origins reveals their organization in specific but flexible sites defined by conserved features. *Genome Res*, 21, 1438–49.

Cui, Y. & Bustamante, C. 2000. Pulling a single chromatin fiber reveals the forces that maintain its higher-order structure. *Proc Natl Acad Sci U S A*, 97, 127–32.

Elgin, S. C. & Reuter, G. 2013. Position-effect variegation, heterochromatin formation, and gene silencing in *Drosophila. Cold Spring Harb Perspect Biol*, 5, a017780.

Finch, J. T., Lutter, L. C., Rhodes, D., Brown, R. S., Rushton, B., Levitt, M. & Klug, A. 1977. Structure of nucleosome core particles of chromatin. *Nature*, 269, 29–36.

Fletcher, T. M. & Hansen, J. C. 1995. Core histone tail domains mediate oligonucleosome folding and nucleosomal DNA organization through distinct molecular mechanisms. *J Biol Chem*, 270, 25359–62.

Fodor, B. D., Shukeir, N., Reuter, G. & Jenuwein, T. 2010. Mammalian Su(var) genes in chromatin control. *Annu Rev Cell Dev Biol*, 26, 471–501.

Grunstein, M. & Gasser, S. M. 2013. Epigenetics in *Saccharomyces cerevisiae. Cold Spring Harb Perspect Biol*, 5, pii: a017491.

Hayes, J. J., Tullius, T. D. & Wolffe, A. P. 1990. The structure of DNA in a nucleosome. *Proc Natl Acad Sci U S A*, 87, 7405–9.

Heitz, E. 1928. [Das heterochrokmatin der Moose.] *Jahrb Wiss Bot*, 69, 762–818.

Hewish, D. R. & Burgoyne, L. A. 1973. Chromatin sub-structure. The digestion of chromatin DNA at regularly spaced sites by a nuclear deoxyribonuclease. *Biochem Biophys Res Commun*, 52, 504–10.

Julienne, H., Zoufir, A., Audit, B. & Arneodo, A. 2013. Human genome replication proceeds through four chromatin states. *PLoS Comput Biol*, 9, e1003233.

Kornberg, R. D. 1974. Chromatin structure: a repeating unit of histones and DNA. *Science*, 184, 868–71.

Kornberg, R. D. & Thomas, J. O. 1974. Chromatin structure; oligomers of the histones. *Science*, 184, 865–8.

Kruithof, M., Chien, F. T., Routh, A., Logie, C., Rhodes, D. & Van Noort, J. 2009. Single-molecule force spectroscopy reveals a highly compliant helical folding for the 30-nm chromatin fiber. *Nat Struct Mol Biol*, 16, 534–40.

Levin, H. L. & Moran, J. V. 2011. Dynamic interactions between transposable elements and their hosts. *Nat Rev Genet*, 12, 615–27.

Lowary, P. T. & Widom, J. 1998. New DNA sequence rules for high affinity binding to histone octamer and sequence-directed nucleosome positioning. *J Mol Biol*, 276, 19–42.

Lu, X., Wontakal, S. N., Kavi, H., Kim, B. J., Guzzardo, P. M., Emelyanov, A. V., Xu, N., Hannon, G. J., Zavadil, J., Fyodorov, D. V. & Skoultchi, A. I. 2013. *Drosophila* H1 regulates the genetic activity of heterochromatin by recruitment of Su(var)3–9. *Science*, 340, 78–81.

Luger, K., Mader, A. W., Richmond, R. K., Sargent, D. F. & Richmond, T. J. 1997. Crystal structure of the nucleosome core particle at 2.8 A resolution. *Nature*, 389, 251–60.

Mcbryant, S. J., Lu, X. & Hansen, J. C. 2010. Multifunctionality of the linker histones: an emerging role for protein-protein interactions. *Cell Res*, 20, 519–28.

Noll, M. 1974. Internal structure of the chromatin subunit. *Nucleic Acids Res*, 1, 1573–8.

Puhl, H. L. & Behe, M. J. 1993. Structure of nucleosomal DNA at high salt concentration as probed by hydroxyl radical. *J Mol Biol*, 229, 827–32.

Scarbrough, K., Hattman, S. & Nur, U. 1984. Relationship of DNA methylation level to the presence of heterochromatin in mealybugs. *Mol Cell Biol*, 4, 599–603.

Schwarz, P. M. & Hansen, J. C. 1994. Formation and stability of higher order chromatin structures. Contributions of the histone octamer. *J Biol Chem*, 269, 16284–9.

Simpson, R. T. & Stafford, D. W. 1983. Structural features of a phased nucleosome core particle. *Proc Natl Acad Sci U S A*, 80, 51–5.

Simpson, R. T. & Whitlock, J. P. 1976. Mapping DNAase l-susceptible sites in nucleosomes labeled at the 5′ ends. *Cell*, 9, 347–53.

Tse, C., Sera, T., Wolffe, A. P. & Hansen, J. C. 1998. Disruption of higher-order folding by core histone acetylation dramatically enhances transcription of nucleosomal arrays by RNA polymerase III. *Mol Cell Biol*, 18, 4629–38.

Williamson, R. 1970. Properties of rapidly labeled deoxyribonucleic acid fragments isolated from the cytoplasm of primary cultures of embryonic mouse liver cells. *J Mol Biol*, 51, 157–68.

CHAPTER 3

The basic mechanism of gene transcription

Transcription is the fundamental mechanism by which a cell retrieves the information that is encrypted in the nucleotide sequence of the genome. The process involves the function of a large number of factors and is regulated from its inception by epigenetic modifications and chromatin remodeling. The nuclear genomes of eukaryotic organisms contain four types of genes or transcription units that are transcribed by different RNA polymerases—multiprotein enzyme complexes that synthesize RNA sequences complementary to specific regions of the DNA. The different types of transcription units include: (1) genes that encode proteins (transcribed by RNA polymerase II, abbreviated RNAPII), (2) transcription units that encode ribosomal RNAs (transcribed by RNAPI), (3) those that encode transfer RNAs (tRNAs) and small RNAs involved in processing other RNAs (transcribed by RNAPIII), and (4) those that encode non-coding RNAs responsible for numerous regulatory steps that impact the structure and function of the genetic material (transcribed by RNAPII or RNAPIII, depending on where they are located in the genome). The different RNA polymerases and the different types of DNA units that they transcribe are summarized in Table 3.1. This chapter will focus on the factors and mechanisms responsible for the transcription of protein-coding genes and of long non-coding RNAs. Pertinent aspects of the synthesis of the other cellular RNAs will be presented in Chapter 6.

The transcription of protein-coding genes and of long non-coding RNAs consists of a series of stages—formation and activation of the initiation complex, transcript elongation and termination—as well as a series of co-transcriptional events that include 5′ capping, 3′ cleavage, polyadenylation and splicing of the transcript.

Table 3.1 Major classes of RNA transcribed by the three eukaryotic RNA polymerases and their functions

Polymerase	RNA transcribed	RNA function
RNA polymerase I	Pre-rRNA (28S, 18S, 5.8S rRNAs)	Ribosome components, protein synthesis
RNA polymerase II	mRNA snRNAs miRNAs	Encodes protein RNA splicing Post-transcriptional gene control
RNA polymerase III	tRNAs 5S rRNA	Protein synthesis Ribosome component, protein synthesis
	snRNA U6	RNA splicing
	Other stable short RNAs	Various functions, unknown for many

(From *Molecular Cell Biology*, sixth edition.)

The pre-initiation complex

The first phase in transcription is the formation of a pre-initiation complex (PIC) at the core promoter, a region of the transcription unit that straddles the DNA base-pair that will be the first to be transcribed and that is designated as the transcription start site (TSS). The promoter region contains several DNA elements that facilitate the binding of various components of the PIC. These elements include the TATA box,

Epigenetics, Nuclear Organization and Gene Function: with implications of epigenetic regulation and genetic architecture for human development and health. John C. Lucchesi, Oxford University Press (2019). © John C. Lucchesi 2019. DOI: 10.1093/oso/9780198831204.001.0001

the initiator element (Inr), the downstream promoter element (DPE), the TFIIB recognition element (BRE), the motif ten element (MTE), the downstream core element (DCE) and the X core promoter element (XCPE1). Different promoters contain different subsets of these elements, which consist of short consensus sequences (Riethoven, 2010). In general, promoters that lack a TATA box must possess conserved copies of the Inr or DPE elements. Conversely, promoters containing optimal TATA sequences do not require Inr and DPE elements for the binding of PIC subunits. The TSS of TATA or DPE-containing promoters usually starts at a specific single nucleotide position. Promoters that lack these elements allow a more variable position for the start of transcription (Anish et al., 2009; Gershenzon and Ioshikhes, 2005; Haberle and Lenhard, 2016; Vo Ngoc et al., 2017).

On some promoters that are transcribed by RNAPII, the PIC is composed of general transcription factors (TFIID, IIA, IIB, IIE, IIF, IIH) that associate in an orchestrated manner with the various core promoter elements. TFIID binds to the TATA box or to the Inr element via one of its subunits—the TATA-binding protein TBP. This association, enhanced and stabilized by the other components of TFIID called TBP-associated factors (TAFs), causes localized bending and partial unwinding of the DNA. While the other general transcription factors are encoded by single copy genes, TFIID is subject to a significant level of variation—TBP-related factors and TAFs have paralogous genes, and these TAF-like proteins are often expressed in a cell type- or tissue-specific manner, reflecting a regulatory role for the PIC in differential gene expression (Muller and Tora, 2004). The TBP plays a role in PIC formation with the other two polymerases. On RNAPI promoters, TBP associates with the Selectivity factor 1 (SL1) complex and five RNAPI-specific TAFs. On RNAPIII promoters, TBP associates with two specific TAFs (BrfI and BdpI) (Akhtar and Veenstra, 2011).

The specific recognition of the TATA box by TFIID requires TFIIA. TFIIB forms a complex with RNAPII and recruits it to the PIC by making DNA-specific contacts and associating with TBP; TFIIB plays a key role in positioning RNAPII, so that the correct start site of transcription is selected (Deng and Roberts, 2007). TFIIF serves as a bridge between RNAPII and TFIID/TFIIB and enables entry of TFIIE and TFIIH.

Other promoters that are transcribed by RNAPII lead to the production of small nuclear RNAs, including U RNAs used for the processing of primary messenger RNA (mRNA) transcripts (see Chapter 6). These promoters contain an essential proximal sequence element (PSE) that is functionally equivalent to the TATA box (Mattaj et al., 1985). Transcription is activated by a five-subunit complex (SNAPc, small nuclear RNA activating polypeptide complex) that binds to the PSE (Sadowski et al., 1993).

In contrast to the formation of the PIC by the orderly and sequential addition of the general transcription factors and RNAPII to the promoter region, a number of cases have been described suggesting that the polymerase and the transcription factors form a preassembled **holoenzyme** that associates with the promoter. First reported in yeast (Koleske and Young, 1994), RNAPII has been purified from human cells in the form of a holoenzyme (Ossipow et al., 1999; Pan et al., 1997). Although the stepwise formation of the PIC on the promoter is widely accepted, whether the preassembly of a holoenzyme occurs in some or in most instances has not been sufficiently refuted or documented.

RNAPII is a 12-subunit enzyme. The large subunit (Rbp1) has a heptapeptide (Tyr_1-Ser_2-Pro_3-Thr_4-Ser_5-Pro_6-Ser_7) that is repeated in the C-terminal domain (CTD). The number of these repeats varies among organisms and appears to be related to phylogenetic complexity—there are 26 repeats in the Rbp1 CTD of yeast, 45 in *Drosophila* and 52 in mammals. In mammals, some of the distal repeats of the CTD show some deviation from the consensus sequence (Eick and Geyer, 2013). Various covalent modifications of the amino acids of the CTD are correlated with the different steps and events of the transcription process. Although it has long been held that RNAPII of the PIC is unmodified, some recent evidence suggests that, as it is recruited, serines 5 and 7 (S_5 and S_7) of the CTD are glycosylated (Ranuncolo et al., 2012). This modification is removed to allow the phosphorylation of these two amino acids that is necessary for transcription to begin. Although CTD glycosylation has been extensively studied *in vitro*, its specific function *in vivo* has not been determined (Lu et al., 2016).

Transcription initiation

Initiation requires the interaction of the PIC with activation factors present at regulatory sites that are upstream of the core promoter region. These promoter-proximal elements, defined by their sequence (e.g. the CAT box and the GC box), are recognized by a specific subset of DNA-binding transcription factors. Additional DNA regulatory modules consist of **enhancers** present on either side of transcription units, sometimes within, but often at substantial, distances from the units (Fig. 3.1). For example, in the mouse, the

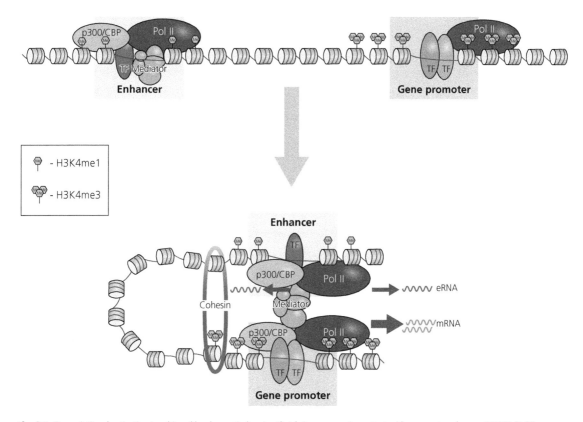

Fig. 3.1 Transcriptional activation is achieved by chromatin looping that brings a gene in contact with a cognate enhancer. RNAPII (PolII) transcribes the enhancer to produce bidirectional RNAs (eRNAs). Some eRNAs interact with cohesin to stabilize the enhancer–promoter association. P300/CBP is a histone acetyl transferase co-activator.

(From Kim et al., 2015.)

enhancer of immunoglobulin H is in the gene itself; in contrast, the wing margin enhancer of the *cut* gene of *Drosophila* is 85 kilobase (kb) upstream of the gene. Enhancers represent binding sites for many regulatory proteins that specify gene activation in relation to the cell cycle, as a response to stress signals or other environmental clues, or confer tissue specificity (Bulger and Groudine, 2011). Enhancer elements are either active or poised, depending on the type of histone modifications that characterize their chromatin. Such modifications also determine the cell specificity of enhancer function and the expression level of the genes that they activate (Zentner et al., 2011).

Enhancers usually interact with the PIC through a large multiprotein complex termed the Mediator complex (Box 3.1). For example, this interaction is found in 60% of actively transcribed genes in mouse embryonic stem (ES) cells (Kagey et al., 2010). The Mediator complex binds to RNAPII and other PIC components, as well as to the activation domains of DNA-bound

factors associated with promoter-proximal and enhancer elements. Mediator exists in functionally distinct forms that differ with respect to particular subunits. While most of these forms activate transcription in a tissue-specific manner, in some cases, they can repress transcription (Conaway and Conaway, 2011). In the case of actively transcribed genes, Mediator is found associated with cohesin, a complex of proteins that form a ring used by cells to hold sister chromatids together following DNA replication. In this case, cohesin's role is likely to maintain the physical association of the enhancer and promoter complexes (Kagey et al., 2010). Active enhancers are transcribed in a bidirectional manner, i.e. from both strands of the DNA (Kim et al., 2010). These enhancer RNAs (eRNAs) are necessary for the activation of target genes (see Chapter 5).

Initiation also requires melting of the double-stranded DNA into a single-stranded bubble to produce an open complex between RNAPII and the DNA sequence to be copied. This is achieved by the

Box 3.1 Mediator

Mediator is a large protein complex that serves as an adaptor between a variety of activators and RNAPII. There is substantial evidence that Mediator is involved in the transcription of the vast majority of genes in yeast and in multicellular animals. Some exceptions may occur in terminally differentiated cells where the transcription machinery appears to function without the help of Mediator.

The numerous Mediator subunits are organized into distinct functional modules (Fig. 3B.1). The head and middle modules are believed to be involved in interactions with the core RNAPII machinery; the tail module interacts with various activators. A Mediator consisting of these three modules has a positive effect on gene transcription. Addition of the kinase module usually leads to a repressive function by interfering

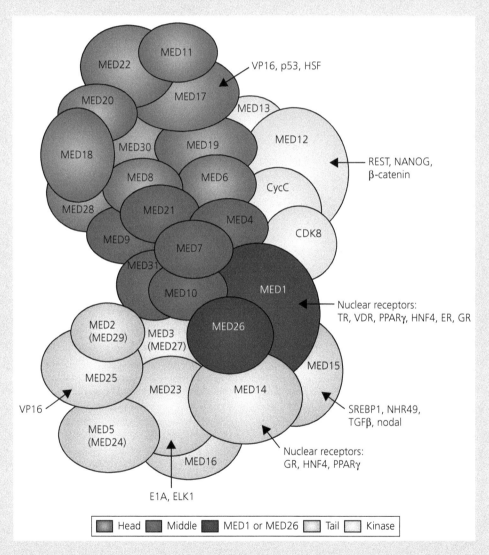

Fig. 3B.1 Diagram of the molecular structure of the Mediator complex. The position of the subunits is based on some published binary interactions. Examples of the transcription factors with which Mediator interacts are indicated: ER, estrogen receptor; GR, glucocorticoid receptor; HNF4, hepatocyte nuclear factor; NHR49, nuclear hormone receptor 49; PPARγ, peroxisome proliferator-activated receptor-γ; SREBP1, sterol regulatory element-binding protein 1; TGFβ, transforming growth factor-β; VDR, vitamin D3 receptor.

(From Malik and Roeder, 2010.)

action of TFIIH, a large multi-subunit complex that contains ATP-dependent helicases (enzymes that partially unwind the two strands of the DNA molecule), and of TFIIE, which regulates the activity of TFIIH. These two transcription factors are added immediately after RNAPII. Following the establishment of the open complex, a subunit of TFIIH—cyclin-dependent kinase 7 (Cdk7) and its associated Cyclin H in metazoans or Kin28 in yeast—phosphorylates the serine at position five (S_5) of the CTD heptad repeats. RNAPII is recruited to the PIC in an unphosphorylated state, and the phosphorylation of S_5 is required to initiate promoter clearance. Transcription initiation is marked by the recruitment of the first two ribonucleoside triphosphates (rNTPs) complementary to the DNA sequence at the transcription start site, and the formation of the first phosphodiester bond.

Another serine at position 7 (S_7) of the CTD repeat is phosphorylated by Kin28 in yeast and CDK7 in metazoans when RNAPII is at the promoter. Unlike phosphorylated S_5 that decreases as the RNAPII complex proceeds along a transcriptional unit (see Productive elongation, p. 22), the level of phosphorylated S_7 is maintained by Bur1 kinase. This covalent modification of the CTD is important for transcription of small nuclear RNA genes (Egloff et al., 2007). In addition to the serine residues, threonine 4 (T_4) can be phosphorylated. This modification is required for viability in mammalian cells where it is achieved by Polo-like kinase 3 (Hintermair et al., 2012).

Promoter clearance and early elongation

This step involves the partial disassembly of the PIC, allowing a subset of the transcription factors (TBP, TFIID, Mediator) to remain behind. These factors serve as the seed for the formation of the next PIC and the next round of transcription.

Following the activation process that allows promoter clearance, RNAPII and its associated general transcription factors constitute the transcription elongation complex (TEC). Initially, this complex is

abortive and can stall and release the DNA; it becomes more stable after successfully synthesizing an 8 to 9 nucleotide pairs (ntp) DNA–RNA hybrid and ejecting TFIIB, and is completely stabilized after the synthesis of a 20 to 25 nucleotide (nt)-long RNA. In a large number—perhaps the majority—of genes in metazoans, the elongating polymerase pauses after having transcribed 25 to 50 (in rare instances up to 100) nt.

Pausing

General RNAPII pausing in eukaryotes was first discovered by John Lis in a *Drosophila* heat-shock gene (*hsp70*) where, in the absence of heat-shock, a polymerase that has transcribed approximately 25 nt is stalled (Rougvie and Lis, 1988). Although Lis thought that this might be a widespread phenomenon, it is only recently that genome-wide studies have shown that pausing occurs broadly in the genomes of flies and mammals. To date, there is no evidence for promoter-proximal polymerase pausing in yeast; this suggests that it may occur only in multicellular organisms. During pausing, the TEC is arrested after having transcribed from 20 to 100 nt. Once arrested, the polymerase can backtrack (Fig. 3.2). The arrest is reversible and involves the action of the transcript cleavage factor TFIIS (Donahue et al., 1994; Izban and Luse, 1992), and pausing is thought to be an important means of transcription regulation (Nechaev and Adelman, 2011).

Pausing is caused by the presence of the negative elongation factor (NELF) and DRB sensitivity-inducing factor (DSIF) complexes. These complexes are highly conserved in eukaryotes. NELF attaches to the polymerase through one of its four subunits (NELF-A), while the heterodimeric DSIF, consisting of the homologues of two yeast transcription factors (Spt4 and Spt5), seems to recognize a particular structure in the nascent transcript.

The DNA sequence has an important influence on the occurrence of pausing and on the specific location where the stalled RNAPII settles (Hendrix et al., 2008; Keene et al., 1999; Lee et al., 1992; Tsai et al, 2016). An A/T-rich

DNA stretch increases the probability that a transcribing TEC will arrest. The TEC then backtracks to a G/C-rich stretch where it pauses. During this reversal, the 3' fragment of the nascent RNA, corresponding to the region of the DNA template extending from the point of the initial arrest to the point where the TEC pauses, is released from the RNAPII active site; this segment of RNA must be cleaved by the general transcription factor TFIIS in order to restore alignment of the active site with the 3'-OH end of the transcript and to prepare for the resumption of transcription.

In higher eukaryotes, the conversion of a paused TEC into a complex engaged in productive elongation involves the recruitment of positive transcription elongation factor b (P-TEFb) that consists of Cdk9 kinase and its activating cyclin. P-TEFb can associate directly with the TEC or be recruited by various transcription factors such as c-Myc and NFκB (Luecke and Yamamoto, 2005; Napolitano et al., 2000). In cells, the level of P-TEFb is regulated by modulating the expression of its two subunits or by its inclusion into an inactive complex that associates with a small nuclear RNA.

The pausing time of the TEC at promoters varies over a wide range, indicating that it is an important means of regulating gene expression. Some genes have a definitive pause that prevents transcription from occurring, unless a specific factor is synthesized. In the more extreme cases, such as the heat-shock (*hsp*) genes, a TEC is released every 10 minutes or so. Induction of rapid transcription requires the heat-shock factor HSF. In other genes, pausing may be minimal and only occurs to allow the recruitment of transcript capping enzymes to cap the 5' end of the nascent RNA (discussed in Co-transcriptional RNA processing events, p. 24). Using the relative ratio of RNAPII density in the promoter region and in the body of genes, the vast majority of transcriptional units in mammalian embryonic stem cells exhibit some level of pausing (Rahl et al., 2010).

Productive elongation

Two events mark the escape from pausing and the resumption of transcript elongation: 5' capping of the nascent RNA and the phosphorylation of Serine 2 (S_2) residues on the heptapeptide repeats of the RNAPII CTD, and of NELF and DSIF. Capping of the transcript occurs as the polymerase moves through the pause region. Nascent RNAs closest to the transcription start site are uncapped, while RNAs longer than 30 nucleotides are capped, suggesting that capping may be necessary for productive elongation (Kwak and Lis, 2013; Rasmussen and Lis, 1993). S_2 phosphorylation is believed to be accomplished in higher eukaryotes by P-TEFb that includes the cyclin-dependent kinase Cdk9 and its cyclin partner Cyclin T (see Box 3.2 for the involvement of other kinases). In general, phosphorylation levels of S_5 on the CTD repeats, which are enriched when RNAPII is at the promoter, decrease and phosphorylation of S_2 increases as the TEC proceeds towards the 3' end of genes, allowing the binding of various complexes to the RNAPII CTD. P-TEFb also phosphorylates NELF and DSIF, disrupting the association of NELF with the polymerase and allowing transcription to resume. There are two parameters that can be used to assess the efficiency of the TEC in synthesizing transcripts: rate of elongation and processivity. The rate of elongation measures the number of nucleotides added to nascent RNA per minute, in other words how fast RNA is synthesized. Processivity reflects the number of nucleotides added per initiation event; it measures the ability of RNAPII to travel to the end of the gene without stopping or prematurely releasing from the template.

Additional factors and complexes associate with the TEC to increase transcription rate and processivity. FACT (facilitates chromatin transcription) is a

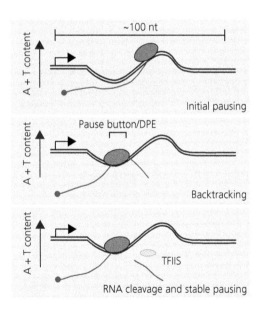

Fig. 3.2   The onset of stable pausing and 3' end cleavage of the nascent RNA to allow realignment of the RNAPII active site with the DNA template. The pause button is a sequence identified by Hendrix et al., 2008, that is frequently present in stalled promoters.

(From Nechaev and Adelman, 2011.)

Box 3.2 Serine 2 (S$_2$) phosphorylation

In yeast, this reaction is carried out predominantly by the catalytic subunit CTK1 of a three-subunit complex (CTDK1). An additional kinase BUR1, in conjunction with its regulatory subunit BUR2, phosphorylates S$_2$ in RNAPII located near the promoter, perhaps facilitating the subsequent phosphorylation by CTDK1 along the whole transcriptional unit (Qiu et al., 2009). In *Drosophila*, CDK12 kinase associated with Cyclin K is responsible for the majority of S$_2$ phosphorylation (Bartkowiak et al., 2010). Although the major kinase involved in phosphorylating S$_2$ in humans is CDK9, two related kinases, both associated with Cyclin K in separate complexes, are present. CDK12–Cyclin K is required for the transcription of a subset of human genes involved in genomic stability such as the *BRCA1* gene associated with breast cancer and the *FANCI* gene associated with acute myelogenous leukemia (Blazek et al., 2011). The kinase function of CDK13–Cyclin K has only been ascertained *in vitro*. Finally, an additional kinase BRD4 has been identified in human cells. BRD4 is a double bromodomain protein that can bind acetylated histones (see Chapter 4) and recruit P-TEFb to promoters (Jang et al., 2005). BRD4 was shown to bind to the CTD and phosphorylate S$_2$ directly (Devaiah et al., 2012).

In *Caenorhabditis*, the phosphorylation of S$_2$ in somatic cells is similar to that in other metazoans. In contrast, most of the phosphorylation in the germline is mediated by CDK12, rather than CDK9, leading to the speculation that the regulation of transcription elongation may differ in different tissues (Bowman et al., 2013).

---

heterodimer that binds to nucleosomal DNA and aids the process of elongation by destabilizing the histone octamers ahead of the TEC; FACT also facilitates the reassembly of octamers following the passage of the TEC (Orphanides et al., 1998; Reinberg and Sims, 2006). Elongin, ELL (eleven-nineteen lysine-rich in leukemia) and EAFs (ELL-associated factors) were identified by searching for proteins that would increase the rate of formation of full-length transcripts in a reconstituted *in vitro* transcription system (Aso et al., 1995; Shilatifard et al., 1996). Spt6 is recruited to the TEC by interacting with RNAPII CTD phosphorylated at S$_2$; it is an essential elongation factor for a number of genes and may play a role in nucleosome reassembly following transcription and in mRNA export (Brès et al., 2008). As an indication that many important elements of the transcription process are still to be discovered, a new

factor, NDF (nucleosome destabilizing factor) was identified recently in *Drosophila* and in human cells. NDF facilitates transcription by associating with the transcribed regions of active genes and destabilizing nucleosomes (Fei et al., 2018).

A number of the factors just listed and additional ones, such as the Elongator or the PAF (RNA polymerase-associating factor) complexes, facilitate transcription by inducing covalent modifications in the histone octamers along transcriptional units. Elongator is a highly conserved multi-subunit complex associated with elongating RNAPII. It functions by acetylating histone H3 along the transcribed portion of different groups of genes in different organisms (Creppe and Buschbeck, 2011). [More recently, though, evidence has accumulated that the primary role of Elongator is to catalyze tRNA modifications (Kolaj-Robin and Seraphin, 2017)]. Compass (complex of proteins associated with Set1) is a complex discovered in yeast that methylates histone H3. In humans, it is one component of a super elongation complex that contains ELL and P-TEFb (Smith et al., 2011). The PAF complex mediates the methylation and ubiquitination of histones H2A and H3, respectively (Moniaux et al., 2009). It interacts with FACT functionally and is also involved in transcript polyadenylation (see Co-transcriptional RNA processing events, p. 24). THO (named for one of its subunits—suppressor of the transcriptional defect of Hpr1 by overexpression) is another conserved multi-subunit factor that appears to play a role in elongation by preventing the pairing of the nascent RNA transcript with the nonsense DNA strand (Gomez-Gonzalez et al., 2011). THO's additional involvement in several post-transcriptional processes, such as termination and transcript release, has made it difficult to understand its precise mechanism of action (Rondon et al., 2010).

Some factors such as FACT are dependent on particular modifications for their activity; others recognize histone modifications characteristic of activated promoters and enhance the association of elongation factors with paused TECs; for example, the bromodomain-containing protein Brd4 participates in the recruitment of P-TEFb (Brès et al., 2008). Some factors facilitate elongation by altering the architectural organization of transcriptional units, rather than by covalently modifying histones. For example, CHD1 (chromo-ATPase/helicase-DNA-binding domain 1) functions as an elongation factor *in vivo*, although very little mechanistic data have been provided. CHD1 physically associates with the PAF elongation complex, DSIF and FACT (Kelley et al. 1999; Krogan et al. 2002;

Simic et al. 2003). The interactions with chromatin of many of these factors and complexes will be discussed in detail in Chapter 5.

Co-transcriptional RNA processing events

These events, which include capping of the 5′ end of the nascent transcript, the successive removal of introns and the termination-coupled processes of cleavage and polyadenylation, are not just coincidental with transcription—they are also functionally coupled with the productive elongation by polymerases. The functional coupling of seemingly independent events can be achieved by different types of biochemical interactions with different consequences, including simple co-localization of factors on the TEC leading to faster rates of reaction, allosteric interactions between proteins of different complexes resulting in novel activities and kinetic coupling ensuring the appropriate order of occurrence of interdependent events (Perales and Bentley, 2009). The location where the majority of the factors and complexes involved in co-transcriptional

processing gather and where the different processing events play out is the RNAPII CTD.

Capping

Capping is necessary to ensure mRNA stability during its transport from the nucleus and its translation. It involves the addition of a methylated guanosine to the first nucleotide triphosphate of the synthesized RNA and requires three enzymatic activities: (1) removal of the terminal phosphate from the first nucleotide of the transcript by a phosphatase, (2) fusion of a guanosine triphosphate (GTP) via a phosphate–phosphate bond by a transferase, a reaction that releases two more phosphates, and (3) methylation of the added guanine by a methyl transferase (Fig. 3.3). In yeast, there are three separate enzymes, while in humans and other metazoans, the phosphatase and transferase activities are present in a single capping enzyme polypeptide. Capping occurs following the phosphorylation of the CTD at S_5, when the TEC is paused and the transcript is 25–50 bases long. Binding of the capping enzyme to

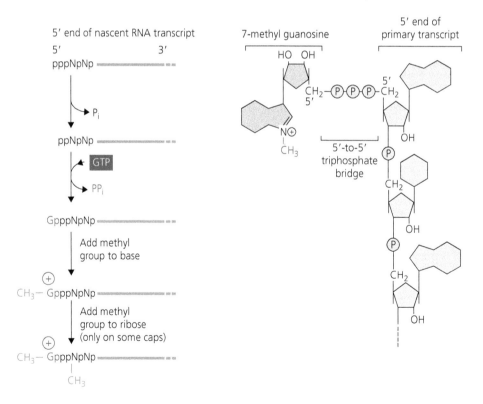

Fig. 3.3 Successive steps resulting in 5′ capping of a nascent RNA transcript.

(From *Molecular Biology of the Cell*, 4th Edition)

the CTD activates guanyl transferase and initiates the capping reaction.

Using a reconstituted transcription system *in vitro*, the capping enzyme was seen to interact with the Spt5 subunit of the DSIF complex and to inactivate the pause-inducing NELF complex (Mandal et al., 2004). This suggests that an additional function of capping may be to serve as a checkpoint ensuring that release from promoter pausing may only occur if the nascent transcript has been capped. A direct *in vivo* role for capping in transcription elongation is provided by the observation that a cap-binding protein complex (CBC) interacts with the P-TEFb complex and ensures optimal phosphorylation of S_2 on the RNAPII CTD (Lenasi et al., 2011). As will be discussed in the next section, the CBC and, indirectly therefore, the 5′ cap play a role in the transcript splicing process.

Transcript splicing

The primary transcript of most genes contains one or more **introns**—DNA segments that interrupt the functional coding sequence and that must be removed (Fig. 3.4). Introns are usually interspersed between functional sections of the transcript referred to as **exons**. The splicing out of introns is the responsibility of the **spliceosome** consisting of several small RNA-containing proteins (snRNPs). Early observations suggested that splicing could occur co-transcriptionally, i.e. while the RNA polymerase is still engaged in transcribing the template. Subsequent investigations have led to the conclusion that co-transcriptional splicing occurs very frequently from yeast to humans (Merkhofer et al., 2014). While various components of the spliceosome can be co-purified with RNAPII, it is not clear whether their interaction is necessary for splicing to occur—the signals for binding of the splicing factors are present in the transcript, and the sequence of events leading to intron removal has been described in detail *in vitro* in the absence of transcription. Two snRNPs (U1 and U2, the latter recruited by the auxiliary factor U2AF) base-pair to the 5′ and 3′ splice sites and to the branch point. A complex of three additional snRNPs (U4–6) joins the others, assumes a catalytically active conformation and executes the phosphodiester bond exchange and the ligation of the two neighboring exons. Once an intron has been removed and the adjacent exons have been stitched together, the spliceosome components are free to be used at other splice sites.

In addition to producing an uninterrupted functional coding sequence in the mature transcript, splicing can represent a regulatory mechanism for differential gene function (Li et al., 2007). The mechanism consists of the use of alternative splicing sites on a given transcript to generate mature transcripts that differ by the presence of one or more exons, resulting in proteins with different functional specificities; alternative splicing can also introduce stop codons that lead to truncated and inactive proteins. The vast majority of human transcripts are alternatively spliced in a tissue-specific and developmental manner. Alternative splicing uses established splicing sites in the transcript sequence or cryptic sites that are made available by the masking of established sites either by RNA-binding regulatory proteins or by the specific folding of the nascent transcript as it exits the transcribing polymerase. Non-coding RNAs play an indirect role in regulating alternative splicing by controlling the level of specific splicing factors (Luco and Misteli, 2011). In addition, the rate of elongation of the TEC plays a role in determining the occurrence of alternative splicing by allowing sufficient time for the splicing machinery to recognize splice sites before they are masked by binding regulatory proteins or by transcript folding. As will be discussed in Chapter 5, the rate of elongation is controlled, in part, by chromatin structure. A classical example of regulatory alternative splicing is provided by the genetic pathway responsible for sex determination in *Drosophila*, and aberrant changes in alternative splicing of histone macroH2A, a histone H2A variant, are correlated with a number of different cancers (Box 3.3).

Transcription termination and processing of the transcript's 3′ end

The termination of transcription ensures that the TEC on a particular gene does not interfere with the promoter element of the next gene transcribed in the same direction, or with the progress of the TEC transcribing the next gene if the latter is proceeding in the opposite direction. In the case of most genes, the TEC continues to transcribe past a signal sequence (AAUAAA in the transcript) and eventually comes to a pause. A large multiprotein complex (CPF in yeast, CPSF in mammals) associates with the AAUAAA sequence; additional proteins (CF1A and CF1B in yeast, CstF, CFI_m and $CFII_m$ in mammals) and poly(A) polymerase (PAP) bind to the CPSF–RNA complex. CstF interacts with a GU-rich (or simply U-rich) sequence that is near the cleavage site. Following cleavage of the RNA molecule by the CFI_m and $CFII_m$ factors, PAP adds the A residues (Fig. 3.5). Polyadenylation of a transcript is necessary

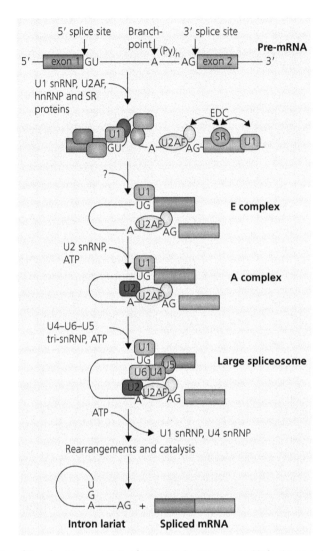

Fig. 3.4 Formation and function of the spliceosome, a complex of snRNPs and proteins responsible for the removal of introns from nascent RNA transcripts.

(From Li et al., 2007.)

Box 3.3 Two examples of regulatory alternative splicing

Sex determination in *Drosophila*. The ratio of the number of X chromosomes to the number of sets of autosomes—equal to 1 in females and to 0.5 in males—initiates the synthesis of a female-specific RNA-binding protein (Sex-lethal, SXL) in females (Fig. 3B.2). SXL regulates the splicing of the transformer (*tra*) gene transcript in such a manner that in females, an exon containing a stop codon is removed, resulting in the synthesis of a functional protein; in males, where SXL is absent, splicing does not remove the stop codon, leading to a truncated non-functional peptide. The TRA protein, in turn, regulates the splicing of the transcript of the doublesex (*dsx*) gene in a female-specific manner, producing a regulatory protein (DSXF) that initiates female sexual differentiation. In males, the absence of a functional TRA protein allows the

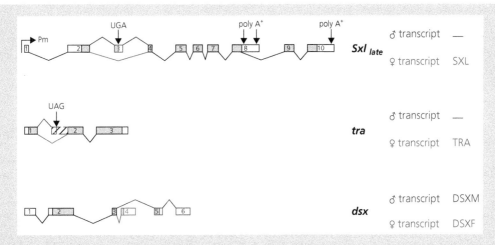

Fig. 3B.2 The sex determination pathway in *Drosophila*.

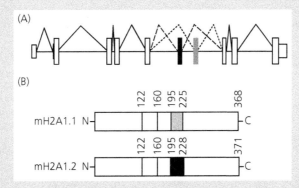

Fig. 3B.3 (A) Diagram illustrating the differential splicing of the macroH2A1 transcript. The two alternatively spliced exons are represented in gray and black, respectively. (B) Diagram of the two macroH2A1 variants.

(Modified from Novikov et al., 2011.)

unregulated or "default" splicing of the *dsx* gene transcript to proceed, leading to a protein (DSXM) that initiates male differentiation (Cline and Meyer, 1996).

A role for alternatively spliced transcripts in oncogenesis. As mentioned in Chapters 2 and 4, variants of some of the core histones can be incorporated into nucleosomes when specific nuclear functions, such as transcription or repair of DNA damage, are carried out. Some H2A variants, collectively called the macroH2A group, have an N-terminal region substantially similar to canonical H2A, and a large, non-histone-like globular C-terminal domain. Although some members of the group are the hallmark of the inactive

X-chromosome (to be discussed in detail in Chapter 9), macroH2As are widely distributed throughout chromatin and localize to other nuclear domains where they are not restricted to silencing but also fine-tune expression levels of specific genes. A macroH2A member of the group (mH2A1) can yield two transcript splice variants mH2A1.1 and mH2A1.2 (Fig. 3B.3). The mH2A1 isoform is involved in regulating cell proliferation in normal tissues. In several types of cancers, splicing of the mH2A1 transcript to yield mH2A1.1 is reduced, leading to a deficiency in the level of the mH2A1.1 protein (Novikov et al., 2011).

for its transport from the nucleus to the site of translation in the cytoplasm; it also ensures its stability by allowing the binding of poly(A)-binding proteins (PABPs) that prevent its degradation; in conjunction with the 5′ end cap, it enhances translation (Mandel et al., 2008). Cleavage and polyadenylation can occur at different sites, although the consensus sequence listed earlier is most frequently used. In mammals, the actual cleavage site is positioned between this sequence and a downstream element (DSE) consisting of a region rich in GU nucleotides. In yeast, four sequence elements have been identified in the polyadenylation region: an AU-rich efficiency element (EE), an A-rich positioning element (PE), the cleavage site and a U-rich element flanking the cleavage site.

Similarly to alternative splicing, polyadenylation can be regulatory. Alternative polyadenylation occurs in approximately 50% of the genes in the human genome and assumes two main forms—alternative poly(A) sites can occur in exons or in introns within coding regions, resulting in proteins that differ in their amino acid sequence; or they can occur in the 3′ untranslated region of transcripts, leading to the same exact protein but yielding mRNAs possessing different lengths of 3′ untranslated sequence. This type of alternative polyadenylation may have a distinct effect on the level of the transcript, and therefore of the protein gene product in particular cell types (Di Giammartino et al., 2011).

In most organisms, mRNAs are polyadenylated, with the exception of the replication-coupled histone mRNAs. The sequence of these transcripts generates a stem-loop structure that is bound by the stem-loop-binding protein (SLBP); cleavage occurs between the stem-loop and a purine-rich downstream element (HDE) that is recognized by a specific snRNP. Following cleavage, the SLBP remains bound to protect the mature histone RNAs from degradation and stimulate their translation (Dominski and Marzluff, 2007). Many non-coding RNAs are not polyadenylated. An intriguing interaction between the assembly of the PIC and the cleavage and polyadenylation complexes CstF and CPSF is mediated by the general transcription factor TFIIB. Phosphorylated TFIIB recruits these complexes at the promoter (PIC), as well as at the termination of transcription when the TEC is poised to allow 3′ end processing (Wang et al., 2010).

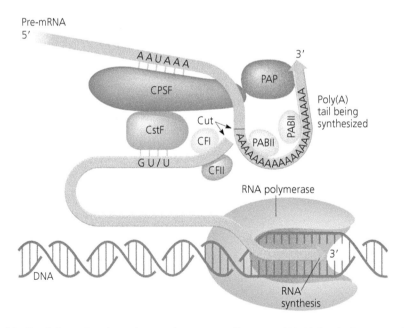

Fig. 3.5 Cleavage of the 3′ end of a terminated transcript occurs between a specific sequence (AAuAAA) and a downstream element rich in GU nucleotides. Following cleavage, a series of As are added to the 3′ end generated by the cleavage.

(From *iGenetics*, Second Edition)

Polymerase recycling and reinitiation of transcription

Clearly, for a polymerase to engage in a new round of transcription, the CTD must be dephosphorylated; this task is accomplished by three phosphatases Fcp1, Ssu72 and Rtr1. Yet, the events that mediate the reinitiation of transcription on an activated gene and the recycling of RNAPII and other factors of the TEC after its release from the DNA template have not been well defined. As the promoter region is cleared, pausing is overcome and productive elongation is initiated, a new PIC can be assembled for the generation of a new transcript. This process can be facilitated in some cases by the formation of a loop that brings the 3' end of the transcriptional unit in contact with the promoter region. First discovered in yeast (O'Sullivan et al., 2004), gene loop formation occurs in a small percentage of very highly expressed genes in mouse and human cells (Grosso et al., 2012). Differences appear to exist between organisms regarding which factors of the PIC remain at the promoter after initiation and which of the general transcription factors travel with the transcribing RNAPII.

A number of genes show bursts of transcription consistent with reinitiation, while on other genes, transcription appears to occur randomly. The latter case is likely due to the dynamic binding of transcription factors and nucleosomes in the promoter region, leading to the observation that, at any given time, some of the cells within a tissue express a particular gene, while others do not (Ko et al., 1990). The first demonstration of transcriptional bursts was reported in the social amoeba *Dictyostelium* by determining the accumulation over time of transcripts from a single gene in single cells (Chubb et al., 2006). More recently, imaging of transcription at high temporal resolution in mammalian cells demonstrated that transcription bursts consist of groups of closely spaced RNAPII complexes, with gaps between the bursts that vary in length in different genes (Tantale et al., 2016).

Chapter summary

Transcription is the fundamental mechanism whereby all of the information contained in the genome is extracted. It is initiated by factors that interact with the RNA polymerases and recruit them to specific sites, unwind the DNA molecules and allow the synthesis of RNA transcripts complementary to one of the single DNA strands. Different polymerases transcribe different regions of the genome: RNA polymerase II (RNAPII) transcribes genes that encode proteins; RNAPI transcribes ribosomal RNA genes; RNAPIII transcribes genes that encode tRNAs and small RNAs involved in processing other RNAs. Genes that encode non-coding RNAs responsible for numerous regulatory steps that impact the structure and function of the genetic material are transcribed by RNAPII or RNAPIII, depending on where they are located in the genome.

The transcription process consists of the formation of a pre-initiation complex (PIC) and its activation, initiation and promoter clearance. Transcriptional activation is achieved by chromatin looping that brings a gene in contact with a cognate enhancer. Enhancers usually interact with the PIC through a large multiprotein complex termed the Mediator complex. Following initiation, RNAPII often makes a promoter-proximal pause that it eventually overcomes to undergo productive elongation of the transcript and eventually termination. Transition through the different phases of transcription is orchestrated by the dynamic phosphorylation of the main subunit of RNAPII and the concomitant recruitment of regulatory factors.

The 5' end of many transcripts is protected by a methylated guanosine "cap," and the 3' end by the addition of a "tail" consisting of a chain of adenosine monophosphates. The transcripts of many protein-coding genes and of long non-coding RNA genes are spliced to remove regions that interrupt the coding sequence or that may interfere with the functional sequence of the RNA. Transcript splicing and polyadenylation can occur at alternate sites, resulting in proteins that differ in their amino acid sequence or in transcripts with different length of untranslated 3' ends.

References

Akhtar, W. & Veenstra, G. J. 2011. TBP-related factors: a paradigm of diversity in transcription initiation. *Cell Biosci*, 1, 23.

Anish, R., Hossain, M. B., Jacobson, R. H. & Takada, S. 2009. Characterization of transcription from TATA-less promoters: identification of a new core promoter element XCPE2 and analysis of factor requirements. *PLoS One*, 4, e5103.

Aso, T., Lane, W. S., Conaway, J. W. & Conaway, R. C. 1995. Elongin (SIII): a multisubunit regulator of elongation by RNA polymerase II. *Science*, 269, 1439–43.

Bartkowiak, B., Liu, P., Phatnani, H. P., Fuda, N. J., Cooper, J. J., Price, D. H., Adelman, K., Lis, J. T. & Greenleaf, A. L. 2010. CDK12 is a transcription elongation-associated CTD kinase, the metazoan ortholog of yeast Ctk1. *Genes Dev*, 24, 2303–16.

Blazek, D., Kohoutek, J., Bartholomeeusen, K., Johansen, E., Hulinkova, P., Luo, Z., Cimermancic, P., Ule, J. & Peterlin, B. M. 2011. The Cyclin K/Cdk12 complex maintains genomic stability via regulation of expression of DNA damage response genes. *Genes Dev*, 25, 2158–72.

Bowman, E. A., Bowman, C. R., Ahn, J. H. & Kelly, W. G. 2013. Phosphorylation of RNA polymerase II is independent of P-TEFb in the *C. elegans* germline. *Development*, 140, 3703–13.

Bres, V., Yoh, S. M. & Jones, K. A. 2008. The multi-tasking P-TEFb complex. *Curr Opin Cell Biol*, 20, 334–40.

Bulger, M. & Groudine, M. 2011. Functional and mechanistic diversity of distal transcription enhancers. *Cell*, 144, 327–39.

Chubb, J. R., Trcek, T., Shenoy, S. M. & Singer, R. H. 2006. Transcriptional pulsing of a developmental gene. *Curr Biol*, 16, 1018–25.

Cline, T. W. & Meyer, B. J. 1996. Vive la différence: males vs females in flies vs worms. *Annu Rev Genet*, 30, 637–702.

Conaway, R. C. & Conaway, J. W. 2011. Function and regulation of the Mediator complex. *Curr Opin Genet Dev*, 21, 225–30.

Creppe, C. & Buschbeck, M. 2011. Elongator: an ancestral complex driving transcription and migration through protein acetylation. *J Biomed Biotechnol*, 2011, 924898.

Deng, W. & Roberts, S. G. 2007. TFIIB and the regulation of transcription by RNA polymerase II. *Chromosoma*, 116, 417–29.

Devaiah, B. N., Lewis, B. A., Cherman, N., Hewitt, M. C., Albrecht, B. K., Robey, P. G., Ozato, K., Sims, R. J., 3RD & Singer, D. S. 2012. BRD4 is an atypical kinase that phosphorylates serine2 of the RNA polymerase II carboxy-terminal domain. *Proc Natl Acad Sci U S A*, 109, 6927–32.

DI Giammartino, D. C., Nishida, K. & Manley, J. L. 2011. Mechanisms and consequences of alternative polyadenylation. *Mol Cell*, 43, 853–66.

Donahue, B. A., Yin, S., Taylor, J. S., Reines, D. & Hanawalt, P. C. 1994. Transcript cleavage by RNA polymerase II arrested by a cyclobutane pyrimidine dimer in the DNA template. *Proc Natl Acad Sci U S A*, 91, 8502–6.

Donner, A. J., Ebmeier, C. C., Taatjes, D. J. & Espinosa, J. M. 2010. CDK8 is a positive regulator of transcriptional elongation within the serum response network. *Nat Struct Mol Biol*, 17, 194–201.

Egloff, S., O'reilly, D., Chapman, R. D., Taylor, A., Tanzhaus, K., Pitts, L., Eick, D. & Murphy, S. 2007. Serine-7 of the RNA polymerase II CTD is specifically required for snRNA gene expression. *Science*, 318, 1777–9.

Eick, D. & Geyer, M. 2013. The RNA polymerase II carboxy-terminal domain (CTD) code. *Chem Rev*, 113, 8456–90.

Fei, J., Ishii, H., Hoeksema, M. A., Meitinger, F., Kassavetis, G. A., Glass, C.K., Ren, B. & Kadonaga, J. T. 2018. NDF, a nucleosome destabilizing factor that facilitates transcription through nucleosomes. *Genes Dev*, 32, 1–13.

Gershenzon, N. I. & Ioshikhes, I. P. 2005. Synergy of human Pol II core promoter elements revealed by statistical sequence analysis. *Bioinformatics*, 21, 1295–300.

Gomez-Gonzalez, B., Garcia-Rubio, M., Bermejo, R., Gaillard, H., Shirahige, K., Marin, A., Foiani, M. & Aguilera, A. 2011. Genome-wide function of THO/TREX in active genes prevents R-loop-dependent replication obstacles. *EMBO J*, 30, 3106–19.

Grosso, A. R., De Almeida, S. F., Braga, J. & Carmo-Fonseca, M. 2012. Dynamic transitions in RNA polymerase II density profiles during transcription termination. *Genome Res*, 22, 1447–56.

Haberle, V. & Lenhard, B. 2016. Promoter architectures and developmental gene regulation. *Semin Cell Dev Biol*, 57, 11–23.

Hendrix, D. A., Hong, J. W., Zeitlinger, J., Rokhsar, D. S. & Levine, M. S. 2008. Promoter elements associated with RNA Pol II stalling in the *Drosophila* embryo. *Proc Natl Acad Sci U S A*, 105, 7762–7.

Hintermair, C., Heidemann, M., Koch, F., Descostes, N., Gut, M., Gut, I., Fenouil, R., Ferrier, P., Flatley, A., Kremmer, E., Chapman, R. D., Andrau, J. C. & Eick, D. 2012. Threonine-4 of mammalian RNA polymerase II CTD is targeted by Polo-like kinase 3 and required for transcriptional elongation. *EMBO J*, 31, 2784–97.

Izban, M. G. & Luse, D. S. 1992. The RNA polymerase II ternary complex cleaves the nascent transcript in a 3′–5′ direction in the presence of elongation factor SII. *Genes Dev*, 6, 1342–56.

Jang, M. K., Mochizuki, K., Zhou, M., Jeong, H. S., Brady, J. N. & Ozato, K. 2005. The bromodomain protein Brd4 is a positive regulatory component of P-TEFb and stimulates RNA polymerase II-dependent transcription. *Mol Cell*, 19, 523–34

Kagey, M. H., Newman, J. J., Bilodeau, S., Zhan, Y., Orlando, D. A., Van Berkum, N. L., Ebmeier, C. C., Goossens, J., Rahl, P. B., Levine, S. S., Taatjes, D. J., Dekker, J. & Young, R. A. 2010. Mediator and cohesin connect gene expression and chromatin architecture. *Nature*, 467, 430–5.

Keene, R. G., Mueller, A., Landick, R. & London, L. 1999. Transcriptional pause, arrest and termination sites for RNA polymerase II in mammalian *N*- and *c-myc* genes. *Nucleic Acids Res*, 27, 3173–82.

Kelley, D. E., Stokes, D. G. & Perry, R. P. 1999. CHD1 interacts with SSRP1 and depends on both its chromodomain and its ATPase/helicase-like domain for proper association with chromatin. *Chromosoma*, 108, 10–25.

Kim, T. K., Hemberg, M. & Gray, J. M. 2015. Enhancer RNAs: a class of long noncoding RNAs synthesized at enhancers. *Cold Spring Harb Perspect Biol*, 7, a018622.

Kim, T. K., Hemberg, M., Gray, J. M., Costa, A. M., Bear, D. M., Wu, J., Harmin, D. A., Laptewicz, M., Barbara-Haley, K., Kuersten, S., Markenscoff-Papadimitriou, E., Kuhl, D., Bito, H., Worley, P. F., Kreiman, G. & Greenberg, M. E. 2010. Widespread transcription at neuronal activity-regulated enhancers. *Nature*, 465, 182–7.

Ko, M. S., Nakauchi, H. & Takahashi, N. 1990. The dose dependence of glucocorticoid-inducible gene expression results from changes in the number of transcriptionally active templates. *EMBO J*, 9, 2835–42.

Kolaj-Robin, O. & Seraphin, B. 2017. Structures and activities of the Elongator complex and its cofactors. *Enzymes*, 41, 117–49.

Koleske, A. J. & Young, R. A. 1994. An RNA polymerase II holoenzyme responsive to activators. *Nature*, 368, 466–9.

Krogan, N. J., Kim, M., Ahn, S. H., Zhong, G., Kobor, M. S., Cagney, G., Emili, A., Shilatifard, A., Buratowski, S. & Greenblatt, J. F. 2002. RNA polymerase II elongation factors of *Saccharomyces cerevisiae*: a targeted proteomics approach. *Mol Cell Biol*, 22, 6979–92.

Kwak, H. & Lis, J. T. 2013. Control of transcriptional elongation. *Annu Rev Genet*, 47, 483–508.

Lee, H., Kraus, K. W., Wolfner, M. F. & Lis, J. T. 1992. DNA sequence requirements for generating paused polymerase at the start of hsp70. *Genes Dev*, 6, 284–95.

Lenasi, T., Peterlin, B. M. & Barboric, M. 2011. Cap-binding protein complex links pre-mRNA capping to transcription elongation and alternative splicing through positive transcription elongation factor b (P-TEFb). *J Biol Chem*, 286, 22758–68.

Li, Q., Lee, J. A. & Black, D. L. 2007. Neuronal regulation of alternative pre-mRNA splicing. *Nat Rev Neurosci*, 8, 819–31.

Lu, L., Fan, D., Hu, C. W., Worth, M., Ma, Z. X. & Jiang, J. 2016. Distributive O-GlcNAcylation on the highly repetitive C-terminal domain of RNA polymerase II. *Biochemistry*, 55, 1149–58.

Luco, R. F. & Misteli, T. 2011. More than a splicing code: integrating the role of RNA, chromatin and non-coding RNA in alternative splicing regulation. *Curr Opin Genet Dev*, 21, 366–72.

Luecke, H. F. & Yamamoto, K. R. 2005. The glucocorticoid receptor blocks P-TEFb recruitment by NFkappaB to effect promoter-specific transcriptional repression. *Genes Dev*, 19, 1116–27.

Malik, S. & Roeder, R. G. 2010. The metazoan Mediator co-activator complex as an integrative hub for transcriptional regulation. *Nat Rev Genet*, 11, 761–72.

Mandal, S. S., Chu, C., Wada, T., Handa, H., Shatkin, A. J. & Reinberg, D. 2004. Functional interactions of RNA-capping enzyme with factors that positively and negatively regulate promoter escape by RNA polymerase II. *Proc Natl Acad Sci U S A*, 101, 7572–7.

Mandel, C. R., Bai, Y. & Tong, L. 2008. Protein factors in pre-mRNA 3′-end processing. *Cell Mol Life Sci*, 65, 1099–122.

Mattaj, I. W., Lienhard, S., Jiricny, J. & De Robertis, E. M. 1985. An enhancer-like sequence within the *Xenopus* U2 gene promoter facilitates the formation of stable transcription complexes. *Nature*, 316, 163–7.

Merkhofer, E. C., Hu, P. & Johnson, T. L. 2014. Introduction to cotranscriptional RNA splicing. *Methods Mol Biol*, 1126, 83–96.

Moniaux, N., Nemos, C., Deb, S., Zhu, B., Dornreiter, I., Hollingsworth, M. A. & Batra, S. K. 2009. The human RNA polymerase II-associated factor 1 (hPaf1): a new regulator of cell-cycle progression. *PLoS One*, 4, e7077.

Muller, F. & Tora, L. 2004. The multicoloured world of promoter recognition complexes. *EMBO J*, 23, 2–8.

Napolitano, G., Majello, B., Licciardo, P., Giordano, A. & Lania, L. 2000. Transcriptional activity of positive transcription elongation factor b kinase *in vivo* requires the C-terminal domain of RNA polymerase II. *Gene*, 254, 139–45.

Nechaev, S. & Adelman, K. 2011. Pol II waiting in the starting gates: regulating the transition from transcription initiation into productive elongation. *Biochim Biophys Acta*, 1809, 34–45.

Novikov, L., Park, J. W., Chen, H., Klerman, H., Jalloh, A. S. & Gamble, M. J. 2011. QKI-mediated alternative splicing of the histone variant MacroH2A1 regulates cancer cell proliferation. *Mol Cell Biol*, 31, 4244–55.

Orphanides, G., Leroy, G., Chang, C. H., Luse, D. S. & Reinberg, D. 1998. FACT, a factor that facilitates transcript elongation through nucleosomes. *Cell*, 92, 105–16.

Ossipow, V., Fonjallaz, P. & Schibler, U. 1999. An RNA polymerase II complex containing all essential initiation factors binds to the activation domain of PAR leucine zipper transcription factor thyroid embryonic factor. *Mol Cell Biol*, 19, 1242–50.

O'sullivan, J. M., Tan-Wong, S. M., Morillon, A., Lee, B., Coles, J., Mellor, J. & Proudfoot, N. J. 2004. Gene loops juxtapose promoters and terminators in yeast. *Nat Genet*, 36, 1014–18.

Pan, G., Aso, T. & Greenblatt, J. 1997. Interaction of elongation factors TFIIS and elongin A with a human RNA polymerase II holoenzyme capable of promoter-specific initiation and responsive to transcriptional activators. *J Biol Chem*, 272, 24563–71.

Perales, R. & Bentley, D. 2009. "Cotranscriptionality": the transcription elongation complex as a nexus for nuclear transactions. *Mol Cell*, 36, 178–91.

Qiu, H., Hu, C. & Hinnebusch, A. G. 2009. Phosphorylation of the Pol II CTD by KIN28 enhances BUR1/BUR2 recruitment and Ser2 CTD phosphorylation near promoters. *Mol Cell*, 33, 752–62.

Rahl, P. B., Lin, C. Y., Seila, A. C., Flynn, R. A., Mccuine, S., Burge, C. B., Sharp, P. A. & Young, R. A. 2010. c-Myc regulates transcriptional pause release. *Cell*, 141, 432–45.

Ranuncolo, S. M., Ghosh, S., Hanover, J. A., Hart, G. W. & Lewis, B. A. 2012. Evidence of the involvement of O-GlcNAc-modified human RNA polymerase II CTD in transcription *in vitro* and *in vivo*. *J Biol Chem*, 287, 23549–61.

Rasmussen, E. B. & Lis, J. T. 1993. *In vivo* transcriptional pausing and cap formation on three *Drosophila* heat shock genes. *Proc Natl Acad Sci U S A*, 90, 7923–7.

Reinberg, D. & Sims, R. J., 3RD 2006. de FACTo nucleosome dynamics. *J Biol Chem*, 281, 23297–301.

Riethoven, J. J. 2010. Regulatory regions in DNA: promoters, enhancers, silencers, and insulators. *Methods Mol Biol*, 674, 33–42.

Rondon, A. G., Jimeno, S. & Aguilera, A. 2010. The interface between transcription and mRNP export: from THO to THSC/TREX-2. *Biochim Biophys Acta*, 1799, 533–8.

Rougvie, A. E. & Lis, J. T. 1988. The RNA polymerase II molecule at the 5′ end of the uninduced *hsp70* gene of *D. melanogaster* is transcriptionally engaged. *Cell*, 54, 795–804.

Sadowski, C. L., Henry, R. W., Lobo, S. M. & Hernandez, N. 1993. Targeting TBP to a non-TATA box cis-regulatory element: a TBP-containing complex activates transcription from snRNA promoters through the PSE. *Genes Dev*, 7, 1535–48.

Shilatifard, A., Lane, W. S., Jackson, K. W., Conaway, R. C. & Conaway, J. W. 1996. An RNA polymerase II elongation factor encoded by the human *ELL* gene. *Science*, 271, 1873–6.

Simic, R., Lindstrom, D. L., Tran, H. G., Roinick, K. L., Costa, P. J., Johnson, A. D., Hartzog, G. A. & Arndt, K. M. 2003. Chromatin remodeling protein Chd1 interacts with transcription elongation factors and localizes to transcribed genes. *EMBO J*, 22, 1846–56.

Smith, E., Lin, C. & Shilatifard, A. 2011. The super elongation complex (SEC) and MLL in development and disease. *Genes Dev*, 25, 661–72.

Tantale, K., Mueller, F., Kozulic-Pirher, A., Lesne, A., Victor, J. M., Robert, M. C., Capozi, S., Chouaib, R., Backer, V., Mateos-Langerak, J., Darzacq, X., Zimmer, C., Basyuk, E. & Bertrand, E. 2016. A single-molecule view of transcription reveals convoys of RNA polymerases and multi-scale bursting. *Nat Commun*, 7, 12248.

Tsai, S. Y., Chang, Y. L., Swamy, K. B., Chiang, R. L. & Huang, D. H. 2016. GAGA factor, a positive regulator of global gene expression, modulates transcriptional pausing and organization of upstream nucleosomes. *Epigenetics Chromatin*, 9, 32.

Vo Ngoc, L., Wang, Y. L., Kassavetis, G. A. & Kadonaga, J. T. 2017. The punctilious RNA polymerase II core promoter. *Genes Dev*, 31, 1289–1301.

Wang, Y., Fairley, J. A. & Roberts, S. G. 2010. Phosphorylation of TFIIB links transcription initiation and termination. *Curr Biol*, 20, 548–53.

Zentner, G. E., Tesar, P. J. & Scacheri, P. C. 2011. Epigenetic signatures distinguish multiple classes of enhancers with distinct cellular functions. *Genome Res*, 21, 1273–83.

PART II

Transcription is Regulated by Epigenetic Mechanisms

.

CHAPTER 4

Chromatin modification and remodeling

Gene transcription in the context of chromatin

In 1976, Harold Weintraub and Mark Groudine made a seminal observation that related gene activity to chromatin structure. Chick red blood cells are nucleated and synthesize large amounts of hemoglobin. Exposure of isolated red blood cell nuclei to DNase I led to the preferential digestion of the globin genes, detected with a probe prepared from adult red cell globin mRNA. In contrast, the globin genes present in fibroblast nuclei, where they are not activated, were refractory to digestion under the same conditions. Weintraub and Groudine arrived to the prescient conclusion that gene activation requires an altered chromosome structure, which, in turn, renders transcription units sensitive to DNase I digestion; the structure of inactive genes protects them from the action of the enzyme. They speculated that "active genes are probably associated with histones in a subunit conformation in which the associated DNA is particularly sensitive to digestion…" These fundamental observations were extended later by the demonstration that a gene in a particular tissue exhibits a clear difference in nuclease sensitivity at low level of expression (in other words, when it is uninduced) and at high level of expression following induction in the same tissue (Eissenberg and Lucchesi, 1983).

RNA polymerase II (RNAPII) transcribes all messenger RNA and most regulatory RNA-encoding sequences and genes. *In vitro*, RNAPII is inhibited by the presence of nucleosomes along the DNA template. Yet, transcription occurs, as required, for differentiation and maintenance in eukaryotic cells. This suggests that cells have the means of overcoming the nucleosome barrier. Of course, transcription involves the implementation of many additional parameters (see Chapter 3). The process begins with the assembly of a complex in the promoter region of transcription units. Simultaneously, various factors are targeted to specific regulatory DNA sequences in the general vicinity of the promoter. These factors interact physically with the promoter complex, which includes an RNA polymerase, and launch it on its path of RNA synthesis. The organization of chromatin described above is unfavorable to these different associations—regulatory sequences can be included in the regions of the DNA molecule that are tightly wrapped around histone octamers, rendering them inaccessible, and the supercoiled arrangement of the 30-nm fiber presents a barrier to the access of large regulatory complexes.

In order to allow transcription to occur, the structure that the chromatin fiber assumes naturally under physiological conditions must be modified—the compaction of the 30-nm fiber must be loosened and the association of nucleosomes with the DNA that is wrapped around them must be altered either by removing the octamers from their position or by diminishing their numerous contacts with the DNA. The usual first step in this process is the binding of specific factors, referred to as **pioneer transcription factors**, to their respective DNA binding sequences. Eukaryotic organisms have evolved an amazing variety of subsequent activities, which can be used individually but most often are used in specific combinations, to modify the architecture of the chromatin fiber. A number of these modifications allow the

Epigenetics, Nuclear Organization and Gene Function: with implications of epigenetic regulation and genetic architecture for human development and health. John C. Lucchesi, Oxford University Press (2019). © John C. Lucchesi 2019.
DOI: 10.1093/oso/9780198831204.001.0001

assembly of the transcriptional machinery, the onset of gene activity and its maintenance; others reverse these effects for the purpose of silencing genes or function to maintain regions of chromatin in a repressed state (see Chapter 7). Most of these modifications do not affect the coding sequence of the DNA; therefore, rather than being considered "genetic" in nature, they are said to be "epigenetic." DNA can be methylated at certain cytosines; histones can be acetylated, methylated, phosphorylated or modified by the addition of ubiquitin or sumo peptides, or of single molecules or chains of ADP ribose. Nucleosomes can be displaced, or their association with DNA can be altered by multiprotein complexes that hydrolyze ATP in order to generate the energy necessary for these functions. The composition of nucleosomes is not static, and core histones are replaced by variants that play roles not only in transcription and in certain aspects of gene silencing, but also in marking sites of double-stranded DNA breaks for repair and in centromere function. Linker histones, which stabilize condensed chromatin, can be covalently modified, leading to their removal and chromatin decompaction. Small non-coding RNA molecules participate in specific DNA or histone modifications, while long non-coding RNA molecules are involved in regulating the activity of individual genes or of entire chromosomes.

In conclusion, the differential expression of the genome in time and space during the development and adult life of eukaryotic organisms involves a complex interplay between DNA sequences, transcription factors, histones and external stimuli. The biochemical nature of this interplay and the special role of nuclear structures on its implementation are the subject of much of the remainder of this book.

Epigenetic modifications are associated with gene activity and genome stability; therefore, they are involved in every step of normal development. The process of transcription is concomitant with the covalent modification of DNA and histones, the replacement of core histones with their variants and the ATP-dependent remodeling of the DNA–histone octamer association. These epigenetic modifications ensure the expression of the genome at the level of individual genes (discussed in Chapter 7). They also control the coordinated expression of gene clusters, as well as the level of expression of individual chromosomes (discussed in Chapter 9). Epigenetic modifications are also involved with DNA replication and with DNA damage repair (discussed in Chapters 14 and 15). Not surprisingly, epigenetic defects are responsible for developmental pathologies (see Box 4.1; Chapter 8, Boxes 8.1 and 8.2; and Chapters 20–22).

Post-translational modifications (PTMs) of histones

The addition of functional groups to specific amino acids within histone molecules can affect chromatin in a number of different ways. Functional groups can attract factors and complexes that recognize and bind to them or that bind to DNA sequences that have been made available, further affecting chromatin structure. If the groups are polar, they may alter the contacts between the DNA and nucleosomes or between neighboring nucleosomes. Certain of these modifications are added preferentially to the amino acids in the N-terminal tails of histone molecules that extend in an unstructured fashion from the globular core of the histone octamers; others occur at specific sites throughout the length of particular histones (Fig. 4.1).

Nucleosomes usually exhibit an amazing array of histone marks. Although some exceptions exist where two specific neighboring modifications collaborate to mediate an effect, in general, these marks are recognized individually by factors and complexes that alter chromatin structure and function.

Acetylation

This was the first covalent modification to be discovered and the first to be most extensively studied with respect to its role in chromatin assembly and gene activation (Allfrey et al., 1964; Pogo et al., 1966). Histone acetylation is the responsibility of a group of lysine acetyltransferases (HKATs); these enzymes use acetyl-coenzyme A (acetyl-CoA) as the donor and exhibit generally broad specificity, including the acetylation of non-histone substrates that is often narrowed to specific lysines or subsets of lysines when the acetyltransferases are present in large multiprotein complexes (Fig. 4.2). The first nuclear HKAT was identified by David Allis in the macronuclei of the ciliate protozoan *Tetrahymena* (Brownell and Allis, 1995); its homology to a known yeast transcriptional activator Gcn5 directly linked histone acetylation with gene activation (Brownell et al., 1996). Acetyl groups can be removed by lysine deacetylases (HKDACs) that often work in conjunction with other factors involved in gene silencing. The first evidence of their existence was reported soon after Allfrey's seminal observations (Inoue and Fujimoto, 1969). A different group of histone deacetylases, the **sirtuins**, perform their reaction differently from HKDACs (discussed in Chapter 21).

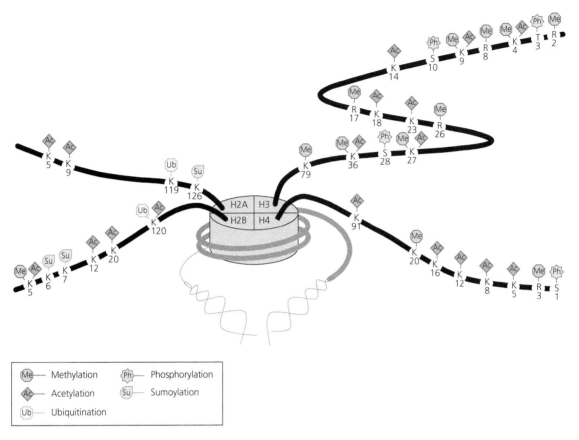

Fig. 4.1 Possible post-translational modifications of the core histones.
(From Araki and Mimura, 2017.)

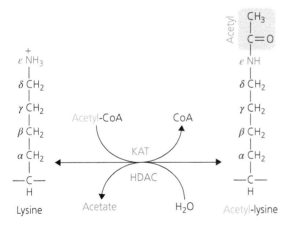

Fig. 4.2 Acetylation and deacetylation of lysine.
(From Kim et al., 2010.)

The genomic distribution of histone acetyltransferases and histone deacetylases in mammalian cells has established their direct correlation with gene expression (Wang et al., 2009). Certain HKATs are primarily associated with enhancers and promoters, while others are elevated specifically on promoters and transcribed regions. Some HKDACs are associated with HKATs on active genes where their primary function is to maintain the appropriate level of histone acetylation; others remove the spurious acetylation of inactive promoters; others still are primed by the previous methylation of histone H3 at lysine 4 (see Methylation, p. 38) to poise genes for potential activation.

Numerous lysine residues, present in the N terminal tails and in the globular portion of the core histones and of linker and variant histone molecules, can be acetylated. This modification neutralizes the positive charge on the lysines, thereby diminishing the charge-dependent attraction of the histones to the negatively charged DNA. For example, the acetylation of histone

H3 at lysine 56 (H3K56ac), which is near the region where DNA enters and exits its association with the histone octamer, or at lysine 122 (H3K122ac), which is located where the interaction between the DNA and the octamer is the strongest, destabilizes the nucleosome and affects transcription (Devaiah et al., 2016; Simon et al., 2011; Tropberger et al., 2013; Xu et al., 2005). In addition to this general effect, particular acetylated lysines perform very specific roles in regulating gene function. For example, the acetylation of histone H3 at lysine 9 (H3K9ac) prevents it from becoming methylated, a mark of silenced chromatin; the acetylation of H4 at lysine 16 (H4K16ac) appears to be especially involved in weakening the interaction between neighboring nucleosomes (Dunlap et al., 2012; Liu et al., 2011; Robinson et al., 2008; Shogren-Knaak et al., 2006). Active enhancers are marked by histone H3 acetylated at lysine 27 (H3K27ac). All of these considerations may explain why histone acetylation is associated with active transcription (discussed in Chapter 5). In contrast, some preliminary evidence suggests that H4 acetylated at lysine 20 (H4K20ac) may play a repressive role; although it is absent from completely silent genes, it is found in the promoter region of minimally expressed genes where it co-localizes with transcriptional repressor complexes (Kaimori et al., 2016). Histone acetylation occurs also in conjunction with DNA replication and DNA repair where, presumably, it facilitates access to the origin of replication sites (Unnikrishnan, et al., 2010) and sites of DNA damage (Bird et al., 2002; Murr et al., 2006).

An analysis of global acetylation (and methylation) motifs in *Drosophila* cells revealed that the loss of a particular modification was often compensated by a gain in a different modification (Feller et al., 2015). For example, reduction in the level of acetylation of a particular lysine (H4K16ac) by ablation of the specific HKAT that is responsible for its deposition [males absent on the first (MOF)] was compensated by a significant increase in the acetylation of the neighboring lysine (H4K12ac) and by an increase in the presence of repressive histone methylation marks. Similar results were obtained with human cells (Feller et al., 2015), leading to the suggestion that, in higher metazoans, in addition to their role in the regulation of individual genes, acetylated histones are modulated to maintain a normal chromatin structure.

In addition to acetylation, a number of different acylations of histone lysines have been reported: formylation (Wisniewski et al., 2008), butyrylation and crotonylation (Tweedie-Cullen et al., 2012) and succinylation and malonylation (Xie et al., 2012). These acylation reactions use coenzyme-A as a group donor and, similarly to acetylation, alter the charge of the lysine residue. Preliminary observations suggest that these modifications may play a role in chromatin regulation (Xie et al., 2012), that is distinct from lysine acetylation (Tan et al., 2011; Xie et al., 2016).

Methylation

This was also discovered several years ago (Allfrey et al., 1964), but its role in the regulation of transcription was suggested much later (Chen et al., 1999), and so was the discovery of the first histone methyl transferase by Thomas Jenuwein (Rea et al., 2000). Spurred by the wealth of observations that linked a group of *Drosophila* genes, identified as suppressors of position–effect variegation (discussed in Chapter 2), with various aspects of chromatin modification, Jenuwein and colleagues cloned the *SUV39h1* human gene and found that its product selectively methylated lysine 9 on histone H3 (H3K9me2). The active site of this enzyme was determined to lie within the highly evolutionarily conserved SET domain, a 130-amino acid-long region (the name is an acronym of the three *Drosophila* genes where it was first discovered) (Jenuwein et al., 1998). Using this molecular fingerprint, numerous additional histone lysine methyltransferases (HKMTs) have been identified and more are predicted. To date, only two HKMTs lack the SET domain (Patel et al., 2011; van Leeuwen et al., 2002). Lysines can be mono-, di- or trimethylated by specific methyl transferases (Fig. 4.3), and the levels of methylation on the same residue have been implicated to affect chromatin structure and transcription differentially. A case in point is the presence of the monomethylated form of lysine 27 on histone H3 (H3K27me1) on the promoters of actively transcribing genes, whereas the trimethylated form H3K27me3 is associated with silenced genes (Barski et al., 2007). In all cases, the methyl group donor is S-adenosylmethionine (SAM).

Histone methylation does not change the overall charge of the histone molecules; it plays a key role in recruiting proteins and complexes that regulate chromatin function. In contrast to acetylation, which is generally associated with gene activation, methylation can have positive or negative effects on transcription, depending on the position of the lysine that is being modified on a particular histone, and on the context in which it occurs. For example, histone H3 monomethylated at lysine 4 (H3K4me1) is present on active enhancers, and the trimethylated form of lysine 4 on histone H3 (H3K4me3) is present at the 5' end of active

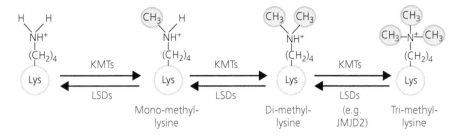

Fig. 4.3 Lysine methylation and demethylation.

(From Pek et al., 2012.)

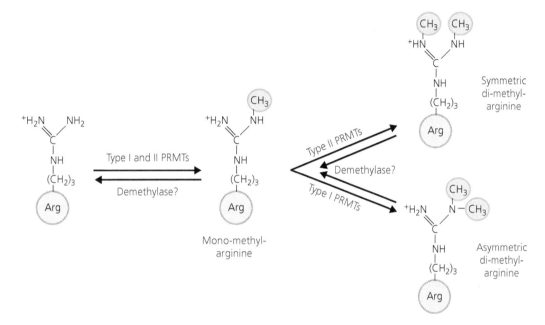

Fig. 4.4 Arginine methylation.

(From Pek et al., 2012.)

genes and is usually associated with transcriptional activity; at DNA damage sites, it can trigger the activation of a checkpoint and interrupt the cell cycle (Shi et al., 2006). Histone H3 methylated at lysine 9 (H3K9me3), the modification laid down by Su(var)3-9 in *Drosophila* and SUV39h1 in humans, is invariably associated with gene silencing, principally through the recruitment of heterochromatin protein 1 (HP1). The presence of H3K9me3 prevents the methylation of H3K4 by its specific HKMT (SET7 and SETDB1 in humans); the reverse is true as well, illustrating that modified amino acids can affect the potential for modification of neighboring sites (Wang et al., 2001a). Histone H3 trimethylated at lysine 27 (H3K27me3) is associated with gene silencing.

Methylation can also occur on arginine. In histones, arginine can be monomethylated, or dimethylated in a symmetric or an asymmetric manner (Fig. 4.4) by protein methyl transferases (PRMTs) that often also methylate non-histone proteins. Whether the dimethylation is symmetric or not influences the interactions of the modified arginine with neighboring modifications. Asymmetrically dimethylated arginine 2 on histone H3 (H3R2me2a) prevents the methylation of H3K4 (Guccione et al., 2007), while symmetrically methylated arginine 2 (H3R2me2s) facilitates it (Yuan et al., 2012).

For many years, following its discovery, histone methylation was considered to be a permanent covalent modification that could only be removed from chromatin by the reassembly of nucleosomes with non-methylated histones or by the conversion of methylated arginine to citrulline. The discovery of the first histone demethylase by Yang Shi (Shi et al., 2004) established the dynamic nature of methylation. The enzyme lysine-specific demethylase 1 (LSD1), which demethylates H3K4me1 and me2, is present in a number of complexes that repress gene activity. Surprisingly, the association of LSD1 with the androgen receptor alters its specificity and targets it to the demethylation of H3K9me2, thereby changing its contribution from gene repression to activation (Metzger et al., 2005). A large number of additional demethylases were discovered, some of which could demethylate H3K4me3. All of these enzymes are characterized by the presence of the Jumonji domain, a 350-amino acid region that recognizes and demethylates modified lysines.

Phosphorylation

Phosphorylation of serine (S), threonine (T) or tyrosine (Y) residues by kinases that use ATP as the phosphate donor leads to a reduction of the overall positive charge of the histones' N-terminal tails. The tails of all four histones can be phosphorylated, and this modification can be removed by phosphatases (Fig. 4.5). As in the case of acetylation, phosphorylation reduces the electrostatic interactions between the histone tails and DNA and promotes a more loose organization of the chromatin fibers that facilitates gene activation during interphase. Surprisingly, during the cell's preparation for mitosis, phosphorylation is instrumental in the process of chromatin condensation (discussed below). The occurrence of phosphorylation at the onset of transcription was first reported in 1991 by L.C. Mahadevan and collaborators in cells arrested in G0 and then induced to grow with mitogen stimulators. In these cells, there was a rapid and transient phosphorylation of histone H3 that was correlated with the activation of immediate early genes, including proto-oncogenes like c-fos and c-jun (Mahadevan et al., 1991); the mitogen-stimulated phosphorylation of H3 occurs at S10. In mammalian cells, a few nucleosomes show phosphorylation of H3 at S28 during interphase. In contrast to mammalian systems, a relatively high level of phosphorylation occurs throughout the genome of *Drosophila* during interphase where it is responsible for the maintenance of open chromatin (Wang et al., 2001b).

Phosphorylation of histone H3 at serine 10 (H3S10ph) is a prototypical example of the functional synergism of covalent modifications occurring on neighboring amino acids. Its presence, in conjunction with the acetylation of lysine 14 or 9 on the same H3 molecule, is necessary for transcriptional activation (Sawicka and Seiser, 2012). In addition to the H3K9ac and S10ph modifications, another pair of modifications of histone H3—H3S28ph and H3K27ac—positively regulates transcription (Gehani et al., 2010; Lau and Cheung, 2011) and is found at some enhancers and promoters in *Drosophila* (Kellner et al., 2012). In euchromatin, H3S10ph and H3S28ph prevent the binding of silencing factors to H3K9me3 and H3K27me3, respectively, leading to the demethylation and acetylation of these sites. Constitutive heterochromatin is marked by both H3S10ph and H3K9me2 (Wang et al., 2014). A number of proteins recognize and bind to phosphorylated histones and either further modify chromatin or serve as bridges for modifying factors. The linker histone H1 can be phosphorylated at multiple residues; in this case, phosphorylation appears to be associated with chromatin decondensation.

During prophase of mitosis, histone H3 phosphorylation occurs at serines 10 and 28, as well as threonine 11.

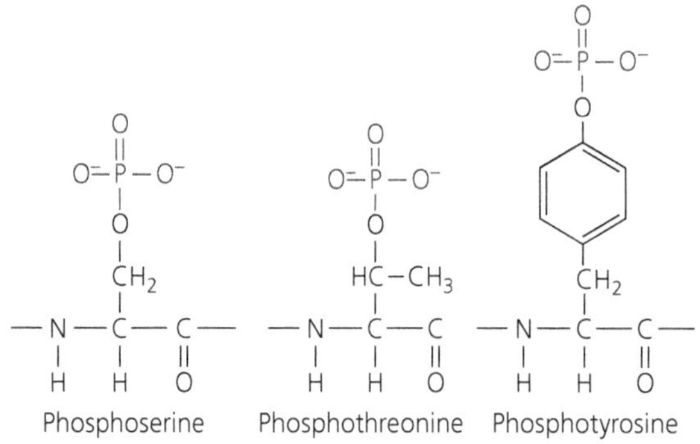

Fig. 4.5 Phosphorylated histone residues.

Phosphorylation begins in the pericentric heterochromatin during prophase where it appears to be required for recruitment of the condensin complex and assembly of the mitotic spindle; it spreads throughout the genome during the G2–M phase transition, peaks during metaphase and decreases during the progression to telophase. Phosphorylation occurring at other sites on histone H3 (threonines 3 and 11) appears to play a role in chromatid segregation, rather than in chromosome condensation (Rossetto et al., 2012).

Ubiquitination

This consists of the addition of ubiquitin, a 76 amino acid-long peptide, to an internal lysine residue. The first ubiquitinated histone was discovered by Harris Busch as an unknown abnormal spot in an electrophoresis gel of nucleolar proteins (Orrick et al., 1973); it was subsequently identified as a modified histone (Goldknopf et al., 1975). In contrast to poly-ubiquitination (the addition of a chain of ubiquitin molecules) that targets proteins for degradation by the proteasome, histones are mono-ubiquitinated and this modification affects chromatin structure and transcription. Three enzymatic reactions are required for ubiquitination— a ubiquitin molecule is activated by one enzyme (E1) and is delivered to a ubiquitin–protein ligase (E3) by a conjugating enzyme (E2); substrate specificity is provided by the E3 enzyme (Fig. 4.6). Ubiquitinated histones H2A and H2B perform very different functions in the regulation of transcription. In metazoans, the ubiquitination of H2B occurs at lysine 120 (H2BK120ub1) and is necessary for the trimethylation

of histone H3 at lysines 4 and 79, two marks that are associated with active transcription (Briggs et al., 2002). While several other lysines of H2B can be ubiquitinated, to date, only one—K34—has a demonstrated function. The presence of H2BK34ub1 greatly enhances the ubiquitination of K120 and the methylation of H3K4/K79 (Wu et al., 2011). A common explanation for the trans-histone effect of ubiquitination on methylation is that the addition of the bulky ubiquitin peptide on one histone alters the structure of the nucleosome in a manner necessary for HKMTs to access their substrates on a neighboring histone. This possibility is not supported by the observation that H2BK120ub1 is not required for the mono- and dimethylation of H3 (Weak and Workman, 2008). An alternate hypothesis is that ubiquitin recruits those subunits of the HKMT complexes that are necessary for trimethylation (Vitaliano-Prunier et al., 2008).

The ubiquitination of H2A occurs at lysine 119. The E3 ubiquitin–protein ligases responsible for this modification are present in silencing complexes and, in general, H2AK119ub1 has the opposite effect on gene function than H2BK120ub1 in that it is required for silencing. Surprisingly, loss of H2AK119ub1 has no effect on H3K9 or H3K27 methylation, emphasizing the differential function of the ubiquitin mark on H2A and H2B.

The removal of the ubiquitin mark is mediated by deubiquitinating enzymes (HDUB) that are targeted to specific regions of the genome where they affect the processes of transcriptional elongation or gene silencing.

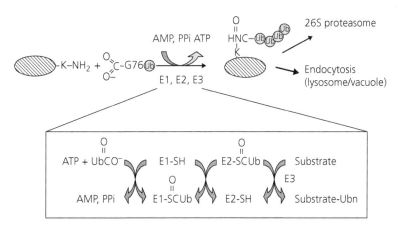

Fig. 4.6 Reactions involved in lysine ubiquitination.
(Modified from Jadhav and Wooten, 2009.)

Sumoylation

This resembles ubiquitination. Sumo (a small ubiquitin-related modifier) is a peptide of approximately 100 amino acids, generated by cleavage of a precursor protein; it is 18% identical to ubiquitin, and its attachment to a histone lysine is mediated by the successive action of activating (E1), conjugating (E2) and ligating (E3) enzymes. Like ubiquitination, sumoylation is reversed by the action of desumoylating enzymes. First discovered as a modification of histone H4 (Shiio and Eisenman, 2003), sumoylation can mark all four core histones. Invertebrates such as nematodes and flies have a single sumo, while vertebrates exhibit four different sumo paralogues. Histone sumoylation is generally thought to repress transcription by recruiting HP1 and co-repressor complexes that include HKDACs (Shiio and Eisenman, 2003).

Glycosylation

Glycosylation of nuclear and cytoplasmic proteins by the addition of single β-N-acetyl-glucosamine residues (O-GlcNAc) was reported over 30 years ago (Torres and Hart, 1984). All core histones are modified by O-GlcNAc at serine and threonine residues, predominantly in the globular region and in the terminal tail of H3. True to the myriad of functions that involve this modification on cellular proteins, histone glycosylation displays unusual complexity. The glycosylation of H2A and H2B may facilitate their dimerization and association with H3 and H4 (Sakabe et al., 2010). Some of the target sites of glycosylation (for example, serine 10 on H3 and serine 112 on H2B) are also sites of phosphorylation. Glycosylated H3S10 is present with chromatin marks characteristic of active transcription (H3K4me3) and repression (H3K9me3) (Zhang et al., 2011). It appears that the ubiquitination of H2B at lysine 120 is enhanced by the glycosylation of serine 112 (Fujiki et al., 2011). Since H2BK120ub is required for the methylation of H3K4/K79, once again, glycosylation is implicated in active transcription.

ADP ribosylation

This is the addition of single or multiple ADP ribose moieties. The reaction is carried out by a family of poly(ADP ribose) polymerases (PARPs), also known as ADP ribose transferases (ARTs), using the co-enzyme nicotinamide adenine dinucleotide (NAD) as the donor (Fig. 4.7). ADP ribosylation occurs at specific lysines on the N-terminal tails of all four histones (Messner et al., 2010). The reaction can be reversed by the activity of ADP ribosyl hydrolases that cleave the

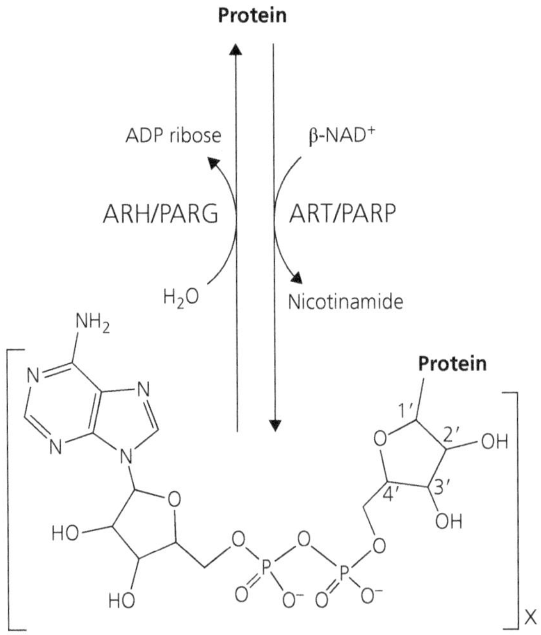

Fig. 4.7 Diagram of the pathway for the ADP ribosylation and ADP ribosyl removal of a protein.
(From Berthold et al., 2009.)

ADP ribose–protein bond (Koch-Nolte et al., 2008). The addition of ADP ribose leads to a negatively charged amino acid target and is therefore believed to diminish the electrostatic interaction between histones and DNA and to loosen the compaction of chromatin fibers.

ADP ribosylation and the PARP enzymes are involved in many different biological processes. The PARP enzymes (PARP 1 and 2) function in DNA repair mechanisms. For example, PARP1 contributes to the non-homologous end-joining (NHEJ) pathway (see Chapter 15) by recruiting a chromatin remodeler (see Remodeling complexes, below) to promote the assembly of core repair factors (Luijsterburg et al., 2016). Under normal genome conditions, PARP-1 has been implicated in the regulation of gene activity. It is present on active promoters where it ADP ribosylates histones, as well as itself, thereby excluding histone H1 and generating its own release (Krishnakumar et al., 2008). In association with macro2HA, PARP-1 is inactive (Nusimow et al., 2007), remains on chromatin and performs its silencing function. PARP-1 has also been implicated in the binding of insulator elements with one another or with the nuclear lamina (Ong et al., 2013; Yu et al., 2004). PARP enzymes and ADP ribosylation also appear to be involved in different aspects of RNA biology, including the formation of nuclear stress bodies (Leung et al., 2011) and the regulation of ribosomal RNA (rRNA) synthesis (Guetg et al., 2012). The occurrence of nuclear stress bodies and the synthesis of rRNA are described in Chapter 13.

Hydroxyisobutyrylation

Hydroxyisobutyrylation of lysine was reported at numerous sites on mammalian histones (Dai et al., 2014). This modification, conserved from yeast to man, is found on all four core histones, as well as on histone H1. Similarly to acetylation, hydroxyisobutyrylation neutralizes the positive lysine charge but confers a much larger change in size. Genome-wide mapping of one specific modified lysine (H4K8hib) revealed a potential depletion of this mark on the mammalian inactive X chromosome; in contrast, H4K8hib is associated with gene activity during spermatogenesis. Both observations strongly suggest a role for lysine hydroxyisobutyrylation in the control of transcription.

Remodeling complexes

In addition to the covalent histone modifications just described, the transcriptional activity of the genome is regulated by the action of remodeling complexes that alter the position of nucleosomes along the chromatin fiber or that modify the association of histone octamers with the DNA (Fig. 4.8). Proposed models involve DNA translocation, nucleosome sliding, intra- and inter-nucleosomal DNA looping and DNA twisting. Each remodeling complex contains an ATPase and uses the energy generated by ATP hydrolysis to perform its task. How exactly ATP hydrolysis is converted into a mechanical force that can break DNA/histone interactions is still unknown.

The first remodeling complex was discovered in yeast by Craig Peterson and Ira Herskowitz (Peterson and Herskowitz, 1992). Some of its 13 components were identified as products of genes that regulate the transcription of a gene encoding an endonuclease required for mating-type switching (*HO*); for this reason, they were named "switch" (*SWI*) genes. Other genes were found to regulate *Suc2*, a gene encoding an invertase involved in sucrose catabolism, and were named "sucrose non-fermenting" (*SNF*) genes. Sequence analysis revealed that *SNF2* and *SWI2* are identical genes. The increasing number of different genes found to be regulated by *SWI* or *SNF* genes and the general similarity of their loss-of-function mutations with defects of RNA polymerase II function led to the conclusion that the different SNF and SWI proteins may be components of a large multiprotein complex with a general role in transcription. Evidence that the SWI/SNF complex interacts with chromatin was provided by observing that mutations in *SWI* genes are suppressed by mutations of genes encoding histones and other chromatin proteins.

Highly related remodeling complexes exist in all eukaryotes. Each of these complexes contains an ATPase subunit that is a member of the SWI2 ATPase superfamily. In any given organism, these ATPases can associate with different groups of proteins and thereby participate in different remodeling functions. In yeast, the RSC complex (**r**emodel the **s**tructure of **c**hromatin) contains Sth1 (SNF two homolog 1) ATPase (Cairns, et al., 1996); in *Drosophila*, Brahma (BRM), the homolog of SWI2, is found in several complexes required for the transcription of most genes; in humans, two orthologs of Brahma—hBRM and BRG1 (brm-related gene 1) (Khavari et al., 1993)—are present in a number of remodeling complexes, generally referred to as BAF (BRG1- or hBRM-associated factors), with tissue and regulatory pathway-specific functions (Kadoch et al., 2013). Another homolog of SWI2 discovered in *Drosophila*—ISWI (**i**mitation **swi**tch)—is found in several remodeling complexes, numbering at least 14 in humans (Oppikofer et al., 2017). Other remodeling ATPases that have been extensively studied to date are

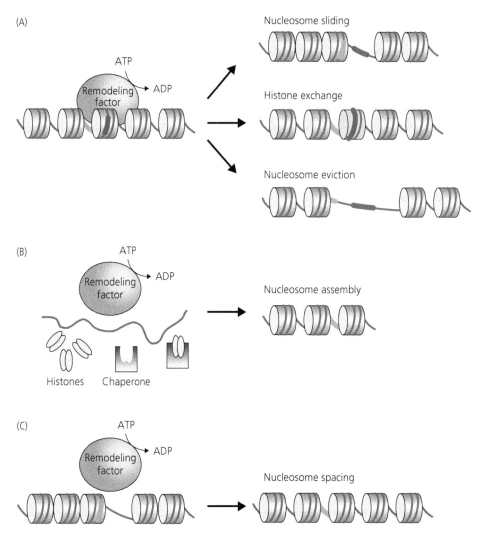

Fig. 4.8 Different actions of ATP-dependent chromatin remodeling complexes. (A) By sliding or evicting nucleosomes to expose previously masked DNA sequences, or by substituting histone variants in a replication-independent manner, remodeling complexes can facilitate transcription. (B) Remodeling complexes collaborate with histone chaperones to assemble nucleosomes (see Chapter 2). (C) Some remodeling complexes position nucleosomes in arrays with particular inter-nucleosomal spacing or linker DNA lengths (see Chapter 5).

(From Becker and Workman, 2013.)

Chd, Mi-2, Ino80 and Swr1. Each of the remodeling enzymes just mentioned represent a distinct group of ATPases that, in addition to the ATPase domain that is common to all, are characterized by the presence of particular additional domains. For example, SWI2/SNF2 ATPases have a bromodomain that recognizes acetylated lysines in histone tails; ISWI and some CHD ATPases contain a SANT domain (acronym of the four proteins where it was first identified) and a SLIDE (SANT-like ISWI) domain that contact extra-nucleosomal or linker DNA; other CHD remodelers also contain a chromodomain that interacts with methylated lysine; the Ino80 group of enzymes is characterized by a long insertion that separates the ATPase domain into two subdomains. In addition to organism-specific subunits, particular remodeling complexes with the same ATPase may have cell-type specific components, allowing them to regulate the transcriptional activity of different groups of genes in different tissues.

Chromatin remodelers are involved in every step of transcription, either to initiate and facilitate it, or to impede it leading to gene repression. Octamer sliding or eviction can uncover promoter regions that are otherwise unavailable to transcription factors and activators (Fig. 4.8). Reversing this process is an integral part of gene repression. In both cases, the remodeling complexes are targeted to their site of action by transcription factors that associate with specific sequences. SWI/SNF complexes are usually involved in transcription initiation via nucleosome displacement, a process that is facilitated by the acetylation of histones. Similarly, INO80 complexes are involved in gene activation. In contrast, ISWI and CHD remodelers restore the regular spacing of nucleosomes; histone acetylation, specifically H4K16ac, interferes with the function of ISWI and Mi-2. It is not surprising that these ATPases are present in complexes that include or that collaborate with histone deacetylases.

An additional, important role of remodeling complexes is to facilitate the exchange of core histones for variant histones. Histones H2A, H2B and H3 can be replaced by histones from which they differ slightly, by a few amino acids, or substantially by the presence of long polypeptide spacers.

Histone variants and nucleosome turnover

Following DNA replication, the newly synthesized DNA molecules are reconstituted into chromatin fibers by associating with preexisting nucleosomal subunits that are supplemented by newly synthesized core histones. The deposition of these histones is said to be replication-dependent (Fig. 4.9A). With the exception of H4, under particular circumstances, all core histones can be replaced by non-allelic isoforms referred to as histone variants that often exhibit the same post-translational modifications as canonical histones. A variety of nuclear processes throughout the cell cycle, ranging from transcription to DNA repair, involve the dynamic incorporation of histone isoforms in a replication-independent manner. Not surprisingly, histone variants play a crucial role in cell differentiation and organism development (Maze et al., 2014).

Histone H3 variants

Although only two variants of histone H3 exist in organisms ranging from yeast to the mouse, several variants have been identified in humans. The most common human H3 variants H3.1 and H3.2 differ in a single amino acid and are deposited in a replication-coupled manner by different dedicated chaperone complexes (Fig. 4.9B)—H3.1 by CAF-1 (chromatin assembly factor 1) and H3.2 by MCM-Asf1 (mini-chromosome maintenance/anti-silencing factor 1) (Latreille et al., 2014).

Surprisingly, the chaperones that deposit the cenH3 (CENP-A in mammals) variant found in the nucleosomes of centromere chromatin are not conserved among multicellular eukaryotes (Chen et al., 2014). The other universal variant H3.3 differs from canonical H3 by only four amino acids. Although initially thought to be uniquely associated with actively transcribed genes, H3.3 is found at the promoter of many silent genes, as well as in telomeric and pericentric heterochromatin (Filipescu et al., 2013; Skene and Henikoff, 2013; Udugama et al., 2015). Its deposition in transcribing regions depends on the HIRA (histone regulator A) chaperone, while the DAXX/ATRX (death domain-associated protein/alpha thalassemia/mental retardation syndrome X-linked) chaperone loads it on telomeric and pericentric heterochromatin and on regions of repeated DNA sequences. Following fertilization, H3.3 appears to play an important role in the remodeling of the male pronucleus—it is incorporated throughout the genome, as the sperm nucleus sheds its protamines and assumes a permissive architecture that is concordant with a state of pluripotency. All of these observations suggest that H3.3 is incorporated at any sites where, following DNA replication, nucleosomes that are absent or have been dislodged are replaced. Surprisingly, in *Drosophila*, deletion of the two genes that encode H3.3 leads to morphologically normal, albeit sterile, individuals (Sakai et al., 2009).

There are two replication-independent H3 variants expressed in human testes—H3.4 and H3.5; their function is not understood.

Histone H2A variants

Three distinct variants of H2A, each with very specific functional roles, are present in chromatin where they replace H2A independently of DNA replication. The most extensively studied variant, to date—H2A.X—plays a key role in DNA damage repair (see Chapter 15). Among the deleterious events that affect genome integrity is the occurrence of double-strand breaks in DNA that can result from exposure to different types of radiation and chemicals, or from mistakes in DNA replication. H2A.X replaces H2A in 5–20% of all nucleosomes with a genome-wide distribution. The exchange of H2A–H2B dimers for H2A.X–H2B dimers is carried out by the FACT (facilitates chromatin transcription) complex (Heo et al., 2008). In response to the

(A) Replication-dependent deposition of canonical histones

(B) Replication-independent deposition of histone variants

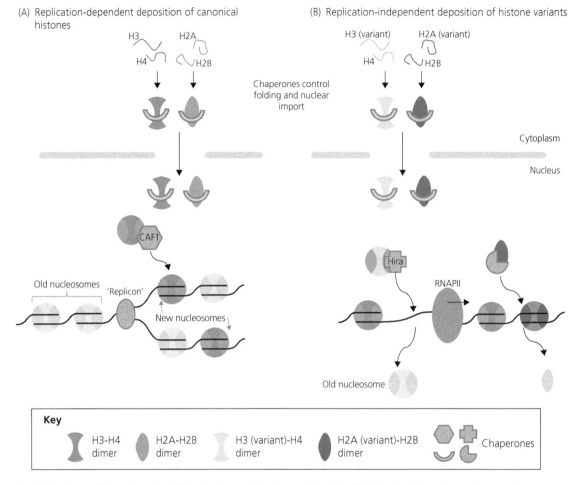

Fig. 4.9 Pathways of histone deposition. (A) Replication-coupled nucleosome formation. (B) Replication-independent replacement of a core hstone with a histone variant.

(From Skene and Henikoff, 2013.)

occurrence of a DNA double-strand break, H2A.X is rapidly phosphorylated by ATM (Ataxia telangiectasia mutated) or ATR (Ataxia telangiectasia and Rad3 related) kinases at a highly conserved serine (S139) residue (Rogakou et al., 1998) and by WSTF-SNF2H (Williams syndrome transcription factor) complex kinase at a conserved tyrosine (Y142) (Xiao et al., 2009). Although the doubly phosphorylated form, termed γH2A.X, initially extends over a distance of a few megabases from the site of the break, it rapidly loses the Y142 phosphorylation. The presence of γH2A.X initiates the recruitment of the damage response/DNA repair proteins (Jasin and Rothstein, 2013).

A second histone H2A variant (H2A.Z) is found near the start of genes and at sites of DNA breaks or aborted replication. It is present in the +1 nucleosome, i.e. in the first nucleosome immediately downstream of the nucleosome-free region that precedes the transcription start site, of the majority of active and inactive genes in yeast cells (Raisner et al., 2005). The exchange of H2A for H2A.Z involves the remodeling complex SWR-C, which, with the help of the chaperone Chz1, promotes H2A.Z incorporation (Luk et al., 2007; Mizuguchi et al., 2004). This function of SWR-C depends on the acetylation of H3K56; in the absence of this modification, SWR-C switches to H2A.Z nucleosome removal and H2A nucleosome deposition (Watanabe et al., 2013). INO80, a remodeling complex that promotes nucleosome turnover, is responsible for the exchange of H2A.Z back to H2A (Brahma et al., 2017; Papamichos-Chronakis et al., 2011), a process that most likely involves the FACT chaperone (Jeronimo et al., 2015).

In mouse and human embryonic stem cells, H2A.Z localizes at distal enhancer elements and at the transcription start sites of active genes and of inactive genes with bivalent promoters [promoters with nucleosomes bearing both histone modifications associated with activation (H3K4me3) and repression (H3K27me3); see Chapter 5] (Ku et al., 2012). In human cells, the deposition of H2A.Z is the responsibility of the SRCAP remodeling complex, and of the p400/TIP60 complex that includes an ATPase of the Ino80 family and a histone acetyl transferase, assisted by the ANP32E chaperone (Obri et al., 2014). A number of experiments, albeit with some apparent exceptions (Taty-Taty et al., 2014), have indicated that H2A is rapidly exchanged for H2A.Z along the site of these breaks, promoting nucleosome acetylation and the formation of open chromatin. In mammalian cells, these last modifications are also executed by the p400/TIP60 complex (Price and D'Andrea, 2013).

A third form of histone H2A (macroH2A) occurs only in vertebrates where it replaces H2A in a few percent of nucleosomes. It contains a large C-terminal non-histone domain connected to the canonical domain of H2A. Splice variants, and some variability in the C-terminal domain, give rise to several macroH2A isoforms. The different macroH2A forms are enriched in inactive regions of the genome, in heterochromatin including the inactive X chromosome (see Chapter 9) and on euchromatic genes silenced by Polycomb-group complexes (Cantarino et al., 2013; Skene and Henikoff, 2013). The latter effect is, in all likelihood, mediated by the macroH2A association with PARP-1. Somewhat surprisingly, macroH2A-containing nucleosomes are also found on the promoters of expressed genes. This dual functional role is explained by the observations that macroH2A nucleosomes mask the binding sites of transcription activators on genes that are silenced and the binding sites of repressors on active genes (Lavigne et al., 2015).

Histone H2B variants

Both known H2B variants are testis-specific. The testis-specific variant TH2B was one of the first histone variants to be discovered. It is incorporated into the chromatin of early spermatocytes in a replication-independent manner and eventually replaces all histone H2B. This switch leads to a shutdown of transcription of the repetitive core histone gene clusters and to the gradual replacement of histones by protamines (Montellier et al., 2013). Discovered more recently,

H2BFWT shows low homology with H2A. Its role in spermatogenesis has not yet been established.

Histone H1 variants

Linker histones stabilize chromatin fibers, resist nucleosome sliding and prevent transcription. Linker histones can be phosphorylated and methylated, modifications that reverse their architectural role and alter their function in gene regulation. In *Drosophila*, one H1 variant is found in somatic cells throughout development; its main function appears to be the maintenance of heterochromatin structure and the silencing of repetitive sequences and transposable elements in the heterochromatic regions of the genome by recruiting the histone H3 lysine 9 methyl transferase Su(var)3-9 (Lu et al., 2013). A new variant (dBigH1) has been recently identified that is present in the early stages of embryogenesis when the somatic H1 is absent (Pérez-Montero et al., 2013). dBigH1 appears to carry out the same functions that are associated with H1.

In contrast to the presence of a single H1 type that is able to participate in all of the developmental requirements of *Drosophila*, many organisms express multiple H1 variants. In humans, there are ten H1 variants—seven somatic and three germline-specific. In mice, mutations that eliminate one or two of the different H1 isoforms have no phenotypic effect, suggesting a substantial level of compensation. The elimination of three isoforms, substantially reducing the level of H1, leads to profound changes in chromatin structure, although only a few genes appear to be affected (Fan et al., 2005).

One of the human's testis-specific variants (H1t) appears to be dispensable, perhaps because it can be replaced by one of the somatic variants; the other testis-specific variant (H1T2) is critical to proper chromatin condensation and to the elongation phase of spermatogenesis (Martianov et al., 2005). The oocyte-specific variant H1oo is expressed throughout oogenesis and is incorporated into sperm chromatin after fertilization where it is thought to facilitate the replacement of protamines with somatic histones (Mizusawa et al., 2010).

DNA modifications

DNA modifications consist predominantly of the methylation or hydroxymethylation of cytosines. The existence of methylcytosines and their prevalence in CpG dinucleotides (where "p" refers to the phosphodiester bond between a cytosine and a guanine nucleotide) was first reported in calf thymus DNA by Robert Sinsheimer (Sinsheimer, 1954). A variable, albeit usually low, level of methylated cytosines that are not present in CpG dinucleotides exists in mammalian genomes (Ramsahoye et al., 2000; Schultz et al., 2015; Varley et al., 2013). These cytosines are referred to as meCpH where H can be either adenine, cytosine or thymine.

In mammals, there are three DNA methyl transferases, two primarily involved in the *de novo* methylation of DNA (DNMTA3A and 3B) and the third (DNMT1) primarily responsible for the methylation of newly replicated DNA, i.e. for the maintenance of specific patterns of methylation of a cell's genome through cell division (see Chapter 16). All methyl transferases use SAM as the donor of the methyl group that is attached to the 5' carbon of cytosine abbreviated as 5mC. While maintenance methylation can be easily explained, the signals that trigger *de novo* methylation are not yet fully understood.

The occurrence of methylated CpGs was initially correlated to developmental gene regulation, because the experimental abrogation of the enzymes responsible for this covalent modification led to early embryonic death (Li et al., 1992). Areas in the genome that exhibit low levels of methylation include CpG-poor distal regulatory regions usually occupied by DNA-binding factors (Stadler et al., 2011). Further studies have produced a more complex picture (see Chapter 8).

Methylated cytosine (5mC) can be converted to hydroxymethylated cytosine (5hmC) by TET (ten eleven translocation) hydroxylases. This DNA modification occurs at promoters that are low in CpG content and is absent in CpG islands, not surprisingly since most of them are unmethylated; it is present along transcribed regions, at enhancers and at the binding sites of the ubiquitous transcription factor CTCF (Szulwach and Jin, 2013). Methyl CpG binding domain (MBD) proteins bind to 5mC and 5hmC with the same affinity (Mellen et al., 2012).

The demethylation of 5mC to convert it back to C can occur passively, by the failure of maintenance methylation following DNA replication. It can also occur enzymatically by a complex process that involves the TET-mediated sequential oxidation of 5mC, the conversion of the final product to thymine and the latter's replacement by the base excision repair (BER) pathway (Schomacher, 2013).

Chapter summary

In its native state, chromatin is refractory to the process of transcription and must be modified to allow access to the DNA. The usual initial step of this modification is the binding of "pioneering" transcription factors to their cognate sequences. These factors then recruit chromatin remodeling and modifying complexes that reposition nucleosomes and induce covalent epigenetic modifications on histones and on the DNA. Nucleosomes can be displaced, or their association with DNA can be altered by multiprotein complexes that hydrolyze ATP in order to generate the energy necessary for their functions. Other complexes replace certain canonical histones with specific histone variants. Histones can undergo acetylation, methylation, phosphorylation, ubiquitination, sumoylation, glycosylation, ADP ribosylation or hydroxyisobutyrylation of particular amino acids along their molecules. DNA nucleotides can be modified by methylation or hydroxymethylation, without altering the coding sequences. All of these chromatin changes are associated with gene function, with sets of modifications usually present on active or repressed genes. Not surprisingly, mutations that affect the timely occurrence or removal of these modifications result in particular diseases.

Chapter 5 describes and discusses the epigenetic changes that are correlated with gene transcription.

Box 4.1 Examples of human diseases associated with histone and DNA modifications

Epigenetic modifications reflect the particular transcriptional programs of the different cell types during the development and aging of multicellular organisms. Through their susceptibility to changes in response to environmental factors, these modifications can impact the processes of differential gene activity responsible for cellular maintenance and differentiation. It is not surprising, therefore, that changes in the epigenetic landscape have been correlated to both disease susceptibility and occurrence, from cancer to nervous system dysfunctions, from metabolic diseases such as diabetes to diseases of the cardiovascular system. Part V is devoted entirely to this broad topic. Below are examples of epigenetic modifications manifested in selected neuropathies.

Two important considerations are to be kept in mind while evaluating the association between epigenetic modifications and disease. The first is whether the data have been collected using experimental protocols that ensure the comparison of appropriately selected tissue samples (that consider, for example, the possible differences in cell composition between patient and control tissues) by experimental methods that are reasonably free of bias (such as ambient conditions and operator influence) and by the use of statistical methods that normalize the data and detect and avoid batch effects. The second consideration is the frequent difficulty in determining if the modifications cause or contribute to the symptoms or are a consequence of the disease process.

Alzheimer's disease (AD) is the most common of a group of brain disorders with progressive behavioral and cognitive deficits. AD involves the formation of extra-cellular plaques consisting primarily of **amyloid β** peptides and of intra-neuronal masses of the protein **tau**. Mutations in three autosomal genes have been associated with development of the disease: *APP* (amyloid precursor protein), and *PSEN1* and *2* (presenilin 1 and presenilin 2) (St. George-Hyslop et al., 1992; Sherrington et al., 1995; Van Broeckhoven et al., 1992). *APP* encodes the protein that is cleaved to give rise to amyloid β peptides, and presenilins are involved in regulating the cleaving enzymes. An additional important genetic factor is a particular allele of the *APOE* (apolipoprotein E) gene that encodes a ubiquitously expressed glycoprotein involved in neuronal growth. While other alleles help clear away amyloid β peptides, the *APOE ε2* allele is much less effective in this process (Deane et al., 2008). Genome-wide association studies (GWAS) use the presence of single-nucleotide polymorphisms in the DNA sequence of individuals who present a disease to identify additional genetic factors that may contribute to the disease. These studies have identified over 20 loci that increase the risk of developing AD (reviewed in Van Cauwenberghe et al., 2016).

Using quantitative immunochemistry with antisera that detected 5-methylcytosine or 5-hydroxymethylcytosine, the levels of DNA methylation in brain tissue of AD patients were found to be significantly less than in healthy controls of the same age (Chouliaras et al., 2013). This general reduction in DNA methylation does not prevent the occurrence of hypermethylation of specific loci that have been shown to contribute to the disease and whose transcription is reduced in diseased brain regions (De Jager et al., 2014). Using mouse models of AD, an anti-epileptic drug known to inhibit histone deacetylase (HDAC) activity was shown to greatly reduce the formation of amyloid β plaques (Qing et al., 2008). In a similar experiment, AD mice treated with a more effective HDAC inhibitor reduced the amount of tau protein in the brain and reversed memory deficits (Ricobaraza et al., 2009).

These results are consistent with the premise that repression of particular genes in the brain by DNA methylation, as well as histone deacetylation, is one of the underlying causes of AD. Evidence for the latter contention was provided by the observations that while H4 acetylated at lysine 16 is increased with normal aging, it is significantly decreased as a function of age in AD patients. Some changes that occur with normal aging are maintained in AD while others are not maintained. In addition, AD brains exhibit some changes that are disease-specific (Nativio et al., 2018).

Parkinson's disease (PD) is a complex neurological disease that involves slowness of movement, tremor, rigidity, postural instability, mood disorders, cognitive disorders and, in some cases, dementia. Mutations in the *SNCA* (synuclein alpha) gene, which normally encodes an abundant, soluble synaptic brain protein, lead to a defective protein that accumulates in neuronal inclusions called Lewy bodies and to the death of dopaminergic neurons in particular regions of the brain (Campbell et al., 2000; Polymeropoulos et al., 1997). In addition, mutations in six additional genes have been shown to contribute to inherited forms of PD, and more than 30 loci are involved in modulating the PD phenotype (reviewed in Labbé et al., 2016). A large analysis of protein and transcripts that compared brain samples from PD patients with healthy individuals indicated that the major pathways affected by the disease involve mitochondria, protein folding and proteins that bind heavy metals (Dumitriu et al., 2016). An analogous comparison of micro RNAs (miRNAs; see Chapter 6) identified several differences in these RNAs between diseased and healthy brains (Hoss et al., 2016). Interestingly, a small number of these miRNAs were also different in Huntington's disease patients.

One of the strong genetic risk factors for developing PD was shown to be loss-of-function mutations in the gene encoding a lysosomal enzyme that degrades glucocerebrosides (Goker-Alpan et al., 2004). These complex lipids directly convert soluble synuclein-α into aggregates (Zunke et al., 2017).

A number of studies have attempted to correlate the methylation status of individual genes in different cultured cell types transfected with *SNCA* transgenes and in tissues from PD individuals, with often conflicting results. On the other hand, epigenome-wide methylation studies revealed a number of genes that were hypomethylated and others that were hypermethylated in the brain and blood of PD patients (Masliah et al., 2013; Moore et al., 2014). Loci with hypomethylation were usually CpG islands associated with promoter regions, while hypermethylation was more often found in intergenic regions and along gene bodies. By measuring the methylation status of a number of CpG sites in the human genome, a correlation can be established between changes

continued

Box 4.1 *Continued*

of DNA methylation—an epigenetic clock—and chronological age (Horvath, 2013). The blood cells of PD patients had a more advanced epigenetic age than their chronological age (Horvath and Ritz, 2015). Synuclein alpha protein over-expressed in cultured human neurons binds to histones and reduces the acetylation of H3 (Kontopoulos et al., 2006). Yet, the *SNCA* gene is controlled by acetylation and levels of histone acetylation are significantly higher in the neurons of affected brain regions in PD patients (Park et al., 2016).

Huntington's disease (HD) is a lethal neurodegenerative disease resulting from a dominant mutation in an autosomal gene that encodes *huntingtin*, a widely expressed protein (The Huntington's Disease Collaboration Research Group, 1993). The mutation causes the expansion of a CAG trinucleotide in the gene. Individuals with 40 or more CAG repeats produce a mutated protein that leads to the selective loss of neurons in particular regions of the brain, resulting in progressive movement disorders, dementia and psychiatric problems. The CAG expansion occurs somatically, so that the length of the repeat increases with age. It remains unclear why the accumulation of the mutant huntingtin protein leads to such profound neurological changes; however, it appears that a group of genes are down-regulated in the brain as a consequence of changes to the epigenome of affected individuals.

The vast majority of the observations on the epigenetic modifications correlated with HD and on potential epigenetic therapeutic interventions have been made with mouse models of the disease. Caution should be exercised in comparing results from different investigations because the mouse lines that exhibit the HD phenotypes have been generated by different genetic engineering techniques and exhibit varying degrees of somatic, as well as germline, expansion. Earlier studies had established that manipulating the level of histone deacetylases would alter the toxic effects

of the mutant huntingtin protein (Saha and Pahan, 2006); other studies reported that the disease was associated with a reduction in the transcription of genes that control neuronal functions (Cha, 2007; Desplats et al., 2006; Luthi-Carter et al., 2000; Zuccato and Cattaneo, 2007). A direct correlation between these different results is provided by the observation that the acetylation of H3 at lysine 27 (H3K27ac), which is mediated by CBP (CREB-binding protein where CREB stands for cAMP response element binding protein) and is a mark of active enhancers, is down-regulated in the affected region of the brain where genes controlling neuronal functions are also down-regulated (Achour et al., 2015). Consistent with the transcriptional repression seen in HD, pharmacological inhibition of DNA methyl transferases, presumably interfering with the methylation of CpG islands that overlap promoter regions, leads to the reactivation of down-regulated genes and protects neurons from huntingtin-induced toxicity (Pan et al., 2016).

Rubinstein–Taybi syndrome (RSTS) is a rare neurological disorder that presents with microcephaly, intellectual disability and postnatal growth retardation. RSTS is caused by an autosomal dominant mutation in the gene that encodes CBP (Petrij et al., 1995). A decade later, mutations in the gene encoding p300, another histone acetyl transferase that shares substantial homology with CBP, were also found to result in RSTS (Roelfsema et al., 2005). Both of these acetyl transferases function as co-activators of transcription, interact with numerous proteins and are involved in a variety of nuclear pathways, ranging from DNA repair and cellular differentiation to tumor suppression. As discussed in Chapter 22, CBP plays an important role in the establishment of long-term memory. Although all the evidence points to dysfunctional chromatin as a result of hypoacetylation, the detailed molecular changes that are the basis for RSTS will be very difficult to establish.

References

Achour, M., Le Gras, S., Keime, C., Parmentier, F., Lejeune, F. X., Boutillier, A. L., Neri, C., Davidson, I. & Merienne, K. 2015. Neuronal identity genes regulated by super-enhancers are preferentially down-regulated in the striatum of Huntington's disease mice. *Hum Mol Genet*, 24, 3481–96.

Allfrey, V. G., Faulkner, R. & Mirsky, A. E. 1964. Acetylation and methylation of histones and their possible role in the regulation of RNA synthesis. *Proc Natl Acad Sci U S A*, 51, 786–94.

Araki, Y. & Mimura, T. 2017. The histone modification code in the pathogenesis of autoimmune diseases. *Mediators Inflamm*, 2017, 2608605.

Barski, A., Cuddapah, S., Cui, K., Roh, T. Y., Schones, D. E., Wang, Z., Wei, G., Chepelev, I. & Zhao, K. 2007. High-resolution profiling of histone methylations in the human genome. *Cell*, 129, 823–37.

Becker, P. B. & Workman, J. L. 2013. Nucleosome remodeling and epigenetics. *Cold Spring Harb Perspect Biol*, 5, pii: a017905.

Berthold, C. L., Wang, H., Nordlund, S. & Hogbom, M. 2009. Mechanism of ADP-ribosylation removal revealed by the structure and ligand complexes of the dimanganese mono-ADP-ribosylhydrolase DraG. *Proc Natl Acad Sci U S A*, 106, 14247–52.

Bird, A. W., Yu, D. Y., Pray-Grant, M. G., Qiu, Q., Harmon, K. E., Megee, P.c., Grant, P. A., Smith, M. M. & Christman, M. F. 2002. Acetylation of histone H4 by Esa1 is required for DNA double-strand break repair. *Nature*, 419, 411–15.

Brahma, S., Udugama, M. I., Kim, J., Hada, A., Bhardwaj, S. K., Hailu, S. G., Lee, T. H. & Bartholomew, B. 2017. INO80 exchanges H2A.Z for H2A by translocating on DNA proximal to histone dimers. *Nat Commun*, 8, 15616.

Briggs, S. D., Xiao, T., Sun, Z. W., Caldwell, J. A., Shabanowitz, J., Hunt, D. F., Allis, C. D. & Strahl, B. D. 2002. Gene silencing: trans-histone regulatory pathway in chromatin. *Nature,* 418, 498.

Brownell, J. E. & Allis C. D. 1995. An activity gel assay detects a single, catalytically active histone acetyltransferase subunit in *Tetrahymena* macronuclei. *Proc Natl Acad Sci U S A*, 92, 6364–8.

Brownell, J. E., Zhou,J., Ranalli, T., Kobayashi, R., Edmondson, D. G., Roth, S. Y. & Allis, C. D. 1996. *Tetrahymena* histone acetyltransferaseA: a homolog to yeast Gcn5p linking histone acetylation to gene activation. *Cell*, 84, 843–51.

Cairns, B. R., Lorch, Y., Li, Y., Zhang, M., Lacomis, L., Erdjument-Bromage, H., Tempst, P., Du, J., Laurent, B. & Kornberg, R. D. 1996. RSC, an essential, abundant chromatin-remodeling complex. *Cell*, 87, 1249–60.

Campbell, B. C., Li, Q. X., Culvenor, J. G., Jakala, P., Cappai, R., Beyreuther, K., Masters, C. L. & Mclean, C. A. 2000. Accumulation of insoluble alpha-synuclein in dementia with Lewy bodies. *Neurobiol Dis*, 7, 192–200.

Cantarino, N., Douet, J. & Buschbeck, M. 2013. MacroH2A—an epigenetic regulator of cancer. *Cancer Lett*, 336, 247–52.

Cha, J. H. 2007. Transcriptional signatures in Huntington's disease. *Prog Neurobiol*, 83, 228–48.

Chen, C. C., Dechassa, M. L., Bettini, E., Ledoux, M. B., Belisario, C., Heun, P., Luger, K. & Mellone, B. G. 2014. CAL1 is the *Drosophila* CENP-A assembly factor. *J Cell Biol*, 204, 313–29.

Chen, D., Ma, H., Hong, H., Koh, S. S., Huang, S. M., Schurter, B. T., Aswad, D. W. & Stallcup, M. R. 1999. Regulation of transcription by a protein methyltransfrase. *Science*, 284, 2174–7.

Chouliaras, L., Mastroeni, D., Delvaux, E., Grover, A., Kenis, G., Hof, P. R., Steinbusch, H. W., Coleman, P. D., Rutten, B. P. & Van Den Hove, D. L. 2013. Consistent decrease in global DNA methylation and hydroxymethylation in the hippocampus of Alzheimer's disease patients. *Neurobiol Aging*, 34, 2091–9.

Dai, L., Peng, C., Montellier, E., Lu, Z., Chen, Y., Ishii, H., Debernardi, A., Buchou, T., Rousseaux, S., Jin, F., Sabari, B. R., Deng, Z., Allis, C. D., Ren, B., Khochbin, S. & Zhao,

Y. 2014. Lysine 2-hydroxyisobutyrylation is a widely distributed active histone mark. *Nat Chem Biol*, 10, 365–70.

De Jager, P. L., Srivastava, G., Lunnon, K., Burgess, J., Schalkwyk, L. C., Yu, L., Eaton, M. L., Keenan, B. T., Ernst, J., Mccabe, C., Tang, A., Raj, T., Replogle, J., Brodeur, W., Gabriel, S., Chai, H. S., Younkin, C., Younkin, S. G., Zou, F., Szyf, M., Epstein, C. B., Schneider, J. A., Bernstein, B. E., Meissner, A., Ertekin-Taner, N., Chibnik, L. B., Kellis, M., Mill, J. & Bennett, D. A. 2014. Alzheimer's disease: early alterations in brain DNA methylation at ANK1, BIN1, RHBDF2 and other loci. *Nat Neurosci*, 17, 1156–63.

Deane, R., Sagare, A., Hamm, K., Parisi, M., Lane, S., Finn, M. B., Holtzman, D. M. & Zlokovic, B. V. 2008. apoE isoform-specific disruption of amyloid beta peptide clearance from mouse brain. *J Clin Invest*, 118, 4002–13.

Desplats, P. A., Kass, K. E., Gilmartin, T., Stanwood, G. D., Woodward, E. L., Head, S. R., Sutcliffe, J. G. & Thomas, E. A. 2006. Selective deficits in the expression of striatal-enriched mRNAs in Huntington's disease. *J Neurochem*, 96, 743–57.

Devaiah, B. N., Case-Borden, C., Gegonne, A., Hsu, C. H., Chen, Q., Meerzaman, D., Dey, A., Ozato, K. & Singer, D. S. 2016. BRD4 is a histone acetyltransferase that evicts nucleosomes from chromatin. *Nat Struct Mol Biol*, 23, 540–8.

Dumitriu, A., Golji, J., Labadorf, A. T., Gao, B., Beach, T. G., Myers, R. H., Longo, K. A. & Latourelle, J. C. 2016. Integrative analyses of proteomics and RNA transcriptomics implicate mitochondrial processes, protein folding pathways and GWAS loci in Parkinson disease. *BMC Med Genomics*, 9, 5.

Dunlap, D., Yokoyama, R., Ling, H., Sun, H. Y., Mcgill, K., Cugusi, S. & Lucchesi, J. C. 2012. Distinct contributions of MSL complex subunits to the transcriptional enhancement responsible for dosage compensation in *Drosophila*. *Nucleic Acids Res*, 40, 11281–91.

Eissenberg, J. C. & Lucchesi, J. C. 1983. Chromatin structure and transcriptional activity of an X-linked heat shock gene in *Drosophila pseudoobscura*. *J Biol Chem*, 258, 13986–91.

Fan, Y., Nikitina, T., Zhao, J., Fleury, T. J., Bhattacharyya, R., Bouhassira, E. E., Stein, A., Woodcock, C. L. & Skoultchi, A. I. 2005. Histone H1 depletion in mammals alters global chromatin structure but causes specific changes in gene regulation. *Cell*, 123, 1199–212.

Feller, C., Forne, I., Imhof, A. & Becker, P. B. 2015. Global and specific responses of the histone acetylome to systematic perturbation. *Mol Cell*, 57, 559–71.

Filipescu, D., Szenker, E. & Almouzni, G. 2013. Developmental roles of histone H3 variants and their chaperones. *Trends Genet*, 29, 630–40.

Fujiki, R., Hashiba, W., Sekine, H., Yokoyama, A., Chikanishi, T., Ito, S., Imai, Y., Kim, J., He, H. H., Igarashi, K., Kanno, J., Ohtake, F., Kitagawa, H., Roeder, R. G., Brown, M. &

Kato, S. 2011. GlcNAcylation of histone H2B facilitates its monoubiquitination. *Nature*, 480, 557–60.

Gehani, S. S., Agrawal-Singh, S., Dietrich, N., Christophersen, N. S., Helin, K. & Hansen, K. 2010. Polycomb group protein displacement and gene activation through MSK-dependent H3K27me3S28 phosphorylation. *Mol Cell*, 39, 886–900.

Goker-Alpan, O., Schiffmann, R., Lamarca, M. E., Nussbaum, R. L., Mcinerney-Leo, A. & Sidransky, E. 2004. Parkinsonism among Gaucher disease carriers. *J Med Genet*, 41, 937–40.

Goldknopf, I. L., Taylor, C. W., Baum, R. M., Yeoman, L. C., Olson, M. O., Prestayko, A. W. & Busch, H. 1975. Isolation and characterization of protein A24, a "histone-like" non-histone chromosomal protein. *J Biol Chem*, 250, 7182–7.

Guccione, E., Bassi, C., Casadio, F., Martinato, F., Cesaroni, M., Schuchlautz, H., Luscher, B. & Amati, B. 2007. Methylation of histone H3R2 by PRMT6 and H3K4 by an MLL complex are mutually exclusive. *Nature*, 449, 933–7.

Guetg, C., Scheifele, F., Rosenthal, F., Hottiger, M. O. & Santoro, R. 2012. Inheritance of silent rDNA chromatin is mediated by PARP1 via noncoding RNA. *Mol Cell*, 45, 790–800.

Heo, K., Kim, H., Choi, S. H., Choi, J., Kim, K., Gu, J., Lieber, M. R., Yang, A. S. & An, W. 2008. FACT-mediated exchange of histone variant H2AX regulated by phosphorylation of H2AX and ADP-ribosylation of Spt16. *Mol Cell*, 30, 86–97.

Horvath, S. 2013. DNA methylation age of human tissues and cell types. *Genome Biol*, 14, R115.

Horvath, S. & Ritz, B. R. 2015. Increased epigenetic age and granulocyte counts in the blood of Parkinson's disease patients. *Aging (Albany NY)*, 7, 1130–42.

Hoss, A. G., Labadorf, A., Beach, T. G., Latourelle, J. C. & Myers, R. H. 2016. MicroRNA profiles in Parkinson's disease prefrontal cortex. *Front Aging Neurosci*, 8, 36.

Inoue, A. & Fujimoto, D. 1969. Enzymatic deacetylation of histone. *Biochem Biophys Res Commun*, 36, 146–50.

Jadhav, T. & Wooten, M. W. 2009. Defining an embedded code for protein ubiquitination. *J Proteomics Bioinform*, 2, 316.

Jasin, M. & Rothstein, R. 2013. Repair of strand breaks by homologous recombination. *Cold Spring Harb Perspect Biol*, 5, a012740.

Jenuwein, T., Laible, G., Dorn, R. & Reuter, G. 1998. SET domain proteins modulate chromatin domains in eu- and heterochromatin. *Cell Mol Life Sci*, 54, 80–93.

Jeronimo, C., Watanabe, S., Kaplan, C. D., Peterson, C. L. & Robert, F. 2015. The histone chaperones FACT and Spt6 restrict H2A.Z from intragenic locations. *Mol Cell*, 58, 1113–23.

Kadoch, C., Hargreaves, D. C., Hodges, C., Elias, L., Ho, L., Ranish, J. & Crabtree, G. R. 2013. Proteomic and bioinformatic analysis of mammalian SWI/SNF complexes identifies extensive roles in human malignancy. *Nat Genet*, 45, 592–601.

Kaimori, J. Y., Maehara, K., Hayashi-Takanaka, Y., Harada, A., Fukuda, M., Yamamoto, S., Ichimaru, N., Umehara, T., Yokoyama, S., Matsuda, R., Ikura, T., Nagao, K., Obuse, C., Nozaki, N., Takahara, S., Takao, T., Ohkawa, Y., Kimura, H. & Isaka, Y. 2016. Histone H4 lysine 20 acetylation is associated with gene repression in human cells. *Sci Rep*, 6, 24318.

Kellner, W. A., Ramos, E., Van Bortle, K., Takenaka, N. & Corces, V. G. 2012. Genome-wide phosphoacetylation of histone H3 at *Drosophila* enhancers and promoters. *Genome Res*, 22, 1081–8.

Khavari, P. A., Peterson, C. L., Tamkun, J. W., Mendel, D. B. & Crabtree, G. R. 1993. BRG1 contains a conserved domain of the SWI2/SNF2 family necessary for normal mitotic growth and transcription. *Nature*, 366, 170–4.

Kim, G. W., Gocevski, G., Wu, C. J. & Yang, X. J. 2010. Dietary, metabolic, and potentially environmental modulation of the lysine acetylation machinery. *Int J Cell Biol*, 2010, 632739.

Koch-Nolte, F., Kernstock, S., Mueller-Dieckmann, C., Weiss, M. S. & Haag, F. 2008. Mammalian ADP-ribosyltransferases and ADP-ribosylhydrolases. *Front Biosci*, 13, 6716–29.

Kontopoulos, E., Parvin, J. D. & Feany, M. B. 2006. Alpha-synuclein acts in the nucleus to inhibit histone acetylation and promote neurotoxicity. *Hum Mol Genet*, 15, 3012–23.

Krishnakumar, R., Gamble, M. J., Frizzell, K. M., Berrocal, J. G., Kininis, M. & Kraus, W. L. 2008. Reciprocal binding of PARP-1 and histone H1 at promoters specifies transcriptional outcomes. *Science*, 319, 819–21.

Ku, M., Jaffe, J. D., Koche, R. P., Rheinbay, E., Endoh, M., Koseki, H., Carr, S. A. & Bernstein, B. E. 2012. H2A.Z landscapes and dual modifications in pluripotent and multipotent stem cells underlie complex genome regulatory functions. *Genome Biol*, 13, R85.

Labbe, C., Lorenzo-Betancor, O. & Ross O.a. 2016. Epigenetic regulation in Parkinson's disease. *Acta Neuropathol*, 132, 515–30.

Latreille, D., Bluy, L., Benkirane, M. & Kiernan, R. E. 2014. Identification of histone 3 variant 2 interacting factors. *Nucleic Acids Res*, 42, 3542–50.

Lau, P. N. & Cheung, P. 2011. Histone code pathway involving H3 S28 phosphorylation and K27 acetylation activates transcription and antagonizes polycomb silencing. *Proc Natl Acad Sci U S A*, 108, 2801–6.

Lavigne, M. D., Vatsellas, G., Polyzos, A., Mantouvalou, E., Sianidis, G., Maraziotis, I., Agelopoulos, M. & Thanos, D. 2015. Composite macroH2A/NRF-1 nucleosomes suppress noise and generate robustness in gene expression. *Cell Rep*, 11, 1090–101.

Leung, A. K., Vyas, S., Rood, J. E., Bhutkar, A., Sharp, P. A. & Chang, P. 2011. Poly(ADP-ribose) regulates stress responses and microRNA activity in the cytoplasm. *Mol Cell*, 42, 489–99.

Li, E., Bestor, T. H. & Jaenisch, R. 1992. Targeted mutation of the DNA methyltransferase gene results in embryonic lethality. *Cell*, 69, 915–26.

Liu, Y., Lu, C., Yang, Y., Fan, Y., Yang, R., Liu, C. F., Korolev, N. & Nordenskiold, L. 2011. Influence of histone tails and H4 tail acetylations on nucleosome-nucleosome interactions. *J Mol Biol*, 414, 749–64.

Lu, X., Wontakal, S. N., Kavi, H., Kim, B. J., Guzzardo, P. M., Emelyanov, A. V., Xu, N., Hannon, G. J., Zavadil, J., Fyodorov, D. V. & Skoultchi, A. I. 2013. *Drosophila* H1 regulates the genetic activity of heterochromatin by recruitment of Su(var)3–9. *Science*, 340, 78–81.

Luijsterburg, M. S., De Krijger, I., Wiegant, W. W., Shah, R. G., Smeenk, G., De Groot, A. J. L., Pines, A., Vertegaal, A. C. O., Jacobs, J. J. L., Shah, G. M. & Van Attikum, H. 2016. PARP1 links CHD-mediated chromatin expansion and H3.3 deposition to DNA repair by non-homologous end-joining. *Mol Cell*, 61, 547–62.

Luk, E., Vu, N. D., Patteson, K., Mizuguchi, G., Wu, W. H., Ranjan, A., Backus, J., Sen, S., Lewis, M., Bai, Y. & Wu, C. 2007. Chz1, a nuclear chaperone for histone H2AZ. *Mol Cell*, 25, 357–68.

Luthi-Carter, R., Strand, A., Peters, N. L., Solano, S. M., Hollingsworth, Z. R., Menon, A. S., Frey, A. S., Spektor, B. S., Penney, E. B., Schilling, G., Ross, C. A., Borchelt, D. R., Tapscott, S. J., Young, A. B., Cha, J. H. & Olson, J. M. 2000. Decreased expression of striatal signaling genes in a mouse model of Huntington's disease. *Hum Mol Genet*, 9, 1259–71.

Mahadevan, L. C., Willis, A. C. & Barratt, M. J. 1991. Rapid histone H3 phosphorylation in response to growth factors, phorbol esters, okadaic acid, and protein synthesis inhibitors. *Cell*, 65, 775–83.

Martianov, I., Brancorsini, S., Catena, R., Gansmuller, A., Kotaja, N., Parvinen, M., Sassone-Corsi, P. & Davidson, I. 2005. Polar nuclear localization of H1T2, a histone H1 variant, required for spermatid elongation and DNA condensation during spermiogenesis. *Proc Natl Acad Sci U S A*, 102, 2808–13.

Masliah, E., Dumaop, W., Galasko, D. & Desplats, P. 2013. Distinctive patterns of DNA methylation associated with Parkinson disease: identification of concordant epigenetic changes in brain and peripheral blood leukocytes. *Epigenetics*, 8, 1030–8.

Maze, I., Noh, K. M., Soshnev, A. A. & Allis, C. D. 2014. Every amino acid matters: essential contributions of histone variants to mammalian development and disease. *Nat Rev Genet*, 15, 259–71.

Mellen, M., Ayata, P., Dewell, S., Kriaucionis, S. & Heintz, N. 2012. MeCP2 binds to 5hmC enriched within active genes and accessible chromatin in the nervous system. *Cell*, 151, 1417–30.

Messner, S., Altmeyer, M., Zhao, H., Pozivil, A., Roschitzki, B., Gehrig, P., Rutishauser, D., Huang, D., Caflisch, A. & Hottiger, M. O. 2010. PARP1 ADP-ribosylates lysine residues of the core histone tails. *Nucleic Acids Res*, 38, 6350–62.

Metzger, E., Wissmann, M., Yin, N., Muller, J. M., Schneider, R., Peters, A. H., Gunther, T., Buettner, R. & Schule, R. 2005. LSD1 demethylates repressive histone marks to promote androgen-receptor-dependent transcription. *Nature*, 437, 436–9.

Mizuguchi, G., Shen, X., Landry, J., Wu, W. H., Sen, S. & Wu, C. 2004. ATP-driven exchange of histone H2AZ variant catalyzed by SWR1 chromatin remodeling complex. *Science*, 303, 343–8.

Mizusawa, Y., Kuji, N., Tanaka, Y., Tanaka, M., Ikeda, E., Komatsu, S., Kato, S. & Yoshimura, Y. 2010. Expression of human oocyte-specific linker histone protein and its incorporation into sperm chromatin during fertilization. *Fertil Steril*, 93, 1134–41.

Montellier, E., Boussouar, F., Rousseaux, S., Zhang, K., Buchou, T., Fenaille, F., Shiota, H., Debernardi, A., Hery, P., Curtet, S., Jamshidikia, M., Barral, S., Holota, H., Bergon, A., Lopez, F., Guardiola, P., Pernet, K., Imbert, J., Petosa, C., Tan, M., Zhao, Y., Gerard, M. & Khochbin, S. 2013. Chromatin-to-nucleoprotamine transition is controlled by the histone H2B variant TH2B. *Genes Dev*, 27, 1680–92.

Moore, K., Mcknight, A. J., Craig, D. & O'neill, F. 2014. Epigenome-wide association study for Parkinson's disease. *Neuromolecular Med*, 16, 845–55.

Murr, R., Loizou, J. I., Yang, Y. G., Cuenin, C., Wang, Z. Q. & Herceg, Z. 2006. Histone acetylation by Trrap-Tip60 modulates loading of repair proteins and repair of DNA double-strand breaks. *Nat Cell Biol*, 8, 91–9.

Nativio, R., Donahue, G., Berson, A., Lan, T., Amlie-Wolf, A., Tuzer, F., Toledo, J. B., Gosai, S. J., Gregory, B. D., Torres, C., Trojanowski, J. Q., Wang, L-S., Johnson, F. B., Bonini, M. N. &Berger, S. L. 2018. Dysregulation of the epigenetic landscape of normal aging in Alzheimer's disease. *Nat Neurosci*, 21, 497-505.

Nusinow, D. A., Hernandez-Munoz, I., Fazzio, T. G., Shah, G. M., Kraus, W. L. & Panning, B. 2007. Poly(ADP-ribose) polymerase is inhibited by a histone H2A variant, MacroH2A, and contributes to silencing of the inactive X chromosome. *J Biol Chem*, 282, 12851–9.

Obri, A., Ouararhni, K., Papin, C., Diebold, M. L., Padmanabhan, K., Marek, M., Stoll, I., Roy, L., Reilly, P. T., Mak, T. W., Dimitrov, S., Romier, C. & Hamiche, A. 2014. ANP32E is a histone chaperone that removes H2A.Z from chromatin. *Nature*, 505, 648–53.

Ong, C. T., Van Bortle, K., Ramos, E. & Corces, V. G. 2013. Poly(ADP-ribosyl)ation regulates insulator function and intrachromosomal interactions in *Drosophila*. *Cell*, 155, 148–59.

Oppikofer, M., Bai, T., Gan, Y., Haley, B., Liu, P., Sandoval, W., Ciferri, C. & Cochran, A. G. 2017. Expansion of the

ISWI chromatin remodeler family with new active complexes. *EMBO Rep*, 18, 1697–706.

Orrick, L. R., Olsn, M. O. J. & Busch, H. 1973. Comparison of nucleolar proteins of normal rat liver and Novikoff hepatoma ascites cells by two-dimensional polyacrylamide gel electrophoresis. *Proc Nat Acad Sci U S A*, 70, 1316–20.

Pan, Y., Daito, T., Sasaki, Y., Chung, Y. H., Xing, X., Pondugula, S., Swamidass, S. J., Wang, T., Kim, A. H. & Yano, H. 2016. Inhibition of DNA methyltransferases blocks mutant huntingtin-induced neurotoxicity. *Sci Rep*, 6, 31022.

Papamichos-Chronakis, M., Watanabe, S., Rando, O. J. & Peterson, C. L. 2011. Global regulation of H2A.Z localization by the INO80 chromatin-remodeling enzyme is essential for genome integrity. *Cell*, 144, 200–13.

Park, G., Tan, J., Garcia, G., Kang, Y., Salvesen, G. & Zhang, Z. 2016. Regulation of histone acetylation by autophagy in Parkinson disease. *J Biol Chem*, 291, 3531–40.

Patel, A., Vought, V. E., Dharmarajan, V. & Cosgrove, M. S. 2011. A novel non-SET domain multi-subunit methyltransferase required for sequential nucleosomal histone H3 methylation by the mixed lineage leukemia protein-1 (MLL1) core complex. *J Biol Chem*, 286, 3359–69.

Pek, J. W., Anand, A. & Kai, T. 2012. Tudor domain proteins in development. *Development*, 139, 2255–66.

Perez-Montero, S., Carbonell, A., Moran, T., Vaquero, A. & Azorin, F. 2013. The embryonic linker histone H1 variant of *Drosophila*, dBigH1, regulates zygotic genome activation. *Dev Cell*, 26, 578–90.

Peterson, C. L. & Herskowitz, I. 1992. Characterization of the yeast SWI1, SWI2, and SWI3 genes, which encode a global activator of transcription. *Cell*, 68, 573–83.

Petrij, F., Giles, R. H., Dauwerse, H. G., Saris, J. J., Hennekam, R. C., Masuno, M., Tommerup, N., Van Ommen, G. J., Goodman, R. H., Peters, D. J. & Et Al. 1995. Rubinstein–Taybi syndrome caused by mutations in the transcriptional co-activator CBP. *Nature*, 376, 348–51.

Pogo, B. G., Allfrey, V. G. & Mirsky, A. E.1996. RNA synthesis and histone acetylation during the course of gene activation in lymphocytes. *Proc Natl Acad Sci U S A*, 55, 805–12.

Polymeropoulos, M. H., Lavedan, C., Leroy, E., Ide, S. E., Dehejia, A., Dutra, A., Pike, B., Root, H., Rubenstein, J., Boyer, R., Stenroos, E. S., Chandrasekharappa, S., Athanassiadou, A., Papapetropoulos, T., Johnson, W. G., Lazzarini, A. M., Duvoisin, R. C., Di Iorio, G., Golbe, L. I. & Nussbaum, R. L. 1997. Mutation in the alpha-synuclein gene identified in families with Parkinson's disease. *Science*, 276, 2045–7.

Price, B. D. & D'andrea, A. D. 2013. Chromatin remodeling at DNA double-strand breaks. *Cell*, 152, 1344–54.

Qing, H., He, G., Ly, P. T., Fox, C. J., Staufenbiel, M., Cai, F., Zhang, Z., Wei, S., Sun, X., Chen, C. H., Zhou, W., Wang, K. & Song, W. 2008. Valproic acid inhibits Abeta production, neuritic plaque formation, and behavioral deficits in Alzheimer's disease mouse models. *J Exp Med*, 205, 2781–9.

Raisner, R. M., Hartley, P. D., Meneghini, M. D., Bao, M. Z., Liu, C. L., Schreiber, S. L., Rando, O. J. & Madhani, H. D. 2005. Histone variant H2A.Z marks the 5' ends of both active and inactive genes in euchromatin. *Cell*, 123, 233–48.

Ramsahoye, B. H., Biniszkiewicz, D., Lyko, F., Clark, V., Bird, A. P. & Jaenisch, R. 2000. Non-CpG methylation is prevalent in embryonic stem cells and may be mediated by DNA methyltransferase 3a. *Proc Natl Acad Sci U S A*, 97, 5237–42.

Rea, S., Eisenhaber, F., O'carroll, D., Strahl, B. D., Sun, Z. W., Schmid, M., Opravil, S., Mechtler, K., Ponting, C. P., Allis, C. D. & Jenuwein, T. 2000. Regulation of chromatin structure by site-specific histone H3 methyltransferases. *Nature*, 406, 593–9.

Ricobaraza, A., Cuadrado-Tejedor, M., Perez-Mediavilla, A., Frechilla, D., Del Rio, J. & Garcia-Osta, A. 2009. Phenylbutyrate ameliorates cognitive deficit and reduces tau pathology in an Alzheimer's disease mouse model. *Neuropsychopharmacology*, 34, 1721–32.

Robinson, P. J., An, W., Routh, A., Martino, F., Chapman, L., Roeder, R. G. & Rhodes, D. 2008. 30 nm chromatin fibre decompaction requires both H4-K16 acetylation and linker histone eviction. *J Mol Biol*, 381, 816–25.

Roelfsema, J. H., White, S. J., Ariyurek, Y., Bartholdi, D., Niedrist, D., Papadia, F., Bacino, C. A., Den Dunnen, J. T., Van Ommen, G. J., Breuning, M. H., Hennekam, R. C. & Peters, D. J. 2005. Genetic heterogeneity in Rubinstein–Taybi syndrome: mutations in both the *CBP* and *EP300* genes cause disease. *Am J Hum Genet*, 76, 572–80.

Rogakou, E. P., Pilch, D. R., Orr, A. H., Ivanova, V. S. & Bonner, W. M. 1998. DNA double-stranded breaks induce histone H2AX phosphorylation on serine 139. *J Biol Chem*, 273, 5858–68.

Rossetto, D., Avvakumov, N. & Cote, J. 2012. Histone phosphorylation: a chromatin modification involved in diverse nuclear events. *Epigenetics*, 7, 1098–108.

Saha, R. N. & Pahan, K. 2006. HATs and HDACs in neurodegeneration: a tale of disconcerted acetylation homeostasis. *Cell Death Differ*, 13, 539–50.

Sakabe, K., Wang, Z. & Hart, G. W. 2010. Beta-N-acetylglucosamine (O-GlcNAc) is part of the histone code. *Proc Natl Acad Sci U S A*, 107, 19915–20.

Sakai, A., Schwartz, B. E., Goldstein, S. & Ahmad, K. 2009. Transcriptional and developmental functions of the H3.3 histone variant in *Drosophila*. *Curr Biol*, 19, 1816–20.

Sawicka, A. & Seiser, C. 2012. Histone H3 phosphorylation – a versatile chromatin modification for different occaions. *Biochimie*, 94, 2193–201.

Schomacher, L. 2013. Mammalian DNA demethylation: multiple faces and upstream regulation. *Epigenetics*, 8, 679–84.

Schultz, M. D., He, Y., Whitaker, J. W., Hariharan, M., Mukamel, E. A., Leung, D., Rajagopal, N., Nery, J. R., Urich, M. A., Chen, H., Lin, S., Lin, Y., Jung, I., Schmitt, A.

D., Selvaraj, S., Ren, B., Sejnowski, T. J., Wang, W. & Ecker, J. R. 2015. Human body epigenome maps reveal noncanonical DNA methylation variation. *Nature*, 523, 212–16.

Sherrington, R., Rogaev, E. I., Liang, Y., Rogaeva, E. A., Levesque, G., Ikeda, M., Chi, H., Lin, C., Li, G., Holman, K., Tsuda, T., Mar, L., Foncin, J. F., Bruni, A. C., Montesi, M. P., Sorbi, S., Rainero, I., Pinessi, L., Nee, L., Chumakov, I., Pollen, D., Brookes, A., Sanseau, P., Polinsky, R. J., Wasco, W., Da Silva, H. A., Haines, J. L., Perkicak-Vance, M. A., Tanzi, R. E., Roses, A. D., Fraser, P. E., Rommens, J. M. & St George-Hyslop, P. H. 1995. Cloning of a gene bearing missense mutations in early-onset familial Alzheimer's disease. *Nature*, 375, 754–60.

Shi, X., Hong, T., Walter, K. L., Ewalt, M., Michishita, E., Hung, T., Carney, D., Pena, P., Lan, F., Kaadige, M. R., Lacoste, N., Cayrou, C., Davrazou, F., Saha, A., Cairns, B. R., Ayer, D. E., Kutateladze, T. G., Shi, Y., Cote, J., Chua, K. F. & Gozani, O. 2006. ING2 PHD domain links histone H3 lysine 4 methylation to active gene repression. *Nature*, 442, 96–9.

Shi, Y., Lan, F., Matson, C., Mulligan, P., Whetstine, J. R., Cole, P. A., Casero, R. A. & Shi, Y. 2004. Histone demethylation mediated by the nuclear amine oxidase homolog LSD1. *Cell*, 119, 941–53.

Shiio, Y. & Eisenman, R. N. 2003. Histone sumoylation is associated with transcriptional repression. *Proc Natl Acad Sci U S A*, 100, 13225–30.

Shogren-Knaak, M., Ishii, H., Sun, J. M., Pazin, M. J., Davie, J. R.& Peterson, C. L. 2006. Histone H4-K16 acetylation controls chromatin structure and protein interactions. *Science*, 311, 844–7.

Simon, M., North, J. A., Shimko, J. C., Forties, R. A., Ferdinand, M. B., Manohar, M., Zhang, M., Fishel, R., Ottesen, J. J. & Poirier, M. G. 2011. Histone fold modifications control nucleosome unwrapping and disassembly. *Proc Natl Acad Sci U S A*, 108, 12711–16.

Sinsheimer, R. L. 1954. The action of pancreatic desoxyribonuclease. I. Isolation of mono- and dinucleotides. *J Biol Chem*, 208, 445–59.

Skene, P. J. & Henikoff, S. 2013. Histone variants in pluripotency and disease. *Development*, 140, 2513–24.

St George-Hyslop, P., Haines, J., Rogaev, E., Mortilla, M., Vaula, G., Pericak-Vance, M., Foncin, J. F., Montesi, M., Bruni, A., Sorbi, S., Rainero, I., Pinessi, L., Pollen, D., Polinsky, R., Nee, L., Kennedy, J., Macciardi, F., Rogaeva, E., Liang, Y., Alexandrova, N., Lukiw, W., Schlumpf, K., Tanzi, R., Tsuda, T., Farrer, L., Cantu, J. M., Duara, R., Amaducci, L., Bergamini, L., Gusella, J., Roses, A., Crapper Mclachlan, D. & Et Al. 1992. Genetic evidence for a novel familial Alzheimer's disease locus on chromosome 14. *Nat Genet*, 2, 330–4.

Stadler, M. B., Murr, R., Burger, L., Ivanek, R., Lienert, F., Scholer, A., Van Nimwegen, E., Wirbelauer, C., Oakeley, E. J., Gaidatzis, D., Tiwari, V. K. & Schubeler, D. 2011. DNA-binding factors shape the mouse methylome at distal regulatory regions. *Nature*, 480, 490–5.

Szulwach, K. E. & Jin, P. 2014. Integrating DNA methylation dynamics into a framework for understanding epigenetic codes. *Bioessays*, 36, 107–17.

Tan, M., Luo, H., Lee, S., Jin, F., Yang, J. S., Montellier, E., Buchou, T., Cheng, Z., Rousseaux, S., Rajagopal, N., Lu, Z., Ye, Z., Zhu, Q., Wysocka, J., Ye, Y., Khochbin, S., Ren, B. & Zhao, Y. 2011. Identification of 67 histone marks and histone lysine crotonylation as a new type of histone modification. *Cell*, 146, 1016–28.

Taty-Taty, G. C., Courilleau, C., Quaranta, M., Carayon, A., Chailleux, C., Aymard, F., Trouche, D. & Canitrot, Y. 2014. H2A.Z depletion impairs proliferation and viability but not DNA double-strand breaks repair in human immortalized and tumoral cell lines. *Cell Cycle*, 13, 399–407.

The Huntington's Disease Collaborative Research Group. 1993. A novel gene containing a trinucleotide repeat that is expanded and unstable on Huntington's disease chromosomes. *Cell*, 72, 971–83.

Torres, C. R. & Hart, G. W. 1984. Topography and polypeptide distribution of terminal N-acetylglucosamine residues on the surfaces of intact lymphocytes. Evidence for O-linked GlcNAc. *J Biol Chem*, 259, 3308–17.

Tropberger, P., Pott, S., Keller, C., Kamieniarz-Gdula, K., Caron, M., Richter, F., Li, G., Mittler, G., Liu, E. T., Buhler, M., Margueron, R. & Schneider, R. 2013. Regulation of transcription through acetylation of H3K122 on the lateral surface of the histone octamer. *Cell*, 152, 859–72.

Tweedie-Cullen, R. Y., Brunner, A. M., Grossmann, J., Mohanna, S., Sichau, D., Nanni, P., Panse, C. & Mansuy, I. M. 2012. Identification of combinatorial patterns of posttranslational modifications on individual histones in the mouse brain. *PLoS One*, 7, e36980.

Udugama, M., Ft, M. C., Chan, F. L., Tang, M. C., Pickett, H. A., Jd, R. M., Mayne, L., Collas, P., Mann, J. R. & Wong, L. H. 2015. Histone variant H3.3 provides the heterochromatic H3 lysine 9 tri-methylation mark at telomeres. *Nucleic Acids Res*, 43, 10227–37.

Unnikrishnan, A., Gafken, P. R. & Tsukiyama, T. 2010. Dynamic changes in histone acetylation regulate origins of DNA replication. *Nat Struct Mol Biol*, 17, 430–7.

Van Broeckhoven, C., Backhovens, H., Cruts, M., De Winter, G., Bruyland, M., Cras, P. & Martin, J. J. 1992. Mapping of a gene predisposing to early-onset Alzheimer's disease to chromosome 14q24.3. *Nat Genet*, 2, 335–9.

Van Cauwenberghe, C., Van Broeckhoven, C. & Sleegers, K. 2016. The genetic landscape of Alzheimer disease: clinical implications and perspectives. *Genet Med*, 18, 421–30.

Van Leeuwen, F., Gafken, P. R. & Gottschling, D. E. 2002. Dot1p modulates silencing in yeast by methylation of the nucleosome core. *Cell*, 109, 745–56.

Varley, K. E., Gertz, J., Bowling, K. M., Parker, S. L., Reddy, T. E., Pauli-Behn, F., Cross, M. K., Williams, B. A.,

Stamatoyannopoulos, J. A., Crawford, G. E., Absher, D. M., Wold, B. J. & Myers, R. M. 2013. Dynamic DNA methylation across diverse human cell lines and tissues. *Genome Res*, 23, 555–67.

Vitaliano-Prunier, A., Menant, A., Hobeika, M., Geli, V., Gwizdek, C. & Dargemont, C. 2008. Ubiquitylation of the COMPASS component Swd2 links H2B ubiquitylation to H3K4 trimethylation. *Nat Cell Biol*, 10, 1365–71.

Wang, C., Li, Y., Cai, W., Bao, X., Girton, J., Johansen, J. & Johansen, K. M. 2014. Histone H3S10 phosphorylation by the JIL-1 kinase in pericentric heterochromatin and on the fourth chromosome creates a composite H3S10phK9me2 epigenetic mark. *Chromosoma*, 123, 273–80.

Wang, H., Cao, R., Xia, L., Erdjument-Bromage, H., Borchers, C., Tempst, P. & Zhang, Y. 2001a. Purification and functional characterization of a histone H3-lysine 4-specific methyltransferase. *Mol Cell*, 8, 1207–17.

Wang, Y., Zhang, W., Jin, Y., Johansen, J. & Johansen, K. M. 2001b. The JIL-1 tandem kinase mediates histone H3 phosphorylation and is required for maintenance of chromatin structure in *Drosophila*. *Cell*, 105, 433–43.

Wang, Z., Zang, C., Cui, K., Schones, D. E., Barski, A., Peng, W. & Zhao, K. 2009. Genome-wide mapping of HATs and HDACs reveals distinct functions in active and inactive genes. *Cell*, 138, 1019–31.

Watanabe, S., Radman-Livaja, M., Rando, O. J. & Peterson, C. L. 2013. A histone acetylation switch regulates H2A.Z deposition by the SWR-C remodeling enzyme. *Science*, 340, 195–9.

Weake, V. M. & Workman, J. L. 2008. Histone ubiquitination: triggering gene activity. *Mol Cell*, 29, 653–63.

Wisniewski, J. R., Zougman, A. & Mann, M. 2008. Nepsilon-formylation of lysine is a widespread post-translational modification of nuclear proteins occurring at residues involved in regulation of chromatin function. *Nucleic Acids Res*, 36, 570–7.

Wu, L., Zee, B. M., Wang, Y., Garcia B. A. & Dou, Y. 2011. The RING finger protein MSL2 in the MOF complex is an E3 ubiquitin ligase or H2B K34 and is involved in crosstalk with H3 K4 and K79 methylation. *Mol Cell*, 43, 132–44.

Xiao, A., Li, H., Shechter, D., Ahn, S. H., Fabrizio, L. A., Erdjument-Bromage, H., Ishibe-Murakami, S., Wang, B., Tempst, P., Hofmann, K., Patel, D. J., Elledge, S. J. & Allis, C. D. 2009. WSTF regulates the H2A.X DNA damage response via a novel tyrosine kinase activity. *Nature*, 457, 57–62.

Xie, Z., Dai, J., Dai, L., Tan, M., Cheng, Z., Wu, Y., Boeke, J. D. & Zhao, Y. 2012. Lysine succinylation and lysine malonylation in histones. *Mol Cell Proteomics*, 11, 100–7.

Xie, Z., Zhang, D., Chung, D., Tang, Z., Huang, H., Dai, L., Qi, S., Li, J., Colak, G., Chen, Y., Xia, C., Peng, C., Ruan, H., Kirkey, M., Wang, D., Jensen, L. M., Kwon, O. K., Lee, S., Pletcher, S. D., Tan, M., Lombard, D. B., White, K. P., Zhao, H., Li, J., Roeder, R. G., Yang, X. & Zhao, Y. 2016. Metabolic regulation of gene expression by histone lysine beta-hydroxybutyrylation. *Mol Cell*, 62, 194–206.

Xu, F., Zhang, K. & Grunstein, M. 2005. Acetylation in histone H3 globular domain regulates gene expression in yeast. *Cell*, 121, 375–85.

Yu, W., Ginjala, V., Pant, V., Chernukhin, I., Whitehead, J., Docquier, F., Farrar, D., Tavoosidana, G., Mukhopadhyay, R., Kanduri, C., Oshimura, M., Feinberg, A. P., Lobanenkov, V., Klenova, E. & Ohlsson, R. 2004. Poly(ADP-ribosyl)ation regulates CTCF-dependent chromatin insulation. *Nat Genet*, 36, 1105–10.

Yuan, C. C., Matthews, A. G., Jin, Y., Chen, C. F., Chapman, B. A., Ohsumi, T. K., Glass, K. C., Kutateladze, T. G., Borowsky, M. L., Struhl, K. & Oettinger, M. A. 2012. Histone H3R2 symmetric dimethylation and histone H3K4 trimethylation are tightly correlated in eukaryotic genomes. *Cell Rep*, 1, 83–90.

Zhang, S., Roche, K., Nasheuer, H. P. & Lowndes, N. F. 2011. Modification of histones by sugar beta-N-acetylglucosamine (GlcNAc) occurs on multiple residues, including histone H3 serine 10, and is cell cycle-regulated. *J Biol Chem*, 286, 37483–95.

Zuccato, C. & Cattaneo, E. 2007. Role of brain-derived neurotrophic factor in Huntington's disease. *Prog Neurobiol*, 81, 294–330.

Zunke, F., Moise, A. C., Belur, N. R., Gelyana, E., Stojkovska, I., Dzaferbegovic, H., Toker, N. J., Jeon, S., Fredriksen, K. & Mazzulli, J. R. 2017. Reversible conformational conversion of alpha-synuclein into toxic assemblies by glucosylceramide. *Neuron*, pii: S0896-6273(17)31135-2.

Epigenetic chromatin changes and the transcription cycle

The fundamental paradigm of epigenetics is that the organization of chromatin—the association of DNA with histone octamers and the resulting compacted physical state—must be modified in order to allow transcription to occur. Transcription is the result of the interaction of activated genomic regulatory regions (enhancers) with the promoter region of poised genomic templates. Differential gene activity in time, during development and adult life, and in space, in different regions of the developing embryo and in different tissues of the adult, is achieved by the activation of particular groups of enhancers. The process is initiated by the binding of cell lineage-specific transcription factors, often called **pioneer factors**, to the DNA region that usually contains their specific binding sequence (Cirillo et al., 2002; Soufi et al., 2015; Zaret and Carroll, 2011). These factors are able to access their binding sites in compacted chromatin that they "open" as the first prerequisite for transcription. Pioneer factors, in turn, recruit chromatin remodelers that can further open up the chromatin, and modifying enzymes that can stabilize the accessibility of the DNA within the region. Some of the initial evidence for this first step in the process of transcription regulation was obtained in yeast where transcription activators were shown to bind to gene promoters and to attract both chromatin remodeling and histone acetylating complexes (Cosma et al., 1999; Gregory et al., 1999). Although the epigenetic changes that modulate transcription at promoters and enhancers are similar in nature, sufficient distinctions exist to warrant their separate consideration.

In eukaryotes, the initial exposure of transcription factor DNA binding sites is achieved by several means that are not epigenetic in nature and that rely on the fact that nucleosomes are dynamic structures, transiently exposing stretches of their DNA (Pollach and Widom, 1995). The transient, partial dissociation of DNA from the histone octamer, often termed nucleosome **breathing**, occurs when the entry and exit ends of the DNA dissociate from the nucleosome (Koopmans et al., 2009; Li et al., 2005). The association of the DNA around the histone octamer may be altered by **sliding** that most likely consists of a rolling motion of the octamer along the DNA. This movement would involve breaking some DNA–octamer contacts at one end and establishing new ones at the other (Eslami-Mossallam et al., 2016; Pennings et al., 1991). A new type of spontaneous remodeling of the DNA–octamer association has been reported; referred to as nucleosome **gaping**, it involves an increase in the distance between the two DNA turns that envelop the octamer. In contrast to breathing or sliding that last for milliseconds, the structural change represented by gaping is thought to last seconds or minutes (Ngo and Ha, 2015).

The dynamic changes in nucleosomal architecture just described can be influenced by the level of affinity of the octamer for particular DNA sequences (Anderson and Widom, 2000; Kaplan et al., 2009). This affinity can be broadly considered to have two related components. Sequences that are enriched in poly dA:dT tracks are refractory to nucleosome formation (Iyer and Struhl, 1995), while the reverse is true in the case of DNA regions with a high number of G:C base-pairs (Valouev et al., 2011). The effects of these sequences, ranging from nucleosome depletion at many enhancers and promoters in yeast (Struhl and Segal, 2013) to destabilization of extant nucleosomes in flies and

Epigenetics, Nuclear Organization and Gene Function: with implications of epigenetic regulation and genetic architecture for human development and health. John C. Lucchesi, Oxford University Press (2019). © John C. Lucchesi 2019.
DOI: 10.1093/oso/9780198831204.001.0001

mammals (Guertin and Lis, 2013), may reflect the facility with which a particular DNA region can bend around the histone octamer (Battistini et al., 2010). Nucleosome destabilization facilitates nucleosome breathing or sliding and increases the transient opportunity that a factor present in the vicinity can access its DNA binding site.

In addition to the usual nucleosomal occlusion of their cognate binding sites, pioneer factors are faced with the problem presented by the random occurrence of similar sequences in the genome. For example, in the human genome, some estimates place the number of transcription factors at approximately 3000 and the total number of their potential binding sites at over 700,000. Differential gene activation, the linchpin of cellular differentiation, requires that particular factors recognize and bind specifically to a selected subset of sites. Binding specificity can be influenced by the degree of affinity of a factor for a precise, invariant sequence; it can also rely on the cooperative binding of the factor preassembled with a co-activator.

Two recent investigations have shed light on the parameters that mediate and affect pioneer transcription factor binding. The FOXA (forkhead box A2), GATA4 (GATA binding protein 4) and OCT4 (octamer binding protein 4) pioneer factors are transiently bound to numerous similar sites in different cell types but are enriched at cell-specific sites; this enrichment is mediated by the interaction of the pioneer factors with co-factors that are present in particular cell types (Donaghey et al., 2018). The Pax7 (paired box 7) pioneer transcription factor that is responsible for the differentiation of particular cell lineages during the development of the pituitary glands binds weakly to a number of genomic sites. In contrast to the transcription factors just discussed that access promoter regions, Pax7 binds to different sets of lineage-specific enhancers. At these high-affinity pioneering sites, Pax7 promotes nucleosome displacement and, after a certain time, allows the subsequent binding of other transcription factors (Mayran et al., 2018).

Role of remodeling complexes

Following the opportunistic association of a transcription factor, a stable DNA platform exposing new binding sites must be established on enhancers and promoter regions to allow the assembly of activating factors and of the transcription machinery (Fig. 5.1). The major regulatory elements of transcription, promoters and enhancers are usually depleted of nucleosomes very soon after the initial induction event. This is usually achieved by the recruitment of remodeling complexes (Becker and Workman, 2013; Neely et al., 1999; Yoshinaga et al., 1992). As previously mentioned, these complexes use the energy of ATP hydrolysis to disrupt histone–DNA interactions, to remove some of the octamer's subunits or to evict entire octamers; they can also slide nucleosomes along the DNA (see Chapter 4). One of the remodeling complexes implicated in the activation of transcription is the NURF (nucleosome remodeling factor) complex. NURF was shown to reposition nucleosomes along a promoter (Hamiche et al., 1999) and to interact with transcription factors (Xiao et al., 2001).

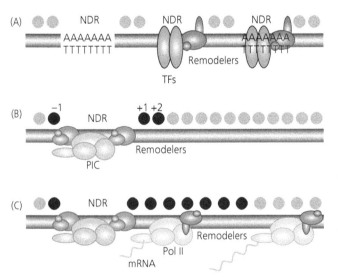

Fig. 5.1 Factors that modulate nucleosome positioning. (A) NDRs are nucleosome-depleted regions that occur either on Poly(dA:dT) tracts in the genome or as a result of transcription factor and remodeling complex binding (the gray circles indicate nucleosomes). (B) Remodeling complexes place nucleosomes at fixed positions (black circles), flanking the NDRs. (C) Downstream of the transcription start sites, the position of nucleosomes depends on the rate of transcription.

(From Struhl and Segal, 2013.)

A whole-genome analysis in *Drosophila* revealed that NURF is associated with specific gene promoters and coding regions, and with distant enhancers where it acts to reposition nucleosomes in a particular arrangement (Kwon et al., 2016). In yeast, two remodeling enzymes—ISWI that is the ATPase enzymatic subunit of NURF, and CHD1 (chromodomain helicase DNA-binding protein 1) that performs the same function in other remodeling complexes—reposition nucleosomes with spacing (linker DNA) of different lengths (Ocampo et al., 2016). Nucleosomal spacing is correlated with the level of activity of particular genes.

The linker histones stabilize nucleosomes and promote chromatin folding and condensation. Removal of these histones permits and facilitates transcription (as well as DNA replication); this function is performed by the histone chaperone SET/TAF1β that binds to H1 and evicts it (Zhang et al., 2015).

There are several subfamilies of remodeling complexes (see Chapter 4). In addition to their role in the very initial steps of gene activation, nucleosome remodeling complexes contribute to the elongation step of the transcription process. The formation and maintenance of the transcription bubble, as transcription proceeds, involve the disassembly of nucleosomes in front of the elongating polymerase complex and their reassembly behind the complex (Kristjuhan and Svejstrup, 2004; Schwabish and Struhl, 2004). These processes are facilitated by the action of ATP-dependent remodeling complexes such as SWI/SNF (BAF, in mammals) or ChD1 that collaborate with chaperones [for example, Asf1 (Adkins et al., 2004)]. As expected, nucleosome remodeling and assembly are distinct and independent functions of a given ATP-dependent complex (Torigoe et al., 2013). In addition, some chaperone complexes can carry out the disassembly and reassembly of nucleosomes in an ATP-independent manner, i.e. they serve both as chaperone and remodeler. One of the more extensively characterized of these complexes—FACT (facilitates chromatin transcription) (Hsieh et al., 2013; Orphanides et al., 1998)—destabilizes nucleosomes and promotes their eviction; by reversing these reactions, it can carry out nucleosome reassembly and stabilization (Formosa, 2013).

Phosphorylation of RNA polymerase

Following its assembly, the RNA polymerase pre-initiation complex is activated by the phosphorylation of serine S_5 on the CTD repeat of RNAPII, and by interacting with factors that are upstream of the core promoter region and with enhancers (see Chapter 3). The activated complex pauses after having transcribed a short segment (from 20 to 100 nt) of the DNA template. Pausing is ubiquitous among multicellular eukaryotes, occurring on approximately 75% of all active genes in *Drosophila* and approximately 50% of active genes in mammals. It is an important means of regulating transcriptional output, as evidenced by the observation that some genes that do not exhibit pausing in some tissues do so in others (Min et al., 2011). Release from pausing is concomitant with the 5´ end capping of the nascent transcript and the phosphorylation by p-TEFb of serines S_2 and S_7 of the RNA polymerase CTD, and of the pausing factors DSIF and NELF. The p-TEFb kinase can be recruited by associating directly with pioneer factors such as c-Myc (named for its homology to the avian myelocytomatosis gene) or NFκB (nuclear factor kappa-light-chain-enhancer of activated B cells), or it can be recruited by co-activators that recognize histone modifications (Kwak and Lis, 2013). The complexity of this latter type of recruitment is illustrated by the cascade of interactions required to eliminate pausing at the *FOSL1* gene promoter (Zippo et al., 2009). At this gene, Myc recruits the kinase PIM1 to phosphorylate serine 10 on histone H3; H3S10ph attracts the 14-3-3 phosphoserine adaptor protein which, in turn, recruits the histone acetyl transferase MOF, leading to the acetylation of H4 at lysine 16; the acetylated lysines of H4K16ac and H3K9ac (presumably mediated by p300/CBP acetyl transferase) are simultaneously recognized by the co-activator BRD4 via its two bromodomains; BDR4 is bound to p-TEFb, ultimately bringing this elongation factor to the promoter. Not surprisingly, MOF can be recruited to paused genes by other means that involve different covalent histone modifications (Kapoor-Vazirani and Vertino, 2014).

Role of histone covalent modifications

The massive body of evidence that transcription is regulated by covalent modifications of histones has revealed an amazing level of diversity and complexity that defy the formulation of precise rules. Some histone modifications alter the compaction of the chromatin fiber; others serve as targets for factors that affect the transcription process; some histone modifications that are present on many active genes are absent on others; the same is true with respect to silenced genes. Perhaps even more perplexing is the presence of modifications that are associated with activation on genes that are repressed.

Histone acetylation

This modification is associated with several functions that affect the transcription process in different ways. In general, the acetylation of lysines on the N-terminal tails of histone H3 and H4 alter the affinity of DNA for the histone octamer in a redundant and additive manner, thereby enhancing the accessibility of DNA sequences to transcription factors (Dion et al., 2005; Lang et al., 2013; Martin et al., 2004). One particular acetylation mark on H4—the acetylation of lysine 16 laid down by the histone acetyl transferase MOF (Hilfiker et al., 1997; Neal et al., 2000)—appears to have a singular effect on transcription. In *Drosophila*, it is highly enriched on the X chromosome in males and is involved in the regulatory mechanism responsible for dosage compensation (see Chapter 9). In addition, H4K16ac is present at numerous promoters of active genes in a variety of organisms, and *in vitro* experiments have demonstrated the unique ability of H4K16ac to disorder nucleosomal arrays and weaken chromatin fibers (Dunlap et al., 2012). These different effects can be explained by the suggested interaction of a basic segment consisting of amino acids 16–20 of the histone H4 tail of one nucleosome, with an acidic patch formed by an H2A/H2B dimer on the surface of a neighboring nucleosome. Such an interaction, which contributes to chromatin condensation (Luger et al., 1997), would be diminished by the acetylation of lysine 16, facilitating access to promoter regions by transcription factors, as well as the progress of RNA polymerase elongation along transcriptional units.

Acetylated lysines are recognized by a particular protein module—the bromodomain—first identified in *Drosophila* and conserved from yeast to humans (Haynes et al., 1992; Tamkun et al., 1992). This module is found in a large number of chromatin-modifying enzymes such as histone acetyl transferases and methyl transferases, remodeling complex subunits and transcription factors, which target nucleosomes and impact gene function in a variety of ways (Filippakopoulos and Knapp, 2012).

Histone methylation

Methylation at lysines or arginines does not change the overall charge; it affects transcription by serving as a target for regulatory factors and structural proteins. Three different lysine methylations—H3K4me, H3K36me and H3K79me—have been correlated with active transcription; two others—H3K9me and H3K27me—are associated with repression and silencing. H3K4 methylation is carried out by the methyl transferase Set1 that is recruited to the initiating RNAPII complex (in humans, additional enzymes such as the mixed lineage leukemia proteins MLL1-4 are capable of methylating H3K4). On nucleosomes that are near the transcription start site, lysine 4 is trimethylated (H3K4me3). As transcription proceeds, the degree of methylation decreases, to H3K4me2, then H3K4me1, and eventually disappears (Fig. 5.2). H3K4me3 is the target of the basal transcription factor TFIID via the PHD (plant homeodomain) finger recognition module of one of its subunits (Vermeulen et al., 2007).

Histone modifications are also involved with the elongation step of the transcription process. Methylated H3K36 is present along gene bodies, with an increased progression from H3K36me1 to H3K36me3 towards the 3' end (Wozniak and Strahl, 2014). In yeast, this modification is laid down by Set2 methyl transferase associated with the phosphorylated CTD of RNAPII. Methylated H3K36 recruits a histone deacetylase complex (Rpd3S) and a remodeling complex (Isw1b). The former complex reduces the acetylation of nucleosomes in the coding region following the passage of RNAPII, while the latter complex assists in re-establishing a condensed chromatin state; the goal of these post-transcriptional chromatin modifications is to prevent the reinitiation of transcription at cryptic start sites (Carrozza et al., 2005; Joshi et al., 2005; Keogh et al., 2005; Maltby et al., 2012; Smolle and Workman., 2012). In metazoans, the role of H3K36 methylation is more complex. Worms and flies possess two enzymes that produce H3K36me2 and H3K36me3, respectively. In addition to a role in transcription elongation, these methylated histones contribute to other regulatory pathways. The functional diversity of H3K36 methylation in humans is most remarkable— eight different histone methyl transferases have, as their target, the mono-, di-, or trimethylation of H3K36 and these isoforms have been associated with germline maintenance, RNA splicing and DNA repair (McDaniel and Strahl, 2017; Venkatesh and Workman, 2013).

The other highly conserved histone modification that is associated with active transcription is the methylation of H3 at lysine 79 by the DOT1 enzyme. The link of H3K79 methylation with transcription was discovered in yeast where it inhibits heterochromatin formation and silencing, and in *Drosophila* where it is associated with many active genes (Schubeler et al., 2004). Elongation rates correlate positively with the frequency of this mark (Jonkers et al., 2014).

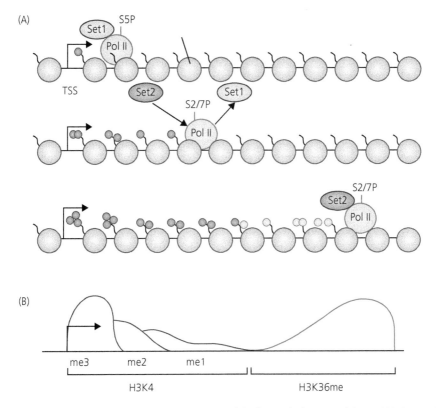

Fig. 5.2 Histone methylation as a consequence of transcriptional processes. (A) Following the formation of the pre-initiation complex, the histone methyl transferase Set1 interacts with Pol II and methylates H3K4 around the promoter region. As transcription elongation begins, Set1 is released and Set2 is recruited; as elongation proceeds, this enzyme methylates lysine 36 on histone H3 (H3K36me) along the gene body. (B) The distribution of these methylated histones is diagrammed at the bottom of the figure.

(From Zentner and Henikoff, 2013.)

Transcription without covalent modification

Some exceptions to the widespread belief that methylation at H3K4, H3K36 and H3K79 correlates with gene active transcription have been noted in particular yeast and *Drosophila* genes (Hodl and Basler, 2012; Zhang et al., 2014). A genome-wide study in *Drosophila* led to the conclusion that, while stably expressed genes bear the canonical histone marks, genes that are regulated and whose expression varies during development lack these marks (Perez-Lluch et al., 2015). Whether active transcription marks are dispensable for the activation of some genes in humans has not yet been demonstrated.

Role of histone variants

As previously discussed (see Chapter 4), many of the canonical histones that are deposited following replication on the daughter DNA molecules are replaced by histone variants. This exchange occurs when nucleosomes are evicted as a result of DNA damage or during transcription, and in both instances, the variants perform specific regulatory roles. The variant H2A.Z replaces approximately 15% of its canonical isoform H2A. In yeast and *Caenorhabditis*, it is present in the promoter region of genes that are poised for activation (bivalent promoters), while in *Drosophila* and mammals, it is present in the promoter of active genes (Henikoff and Smith, 2015). H2A.Z is removed prior to the assembly of the pre-initiation RNAPII complex, suggesting that it prepares the region for activation (Hardy et al., 2009). In yeast and *Drosophila*, H2A.Z is also present along the coding regions of transcribed genes where it facilitates elongation (Santisteban et al., 2011; Weber and Henikoff, 2014). Both effects can be explained if the presence of H2A.Z renders nucleosomes more easily disassembled. H2A.Z is also found at the promoter of silenced genes where, by destabilizing nucleosomes, it presumably facilitates the binding of repressive complexes (Weber and Henikoff, 2014).

Another histone variant that is ubiquitously deposited at active enhancers, promoters and along gene bodies is H3.3. The function of this variant is not yet fully understood, although some evidence exists that it may destabilize nucleosomes (Jin and Felsenfeld, 2007), explaining, once again, its dual role on transcription activation and repression. At active promoters, individual nucleosomes contain both H2A.Z and H3.3; this dual modification renders nucleosomes so unstable that it may lead to the conclusion that promoters are nucleosome-free (Jin et al., 2009).

A third histone variant (macroH2A) that is found in nucleosomes associated with heterochromatic regions and repressed genes is also found at the promoter of expressed genes. It carries out a functionally conflicting role by masking the binding sites of transcription activators on genes that are silenced and the binding sites of repressors on active genes (Lavigne et al., 2015).

Interactions among transcription regulators

Many chromatin modifications correlated with transcriptional modulation are interdependent. In human cells, H4K16ac occurs with H3K4me3 on the same nucleosomes in the promoter region of actively transcribed genes (Ruthenburg et al., 2011). DNA methylation, H3K9 methylation and unmethylated H3K4 are present in silent chromatin. The reverse situation, i.e. unmethylated CpGs, unmethylated H3K9 and methylated H3K4, is associated with expressed genes (Meissner et al., 2008).

The ubiquitination of histone H2B is required for the methylation of H3K4 and H3K79 in the same octamer (see Chapter 4). Different, bead-based, approaches have been used to determine the affinity of binding domains of chromatin proteins for single or combinations of histone modifications and to map their genomic location (Garske et al., 2010; Su et al., 2014). Covalent histone modifications mediate the recruitment of chromatin remodeling complexes and other histone-modifying enzymes; these interactions are usually gene-specific. For example, H3K4me3 can recruit the NURF and CHD1 remodeling complexes, as well as histone acetyl transferase complexes, such as SAGA, or histone demethylases that remove repressive methylation marks (Wozniak and Strahl, 2014). Covalent histone modifications can also modulate the replication-independent exchange of canonical histones for histone variants. For example, in *Drosophila*, the pioneer heat-shock

factor (HSF) recruits Set1 methyltransferase and Tip60 acetyl transferase complexes to the heat-shock genes' promoters; the methylation of H3K4 stimulates the incorporation and acetylation of H2Av, the *Drosophila* H2A.Z (Kusch et al., 2014).

Epigenetic changes at enhancers

The seminal events that activate enhancers are similar to those that activate promoters. Pioneer transcription factors bind to their cognate DNA sequences and initiate the accessibility of enhancer regions to remodeling factors and histone-modifying enzymes. Active enhancers are marked by the presence of H3K4me1 and H3K27ac, of the histone variant H2AZ and of p300/CBP acetyl transferase. A very low level of H3K4me3 may also be present. Enhancers that are poised for activation are marked by H3K4me1, lack H3K27ac and instead, at least in embryonic stem cells, are associated with H3K27me3 (Natoli and Andrau, 2012; Rada-Iglesias et al., 2011) and may display lower levels of H2AZ (Chen et al., 2014). Once activated, poised enhancers lose the H3K27me3 mark and replace it with H3K27ac. The H3K4me1 mark appears to function as a target for chromatin remodeling factors such as BAF and other complexes (Local et al., 2018). Yet, the presence of H3K4me1 in *Drosophila* enhancers (Rickels et al., 2017) and in mouse embryonic stem cells (Dorighi et al., 2017) appears to be dispensable. Surprisingly, the methyl transferases responsible for this modification play a methylation-independent role in regulating enhancer-mediated gene activation.

In certain instances, groups of enhancers form clusters that are referred to as super-enhancers. These special regulatory regions were first described in mouse embryonic stem cells where they are present at the sites of binding of the pioneer transcription factors—Oct4, Sox2 and Nanog (see Chapter 17)—responsible for the maintenance of the pluripotent state of these cells (Whyte et al., 2013). Subsequently, super-enhancers were identified in other cell types near genes that have cell-type specific functions (Hnisz et al., 2013). Diagnostic features for the identification and localization of super-enhancers in the genome are enrichment in H3K27ac and a high level of Mediator complex. It should be noted that these features are sometimes associated with individual enhancers (Pott and Lieb, 2015).

Active enhancers are transcribed by traditional RNAPII complexes (Kim et al., 2010). Although it was initially suggested that enhancer RNAs represented transcriptional noise, numerous observations indicate that enhancer transcription has a functional role. Enhancer transcripts, termed eRNAs, can be transcribed in one direction or bidirectionally from the transcription start site. They have a 5′ cap and can be polyadenylated and spliced; during transcription, RNAPII undergoes pausing, as it does on protein-coding genes. Highly active enhancers display H3K4me3, rather than H3K4me1 (Henriques et al., 2018). The tissue-specific expression of eRNAs is correlated with changes in the expression of nearby genes, perhaps by evicting repressive complexes (Lam et al., 2014). eRNAs and other non-coding RNAs (see Chapter 6) appear to function by recruiting transcription activators, by helping to establish enhancer–promoter association via the Mediator complex or cohesin, or by titrating away the NELF complex to allow productive elongation of the gene's transcript (Li et al., 2016). In other cases, although the enhancer is transcribed, the RNA that it produces does not participate in the activation of the cognate gene and should not be referred to as an eRNA; the enhancer function of these regulatory elements resides in their DNA sequence (Engreitz et al., 2016; Paralkar et al., 2016). It is also theoretically possible that, in some instances, it is the actual occurrence of transcription of the enhancer that stimulates the activity of the gene with which it associates by increasing the local concentration of transcription factors, of polymerase and of remodeling complexes.

Intragenic enhancers can have a negative regulatory effect on the genes wherein they are located by interfering with transcript elongation. These enhancers' own transcription is responsible for this effect (Cinghu et al., 2017).

Bivalent promoters

Bivalent promoters display simultaneously histone covalent modifications that are characteristic of active genes (H3K4me3) and of repressed genes (H3K27me3). First described in mouse embryonic stem cells where they are relatively prevalent (Bernstein et al., 2006), bivalent promoters are retained in a subset of genes in adult mammalian cell types (Roh et al., 2006). In addition, a small number of promoters in differentiated cells acquire bivalent histone marks (Mohn et al., 2008). In bivalent domains, the K4me3 and K27me3 modifications occur on opposite histone H3 molecules of individual nucleosomes (Voigt et al., 2012). The occurrence of bivalent domains appears to be evolutionarily restricted—bivalent promoters are absent in *Drosophila*; they are present in some non-mammalian vertebrates such as the Zebrafish, but not in others

(Voigt and Reinberg, 2013). Initially interpreted as a means of poising genes for rapid activation or permanent silencing, the low level of transcription allowed by bivalent promoters suggest that they ensure against inappropriately high transcription levels of key developmental genes or of genes responsible for differentiated cell function (Voigt and Reinberg, 2013).

Many genes are subjected to monoallelic expression

Mammals exhibit three different instances where only one allele of a gene is expressed. The first occurs during **random X chromosome inactivation** in female somatic cells—beginning around the time of implantation, approximately half of the embryo's cells express the alleles on the paternal X chromosome, while the other half express the maternal alleles (see Chapter 9). The second instance of monoallelic expression is **genomic imprinting**, a regulatory mechanism that directs the expression of a small number of autosomal genes from either the maternal or the paternal alleles (see Chapter 8). Finally, the third instance is a tissue-specific version of monoallelic inheritance. First discovered for gamma globulins (Pernis et al., 1965), and later in olfactory receptors (Chess et al., 1994), an estimated 10% to 20% of mammalian protein-encoding genes are expressed from the maternal copy in some cells and from the paternal copy in others. A number of widely expressed genes that are normally transcribed from both alleles exhibit monoallelic expression in a particular tissue, while a significant proportion of tissue-specific genes exhibit monoallelic expression. Not surprisingly, the alleles of monoallelically expressed genes differ in the histone epigenetic marks that they exhibit; for example, H3K36me3 is distributed throughout the body of the active allele, while H3K27me3 is found throughout the silent allele (Nag et al., 2013).

Heterogeneous gene expression among the cells of a particular tissue

This regulatory aspect of transcription, whereby some cells in a particular tissue type express a gene product while others do not, is of fundamental importance, given its occurrence in prokaryotes and eukaryotes. Heterogeneous gene expression is a largely stochastic process that, nevertheless, involves classical modes of transcriptional or post-transcriptional regulation (Johnston and Desplan, 2010; 2014). In bacteria, it provides the diversity necessary for the population to survive in changing environmental conditions; in multicellular organisms, it may prepare pluripotent cells to respond to various differentiation signals (Torres-Padilla and Chambers, 2014) and generate cellular diversity in groups of cells that are undergoing differentiation during development.

Repression of transcription

When physiological circumstances require it, transcription can be abrogated by reversing the mechanisms that initiate and maintain it. Gene silencing is largely the responsibility of specific complexes that not only erase some of the chromatin marks associated with gene activation, but also lay down specifically repressive epigenetic marks on histones and the DNA. Two major multiprotein complexes PRC1 and PRC2 (Polycomb repressive complex 1 and 2) are present, from worms to mammals. PRC2 methylates lysine 27 of histone H3, while PRC1 ubiquitynates lysine 119 of histone H2A, marks that are associated with inactive and condensed chromatin (see Chapter 7). In some instances, gene repression is initiated by remodeling complexes that act prior to the establishment of the histone modifications characteristic of repressed chromatin.

Transcriptional configuration of the genome

The thought that the genomes of eukaryotes may be divided into domains with differential functions is rooted in early cytological observations with *Drosophila* interphase polytene chromosomes. These chromosomes exhibit regions of extensive DNA folding and condensation that alternate with uncondensed regions along individual chromatids. In mammals, an analogous banded pattern of mitotic chromosomes is obtained with a variety of stains. With the advent of high-resolution methods to locate the position of chromatin-associated factors and modifications, genome-wide maps of functional domains have been generated. In the original study, 53 different non-histone protein components of chromatin were mapped and five principal chromatin types were identified on the basis of particular combinations of proteins: three different types of repressed chromatin corresponding to constitutive, Polycomb group-mediated and facultative heterochromatin, and two types of active chromatin (Filion et al., 2010). From a different perspective, by comparing the genomic distribution of eight common histone modifications and of CTCF, a protein with diverse nuclear functions (see Chapter 10), in nine mammalian cell types, six functional classes of chro-

matin were distinguished that included promoters, enhancers, insulators and transcribed, repressed and inactive states. Within these classes, further distinctions could be made between active and poised promoters, strong and weakly transcribed regions and constitutively repressed vs. Polycomb-repressed regions (Ernst et al., 2011). Similar studies have defined combinatorial chromatin states at the level of individual regulatory regions or of whole chromosomes, and correlated chromatin modifications of active genes with transcriptional unit length, exon patterns, regulatory functions and genomic contexts (Kharchenko et al., 2011). An analysis of over 300 genomic maps of major histone modifications from a variety of human tissues and stem cells partitioned the genome into four types of chromatin that contain: (1) active loci, marked by high levels of H3K4me3 and H3K36me3, (2) loci inactivated by Polycomb group repressive complexes with high levels of H3K27me3, (3) heterochromatic loci, with high levels of H3K9me3 and (4) null loci, with few, if any, histone modifications (Zhu et al., 2013).

A caveat to these and all other mapping studies of histone modifications that rely on chromatin immunoprecipitation is the potential cross-reactivity of antibodies with other residues (Egelhofer et al., 2011). Further, the binding of an antibody to the epitope for which it was generated may be inhibited by the presence of a neighboring histone mark. The *in vivo* specificity of antisera can be easily validated in yeast where individual histone residues can be mutated; unfortunately, given the multiple copies of histone genes in the genome, this type of control is not available in flies or mammals.

All of the above studies have revealed that the genome is subdivided into a very large number of regions with different protein compositions and functional characteristics. Superimposed on this linear domain differentiation is a complex level of long-distance interactions among domains and between domains and different nuclear structures. This three-dimensional organization of the genome will be discussed in Chapters 10–14.

Chapter summary

The association of DNA with histone octamers and the compacted physical state of the chromatin fiber must be modified in order to allow transcription to occur. The process is initiated by the opportunistic binding of pioneer transcription factors to their cognate DNA binding sites, which is facilitated by various stochastic events that relax, momentarily, the nucleosome–DNA association.

Once bound, pioneer factors recruit chromatin remodelers and histone-modifying enzymes for the purpose of repositioning nucleosomes and exposing regulatory regions (enhancers and gene promoters) to the components necessary for the initiation of transcription. Histone modifications, such as acetylation, methylation and ubiquitination, and the dynamic phosphorylation of specific amino acids on the major RNAPII subunit activate transcription and attract the factors necessary to eliminate the pausing that normally occurs soon after initiation. Further histone modifications and the replacement of certain core histones by histone variants facilitate transcript elongation and termination.

Two additional major epigenetic modifications that impact the process of transcription consist of the action of non-coding RNAs and DNA methylation. The importance of these modifications and their functional complexity are highlighted by discussing them in separate chapters (see Chapters 6 and 8).

References

Adkins, M. W., Howar, S. R. & Tyler, J. K. 2004. Chromatin disassembly mediated by the histone chaperone Asf1 is essential for transcriptional activation of the yeast PHO5 and PHO8 genes. *Mol Cell*, 14, 657–66.

Anderson, J. D. & Widom, J. 2000. Sequence and position-dependence of the equilibrium accessibility of nucleosomal DNA target sites. *J Mol Biol*, 296, 979–87.

Battistini, F., Hunter, C. A., Gardiner, E. J. & Packer, M. J. 2010. Structural mechanics of DNA wrapping in the nucleosome. *J Mol Biol*, 396, 264–79.

Becker, P. B. & Workman, J. L. 2013. Nucleosome remodeling and epigenetics. *Cold Spring Harb Perspect Biol*, 5, pii: a017905.

Bernstein, B. E., Mikkelsen, T. S., Xie, X., Kamal, M., Huebert, D. J., Cuff, J., Fry, B., Meissner, A., Wernig, M., Plath, K., Jaenisch, R., Wagschal, A., Feil, R., Schreiber, S. L. & Lander, E. S. 2006. A bivalent chromatin structure marks key developmental genes in embryonic stem cells. *Cell*, 125, 315–26.

Carrozza, M. J., Li, B., Florens, L., Suganuma, T., Swanson, S. K., Lee, K. K., Shia, W. J., Anderson, S., Yates, J., Washburn, M. P. & Workman, J. L. 2005. Histone H3 methylation by Set2 directs deacetylation of coding regions by Rpd3S to suppress spurious intragenic transcription. *Cell*, 123, 581–92.

Chen, P., Wang, Y. & Li, G. 2014. Dynamics of histone variant H3.3 and its coregulation with H2A.Z at enhancers and promoters. *Nucleus*, 5, 21–7.

Chess, A., Simon, I., Cedar, H. & Axel, R. 1994. Allelic inactivation regulates olfactory receptor gene expression. *Cell*, 78, 823–34.

Cinghu, S., Yang, P., Kosak, J. P., Conway, A. E., Kumar, D., Oldfield, A. J., Adelman, K. & Jothi, R. 2017. Intragenic enhancers attenuate host gene expression. *Mol Cell*, 68, 104–17 e6.

Cirillo, L. A., Lin, F. R., Cuesta, I., Friedman, D., Jarnik, M. & Zaret, K. S. 2002. Opening of compacted chromatin by early developmental transcription factors HNF3 (FoxA) and GATA-4. *Mol Cell*, 9, 279–89.

Cosma, M. P., Tanaka, T. & Nasmyth, K. 1999. Ordered recruitment of transcription and chromatin remodeling factors to a cell cycle- and developmentally regulated promoter. *Cell*, 97, 299–311.

Dion, M. F., Altschuler, S. J., Wu, L. F. & Rando, O. J. 2005. Genomic characterization reveals a simple histone H4 acetylation code. *Proc Natl Acad Sci U S A*, 102, 5501–6.

Donaghey, J., Thakurela, S., Charlton, J., Chen, J. S., Smith, Z. D., Gu, H., Pop, R., Clement, K., Stamenova, E. K., Karnik, R., Kelley, D. R., Gifford, C. A., Cacchiarelli, D., Rinn, J. L., Gnirke, A., Ziller, M. J. & Meissner, A. 2018. Genetic determinants and epigenetic effects of pioneer-factor occupancy. *Nat Genet*, 50, 250–8.

Dorighi, K. M., Swigut, T., Henriques, T., Bhanu, N. V., Scruggs, B. S., Nady, N., Still, C. D., 2ND, Garcia, B. A., Adelman, K. & Wysocka, J. 2017. Mll3 and Mll4 facilitate enhancer RNA synthesis and transcription from promoters independently of H3K4 monomethylation. *Mol Cell*, 66, 568–576 e4.

Dunlap, D., Yokoyama, R., Ling, H., Sun, H. Y., Mcgill, K., Cugusi, S. & Lucchesi, J. C. 2012. Distinct contributions of MSL complex subunits to the transcriptional enhancement responsible for dosage compensation in *Drosophila*. *Nucleic Acids Res*, 40, 11281–91.

Egelhofer, T. A., Minoda, A., Klugman, S., Lee, K., Kolasinska-Zwierz, P., Alekseyenko, A. A., Cheung, M. S., Day, D. S., Gadel, S., Gorchakov, A. A., Gu, T., Kharchenko, P. V., Kuan, S., Latorre, I., Linder-Basso, D., Luu, Y., Ngo, Q., Perry, M., Rechtsteiner, A., Riddle, N. C., Schwartz, Y. B., Shanower, G. A., Vielle, A., Ahringer, J., Elgin, S. C., Kuroda, M. I., Pirrotta, V., Ren, B., Strome, S., Park, P. J., Karpen, G. H., Hawkins, R. D. & Lieb, J. D. 2011. An assessment of histone-modification antibody quality. *Nat Struct Mol Biol*, 18, 91–3.

Engreitz, J. M., Haines, J. E., Perez, E. M., Munson, G., Chen, J., Kane, M., Mcdonel, P. E., Guttman, M. & Lander, E. S. 2016. Local regulation of gene expression by lncRNA promoters, transcription and splicing. *Nature*, 539, 452–5.

Ernst, J., Kheradpour, P., Mikkelsen, T. S., Shoresh, N., Ward, L. D., Epstein, C. B., Zhang, X., Wang, L., Issner, R., Coyne, M., Ku, M., Durham, T., Kellis, M. & Bernstein, B. E. 2011. Systematic analysis of chromatin state dynamics in nine human cell types. *Nature*, 473, 43–9.

Eslami-Mossallam, B., Schiessel, H. & Van Noort, J. 2016. Nucleosome dynamics: sequence matters. *Adv Colloid Interface Sci*, 232, 101–13.

Filion, G. J., Van Bemmel, J. G., Braunschweig, U., Talhout, W., Kind, J., Ward, L. D., Brugman, W., De Castro, I. J., Kerkhovn, R. M., Bussemaker, H. J. & Van Steensel, B. 2010. Systematic protein location mapping reveals five principal chromatin types in *Drosophila* cells. *Cell*, 143, 212–24.

Filippakopoulos, P. & Knapp, S. 2012. The bromodomain interaction module. *FEBS Lett,* 586, 2692–704.

Formosa, T. 2013. The role of FACT in making and breaking nucleosomes. *Biochim Biophys Acta,* 1819, 247–55.

Garske, A. L., Oliver, S. S., Wagner, E. K., Musselman, C. A., Leroy, G., Garcia, B. A., Kutateladze, T. G. & Denu, J. M. 2010. Combinatorial profiling of chromatin binding modules reveals multisite discrimination. *Nat Chem Biol,* 6, 283–90.

Gregory, P. D., Schmid, A., Zavari, M., Munsterkotter, M. & Horz, W. 1999. Chromatin remodelling at the PHO8 promoter requires SWI-SNF and SAGA at a step subsequent to activator binding. *EMBO J,* 18, 6407–14.

Guertin, M. J. & Lis, J. T. 2013. Mechanisms by which transcription factors gain access to target sequence elements in chromatin. *Curr Opin Genet Dev,* 23, 116–23.

Hamiche, A., Sandaltzopoulos, R., Gdula, D. A. & Wu, C. 1999. ATP-dependent histone octamer sliding mediated by the chromatin remodeling complex NURF. *Cell,* 97, 833–42.

Hardy, S., Jacques, P. E., Gevry, N., Forest, A., Fortin, M. E., Laflamme, L., Gaudreau, L. & Robert, F. 2009. The euchromatic and heterochromatic landscapes are shaped by antagonizing effects of transcription on H2A.Z deposition. *PLoS Genet,* 5, e1000687.

Haynes, S. R., Dollard, C., Winston, F., Beck, S., Trowsdale, J. & Dawid, I. B. 1992. The bromodomain: a conserved sequence found in human, *Drosophila* and yeast proteins. *Nucleic Acids Res,* 20, 2603.

Henikoff, S. & Smith, M. M. 2015. Histone variants and epigenetics. *Cold Spring Harb Prospect Biol,* 7. doi: 10.1101/cshperspect.a019364.

Henriques, T., Scruggs, B. S., Inouye, M. O., Muse, G. W., Williams, L. H., Burkholder, A. B., Lavender, C. A., Fargo, D. C. & Adelman, K. 2018. Widespread transcriptional pausing and elongation control at enhancers. *Genes Dev,* 32, 26–41.

Hilfiker, A., Hilfiker-Kleiner, D., Pannuti, A. & Lucchesi, J. C. 1997. MOF, a putative acetyl transferase gene related to the Tip60 and MOZ human genes and to the SAS genes of yeast, is required for dosage compensation in *Drosophila*. *EMBO J,* 16, 2054–60.

Hnisz, D., Abraham, B. J., Lee, T. I., Lau, A., Saint-Andre, V., Sigova, A. A., Hoke, H. A. & Young, R. A. 2013. Super-enhancers in the control of cell identity and disease. *Cell,* 155, 934–47.

Hodl, M. & Basler, K. 2012. Transcription in the absence of histone H3.2 and H3K4 methylation. *Curr Biol,* 22, 2253–7.

Hsieh, F. K., Kulaeva, O. I., Patel, S. S., Dyer, P. N., Luger, K., Reinberg, D. & Studitsky, V. M. 2013. Histone chaperone FACT action during transcription through chromatin by RNA polymerase II. *Proc Natl Acad Sci U S A,* 110, 7654–9.

Iyer, V. & Struhl, K. 1995. Poly(dA:dT), a ubiquitous promoter element that stimulates transcription via its intrinsic DNA structure. *EMBO J,* 14, 2570–9.

Jin, C. & Felsenfeld, G. 2007. Nucleosome stability mediated by histone variants H3.3 and H2A.Z. *Genes Dev,* 21, 1519–29.

Jin, C., Zang, C., Wei, G., Cui, K., Peng, W., Zhao, K. & Felsenfeld, G. 2009. H3.3/H2A.Z double variant-containing nucleosomes mark "nucleosome-free regions" of active promoters and other regulatory regions. *Nat Genet,* 41, 941–5.

Johnston, R. J.,JR. & Desplan, C. 2010. Stochastic mechanisms of cell fate specification that yield random or robust outcomes. *Annu Rev Cell Dev Biol,* 26, 689–719.

Johnston, R. J.,JR. & Desplan, C. 2014. Interchromosomal communication coordinates intrinsically stochastic expression between alleles. *Science,* 343, 661–5.

Jonkers, I., Kwak, H. & Lis, J. T. 2014. Genome-wide dynamics of Pol II elongation and its interplay with promoter proximal pausing, chromatin, and exons. *Elife,* 3, e04207.

Joshi, A. A. & Struhl, K. 2005. Eaf3 chromodomain interaction with methylated H3-K36 links histone deacetylation to Pol II elongation. *Mol Cell,* 20, 971–8.

Kaplan, N., Moore, I. K., Fondufe-Mittendorf, Y., Gossett, A. J., Tillo, D., Field, Y., Leproust, E. M., Hughes, T. R., Lieb, J. D., Widom, J. & Segal, E. 2009. The DNA-encoded nucleosome organization of a eukaryotic genome. *Nature,* 458, 362–6.

Kapoor-Vazirani, P. & Vertino, P. M. 2014. A dual role for the histone methyltransferase PR-SET7/SETD8 and histone H4 lysine 20 monomethylation in the local regulation of RNA polymerase II pausing. *J Biol Chem,* 289, 7425–37.

Keogh, M. C., Kurdistani, S. K., Morris, S. A., Ahn, S. H., Podolny, V., Collins, S. R., Schuldiner, M., Chin, K., Punna, T., Thompson, N. J., Boone, C., Emili, A., Weissman, J. S., Hughes, T. R., Strahl, B. D., Grunstein, M., Greenblatt, J. F., Buratowski, S. & Krogan, N. J. 2005. Cotranscriptional set2 methylation of histone H3 lysine 36 recruits a repressive Rpd3 complex. *Cell,* 123, 593–605.

Kharchenko, P. V., Alekseyenko, A. A., Schwartz, Y. B., Minoda, A., Riddle, N. C., Ernst, J., Sabo, P. J., Larschan, E., Gorchakov, A. A., Gu, T., Linder-Basso, D., Plachetka, A., Shanower, G., Tolstorukov, M. Y., Luquette, L. J., Xi, R., Jung, Y. L., Park, R. W., Bishop, E. P., Canfield, T. K., Sandstrom, R., Thurman, R. E., Macalpine, D. M., Stamatoyannopoulos, J. A., Kellis, M., Elgin, S. C., Kuroda, M. I., Pirrotta, V., Karpen, G. H. & Park, P. J. 2011. Comprehensive analysis of the chromatin landscape in *Drosophila melanogaster*. *Nature,* 471, 480–5.

Kim, T. K., Hemberg, M., Gray, J. M., Costa, A. M., Bear, D. M., Wu, J., Harmin, D. A., Laptewicz, M., Barbara-Haley, K., Kuersten, S., Markenscoff-Papadimitriou, E., Kuhl, D., Bito, H., Worley, P. F., Kreiman, G. & Greenberg, M. E. 2010. Widespread transcription at neuronal activity-regulated enhancers. *Nature,* 465, 182–7.

Koopmans, W. J., Buning, R., Schmidt, T. & Van Noort, J. 2009. spFRET using alternating excitation and FCS reveals progressive DNA unwrapping in nucleosomes. *Biophys J,* 97, 195–204.

Kristjuhan, A. & Svejstrup, J. Q. 2004. Evidence for distinct mechanisms facilitating transcript elongation through chromatin in vivo. *EMBO J,* 23, 4243–52.

Kusch, T., Mei, A. & Nguyen, C. 2014. Histone H3 lysine 4 trimethylation regulates cotranscriptional H2A variant exchange by Tip60 complexes to maximize gene expression. *Proc Natl Acad Sci U S A,* 111, 4850–5.

Kwak, H. & Lis, J. T. 2013. Control of transcriptional elongation. *Annu Rev Genet,* 47, 483–508.

Kwon, S. Y., Grisan, V., Jang, B., Herbert, J. & Badenhorst, P. 2016. Genome-wide mapping targets of the metazoan chromatin remodeling factor NURF reveals nucleosome remodeling at enhancers, core promoters and gene insulators. *PLoS Genet,* 12, e1005969.

Lam, M. T., Li, W., Rosenfeld, M. G. & Glass, C. K. 2014. Enhancer RNAs and regulated transcriptional programs. *Trends Biochem Sci,* 39, 170–82.

Lang, D., Schumann, M., Gelato, K., Fischle, W., Schwarzer, D. & Krause, E. 2013. Probing the acetylation code of histone H4. *Proteomics,* 13, 2989–97.

Lavigne, M. D., Vatsellas, G., Polyzos, A., Mantouvalou, E., Sianidis, G., Maraziotis, I., Agelopoulos, M. & Thanos, D. 2015. Composite macroH2A/NRF-1 nucleosomes suppress noise and generate robustness in gene expression. *Cell Rep,* 11, 1090–101.

Li, G., Levitus, M., Bustamante, C. & Widom, J. 2005. Rapid spontaneous accessibility of nucleosomal DNA. *Nat Struct Mol Biol,* 12, 46–53.

Li, W., Notani, D. & Rosenfeld, M. G. 2016. Enhancers as non-coding RNA transcription units: recent insights and future perspectives. *Nat Rev Genet,* 17, 207–23.

Local, A., Huang, H., Albuquerque, C. P., Singh, N., Lee, A. Y., Wang, W., Wang, C., Hsia, J. E., Shiau, A. K., Ge, K., Corbett, K. D., Wang, D., Zhou, H. & Ren, B. 2018. Identification of H3K4me1-associated proteins at mammalian enhancers. *Nat Genet,* 50, 73–82.

Luger, K., Mader, A. W., Richmond, R. K., Sargent, D. F. & Richmond, T. J. 1997. Crystal structure of the nucleosome core particle at 2.8 A resolution. *Nature,* 389, 251–60.

Maltby, V. E., Martin, B. J., Schulze, J. M., Johnson, I., Hentrich, T., Sharma, A., Kobor, M. S. & Howe, L. 2012. Histone H3 lysine 36 methylation targets the Isw1b remodeling complex to chromatin. *Mol Cell Biol,* 32, 3479–85.

Martin, A. M., Pouchnik, D. J., Walker, J. L. & Wyrick, J. J. 2004. Redundant roles for histone H3 N-terminal lysine residues in subtelomeric gene repression in *Saccharomyces cerevisiae. Genetics,* 167, 1123–32.

Mayran, A., Khetchoumian, K., Hariri, F., Pastinen, T., Gauthier, Y., Balsalobre, A. & Drouin, J. 2018. Pioneer factor Pax7 deploys a stable enhancer repertoire for specification of cell fate. *Nat Genet,* 50, 259–69.

Mcdaniel, S. L. & Strahl, B. D. 2017. Shaping the cellular landscape with Set2/SETD2 methylation. *Cell Mol Life Sci,* 74, 3317–34.

Meissner, A., Mikkelsen, T. S., Gu, H., Wernig, M., Hanna, J., Sivachenko, A., Zhang, X., Bernstein, B. E., Nusbaum, C., Jaffe, D. B., Gnirke, A., Jaenisch, R. & Lander, E. S. 2008. Genome-scale DNA methylation maps of pluripotent and differentiated cells. *Nature,* 454, 766–70.

Min, I. M., Waterfall, J. J., Core, L. J., Munroe, R. J., Schimenti, J. & Lis, J. T. 2011. Regulating RNA polymerase pausing and transcription elongation in embryonic stem cells. *Genes Dev,* 25, 742–54.

Mohn, F., Weber, M., Rebhan, M., Roloff, T. C., Richter, J., Stadler, M. B., Bibel, M. & Schubeler, D. 2008. Lineage-specific polycomb targets and *de novo* DNA methylation define restriction and potential of neuronal progenitors. *Mol Cell,* 30, 755–66.

Nag, A., Savova, V., Fung, H. L., Miron, A., Yuan, G. C., Zhang, K. & Gimelbrant, A. A. 2013. Chromatin signature of widespread monoallelic expression. *Elife,* 2, e01256.

Natoli, G. & Andrau, J. C. 2012. Noncoding transcription at enhancers: general principles and functional models. *Annu Rev Genet,* 46, 1–19.

Neal, K. C., Pannuti, A., Smith, E. R. & Lucchesi, J. C. 2000. A new human member of the MYST family of histone acetyl transferases with high sequence similarity to *Drosophila* MOF. *Biochim Biophys Acta,* 1490, 170–4.

Neely, K. E., Hassan, A. H., Wallberg, A. E., Steger, D. J., Cairns, B. R., Wright, A. P. & Workman, J. L. 1999. Activation domain-mediated targeting of the SWI/SNF complex to promoters stimulates transcription from nucleosome arrays. *Mol Cell,* 4, 649–55.

Ngo, T. T. & Ha, T. 2015. Nucleosomes undergo slow spontaneous gaping. *Nucleic Acids Res,* 43, 3964–71.

Ocampo, J., Chereji, R. V., Eriksson, P. R. & Clark, D. J. 2016. The ISW1 and CHD1 ATP-dependent chromatin remodelers compete to set nucleosome spacing *in vivo. Nucleic Acids Res,* 44, 4625–35.

Orphanides, G., Leroy, G., Chang, C. H., Luse, D. S. & Reinberg, D. 1998. FACT, a factor that facilitates transcript elongation through nucleosomes. *Cell,* 92, 105–16.

Paralkar, V. R., Taborda, C. C., Huang, P., Yao, Y., Kossenkov, A. V., Prasad, R., Luan, J., Davies, J. O., Hughes, J. R., Hardison, R. C., Blobel, G. A. & Weiss, M. J. 2016. Unlinking an lncRNA from its associated cis element. *Mol Cell,* 62, 104–10.

Pennings, S., Meersseman, G. & Bradbury, E. M. 1991. Mobility of positioned nucleosomes on 5 S rDNA. *J Mol Biol,* 220, 101–10.

Perez-Lluch, S., Blanco, E., Tilgner, H., Curado, J., Ruiz-Romero, M., Corominas, M. & Guigo, R. 2015. Absence of canonical marks of active chromatin in developmentally regulated genes. *Nat Genet,* 47, 1158–67.

Pernis, B., Chiappino, G., Kelus, A. S. & Gell, P. G. 1965. Cellular localization of immunoglobulins with different allotypic specificities in rabbit lymphoid tissues. *J Exp Med,* 122, 853–76.

Polach, K. J. & Widom, J. 1995. Mechanism of protein access to specific DNA sequences in chromatin: a dynamic equilibrium model for gene regulation. *J Mol Biol,* 254, 130–49.

Pott, S. & Lieb, J. D. 2015. What are super-enhancers? *Nat Genet,* 47, 8–12.

Rada-Iglesias, A., Bajpai, R., Swigut, T., Brugmann, S. A., Flynn, R. A. & Wysocka, J. 2011. A unique chromatin signature uncovers early developmental enhancers in humans. *Nature,* 470, 279–83.

Rickels, R., Herz, H. M., Sze, C. C., Cao, K., Morgan, M. A., Collings, C. K., Gause, M., Takahashi, Y. H., Wang, L., Rendleman, E. J., Marshall, S. A., Krueger, A., Bartom, E. T., Piunti, A., Smith, E. R., Abshiru, N. A., Kelleher, N. L., Dorsett, D. & Shilatifard, A. 2017. Histone H3K4 monomethylation catalyzed by Trr and mammalian COMPASS-like proteins at enhancers is dispensable for development and viability. *Nat Genet,* 49, 1647–1653.

Roh, T. Y., Cuddapah, S., Cui, K. & Zhao, K. 2006. The genomic landscape of histone modifications in human T cells. *Proc Natl Acad Sci U S A,* 103, 15782–7.

Ruthenburg, A. J., Li, H., Milne, T. A., Dewell, S., Mcginty, R. K., Yuen, M., Ueberheide, B., Dou, Y., Muir, T. W., Patel, D. J. & Allis C. D. 2011. Recognition of a mononcleosomal histone modification pattern by BPTF via multivalent interactions. *Cell,* 145, 692–706.

Santisteban, M. S., Hang, M. & Smith, M. M. 2011. Histone variant H2A.Z and RNA polymerase II transcription elongation. *Mol Cell Biol,* 31, 1848–60.

Schubeler, D., Macalpine, D. M., Scalzo, D., Wirbelauer, C., Kooperberg, C., Van Leeuwen, F., Gottschling, D. E., O'neill, L. P., Turner, B. M., Delrow, J., Bell, S. P. & Groudine, M. 2004. The histone modification pattern of active genes revealed through genome-wide chromatin analysis of a higher eukaryote. *Genes Dev,* 18, 1263–71.

Schwabish, M. A. & Struhl, K. 2004. Evidence for eviction and rapid deposition of histones upon transcriptional elongation by RNA polymerase II. *Mol Cell Biol,* 24, 10111–17.

Smolle, M. & Workman, J. L. 2012. Transcription-associated histone modifications and cryptic trasciption. *Biochim Biophys Acta,* 1829, 84–97.

Soufi, A., Garcia, M. F., Jaroszewicz, A., Osman, N., Pellegrini, M. & Zaret, K. S. 2015. Pioneer transcription factors target partial DNA motifs on nucleosomes to initiate reprogramming. *Cell,* 161, 555–68.

Struhl, K. & Segal, E. 2013. Determinants of nucleosome positioning. *Nat Struct Mol Biol,* 20, 267–73.

Su, Z., Boersma, M. D., Lee, J. H., Oliver, S. S., Liu, S., Garcia, B. A. & Denu, J. M. 2014. ChIP-less analysis of chromatin states. *Epigenetics Chromatin,* 7, 7.

Tamkun, J. W., Deuring, R., Scott, M. P., Kissinger, M., Pattatucci, A. M., Kaufman, T. C. & Kennison, J. A. 1992. Brahma: a regulator of *Drosophila* homeotic genes structurally related to the yeast transcriptional activator SNF2/SWI2. *Cell,* 68, 561–72.

Torigoe, S. E., Patel, A., Khuong, M. T., Bowman, G. D. & Kadonaga, J. T. 2013. ATP-dependent chromatin assembly is functionally distinct from chromatin remodeling. *Elife,* 2, e00863.

Torres-Padilla, M. E. & Chambers, I. 2014. Transcription factor heterogeneity in pluripotent stem cells: a stochastic advantage. *Development,* 141, 2173–81.

Valouev, A., Johnson, S. M., Boyd, S. D., Smith, C. L., Fire, A. Z. & Sidow, A. 2011. Determinants of nucleosome organization in primary human cells. *Nature,* 474, 516–20.

Vermeulen, M., Mulder, K. W., Denissov, S., Pijnappel, W. W., Van Schaik, F. M., Varier, R. A., Baltissen, M. P., Stunnenberg, H. G., Mann, M. & Timmers, H. T. 2007. Selective anchoring of TFIID to nucleosomes by trimethylation of histone H3 lysine 4. *Cell,* 131, 58–69.

Voigt, P., Leroy, G., Drury, W. J., 3RD, Zee, B. M., Son, J., Beck, D. B., Young, N. L., Garcia, B. A. & Reinberg, D. 2012. Asymmetrically modified nucleosomes. *Cell,* 151, 181–93.

Voigt, P., Tee, W. W. & Reinberg, D. 2013. A double take on bivalent promoters. *Genes Dev,* 27, 1318–38.

Weber, C. M. & Henikoff, S. 2014. Histone variants: dynamic punctuation in transcription. *Genes Dev,* 28, 672–82.

Whyte, W. A., Orlando, D. A., Hnisz, D., Abraham, B. J., Lin, C. Y., Kagey, M. H., Rahl, P. B., Lee, T. I. & Young, R. A. 2013. Master transcription factors and mediator establish super-enhancers at key cell identity genes. *Cell,* 153, 307–19.

Wozniak, G. G. & Strahl, B. D. 2014. Hitting the "mark:" interpreting lysine methylation in the context of active transcription. *Biochim Biophys Acta,* 1839, 1353–61.

Xiao, H., Sandaltzopoulos, R., Wang, H. M., Hamiche, A., Ranallo, R., Lee, K. M., Fu, D. & Wu, C. 2001. Dual functions of largest NURF subunit NURF301 in nucleosome sliding and transcription factor interactions. *Mol Cell,* 8, 531–43.

Yoshinaga, S. K., Peterson, C. L., Herskowitz, I. & Yamamoto, K. R. 1992. Roles of SWI1, SWI2, and SWI3 proteins for transcriptional enhancement by steroid receptors. *Science,* 258, 1598–604.

Zaret, K. S. & Carroll, J. S. 2011. Pioneer transcription factors: establishing competence for gene expression. *Genes Dev,* 25, 2227–41.

Zentner, G. E. & Henikoff, S. 2013. Regulation of nucleosome dynamics by histone modifications. *Nat Struct Mol Biol,* 20, 259–66.

Zhang, H., Gao, L., Anandhakumar, J. & Gross, D. S. 2014. Uncoupling transcription from covalent histone modification. *PLoS Genet,* 10, e1004202.

Zhang, Q., Giebler, H. A., Isaacson, M. K. & Nyborg, J. K. 2015. Eviction of linker histone H1 by NAP-family histone chaperones enhances activated transcription. *Epigenetics Chromatin,* 8, 30.

Zhu, J., Adli, M., Zou, J. Y., Verstappen, G., Coyne, M., Zhang, X., Durham, T., Miri, M., Deshpande, V., De Jager, P. L., Bennett, D. A., Houmard, J. A., Muoio, D. M., Onder, T. T., Camahort, R., Cowan, C. A., Meissner, A., Epstein, C. B., Shoresh, N. & Bernstein, B. E. 2013. Genome-wide chromatin state transitions associated with developmental and environmental cues. *Cell,* 152, 642–54.

Zippo, A., Serafini, R., Rocchigiani, M., Pennacchini, S., Krepelova, A. & Oliviero, S. 2009. Histone crosstalk between H3S10ph and H4K16ac generates a histone code that mediates transcription elongation. *Cell,* 138, 1122–36.

CHAPTER 6

The role of non-coding RNAs

It had long been known that, in addition to messenger RNA (mRNA), the genome of eukaryotic organisms produces two additional classes of RNA—transfer RNAs (tRNAs) and ribosomal RNAs (rRNAs)—and that all three classes are exported to the cytoplasm where they participate in the translation process. In 1959, Henry Harris reported that only a small fraction of the nuclear RNA in connective tissue cells and in macrophages actually passes into the cytoplasm, leading him to conclude that most of the RNA is not engaged in the function of "carrying 'information' from the genes to the cytoplasm" (Harris, 1959). This historical insight has been fully validated—modern analyses of the total transcriptional output by hybridization to genomic arrays or high-throughput nucleic acid sequencing have revealed that, while only a few percent of the sequences of the genomes of higher metazoans encode proteins, the majority of their DNA is, in fact, transcribed (Venters and Pugh, 2013). Generated as intergenic transcripts or transcripts that partially overlap coding segments, these RNAs can be divided into small or long non-coding RNAs, which can, in turn, be subdivided into different subclasses with specific functions. Non-coding RNAs, usually in the form of RNA–protein complexes, regulate all aspects of gene expression, confer stability to the genome and defend against invading foreign nucleic acids.

Long non-coding RNAs

Long non-coding RNAs (lncRNAs) represent the majority of the transcripts of the mammalian genome. They were first recognized as polyadenylated transcripts that remained in the nucleus and were distinct from mRNAs in sequence (Herman et al., 1976; Perry et al., 1974). Non-coding RNAs are assigned to this group if they are at least 200 nucleotides in length; they are transcribed by RNAPII from the coding or non-coding strands of regions that overlap protein-coding genes, from the coding strand of promoter regions in the opposite direction of the associated protein-coding gene (by far the most abundant class), from the coding or non-coding strands of enhancer regions (eRNAs, the second most abundant class) or from intergenic regions (lincRNAs). Many lncRNAs undergo splicing and are capped and polyadenylated at their 5' and 3' ends, respectively. Of course, their transcription is subject to regulation. This is made obvious by the fact that the majority of the lncRNAs identified to date are transcribed in a cell type-specific manner. Once transcribed, these RNAs can repress genes or participate in their activation; they can also help to organize the genome into functionally different topological regions. Their sequence usually contains motifs that are recognized by specific proteins and sometimes provides the complementarity necessary for recognizing remote genomic target sites. Although the functional inventory of lncRNAs is in its initial phases, a growing number of them have been characterized in plants and animals. Below are some examples of lncRNAs that illustrate the many ways in which this class of RNAs affects gene expression.

lncRNAs and lincRNAs repress or enhance the transcription of target genes

Some lncRNAs interact with Polycomb group complexes (PRCs), recruit other histone-modifying enzymes and

Epigenetics, Nuclear Organization and Gene Function: with implications of epigenetic regulation and genetic architecture for human development and health. John C. Lucchesi, Oxford University Press (2019). © John C. Lucchesi 2019.
DOI: 10.1093/oso/9780198831204.001.0001

cause gene repression in *cis* in the same regions of the genome from which they were transcribed (Fig. 6.1A). ANRIL (antisense RNA in the INK4 locus) is a lncRNA that is complementary to a segment of DNA in the tumor suppressor gene cluster *INK4-ARF* (Pasmant et al., 2007). ANRIL binds to a subunit (CBX) of the PRC1 repressive complex and specifically targets it to the *INK4-ARF* locus where it interacts with the methylated H3K27 mark laid down by PRC2 (Yap et al., 2010). Several other cases of recruitment of PRCs to chromatin by lincRNAs have been reported, among them Kcnq1ot1 (KCNQ1 opposite strand/antisense transcript 1), HOTAIR (HOX transcript antisense RNA) and Xist (X inactive specific transcript). Kcnq1ot1 is involved in genomic imprinting (see Chapter 8); it is transcribed as an antisense RNA only on the paternal chromosome, and it is involved in the silencing of several protein-encoding genes on this chromosome by interacting with G9, the enzyme responsible for H3K9 methylation characteristic of inactive promoters, and with the PRC2 complex (Pandey et al., 2008). HOTAIR is a lincRNA transcribed from the homeobox gene cluster *HOXC*; it represses the expression of genes in the *HOXD* cluster by interacting with the PRC2 complex and with LSD1, the enzyme responsible for

demethylating the H3K4me mark found on active genes. Since the two clusters are located in different chromosomes, HOTAIR exercises its repressive effect in *trans*.

Xist was the first lincRNA identified (Brown et al., 1991). Because Xist and PRC2 co-localize along the inactive X chromosome in females, this RNA was reported to achieve its repressive effect by direct interaction with PRC2 and recruitment to the X (Mak et al., 2002). This notion has been extensively modified by recent data (Almeida et al., 2017). This case and circumstantial observations relating to the promiscuous binding of PRC2 to any RNA (Davidovich et al., 2013) suggest that the association of PRC2 with lncRNAs may occur through intermediate linker proteins or the action of intermediate complexes more often than was initially suspected.

Some lincRNAs can have a positive effect on the expression of the genes that they target. In *Drosophila*, roX1 and roX2 are two lincRNAs that provide the scaffold for the assembly of the MSL dosage compensation complex in males (Amrein and Axel, 1997; Meller et al., 1997). These RNAs are inactive upon transcription, and their secondary structure must be remodeled by a helicase in order to permit the cascade of associations of the different protein components that form the

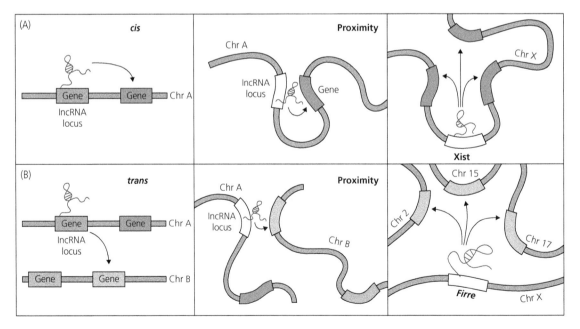

Fig. 6.1 lncRNAs can regulate genes in *cis* (A) or in *trans* (B). Regulation is facilitated when the lncRNA locus and its targets are in close proximity (middle panels).

(From Quinodoz and Guttman, 2014.)

MSL complex (Maenner et al., 2013). In contrast to the repressive effect of Xist along the X chromosome in mammals, the roX RNAs and associated MSL proteins increase the overall transcription rates of most X-linked genes in male *Drosophila* (see Chapter 9).

HOTTIP (HOXA transcript at the distal tip) is the product of a gene located at the 3' end of the *HOXA* gene cluster. Absence of this RNA leads to the silencing of the *HOXA* genes, with those that are closest to the HOTTIP transcriptional unit being most affected. To mediate the activation of *HOXA* genes, HOTTIP RNA associates directly with the MLL complex that is responsible for the methylation of H3K4 (Wang et al., 2011). Some examples exist of genes that are activated by a particular lncRNA and repressed by a different one (Yang et al., 2011).

Some lncRNAs may modulate gene activity by sequestering particular micro RNAs (miRNAs). As discussed in Short non-coding RNAs, p. 72, miRNAs are negative regulators of gene expression and their association with a lncRNA would limit their ability to target their cognate genes (Wang et al., 2013). Few examples of this type of regulation exist and, given the low nuclear concentration of most lncRNAs, they may exert only a minor regulatory effect of this type on gene function (Thomson and Dinger, 2016).

Recently, a systematic study of lncRNA promoters whose genomic position is conserved between mice and humans has revealed the presence of a subset of these RNAs located at the ends of loops that form topological chromatin boundaries, described in Chapter 10 (Amaral et al., 2018). As is the case for these loop ends, the promoters of most of the lncRNA genes are enriched in the architectural protein CTCF.

Some lncRNAs transcribed from enhancers promote the association of these regulatory modules with their target genes

Enhancers are marked by H3K4me1, H3K27ac and often by the presence of co-activators such as the histone acetyl transferases p300/CBP. These associations suggested that enhancers may be transcribed and led to the discovery that both strands of active enhancers are, in fact, transcribed by divergent RNAPII complexes (De Santa et al., 2010; Kim et al., 2010). In particular instances, the functionality of the resulting eRNAs was demonstrated by inducing their degradation with RNA interference (a process that is discussed in Short non-coding RNAs, p. 72) or by targeting and tethering them to a reporter gene promoter or enhancer sequence on a plasmid (Li et al., 2013; Melo et al., 2013). These experiments demonstrated that the eRNAs produced by enhancers following activation are responsible for the expression of the enhancer-targeted genes. This effect is achieved by helping to establish or to stabilize the physical association that enhancers undergo with their target gene promoters (see Chapters 3, 5 and 10). In other instances, the lncRNAs produced by enhancer transcription appear to have no direct effect on the activation of the gene with which the enhancer associates and can be considered to constitute transcriptional noise. To determine whether a lncRNA is, in fact, an eRNA simply requires deleting the promoter region of the enhancer that produces it and/or forcing the early termination of its transcript. If the first modification abrogates the activity of the regulated gene, while the second modification does not, the RNA can be considered to be an eRNA (Engreitz et al., 2016; Paralkar et al., 2016).

lincRNAs can affect gene activity by interacting directly with transcription co-factors

Glucocorticoids are a class of hormones that modulate a number of physiological pathways involved in development, metabolism and immune response. These hormones act by associating with, and activating, a receptor molecule (GR). The activated receptor–hormone complex binds to genomic sequences termed glucocorticoid response elements (GREs) present in or near the promoter of target genes and stimulates transcription. A lincRNA—Gas5 (growth arrest specific 5) (Schneider et al., 1988)—can bind to the DNA-binding domain of the hormone-activated GR and prevents it from associating with the GREs. In this manner, the Gas5 RNA acts as a "decoy" to block the activation of glucocorticoid-responsive genes (Kino et al., 2010).

lincRNAs contribute to the topological organization of chromatin

Given their ubiquitous roles in the regulation of gene expression, it is not surprising that lncRNAs contribute to the nuclear organization of the genome into its various topological regions. Inhibition of transcription and low-resolution cytology had led to the thought that RNA plays a role in the maintenance of the nuclear matrix and that the latter may organize the genome into higher-order functional structures (Nickerson et al., 1989). Recent experiments have provided a striking example of the role played by a particular lincRNA—Firre (functional intergenic repeating RNA element)—in organizing a group of genes involved in the same physiological pathway, located on different chromosomes, into a cluster associated with its own locus (Fig. 6.1B). Firre is located on the X chromosome but is one of the genes that escape inactivation (see Chapter 9). The formation of a gene cluster by Firre requires the binding of Firre to a scaffold attachment protein. In the absence of the RNA or of its associated protein, the cross-chromosomal clustering is abrogated (Hacisuleyman et al., 2014). Another role for the Firre RNA is to anchor the mouse inactive X chromosome to the nucleolus, an association mediated by binding CTCF (CCCTC binding factor); this association appears to be necessary for the maintenance of the repressive H3K27me3 mark on this chromosome (Yang et al., 2015).

In addition to these functions in establishing intra- and inter-chromosomal contacts, lincRNAs have a surprising role in the formation of some nuclear bodies. A specific example is the NEAT1 RNA that is required for the formation of paraspeckles (Clemson et al., 2009) (see Chapter 13).

The involvement of lncRNAs, and especially of lincRNAs, in a variety of nuclear functions is now solidly established; yet, a number of fundamental questions regarding their regulatory roles remain unanswered, to wit: what is the total number of lncRNAs in a given cell type? How does the lncRNA repertoire of a stem cell change as a function of differentiation? What is the complete inventory of lncRNAs' regulatory activities? What are the mechanisms involved in their targeting? What are the molecular and biophysical means by which they carry out their regulatory role? Answers to these questions will most likely continue to require the painstaking process of identifying new RNA molecules and characterizing their individual functions.

Short non-coding RNAs

The first endogenous antisense RNA that interfered with the expression of a gene was reported in *Caenorhabditis elegans* (Lee et al., 1993). This RNA is the first identified example of a miRNA [see Micro RNAs (miRNAs), below]. In prior years, injection of antisense RNA for the purpose of disrupting the post-transcriptional function of specific mRNAs had been used as an experimental tool for the study of gene function during embryonic development (Harland and Weintraub, 1985; Rosenberg et al., 1985). Although it yielded the expected results, this approach was inefficient in that it required large amounts of injected RNA. A major improvement was obtained by injecting double-stranded RNA (Fire et al., 1998). This breakthrough observation paved the way for the characterization of the biochemical pathways responsible for the synthesis of endogenous interfering antisense RNAs and of other short non-coding RNAs (sncRNAs).

The main classes of short non-coding RNAs are: miRNAs (micro RNAs), siRNAs (small interfering RNAs), piRNAs (Piwi-interacting RNAs), snRNAs (small nuclear RNAs) and tRNA-derived non-coding RNAs. Additional short RNAs are rasiRNAs (repeat-associated small interfering RNAs) that appear to be involved in centromeric heterochromatin formation. First reported in the fission yeast *Schizosaccharomyces* (Verdel et al., 2004; Volpe et al., 2002) and in *Drosophila* (Pal-Bhadra et al., 2002), sncRNAs play a role in the formation of heterochromatin in vertebrates as well (Fukagawa et al., 2004). The various types of sncRNAs are of slightly different sizes, are found in different cellular locations and differ in the proteins involved in their production and in the mechanisms by which they effect their regulation.

Micro RNAs (miRNAs)

These RNAs are 20–22 nucleotides in length. In general, they are negative regulators of gene expression by inhibiting translation or promoting mRNA decay. Individual miRNAs can regulate many genes in response to developmental signals or environmental stimuli. They are transcribed by RNAPII as long RNA molecules, termed primary miRNA transcripts or pri-miRNAs, with an inverted repeat that generates a hairpin (Fig. 6.2B). RNA hairpins are secondary structures formed by double-stranded regions, known as stems, with the paired strands connected by a terminal loop. The miRNA sequence is contained within the stem. An RNase (Drosha), present in a complex called the Microprocessor complex, releases the hairpin from the rest of the molecule. This process is facilitated by another subunit of the complex DGCR8, which guides Drosha to cut the hairpin at the junction between the double-stranded and single-stranded regions. The hairpin, now referred to as pre-miRNA, is exported to the cytoplasm by the Exportin 5 protein, where it is further cleaved by another RNase (Dicer). The remainder of the stem duplex is loaded onto AGO proteins, members of the Argonaute family, leading to the formation of the RNA-induced silencing complex (RISC). In this complex, the duplex RNA is unwound and the strand that is complementary to some particular mRNA (guide strand) is retained. The target site of the miRNA is usually in the 3' untranslated region of the mRNA, and a short nucleotide stretch of the miRNA, referred

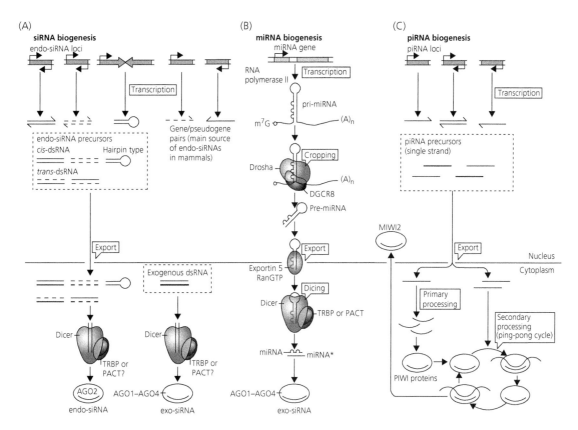

Fig. 6.2 The biogenesis of sncRNAs. See the text for a discussion of the various steps.

(From Huang et al., 2013.)

to as the "seed," is required for association. The AGO proteins recruit factors that repress the translation of the target RNA (Fukao et al., 2014; Petersen et al., 2006) or that destabilize it and lead to its degradation (Behm-Ansmant et al., 2006; Eulalio et al., 2009).

Endogenous small interfering RNAs (siRNAs)

These RNAs are produced from double-stranded lncR-NAs formed by bidirectional transcription of one gene, by the transcription of opposite strands of a gene and one of its pseudogenes, or the transcription of genomic sequences containing inverted repeats (Fig. 6.2A). Similarly to the biogenesis of miRNAs, these double-stranded RNAs are cleaved by Dicer and one strand of the resulting short duplex is incorporated into the RISC complex. In contrast to miRNAs, which tolerate some non-complementary base associations in binding to their targets, siRNAs require full complementarity to induce gene repression.

Piwi-interacting RNAs (piRNAs)

These RNAs were first discovered in mouse testes (Aravin et al., 2006; Girard et al., 2006; Grivna et al., 2006; Watanabe et al., 2006). They are derived from long transcripts of clusters of transposon sequences (Brennecke et al., 2007). The transcripts are transported to the cytoplasm where they are processed into smaller fragments (24–31 nucleotides in length) by some Dicer-independent process (Fig. 6.2C); these fragments associate with Piwi proteins, another group of the Argonaute family of proteins, to form piRISCs. Some of the piRISCs are returned to the nucleus, while others act in the cytoplasm. Their main function is to repress the action of transposable elements (Box 6.1) in the germline by cleaving transcripts or by inducing epigenetic changes, such as H3K9me3 deposition, that interfere with the elements' expression and transposition. Transposable elements can disrupt the integrity of genomic information in a variety of ways, ranging from the induction of inter- or intra-chromosomal rearrangements to gene inactivation through insertion or the spread of heterochromatin beyond the transposon sequence. New piRNAs are generated when, by chance, a copy of an invading transposon jumps into one of the existing piRNA clusters.

In addition to their biogenesis from transposable elements, a substantial proportion of piRNAs are produced from the 3' untranslated region of a variety of protein-coding mRNAs transcribed in the gonads (Robine et al., 2009). The mechanism that identifies which genes will give rise to these piRNAs or the nature of the biological role that they play have not yet been determined.

Small nuclear RNAs (snRNAs)

snRNAs can be subdivided into three subgroups: U RNAs (spliceosomal RNAs), snoRNAs (small nucleolar RNAs) and scaRNAs (small Cajal body-specific RNAs). The snoRNAs guide site-specific ribosomal RNA modifications (Jady and Kiss, 2001) or contribute to the regulation of alternative splicing of some specific gene transcripts (Kishore and Stamm, 2006). The scaRNAs are found exclusively in the Cajal bodies of the nucleus (see Chapters 12 and 13). U RNAs associate with specific proteins to form a set of U snRNPs (small nuclear ribonucleoproteins). Five of these U snRNPs (U1, 2, 4, 5 and 6) are the building blocks of a large complex—the spliceosome—responsible for the removal of introns from the vast majority of primary transcripts (Frendewey and Keller, 1985). As the components of the spliceosome are delivered to the sites on the pre-mRNA where the excision of introns will occur, they undergo dynamic interactions and conformational changes. These changes eventually lead to the timely activation of the spliceosome and ensure that cleavage of the pre-mRNA does not occur prematurely.

tRNA-derived non-coding RNAs

These RNAs are produced from tRNA primary transcripts or mature sequences (Fu et al., 2009; Lee et al., 2009; Pekarsky et al., 2016). There are two major classes: tRNA halves, known as tRNA-derived stress-induced RNAs (tiRNAs); and tRNA-derived fragments (tRFs). tiRNAs are the result of cleavage at the anti-codon loop of tRNA molecules by the RNase angiogenin that is activated under conditions of stress (nutritional deficiency, hypoxia or hypothermia, etc.). tiRNAs act in the cytoplasm to inhibit the translation initiation of most genes, except for those that encode stress-related functions (Ivanov et al., 2011); they are responsible for the formation of stress granules (Aulas

et al., 2017). tRFs are smaller fragments derived from the extreme 5' or 3' ends of mature tRNAs, as well as from the sequence that is immediately downstream of the primary tRNA transcripts. 5' tRFs are nuclear, while the other two types are cytoplasmic (Kumar et al., 2014). tRFs have similar properties to miRNAs and play a major role in RNA interference.

Antisense non-coding RNAs are a common byproduct of transcription

In contrast to the common belief that RNA transcription proceeded in one direction only, along the 3' → 5' non-coding template strand, it is now evident that most active protein-coding genes are transcribed bidirectionally (Core et al., 2008; Seila et al., 2008). The upstream antisense transcript (uaRNA) is generated by a different promoter, is short, lacks polyadenylation and is quickly degraded. Of course, transcription in the opposite direction can give rise to a stable RNA if two protein-coding genes are adjacent to each other and are divergently transcribed, or if the upstream DNA sequence encodes a tRNA or a lincRNA (Oler et al., 2010; Trinklein et al., 2004). Divergent promoters, whether they produce uaRNAs or stable RNAs, are located very close to protein-encoding gene promoters and exhibit classic TATA box or Inr elements (Core et al., 2014). As expected, all promoters contain the TATA-binding protein TBP and the general transcription factor TFIIB (Pugh and Venters, 2016).

RNA editing

All of the different types of RNA found in eukaryotic cells—rRNA, tRNA and mRNA, as well as sncRNA and lncRNA—are subject to a variety of chemical modifications (Machnicka et al., 2013). The most extensively modified RNA is tRNA, with up to 25% of the nucleotides undergoing some change. A common modification is the deamination of adenosine (A) to form inosine (I) (Fig. 6.3), mediated by ADAR (adenosine deaminases acting on RNA) enzymes (Yang et al., 2006). During mRNA translation, inosine is treated as a guanine (G). Depending on where it is located, the new inosine can affect alternative splicing (Mazloomian and Meyer, 2015) or the stability of a primary transcript; upon translation, it can lead to a protein that differs from its encoded amino acid sequence. The presence of inosine can also alter the biogenesis or the specificity of miRNA (Cui et al., 2015; Yang et al., 2006).

In mRNA, lncRNAs, sncRNAs, rRNA and tRNA, adenosine can also be methylated at either one of two positions to yield N6- or N1-methyladenosine (m1A or

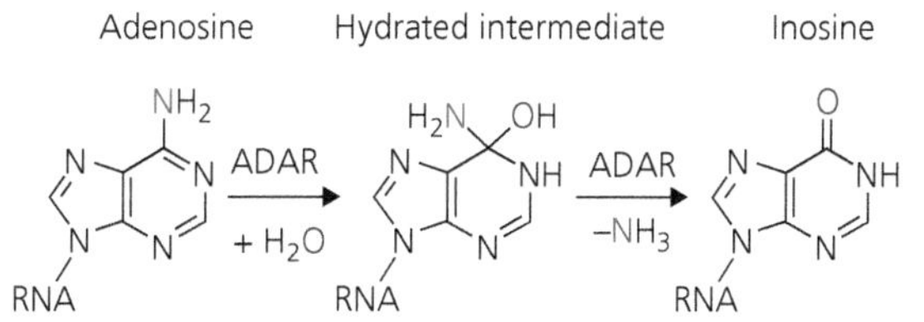

Fig. 6.3 Enzymatic conversion of adenosine to inosine.

(From Thomas and Beal, 2017.)

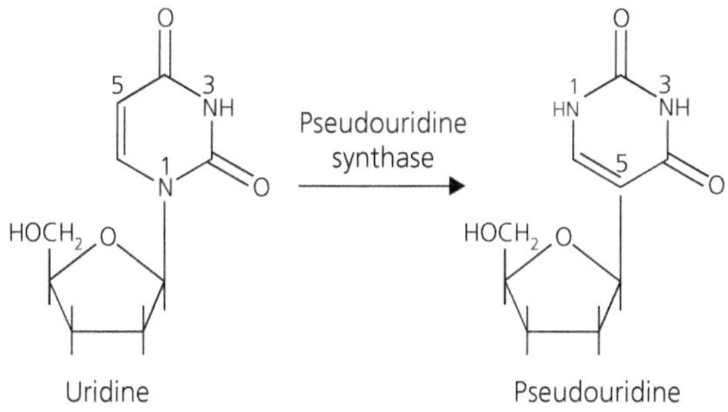

Fig. 6.4 Enzymatic conversion of uridine to its isomer pseudouridine.

(From Ni et al., 1997.)

m6A) by a methyl transferase complex. The presence of m6A alters the stability and export to the cytoplasm of mRNAs and enhances their translation (Wang et al., 2014; Wang et al., 2015). As mentioned earlier, RNA molecules can form secondary structures that are either an integral part of their function or that must be modified. The conversion of A to m6A appears to impact RNA structure by favoring a paired to an unpaired transition (Spitale et al., 2015). Another common modification that once again occurs in different RNA types is pseudouridine (denoted by the Greek letter psi Ψ) that is formed by a different link between uracil and ribose

(Fig. 6.4). This modification appears to enhance the thermodynamic stability of RNA duplexes (Kierzek et al., 2014).

A modification that occurs in tRNA, rRNA and mRNA, but at a relatively low level, is 5-methylcytosine (Squires et al., 2012). To date, the function of this modification has not been established.

Chapter summary

Modern sequencing technologies have revealed that most of the genome is transcribed into non-coding transcripts that far exceed in number the transcripts of coding genes. While the function of most of these RNAs has yet to be established, sufficient examples have been characterized to allow their subdivision into different classes with different regulatory roles.

Long non-coding RNAs (lncRNAs) are at least 200 nucleotides in length and are transcribed from promoter, coding, intergenic or enhancer regions. These RNAs repress or enhance the transcription of target genes by facilitating the interaction between promoters and enhancers or by interacting with transcription factors and targeting histone-modifying enzymes.

Short non-coding RNAs include a diverse group of functional types—miRNAs (micro RNAs) and siRNAs (small interfering RNAs) are negative regulators of gene expression; piRNAs (Piwi-interacting RNAs) suppress the action of transposable elements in the germline; snRNAs (small nuclear RNAs) are involved in mRNA splicing and rRNA maturation; tRNA-derived non-coding RNAs are involved in the cellular reaction to stress and in the repression of gene function. Additional short RNAs are rasiRNAs (repeat-associated small interfering RNAs) that appear to be involved in centromeric heterochromatin formation.

The role of non-coding RNAs is so pervasive in nuclear function that reference is made to them in most of the subsequent chapters of the book.

Box 6.1 Transposable elements

The existence of transposable elements was first reported in maize by Barbara McClintock who realized that the somatic inactivation of a particular gene was not the result of a mutation in the gene itself; rather it was the result of the insertion of a "chromatin element" adjacent to the gene (McClintock, 1950). Since that seminal observation, genomic sequencing has revealed that, as the result of insertional events that occurred throughout evolution, the genomes of prokaryotes and eukaryotes contain a variety of transposable elements (TEs). In humans, these elements make up almost half of the nuclear DNA. Over the course of time, mutations have abrogated the ability of most TEs to transpose. Nevertheless, some families of TEs have retained this function and are, therefore, a source of genetic variation. Among these, autonomous TEs have maintained the necessary factors for their transposition,

continued

Box 6.1 *Continued*

while others are non-autonomous in that they use the factors produced by other TEs to transpose.

There are two major classes of TEs—Class I are TEs that transpose using an RNA intermediate; Class II are TEs that transpose directly as DNA, leaving their location and inserting themselves elsewhere in the genome (Fig. 6B.1). Class I elements, referred to as retrotransposons, are of two types—those that have long terminal repeated sequences (LTR retrotransposons) and those that lack this feature (non-LTR retrotransposons, or LINEs). Both types initiate the process of transposition by the transcription of their genomic sequences using the cell's RNAPII (or RNAPI, in the case of some non-LTR transposons that target rDNA genes). In the case of LTR retrotransposons,

the next steps in the sequence, discovered in yeast, involve the cytoplasmic encapsulation of the RNA into virus-like particles, and the synthesis within the particles of complementary DNA (cDNA) strands by a reverse transcriptase (Garfinkel et al., 1985). The transcript of non-LTR retrotransposons is copied, using as primer one end of a DNA strand of the target site, nicked by an endonuclease (Bucheton, 1990). The integration reactions of DNA transposons and LTR retrotransposons share some similarities. The former use a transposase, while the latter use a related enzyme—an integrase—in a "cut-and-paste" mechanism. The integration of newly synthesized non-LTR elements into the genome has not yet been fully elucidated.

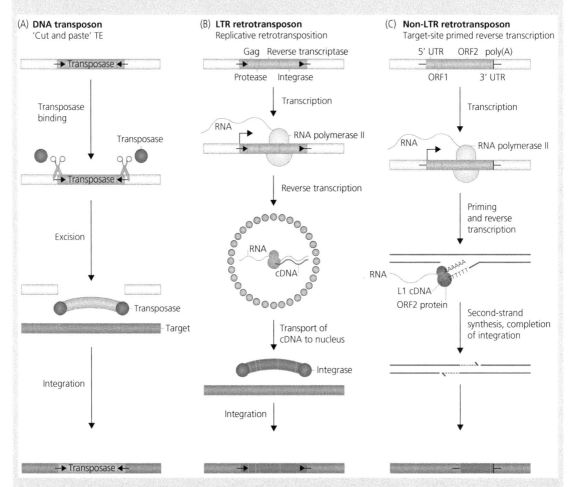

Fig. 6B.1 Mechanisms of transposition of DNA transposon and of LTR and non-LTR retrotransposons.
(Modified from Levin and Moran, 2011.)

References

Almeida, M., Pintacuda, G., Masui, O., Koseki, Y., Gdula, M., Cerase, A., Brown, D., Mould, A., Innocent, C., Nakayama, M., Schermelleh, L., Nesterova, T. B., Koseki, H. & Brockdorff, N. 2017. PCGF3/5-PRC1 initiates Polycomb recruitment in X chromosome inactivation. *Science,* **356,** 1081–4.

Amaral, P. P., Leonardi, T., Han, N., Vire, E., Gascoigne, D. K., Arias-Carrasco, R., Buscher, M., Pandolfini, L., Zhang, A., Pluchino, S., Maracaja-Coutinho, V., Nakaya, H. I., Hemberg, M., Shiekattar, R., Enright, A. J. & Kouzarides, T. 2018. Genomic positional conservation identifies topological anchor point RNAs linked to developmental loci. *Genome Biol,* 19, 32-53.

Amrein, H. & Axel, R. 1997. Genes expressed in neurons of adult male *Drosophila. Cell,* **88,** 459–69.

Aravin, A., Gaidatzis, D., Pfeffer, S., Lagos-Quintana, M., Landgraf, P., Iovino, N., Morris, P., Brownstein, M. J., Kuramochi-Miyagawa, S., Nakano, T., Chien, M., Russo, J. J., Ju, J., Sheridan, R., Sander, C., Zavolan, M. & Tuschl, T. 2006. A novel class of small RNAs bind to MILI protein in mouse testes. *Nature,* **442,** 203–7.

Aulas, A., Fay, M. M., Lyons, S. M., Achorn, C. A., Kedersha, N., Anderson, P. & Ivanov, P. 2017. Stress-specific differences in assembly and composition of stress granules and related foci. *J Cell Sci,* 130, 927–37.

Behm-Ansmant, I., Rehwinkel, J., Doerks, T., Stark, A., Bork, P. & Izaurralde, E. 2006. mRNA degradation by miRNAs and GW182 requires both CCR4:NOT deadenylase and DCP1:DCP2 decapping complexes. *Genes Dev,* 20, 1885–98.

Brennecke, J., Aravin, A. A., Stark, A., Dus, M., Kellis, M., Sachidanandam, R. & Hannon, G. J. 2007. Discrete small RNA-generating loci as master regulators of transposon activity in *Drosophila. Cell,* **128,** 1089–103.

Brown, C. J., Ballabio, A., Rupert, J. L., Lafreniere, R. G., Grompe, M., Tonlorenzi, R. & Willard, H. F. 1991. A gene from the region of the human X inactivation centre is expressed exclusively from the inactive X chromosome. *Nature,* 349, 38–44.

Bucheton, A. 1990. I transposable elements and I-R hybrid dysgenesis in *Drosophila. Trends Genet,* 6, 16–21.

Clemson, C. M., Hutchinson, J. N., Sara, S. A., Ensminger, A. W., Fox, A. H., Chess, A. & Lawrence, J. B. 2009. An architectural role for a nuclear noncoding RNA: NEAT1 RNA is essential for the structure of paraspeckles. *Mol Cell,* 33, 717–26.

Core, L. J., Martins, A. L., Danko, C. G., Waters, C. T., Siepel, A. & Lis, J. T. 2014. Analysis of nascent RNA identifies a unified architecture of initiation regions at mammalian promoters and enhancers. *Nat Genet,* 46, 1311–20.

Core, L. J., Waterfall, J. J. & Lis, J. T. 2008. Nascent RNA sequencing reveals widespread pausing and divergent initiation at human promoters. *Science,* 322, 1845–8.

Cui, Y., Huang, T. & Zhang, X. 2015. RNA editing of microRNA prevents RNA-induced silencing complex recognition of target mRNA. *Open Biol,* 5, 150126.

Davidovich, C., Zheng, L., Goodrich, K. J. & Cech, T. R. 2013. Promiscuous RNA binding by Polycomb repressive complex 2. *Nat Struct Mol Biol,* 20, 1250–7.

De Santa, F., Barozzi, I., Mietton, F., Ghisletti, S., Polletti, S., Tusi, B. K., Muller, H., Ragoussis, J., Wei, C. L. & Natoli, G. 2010. A large fraction of extragenic RNA pol II transcription sites overlap enhancers. *PLoS Biol,* 8, e1000384.

Engreitz, J. M., Haines, J. E., Perez, E. M., Munson, G., Chen, J., Kane, M., Mcdonel, P. E., Guttman, M. & Lander, E. S. 2016. Local regulation of gene expression by lncRNA promoters, transcription and splicing. *Nature,* 539, 452–5.

Eulalio, A., Huntzinger, E., Nishihara, T., Rehwinkel, J., Fauser, M. & Izaurralde, E. 2009. Deadenylation is a widespread effect of miRNA regulation. *RNA,* 15, 21–32.

Fire, A., Xu, S., Montgomery, M. K., Kostas, S. A., Driver, S. E. & Mello, C. C. 1998. Potent and specific genetic interference by double-stranded RNA in *Caenorhabditis elegans. Nature,* 391, 806–11.

Frendewey, D. & Keller, W. 1985. Stepwise assembly of a pre-mRNA splicing complex requires U-snRNPs and specific intron sequences. *Cell,* 42, 355–67.

Fu, H., Feng, J., Liu, Q., Sun, F., Tie, Y., Zhu, J., Xing, R., Sun, Z. & Zheng, X. 2009. Stress induces tRNA cleavage by angiogenin in mammalian cells. *FEBS Lett,* 583, 437–42.

Fukagawa, T., Nogami, M., Yoshikawa, M., Ikeno, M., Okazaki, T., Takami, Y., Nakayama, T. & Oshimura, M. 2004. Dicer is essential for formation of the heterochromatin structure in vertebrate cells. *Nat Cell Biol,* 6, 784–91.

Fukao, A., Mishima, Y., Takizawa, N., Oka, S., Imataka, H., Pelletier, J., Sonenberg, N., Thoma, C. & Fujiwara, T. 2014. MicroRNAs trigger dissociation of eIF4AI and eIF4AII from target mRNAs in humans. *Mol Cell,* 56, 79–89.

Garfinkel, D. J., Boeke, J. D. & Fink, G. R. 1985. Ty element transposition: reverse transcriptase and virus-like particles. *Cell,* 42, 507–17.

Girard, A., Sachidanandam, R., Hannon, G. J. & Carmell, M. A. 2006. A germline-specific class of small RNAs binds mammalian Piwi proteins. *Nature,* 442, 199–202.

Grivna, S. T., Pyhtila, B. & Lin, H. 2006. MIWI associates with translational machinery and PIWI-interacting RNAs (piRNAs) in regulating spermatogenesis. *Proc Natl Acad Sci U S A,* 103, 13415–20.

Harland, R. & Weintraub, H. 1985. Translation of mRNA injected into *Xenopus* oocytes is specifically inhibited by antisense RNA. *J Cell Biol,* 101, 1094–9.

Harris, H. 1959. Turnover of nuclear and cytoplasmic ribonucleic acid in two types of animal cell, with some further observations on the nucleolus. *Biochem J,* 73, 362–9.

Hasisuleyman, E., Goff, L. A., Trapnell, C., Williams, A., Henao-Meila, J., Sun, L., Mcclanahan, P., Hendrickson, D. J., Sauvageau, M., Kelley, D. R., Morse, M., Engreitz, J., Lander, E. S., Uttman, M., Lodish, H. F., Flavell, R., Raj, A. & Rinn, J. L. 2014. Topological organization of multichromosomal regions by the long intergenc noncoding RNA Fire. *Nat Struct Mol Biol,* 21, 198–206.

Herman, R. C., Williams, J. G. & Penman, S. 1976. Message and non-message sequences adjacent to poly(A) in steady state heterogeneous nuclear RNA of HeLa cells. *Cell, 7*, 429–37.

Huang, Y., Zhang, J. L., Yu, X. L., Xu, T. S., Wang, Z. B. & Cheng, X. C. 2013. Molecular functions of small regulatory noncoding RNA. *Biochemistry (Mosc), 78*, 221–30.

Ivanov, P., Emara, M. M., Villen, J., Gygi, S. P. & Anderson, P. 2011. Angiogenin-induced tRNA fragments inhibit translation initiation. *Mol Cell, 43*, 613–23.

Jady, B. E. & Kiss, T. 2001. A small nucleolar guide RNA functions both in 2′-O-ribose methylation and pseudouridylation of the U5 spliceosomal RNA. *EMBO J, 20*, 541–51.

Kierzek, E., Malgowska, M., Lisowiec, J., Turner, D. H., Gdaniec, Z. & Kierzek, R. 2014. The contribution of pseudouridine to stabilities and structure of RNAs. *Nucleic Acids Res, 42*, 3492–501.

Kim, T. K., Hemberg, M., Gray, J. M., Costa, A. M., Bear, D. M., Wu, J., Harmin, D. A., Laptewicz, M., Barbara-Haley, K., Kuersten, S., Markenscoff-Papadimitriou, E., Kuhl, D., Bito, H., Worley, P. F., Kreiman, G. & Greenberg, M. E. 2010. Widespread transcription at neuronal activity-regulated enhancers. *Nature, 465*, 182–7.

Kino, T., Hurt, D. E., Ichijo, T., Nader, N. & Chrousos, G. P. 2010. Noncoding RNA gas5 is a growth arrest- and starvation-associated repressor of the glucocorticoid receptor. *Sci Signal, 3*, ra8.

Kishore, S. & Stamm, S. 2006. Regulation of alternative splicing by snoRNAs. *Cold Spring Harb Symp Quant Biol, 71*, 329–34.

Kumar, P., Anaya, J., Mudunuri, S. B. & Dutta, A. 2014. Meta-analysis of tRNA derived RNA fragments reveals that they are evolutionarily conserved and associate with AGO proteins to recognize specific RNA targets. *BMC Biol, 12*, 78.

Lee, R. C., Feinbaum, R. L. & Ambros, V. 1993. The *C. elegans* heterochronic gene lin-4 encodes small RNAs with antisense complementarity to lin-14. *Cell, 75*, 843–54.

Lee, Y. S., Shibata, Y., Malhotra, A. & Dutta, A. 2009. A novel class of small RNAs: tRNA-derived RNA fragments (tRFs). *Genes Dev, 23*, 2639–49.

Levin, H. L. & Moran, J. V. 2011. Dynamic interactions between transposable elements and their hosts. *Nat Rev Genet, 12*, 615–27.

Li, W., Notani, D., Ma, Q., Tanasa, B., Nunez, E., Chen, A. Y., Merkurjev, D., Zhang, J., Ohgi, K., Song, X., Oh, S., Kim, H. S., Glass, C. K. & Rosenfeld, M. G. 2013. Functional roles of enhancer RNAs for oestrogen-dependent transcriptional activation. *Nature, 498*, 516–20.

Machnicka, M. A., Milanowska, K., Osman Oglou, O., Purta, E., Kurkowska, M., Olchowik, A., Januszewski, W., Kalinowski, S., Dunin-Horkawicz, S., Rother, K. M., Helm, M., Bujnicki, J. M. & Grosjean, H. 2013. MODOMICS: a database of RNA modification pathways–2013 update. *Nucleic Acids Res, 41*, D262–7.

Maenner, S., Muller, M., Frohlich, J., Langer, D. & Becker, P. B. 2013. ATP-dependent roX RNA remodeling by the helicase maleless enables specific association of MSL proteins. *Mol Cell, 51*, 174–84.

Mak, W., Baxter, J., Silva, J., Newall, A. E., Otte, A. P. & Brockdorff, N. 2002. Mitotically stable association of polycomb group proteins eed and enx1 with the inactive x chromosome in trophoblast stem cells. *Curr Biol, 12*, 1016–20.

Mazloomian, A. & Meyer, I. M. 2015. Genome-wide identification and characterization of tissue-specific RNA editing events in *D. melanogaster* and their potential role in regulating alternative splicing. *RNA Biol, 12*, 1391–401.

Mcclintock, B. 1950. The origin and behavior of mutable loci in maize. *Proc Natl Acad Sci U S A, 36*, 344–55.

Meller, V. H., Wu, K. H., Roman, G., Kuroda, M. I. & Davis, R. L. 1997. roX1 RNA paints the X chromosome of male *Drosophila* and is regulated by the dosage compensation system. *Cell, 88*, 445–57.

Melo, C. A., Drost, J., Wijchers, P. J., Van De Werken, H., De Wit, E., Oude Vrielink, J. A., Elkon, R., Melo, S. A., Leveille, N., Kalluri, R., De Laat, W. & Agami, R. 2013. eRNAs are required for p53-dependent enhancer activity and gene transcription. *Mol Cell, 49*, 524–35.

Ni, J., Tien, A. L. & Fournier, M. J. 1997. Small nucleolar RNAs direct site-specific synthesis of pseudouridine in ribosomal RNA. *Cell, 89*, 565–73.

Nickerson, J. A., Krochmalnic, G., Wan, K. M. & Penman, S. 1989. Chromatin architecture and nuclear RNA. *Proc Natl Acad Sci U S A, 86*, 177–81.

Oler, A. J., Alla, R. K., Roberts, D. N., Wong, A., Hollenhorst, P. C., Chandler, K. J., Cassiday, P. A., Nelson, C. A., Hagedorn, C. H., Graves, B. J. & Cairns, B. R. 2010. Human RNA polymerase III transcriptomes and relationships to Pol II promoter chromatin and enhancer-binding factors. *Nat Struct Mol Biol, 17*, 620–8.

Pal-Bhadra, M., Bhadra, U. & Birchler, J. A. 2002. RNAi related mechanisms affect both transcriptional and post-transcriptional transgene silencing in *Drosophila*. *Mol Cell, 9*, 315–27.

Pandey, R. R., Mondal, T., Mohammad, F., Enroth, S., Redrup, L., Komorowski, J., Nagano, T., Mancini-Dinardo, D. & Kanduri, C. 2008. Kcnq1ot1 antisense noncoding RNA mediates lineage-specific transcriptional silencing through chromatin-level regulation. *Mol Cell, 32*, 232–46.

Paralkar, V. R., Taborda, C. C., Huang, P., Yao, Y., Kossenkov, A. V., Prasad, R., Luan, J., Davies, J. O., Hughes, J. R., Hardison, R. C., Blobel, G. A. & Weiss, M. J. 2016. Unlinking an lncRNA from its associated *cis* element. *Mol Cell, 62*, 104–10.

Pasmant, E., Laurendeau, I., Heron, D., Vidaud, M., Vidaud, D. & Bieche, I. 2007. Characterization of a germ-line deletion, including the entire INK4/ARF locus, in a melanoma-neural system tumor family: identification of ANRIL, an antisense noncoding RNA whose expression coclusters with ARF. *Cancer Res, 67*, 3963–9.

Pekarsky, Y., Balatti, V., Palamarchuk, A., Rizzotto, L., Veneziano, D., Nigita, G., Rassenti, L. Z., Pass, H. I., Kipps, T. J., Liu, C. G. & Croce C. M. 2016. Dysregulation of a family of short noncoding RNAs, tsRNAs, in human cancer. *Proc Natl Acad Sci U S A,* 113, 5071–6.

Perry, R. P., Kelley, D. E. & Latorre, J. 1974. Synthesis and turnover of nuclear and cytoplasmic polyadenylic acid in mouse L cells. *J Mol Biol,* 82, 315–31.

Petersen, C. P., Bordeleau, M. E., Pelletier, J. & Sharp, P. A. 2006. Short RNAs repress translation after initiation in mammalian cells. *Mol Cell,* 21, 533–42.

Pugh, B. F. & Venters, B. J. 2016. Genomic organization of human transcription initiation complexes. *PLoS One,* 11, e0149339.

Quinodoz, S. & Guttman, M. 2014. Long noncoding RNAs: an emerging link between gene regulation and nuclear organization. *Trends Cell Biol,* 24, 651–63.

Robine, N., Lau, N. C., Balla, S., Jin, Z., Okamura, K., Kuramochi-Miyagawa, S., Blower, M. D. & Lai, E. C. 2009. A broadly conserved pathway generates 3´UTR-directed primary piRNAs. *Curr Biol,* 19, 2066–76.

Rosenberg, U. B., Preiss, A., Seifert, E., Jackle, H. & Knipple, D. C. 1985. Production of phenocopies by Kruppel antisense RNA injection into *Drosophila* embryos. *Nature,* 313, 703–6.

Schneider, C., King, R. M. & Philipson, L. 1988. Genes specifically expressed at growth arrest of mammalian cells. *Cell,* 54, 787–93.

Seila, A. C., Calabrese, J. M., Levine, S. S., Yeo, G. W., Rahl, P. B., Flynn, R. A., Young, R. A. & Sharp, P. A. 2008. Divergent transcription from active promoters. *Science,* 322, 1849–51.

Spitale, R. C., Flynn, R. A., Zhang, Q. C., Crisalli, P., Lee, B., Jung, J. W., Kuchelmeister, H. Y., Batista, P. J., Torre, E. A., Kool, E. T. & Chang, H. Y. 2015. Structural imprints *in vivo* decode RNA regulatory mechanisms. *Nature,* 519, 486–90.

Squires, J. E., Patel, H. R., Nousch, M., Sibbritt, T., Humphreys, D. T., Parker, B. J., Suter, C. M. & Preiss, T. 2012. Widespread occurrence of 5-methylcytosine in human coding and noncoding RNA. *Nucleic Acids Res,* 40, 5023–33.

Thomas, J. M. & Beal, P. A. 2017. How do ADARs bind RNA? New protein-RNA structures illuminate substrate recognition by the RNA editing ADARs. *Bioessays,* 39. doi: 10.1002/bies.201600187.

Thomson, D. W. & Dinger, M. E. 2016. Endogenous microRNA sponges: evidence and controversy. *Nat Rev Genet,* 17, 272–83.

Trinklein, N. D., Aldred, S. F., Hartman, S. J., Schroeder, D. I., Otillar, R. P. & Myers, R. M. 2004. An abundance of bidirectional promoters in the human genome. *Genome Res,* 14, 62–6.

Venters, B. J. & Pugh, B. F. 2013. Genomic organization of human transcription initiation complexes. *Nature,* 502, 53–8.

Verdel, A., Jia, S., Gerber, S., Sugiyama, T., Gygi, S., Grewal, S. I. & Moazed, D. 2004. RNAi-mediated targeting of heterochromatin by the RITS complex. *Science,* 303, 672–6.

Volpe, T. A., Kidner, C., Hall, I. M., Teng, G., Grewal, S. I. & Martienssen, R. A. 2002. Regulation of heterochromatic silencing and histone H3 lysine-9 methylation by RNAi. *Science,* 297, 1833–7.

Wang, K. C., Yang, Y. W., Liu, B., Sanyal, A., Corces-Zimmerman, R., Chen, Y., Lajoie, B. R., Protacio, A., Flynn, R. A., Gupta, R. A., Wysocka, J., Lei, M., Dekker, J., Helms, J. A. & Chang, H. Y. 2011. A long noncoding RNA maintains active chromatin to coordinate homeotic gene expression. *Nature,* 472, 120–4.

Wang, X., Lu, Z., Gomez, A., Hon, G. C., Yue, Y., Han, D., Fu, Y., Parisien, M., Dai, Q., Jia, G., Ren, B., Pan, T. & He, C. 2014. N6-methyladenosine-dependent regulation of messenger RNA stability. *Nature,* 505, 117–20.

Wang, X., Zhao, B. S., Roundtree, I. A., Lu, Z., Han, D., Ma, H., Weng, X., Chen, K., Shi, H. & He, C. 2015. N(6)-methyladenosine modulates messenger RNA translation efficiency. *Cell,* 161, 1388–99.

Wang, Y., Xu, Z., Jiang, J., Xu, C., Kang, J., Xiao, L., Wu, M., Xiong, J., Guo, X. & Liu, H. 2013. Endogenous miRNA sponge lincRNA-RoR regulates Oct4, Nanog, and Sox2 in human embryonic stem cell self-renewal. *Dev Cell,* 25, 69–80.

Watanabe, T., Takeda, A., Tsukiyama, T., Mise, K., Okuno, T., Sasaki, H., Minami, N. & Imai, H. 2006. Identification and characterization of two novel classes of small RNAs in the mouse germline: retrotransposon-derived siRNAs in oocytes and germline small RNAs in testes. *Genes Dev,* 20, 1732–43.

Yang, F., Deng, X., Ma, W., Berletch, J. B., Rabaia, N., Wei, G., Moore, J. M., Filippova, G. N., Xu, J., Liu, Y., Noble, W. S., Shendure, J. & Disteche, C. M. 2015. The lncRNA Firre anchors the inactive X chromosome to the nucleolus by binding CTCF and maintains H3K27me3 methylation. *Genome Biol,* 16, 52.

Yang, L., Lin, C., Liu, W., Zhang, J., Ohgi, K. A., Grinstein, J. D., Dorrestein, P. C. & Rosenfeld, M. G. 2011. ncRNA- and Pc2 methylation-dependent gene relocation between nuclear structures mediates gene activation programs. *Cell,* 147, 773–88.

Yang, W., Chendrimada, T. P., Wang, Q., Higuchi, M., Seeburg, P. H., Shiekhattar, R. & Nishikura, K. 2006. Modulation of microRNA processing and expression through RNA editing by ADAR deaminases. *Nat Struct Mol Biol,* 13, 13–21.

Yap, K. L., Li, S., Munoz-Cabello, A. M., Raguz, S., Zeng, L., Mujtaba, S., Gil, J., Walsh, M. J. & Zhou, M. M. 2010. Molecular interplay of the noncoding RNA ANRIL and methylated histone H3 lysine 27 by polycomb CBX7 in transcriptional silencing of INK4a. *Mol Cell,* 38, 662–74.

CHAPTER 7

Maintenance of the active and inactive states

Cellular differentiation during development and the maintenance of tissue identity and function in postnatal through adult stages require the stabilization of differential gene expression. This is achieved by the action of two highly conserved groups of regulatory genes, named after the first member of each group to be identified in *Drosophila*: the trithorax group (TrxG) and the Polycomb group (PcG). TrxG gene products maintain the active state, while PcG genes are responsible for repressing gene function; both actions are perpetuated through cell division and constitute the basis of cellular memory (see Chapter 16).

Drosophila fertilized eggs divide rapidly to produce a syncytial mass of nuclei that eventually migrate to the surface of the egg and, following cell membrane formation, give rise to a cellular blastoderm, the equivalent of the blastula in vertebrate development. During this time, transcription factors and mRNAs that had been deposited in an asymmetrical fashion during egg formation, and external signals from the ovarian environment establish the axes of the developing embryo. As development proceeds, the embryo becomes divided along its length into a series of repeating segments that must become different from one another in order to give rise to the many structures of the adult body. The identity of individual segments is determined by the expression of **homeotic genes**, most of which encode DNA-binding transcription factors (Denell, 1973; Kaufman et al., 1980; Lewis, 1978). It is crucial to normal development that specific homeotic gene products be present only in the appropriate embryonic segments—loss-of-function mutations of homeotic

genes in a segment where they are normally expressed, or ectopic expression in a segment where they are normally silent, cause the development of the segment into a different adult structure (referred to as a homeotic transformation). TrxG genes maintain specific homeotic genes in an active state, while PcG genes maintain them in a repressed state in those segments where their expression would be inappropriate (Bowman et al., 2014). Homeotic genes contain a 180-nucleotide sequence, called the *homeobox*, that is translated into a 60-amino acid DNA-binding domain. Homeobox-containing genes, generally referred to as *HOX* genes, are found in plants and are conserved from worms to mammals. With a few exceptions (Schiemann et al., 2017), in higher metazoans such as fruit flies and mammals, they exhibit the fascinating feature that they are present in the genome as a cluster of genes ordered in a manner that reflects the order of the adult structures along the organism's anterior–posterior axis (Fig. 7.1). Numerous hypotheses have been offered to explain the origin and evolutionary maintenance of the *HOX* genes collinearity (Gaunt, 2015).

Altough the role of PcG-mediated repression of gene activity is widespread and crucial during development and differentiation, other complexes exist, with silencing functions targeted to specific subsets of genes. An example is the NoRC (nuclear remodeling complex) that is responsible for the silent genes that encode ribosomal RNA, the so-called rDNA genes (Strohner et al., 2004). This complex will be discussed in Chapter 9.

Epigenetics, Nuclear Organization and Gene Function: with implications of epigenetic regulation and genetic architecture for human development and health. John C. Lucchesi, Oxford University Press (2019). © John C. Lucchesi 2019.
DOI: 10.1093/oso/9780198831204.001.0001

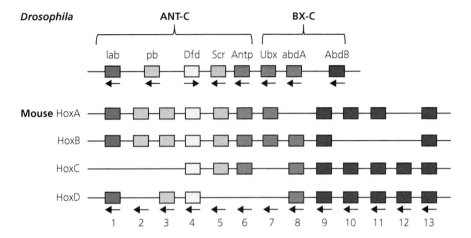

Fig. 7.1 *Hox* gene clusters of *Drosophila* and mouse. Orthologous genes are represented by the same color. Although duplication of individual genes and of the entire *Hox* set have occurred during mammalian evolution, the linear arrangement of the different *Hox* genes is maintained between the two organisms representing different phyla.

(Modified from Gaunt, 2015.)

PcG gene products guard against improper *HOX* gene expression

The *HOX* genes in *Drosophila melanogaster* are found in two complexes: the Antennapedia complex (ANT-C) responsible for the differentiation of the head and the anterior thoracic region (Kaufman et al., 1980), and the Bithorax complex (BX-C) that controls the differentiation of the posterior thoracic region and of the abdomen (Lewis, 1978). The expression of the genes in these two complexes must be spatially restricted during development in order to lead to the eventual differentiation of embryonic segments into adult structures. PcG genes were identified by mutations that allowed the expression of *HOX* genes in embryonic regions where they should remain silent. Historically, geneticists working with *Drosophila* have named newly discovered genes according to the phenotypes of their loss-of-function mutations. According to this custom, the first unlinked, transacting modifier of one of the *HOX* genes in the ANT-C was named Polycomb (*Pc*)—in males, the anterior legs are morphologically different from the others because of the presence of a structure called the sex comb, and the *Pc* mutation caused the transformation of the second and third pairs of legs into first legs (Lewis, 1978).

Genetic screens based on homeotic transformations subsequently identified additional *HOX* gene expression modifiers. Some, such as Polycomb-like (*Pcl*) and Sex comb on midleg (*Scm*), were named after their homeotic phenotypes; others, such as Enhancer of zeste [*E(z)*] or Suppressor of zeste [*Su(z)12*], were named after a previously discovered, additional effect on eye pigment formation (Jurgens, 1985; Wu et al., 1989). To date, the PcG numbers well over a dozen protein-coding genes in *Drosophila*, and over three dozens in mammals; these genes not only control the expression of the ANT-C and BX-C or the HOXa–d clusters of genes, respectively, in these organisms, but also exert a broad regulatory influence on genome expression in all metazoans.

PcG activity is mediated through the action of repressive protein complexes

In flies and in mammals, two major protein complexes PRC1 and PRC2 (Fig. 7.2) mediate transcriptional repression (Shao et al., 1999; Tie et al., 2001). *Drosophila* PRC1 consists of a core of four proteins that include a histone H2A ubiquitin ligase (RING1). Several paralogues of the core subunits are present in the genome, leading to a number of isoforms of the complex. A greater number of homologues of the core proteins exist in mammals, resulting in an even more extensive variety of isoforms. The *Drosophila* PRC2 complex includes, among its core subunits, a histone methyl transferase [*E(z)*] that trimethylates H3K27 (Müller et al., 2002); an identical function occurs in mammals (Kuzmichev et al., 2002). This modification, which acts as a target for the PRC1 complex, is the historical linchpin of PcG repression (but see below). As in the case of PRC1, several isoforms of PRC2 exist in flies and mammals, resulting in distinct classes of complexes. In

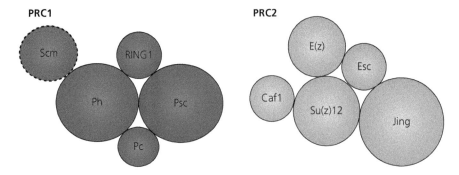

Fig. 7.2 In *Drosophila*, the PRC1 core subunits are Polycomb (Pc), which binds to H3K27me3; Polyhomeotic (Ph); RING1, a histone H2A ubiquitin ligase; and Posterior sex combs (Psc). Sex comb on midleg (Scm) appears to be associated with the core complex. PRC2 contains the methyl transferase Enhancer of zeste [E(z)]; Extra sexcombs (Esc) and Su(z)12; and the histone chaperone Caf1. Jing stabilizes the complex, but it is not essential for function.

(From Schwartz and Pirrotta, 2013.)

addition, several PRC1- and PRC2-related complexes are formed by the absence of one of the core proteins and the inclusion of new subunits. In *Drosophila*, all of these complexes are targeted to specific DNA sequences referred to as Polycomb response elements, or PREs (Simon et al., 1993), although, surprisingly, none of their protein subunits have DNA-binding motifs. Genome-wide studies have identified hundreds of PREs located adjacent to promoters, within introns or tens of kilobases away from the genes that they regulate. PREs are made up of several segments of a few hundred base-pairs that have cumulative effects in mediating gene repression. As the two PRC complexes appear to function together in order to achieve gene repression, it was reasonable to assume that PRC2 binds first and methylates H3K27, which serves as a target for PRC1 through the chromodomain of its Pc subunit. Somewhat surprisingly, H3K27me3-marked nucleosomes extend beyond the PREs over a relatively large area, and although PRC1 peaks at the PRE, it spreads across PcG domains. A number of PRC1-like variant complexes exist in mammals. Some of these complexes contain orthologs of the canonical Pc subunit or of HP1 (heterochromatin protein 1); these different proteins are referred to as CBX (chromobox) proteins. In other complexes, the CBX subunit is absent and is replaced by some other protein. These different complexes are targeted by interacting with traditional DNA-binding factors. Recently, a new order for the binding of PcG complexes has been discovered in mouse cells (Blackledge et al., 2014; Cooper et al., 2014; Kalb et al., 2014). Variant PRC1 complexes are targeted first and mono-ubiquitinate histone H2A on genes that they will repress; this nucleosomal mark recruits PRC2

complexes that methylate H3K27; H3K7me3 then serves as the traditional target for canonical PRC1 complexes. Similar studies in *Drosophila* have shown that, at some PREs, PRC2 binding is PRC1-dependent while at other PRE sites, it is not (Kahn et al., 2016).

PREs are not present in mammalian genomes. Associated with a few genes are sequences that resemble *Drosophila* PREs in that they are targeted by PcG proteins and mediate PcG-dependent repression when they are inserted near reporter genes (Kassis and Brown, 2013). The only general DNA features that have been shown to recruit PRC complexes in mammals are unmethylated CpG islands that lack activator factor binding sites (Ku et al., 2008; Lynch et al., 2012). Recruitment at CpG islands is initiated by the targeting of PRC1-like variant complexes via their histone demethylase subunit (KDM2B) and is followed by the nucleation of PRC2 and canonical PRC1 complexes (Blackledge et al., 2014; Cooper et al., 2014; Kalb et al., 2014). The presence of histone marks characteristic of active transcription within CpG islands prevents PCR binding.

In some instances, PRC2 can be targeted to sites of repression by associating directly or indirectly with long non-coding RNAs (lncRNAs). Examples of this alternate mode of recruitment are the processes that lead to parental imprinting (see Chapter 8) and to X chromosome inactivation in mammals (see Chapter 9). Genome-wide analyses have demonstrated that PRC2 is bound to thousands of RNAs in different cell lines, principally through its Ezh2 [E(z) in *Drosophila*] subunit (Zhao et al., 2010). Whether non-coding RNAs directly recruit PcG complexes, as in the case of X chromosome inactivation, or whether PcG recruiting is the result of non-coding RNA silencing remains to be determined. PRC2 has been found to associate with both active and inactive genes. In the former, PRC2 is not catalytically active; in the latter, it methylates H3K27, suggesting that PRC2 function depends on the chromatin state of the site where it binds and that its function is to maintain a previously and independently established inactive state (Davidovitch et al., 2013).

In both flies and mammals, many PcG gene products are not included in PRC1 or PRC2 and form additional complexes (Schwartz and Pirrotta, 2013). One of these complexes, called the PhoRC, because it includes the DNA-binding protein pleiohomeotic (Pho; YY1 in mammals), is present at PREs that are simultaneously occupied by PRC1 and PRC2 and is thought to facilitate the recruitment of these two complexes (Oktaba et al., 2008). Another complex found at PcG targets contains a hydrolase that removes monoubiquitin from histone H2A. The presence of this complex, named Polycomb repressive deubiquitinase (PR-DUB), is counterintuitive, given the ubiquitination activity of PRC1, and suggests that repression is sensitive to a dynamically established level of this H2A covalent modification (Scheuermann et al., 2010).

Neighboring PREs and their bound PcG repressive complexes, as well as remote PREs, associate into clusters referred to as PcG bodies (discussed in PcG proteins alter the three-dimensional structure of the genome, p. 84 and in Chapter 10).

The mechanistic basis of PcG repression

Before addressing this topic, it is important to realize that different PcG complexes have different activities that they exert at different times and in different tissues during differentiation and development. For example, in mammals, some variants of the canonical PRC1 complex do not require the presence of PRC2 and H3K27me3 to ubiquitinate H2A (Tavares et al., 2012). Also, starting in early embryonic stages, the assembly and disassembly of different PRC1 complexes orchestrate the balance between maintaining pluripotency (the potential to differentiate into several related cell types) and allowing cellular differentiation (Morey et al., 2012). Finally, canonical and variant PRC1 complexes regulate different classes of genes (Morey et al., 2013). These caveats notwithstanding, a number of specific chromatin modifications associated with PcG-mediated repression can be listed. Whether these modifications are causative or are only concomitant to repression (Henikoff and Shilatifard, 2011) has not yet been fully resolved.

Genes that are silenced by PcG complexes are often marked by nucleosomes containing H3K4me3, a modification characteristic of active genes (Bernstein et al., 2006; Roh et al., 2006), and by non-elongating RNAPII (Chopra et al., 2009). These modifications are presumed to poise these genes for rapid reactivation.

PRC2

As mentioned earlier, the major function of PRC2 is the sequential methylation of H3K27 to the trimethylated state. The final step of this reaction is stimulated by Polycomb-like (PCL) co-factors that can also aid in the recruitment of PRC2 (Walker et al., 2010). In mammals, PCLs can associate with H3K36me3 present in the coding region of transcribing genes and recruit a demethylase (No66) that will specifically remove this active gene mark (Brien et al., 2012). Concomitant with this effect, genes that are targeted for repression often exhibit the presence of H3K27me3 throughout their length. Recently, the H3K36me3 and H3K27me3 modifications were found to overlap, with the former establishing the limit of the extent of the latter (Streubel et al., 2018). The presence of PRC2 and of H3K27me3 also interferes with the process of RNAPII elongation (Chopra et al., 2011), presumably due to the fact that H3K27me3 recruits those PRC1 complexes that include one of the CBX subunits.

PRC1

These complexes ubiquitinate histone H2A (Zhang et al., 2004). This histone modification appears to be required for the repression of many, but not all, PRC1-targeted genes. H2A ubiquitination could alter chromatin structure, interfere directly with the transcription process or recruit other repressing factors; each of these possibilities that, of course, are not mutually exclusive have substantial experimental support (Simon and Kingston, 2013). H2A ubiquitination also appears to facilitate H3K27 methylation, establishing an interesting feedback loop along repressed genes that facilitates the spreading of H3K27me3 (Kalb et al., 2014).

In *Drosophila*, PRC1 represses canonical targets in the absence of H2A ubiquitylation (Pengelly et al., 2015), confirming that PRC1 has a repressive effect that is independent of its ability to ubiquitinate histone H2A—its presence induces chromatin compaction that physically inhibits transcription (Eskeland et al., 2010). This action is very likely the result of the cooperation of the H3K27me3 mark with the H3K9me3 mark laid down by SUV 39H methyl transferase, in attracting and anchoring the heterochromatin protein HP1 isoforms (Boros et al., 2014). A more direct effect on chromatin compaction has been ascribed to a highly positively charged domain containing numerous basic amino acids present in one of the PRC1 components (Grau et al., 2011). Mutation of this region, which, in mammals, is present on the CBX2 subunit of PRC1 (a homolog of the *Drosophila Pc* gene product), results in homeotic transformations, i.e. developmental changes in the normal axial pattern (Lau et al., 2017).

PcG proteins alter the three-dimensional structure of the genome

The association of clusters of PRCs, bound to PREs that regulate different genes, was first noted at the cytological level in *Drosophila* by the appearance of the Pc protein in distinct punctate foci in the interphase nuclei of diploid cells (Messner et al., 1992). Using a variety of techniques that included fluorescence *in situ* hybridization (FISH) and different versions of chromosome conformation capture (see Chapter 10, Box 10.1), several characteristics of these foci, also known as PcG bodies, have come to light. The PREs controlling a particular *HOX* gene are present in the PcG body in those regions of the embryo where the gene is repressed, but they are outside of the body in regions where the *HOX* gene is expressed (Bantignies et al., 2011). Genes of different clusters are present in the same nuclear PcG body if they are repressed (Lanzuolo et al., 2007), and removal of one particular PRE-containing element from a PcG body weakens the repression of the other genes associated with the body, indicating a stabilizing function. The interaction of Pc domains and the formation of PcG bodies depend on the presence of insulator proteins that define functional chromatin domains throughout the genome (Van Bortle et al., 2014). Similarly, loci marked with H3K27me3 form associations in mammalian cells; these associations include the members of the *HOX* gene clusters that have been inactivated by the action of PcG complexes and occur in the active compartment of the nucleus (Vieux-Rochas et al., 2015). Rather than locating these PcG-inactivated gene clusters to the nuclear lamina, their presence in a more active central location may facilitate their activation in other cell types.

Repression can occur without the intervention of PRC complexes

Recent evidence has been presented for the silencing of genes in mouse B lymphocyte precursors by the nucleosome remodeling and deacetylase (NuRD)

complex that was targeted by the transcription factor Ikaros. Although the silencing of genes by Ikaros in other hematopoietic cell lines is mediated by its recruitment of PRC2 (Oravecz et al., 2015), in pre-B cells, immediately following Ikaros binding and NuRD recruitment, promoters of targeted genes are invaded by nucleosomes, preventing the binding of RNAPII (Liang et al., 2017). It is only subsequently that histones are deacetylated.

Normal levels of homeotic gene expression are maintained by the TrxG genes

The first gene of this group—trithorax (*Trx*)—was identified by a mutation that caused the homeotic transformation of the posterior abdominal segments of *Drosophila* embryos to anterior abdominal segments (Ingham and Whittle, 1980). Additional TrxG genes were identified by mutations that mimicked the loss of function of *HOX* genes or that suppressed the homeotic transformations caused by *PcG* gene mutations. The rationale for this latter type of genetic screen was that reducing the level of a factor involved in maintaining active transcription should compensate for the reduction of a repressor. The main TrxG proteins found in *Drosophila* are TRX, ASH1 (absent, small or homeotic discs 1), FSH [fs(1)h, female sterile homeotic] and BRM (brahma); their homologs in vertebrates are MLL (myeloid/ lymphoid or mixed-lineage leukemia), ASH1L, BRD4 (bromodomains containing 4) and BRG1 (brahma-related gene 1, renamed SMARCA1), respectively. Isoforms of these gene products are present in a variety of complexes and depend on their interaction with the other subunits to carry out their function. Since loss-of-function of the subunits with which they associate leads to phenotypes that are similar to mutations in the genes that encode the main TrxG proteins themselves, many of the subunits have been added to the growing list of TrxG proteins.

TrxG complexes function by preventing PcG silencing

In order to prevent the repressive action of PRCs, TrxG complexes bind PREs (for this reason, it may be more logical to refer to these genomic sites as PRE/TREs). The mechanistic basis of this antagonistic interaction was provided by the *in vitro* observations that PRC2 complexes can bind to nucleosomes bearing the H3K4me3 and H3K36me3 marks of actively transcribed genes, but this binding causes an allosteric inhibition of the

E(z) subunit's methyl transferase activity (Schmitges et al., 2011; Yuan et al., 2011). Additional modifications that are common on active gene promoters—H3K27 acetylated by CBP [CREB (cAMP response element-binding protein) binding protein] and H3S28 phosphorylated by MSKs (mitogen and stress-activated protein kinases)—antagonize PRC2 function (Gehani et al., 2010; Tie et al., 2009).

In *Drosophila*, BRM (the yeast SWI2 homolog) is associated with the majority of transcriptionally active chromatin (Armstrong et al., 2002). BRM and its homologs are the ATPase subunits of *Drosophila* and mammalian SWI/SNF complexes that alter the position of nucleosomes along the chromatin fiber or that modify the association of histone octamers with the DNA (see Chapter 4). These complexes contain a set of core subunits and a large number of additional subunits that vary in different cell lineages, endowing them with context-specific functions (Fig. 7.3). In humans, the BAF (BRG1- or hBRM-associated factors) complexes contain over a dozen subunits and are highly polymorphic. These complexes activate loci in repressed facultative heterochromatin by rapidly evicting PRC complexes. The experimental removal of BAF complexes re-establishes heterochromatin by the rapid reappearance of PRCs (Kadoch et al., 2017).

Another ATPase closely related to SWI2 is ISWI (imitation switch), found in a number of remodeling complexes: CHRAC (chromatin accessibility complex), ACF (assembly of core histones factor) and NURF (nucleosome remodeling factor). Based on functional similarity, members of the other families of ATPases—Chd1, Mi-2, Ino80 and Swr1 (see Chapter 4)—can be designated as TrxG proteins and the chromatin remodeling complexes in which they participate considered to be TrxG complexes.

TrxG proteins participate in the general transcription process

The common function of all TrxG protein complexes is to maintain and enhance gene expression by participating in every major step of the transcription process (Kingston and Tamkun, 2014). Three different lysine methylations—H3K4me, H3K36me and H3K79me—have been correlated with active transcription (see Chapter 5). All of these methylations are effected by TrxG proteins. MLL, and by inference its *Drosophila* homolog TRX, are methyl transferases responsible for H3K4 methylation (Milne et al., 2002); other members of the MLL family (Trr and LPT in *Drosophila*) carry out the same reaction. These enzymes, present in plants

and conserved from yeast to mammals, function as subunits of the Compass (complex of proteins associated with Set 1) family of complexes (Mohan et al., 2011). ASH1 and ASH1L methylate H3K36 (Tanaka et al., 2007). The methylation of H3K79 is the responsibility of a methyl transferase that is conserved from yeast (Dot1) to *Drosophila* (GPP) and mammals (Dot1L) (Feng et al., 2002).

Human BRD4 associates with acetylated histones in the promoter region of activated genes and recruits the positive transcription elongation factor P-TEFb that releases paused polymerases (Yang et al., 2005); it also directly phosphorylates serine 2 in the RNAPII C-terminal domain (Devaiah et al., 2012). Similar functions are assumed for the *Drosophila* homolog FSH, which is found at enhancers and promoters (Kellner et al., 2013).

Another connection of TrxG genes with general transcription is their role in the assembly of Mediator, the multiprotein complex that provides the functional bridge between most eukaryotic genes and their enhancers (see Chapter 3). Two Mediator subunits in *Drosophila* are encoded by genes that were discovered because their loss-of-function alleles suppressed *Pc* mutations (Treisman, 2001).

It has been known for some time that PcG and TrxG proteins bind RNA, suggesting that targeting of these complexes to the genomic sites where they exert their function is mediated by RNA intermediaries (Rinn et al., 2007). This hypothesis has been validated in the case of the *Drosophila* developmental gene vestigial (*vg*). This gene is active in some tissues, at certain times, and is repressed in others. Activation correlates with the transcription of the linked PRE/TRE along one strand, while repression is correlated with transcription of the element from the opposite strand (Herzog et al., 2014).

The role of cohesin

Cohesin is a complex of four different subunits that is loaded onto chromosomes to ensure chromatid cohesion following DNA replication (Haering et al., 2002). In addition, cohesin can modulate transcription by stimulating the release of paused RNAPII complexes, either by stabilizing the association of distant enhancers with their cognate promoters or by acting directly on the promoters of the many genes that lack enhancers (Schaaf et al., 2013). The role of cohesin in the maintenance of active transcription was first noted in a screen designed to uncover new genes that determined segment identity during *Drosophila* development (Kennison

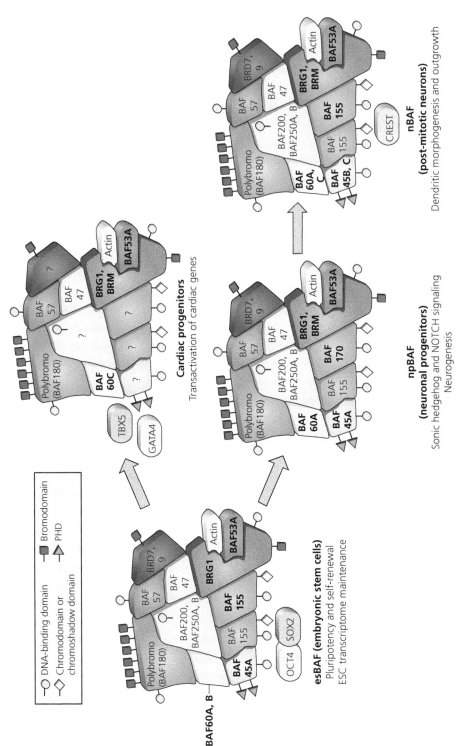

Fig. 7.3 Assembly of mammalian SWI/SNF complexes with cell lineage specificity. BAF designates Brahma (BRM)-associated factor subunits, and BRG1 is the Brahma-related gene 1. These complexes are sometimes referred to as BAP (Brahma-associated protein) complexes.

(From Ho and Crabtree, 2010.)

and Tamkun, 1988). Cohesin and the complex that loads it onto chromatin are associated predominantly with active genes and are absent from PcG-repressed genes (Misulovin et al., 2008). Surprisingly, cohesin interacts biochemically with both TrxG and PcG proteins (Strübbe et al., 2011) and is found concomitantly with PRC1 on active genes (Schaaf et al., 2013). At these genes, the presence of cohesin facilitates PRC1 binding and PRC1 blocks the release of paused RNAPII while cohesin counteracts this effect (Schaaf et al., 2013).

Transvection

Once again discovered in *Drosophila* (Lewis, 1954), the phenomenon of transvection consists of the influence of one allele of a gene on the expression of the other allele. This effect is usually dependent on the intimate physical juxtaposition of the two alleles that occurs in those organisms where somatic chromosomes are paired (synapsed) during interphase (Fig. 7.4). Somatic pairing allows the regulatory region associated with one allele of a gene on one chromosome to influence the activity of the promoter of the allele on the homologous chromosome (Gelbart, 1982; Geyer et al., 1990; Korge, 1977). This association has been visualized cytologically by FISH, a technique that allows the measurement of the distance separating fluorescent probes associated with specific genomic sites by nucleic acid hybridization (Ronshaugen and Levine, 2004). Transvection appears to be an option for the activation of a number of genes throughout the *Drosophila* genome. In all of the cases that have been studied individually, transvection requires the appropriate dosage of the *zeste* (*z*) gene product (Gelbart and Wu, 1982; Juni and Yamamoto, 2009). The Z protein has DNA- and protein-binding modules that allow it to stabilize long-range interactions between enhancers and promoters. The *Drosophila* genome has approximately 300 regions that are bound by Zeste; these regions include around 1500 potential Z binding sites (Moses et al., 2006).

The first observation suggesting that transvection may occur in mammals was made in human cells. A β-globin gene carried by a plasmid was activated by transfection into non-hematopoietic cells; the plasmid was seen to co-localize with the endogenous quiescent β-globin locus in the majority of the cells examined, suggesting a mechanism of "transinduction" (Ashe et al., 1997). Another observation was made in mice carrying the LoxP–Cre system. This system uses two copies of the Lox sequence and the gene encoding the Cre recombinase from the P1 bacteriophage, introduced by transgenic technology into a mammalian genome. Depending on the orientation of the two Lox sequences, the Cre recombinase can mediate the deletion of the DNA segment that separates them (Sauer and Henderson, 1989). In the male germline, following recombination, the LoxP sequence becomes methylated and, in the next generation, it induces the methylation of an allelic LoxP sequence inherited from the female, preventing its Cre-mediated excision (Rassoulzadegan et al., 2002). To date, three endogenous genes have been shown to undergo classical transvection in mice: *Kit* (encodes a tyrosine kinase receptor), *Cdk9* (cyclin-dependent kinase) and *Sox9* (a transcription factor). The transvection mechanism has been correlated with small non-coding RNAs, produced by the transvecting allele, that enhance the transcription of the transvected allele (Cuzin and Rassoulzadegan, 2010), and with the methyl transferase DNMT2 that, in contrast to the other members of the DNMT family of DNA methyl transferases, methylates RNA (Kiani et al., 2013).

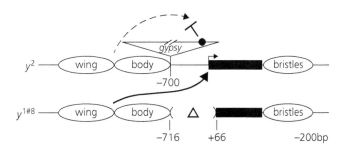

Fig. 7.4 Transvection at the yellow locus of *Drosophila*. The two upstream enhancers (ovals) are responsible for the dark pigmentation of the body cuticle and wing blades, respectively. The presence of the gypsy insulator prevents activation of the y2 allele by the wing and body enhancers in *cis*. The y2 allele is activated in *trans* by the wing enhancer present on the paired homologous chromosome.
(From Lee and Wu, 2006.)

Paramutations: a special case of transvection in plants

First discovered by Alexander Brink in corn, paramutation refers to the repression of the activity of one allele by the other allele of the gene; in subsequent generations, both silencing and silenced alleles are able to induce the repression of active alleles with which they are paired (Brink, 1956). Newly generated silencing alleles revert to a non-silencing status if they are maintained in a hemizygous condition, i.e. with a homologous chromosome deficient for the locus (Styles and Brink, 1969). This observation clearly indicates that paramutation represents a type of heritable epigenetic change affecting transcription.

One of the best studied cases of paramutation is that of the *b1* gene that encodes a transcription factor involved in anthocyanin pigment production. The silencing of a *b1* paramutable allele by a paramutagenic allele requires the presence of a tandemly repeated sequence, located at some distance upstream of the gene; the tandem repeat is transcribed into small interfering RNAs (siRNAs) that direct the RNA-dependent DNA methylation of the repeat in both the silencing and silenced alleles. The repeats no longer exhibit the open chromatin marks that characterize them in non-paramutating genotypes, suggesting that they no longer act as *b1* enhancers. The promoters and coding regions of both silencing and silenced alleles bear the activating and repressing histone modifications present on mammalian bivalent chromatin domains (Haring et al., 2010).

That paramutation may be a relatively significant aspect of gene regulation in flowering plants is indicated by genome-wide methylation patterns (methylomes) in *Arabidopsis*. Hybrid progenies produced by parent plants of different strains exhibit the transchromosomal methylation or demethylation characteristic of one or the other parent (Greaves et al., 2014). Similar results have been obtained in corn (Eichten et al., 2013; Li et al., 2013; Regulski et al., 2013).

Chapter summary

A corollary to the tenet that cellular differentiation and organismal development are based on differential gene activity is that certain genes, once activated, must remain active while others, once repressed, must remain inactive. Although the initial step in gene activation is carried out by transcription factors, the maintenance of a gene in the active state or its repression and maintenance in an inactive state are carried out by epigenetic modifications of the histones and of the DNA itself.

Two major classes of complexes (PCR1 and PCR2), containing members of an evolutionarily conserved group of proteins (Polycomb group or PcG), mediate transcriptional repression. PRC2 trimethylates histone H3 at lysine 27 (H3K27me3); in many cases, this modification attracts PRC1, leading to the ubiquitination of histone H2A. A number of variations on this theme can occur; for example, variant PRC1 complexes can be targeted first, and mono-ubiquitinated histone H2A recruits PRC2 complexes that methylate H3K27 that, in turn, serves as the traditional target for canonical PRC1 complexes. In some instances, PRC2 can be targeted to sites of repression by associating directly or indirectly with long non-coding RNAs (lncRNAs).

Trithorax group (TrxG) proteins form complexes that counteract PcG-mediated repression. Some of the subunits present in these complexes maintain and enhance transcription by carrying out different lysine methylations (H3K4me, H3K36me and H3K79me) that are associated with active gene function; other subunits remodel chromatin by displacing and repositioning nucleosomes.

Additional effects on transcription are rare instances of transvection, whereby somatic pairing allows the regulatory region associated with one allele of a gene on one chromosome to influence the activity of the promoter of the allele on the homologous chromosome.

The propagation of active or repressed gene expression patterns through cell divisions during development by PcG and TrxG proteins implicates them in processes of somatic and transgenerational epigenetic inheritance. These topics are discussed in Chapters 16 and 19 in Part IV.

References

Armstrong, J. A., Papoulas, O., Daubresse, G., Sperling, A. S., Lis, J. T., Scott, M. P. & Tamkun, J. W. 2002. The *Drosophila* BRM complex facilitates global transcription by RNA polymerase II. *EMBO J*, 21, 5245–54.

Ashe, H. L., Monks, J., Wijgerde, M., Fraser, P. & Proudfoot, N. J. 1997. Intergenic transcription and transinduction of the human beta-globin locus. *Genes Dev*, 11, 2494–509.

Bantignies, F., Roure, V., Comet, I., Leblanc, B., Schuettengruber, B., Bonnet, J., Tixier, V., Mas, A. & Cavalli, G. 2011. Polycomb-dependent regulatory contacts between distant Hox loci in *Drosophila*. *Cell*, 144, 214–26.

Bartolomei, M. S., Zemel, S. & Tilghman, S. M. 1991. Parental imprinting of the mouse H19 gene. *Nature*, 351, 153–5.

Bernstein, B. E., Mikkelsen, T. S., Xie, X., Kamal, M., Huebert, D. J., Cuff, J., Fry, B., Meissner, A., Wernig, M., Plath, K., Jaenisch, R., Wagschal, A., Feil, R., Schreiber, S. L. & Lander, E, S. 2006. A bivalent chromatin structure marks key developmental genes in embryonic stem cells. *Cell,* 125, 315–26.

Blackledge, N. P., Farcas, A. M., Kondo, T., King, H. W., Mcgouran, J. F., Hanssen, L. L., Ito, S., Cooper, S., Kondo, K., Koseki, Y., Ishikura, T., Long, H. K., Sheahan, T. W., Brockdorff, N., Kessler, B. M., Koseki, H. & Klose, R. J. 2014. Variant PRC1 complex-dependent H2A ubiquityla- tion drives PRC2 recruitment and polycomb domain formation. *Cell,* 157, 1445–59.

Boros, J., Arnoult, N., Stroobant, V., Collet, J. F. & Decottignies, A. 2014. Polycomb repressive complex 2 and H3K27me3 cooperate with H3K9 methylation to maintain hetero- chromatin protein 1alpha at chromatin. *Mol Cell Biol,* 34, 3662–74.

Bowman, S. K., Deaton, A. M., Domingues, H., Wang, P. I., Sadreyev, R. I., Kingston, R. E. & Bender, W. 2014. H3K27 modifications define segmental regulatory domains in the *Drosophila* bithorax complex. *Elife,* 3, e02833.

Brien, G. L., Gambero, G., O'connell, D. J., Jerman, E., Turner, S. A., Egan, C. M., Dunne, E. J., Jurgens, M. C., Wynne, K., Piao, L., Lohan, A. J., Ferguson, N., Shi, X., Sinha, K. M., Loftus, B. J., Cagney, G. & Bracken, A. P. 2012. Polycomb PHF19 binds H3K36me3 and recruits PRC2 and demethy- lase NO66 to embryonic stem cell genes during differen- tiation. *Nat Struct Mol Biol,* 19, 1273–81.

Brink, R. A. 1956. A genetic change associated with the R locus in maize which is directed and potentially reversible. *Genetics,* 41, 872–89.

Carrera, I., Janody, F., Leeds, N., Duveau, F. & Treisman, J. E. 2008. Pygopus activates Wingless target gene transcrip- tion through the mediator complex subunits Med12 and Med13. *Proc Natl Acad Sci U S A,* 105, 6644–9.

Chopra, V. S., Hendrix, D. A., Core, L. J., Tsui, C., Lis, J. T. & Levine, M. 2011. The polycomb group mutant esc leads to augmented levels of paused Pol II in the *Drosophila* embryo. *Mol Cell,* 42, 837–44.

Chopra, V. S., Hong, J. W. & Levine, M. 2009. Regulation of *Hox* gene activity by transcriptional elongation in *Drosophila. Curr Biol,* 19, 688–93.

Cooper, S., Dienstbier, M., Hassan, R., Schermelleh, L., Sharif, J., Blackledge, N. P., De Marco, V., Elderkin, S., Koseki, H., Klose, R., Heger, A. & Brockdorff, N. 2014. Targeting poly- comb to pericentric heterochromatin in embryonic stem cells reveals a role for H2AK119u1 in PRC2 recruitment. *Cell Rep,* 7, 1456–70.

Cuzin, F. & Rassoulzadegan, M. 2010. Non-Mendelian epi- genetic heredity: gametic RNAs as epigenetic regulators and transgenerational signals. *Essays Biochem,* 48, 101–6.

Davidovich, C. & Cech, T. R. 2015. The recruitment of chro- matin modifiers by long noncoding RNAs: lessons from PRC2. *RNA,* 21, 2007–22.

Davidovich, C., Zheng, L., Goodrich, K. J. & Cech, T. R. 2013. Promiscuous RNA binding by Polycomb repressive com- plex 2. *Nat Struct Mol Biol,* 20, 1250–7.

Denell, R. E. 1973. Homoeosis in *Drosophila.* I. Complemen- tation studies with revertants of Nasobemia. *Genetics,* 75, 279–97.

Devaiah, B. N., Lewis, B. A., Cherman, N., Hewitt, M. C., Albrecht, B. K., Robey, P. G., Ozato, K., Sims, R. J., 3RD & Singer, D. S. 2012. BRD4 is an atypical kinase that phos- phorylates serine2 of the RNA polymerase II carboxy-ter- minal domain. *Proc Natl Acad Sci U S A,* 109, 6927–32.

Eichten, S. R., Briskine, R., Song, J., Li, Q., Swanson-Wagner, R., Hermanson, P. J., Waters, A. J., Starr, E., West, P. T., Tiffin, P., Myers, C. L., Vaughn, M. W. & Springer, N. M. 2013. Epigenetic and genetic influences on DNA methyla- tion variation in maize populations. *Plant Cell,* 25, 2783–97.

Eskeland, R., Leeb, M., Grimes, G. R., Kress, C., Boyle, S., Sproul, D., Gilbert, N., Fan, Y., Skoultchi, A. I., Wutz, A. & Bickmore, W. A. 2010. Ring1B compacts chromatin struc- ture and represses gene expression independent of his- tone ubiquitination. *Mol Cell,* 38, 452–64.

Farcas, A. M., Blackledge, N. P., Sudbery, I., Long, H. K., Mcgouran, J. F., Rose, N. R., Lee, S., Sims, D., Cerase, A., Sheahan, T. W., Koseki, H., Brockdorff, N., Ponting, C. P., Kessler, B. M. & Klose, R. J. 2012. KDM2B links the Polycomb Repressive Complex 1 (PRC1) to recognition of CpG islands. *Elife,* 1, e00205.

Feng, Q., Wang, H., Ng, H. H., Erdjument-Bromage, H., Tempst, P., Struhl, K. & Zhang, Y. 2002. Methylation of H3-lysine 79 is mediated by a new family of HMTases without a SET domain. *Curr Biol,* 12, 1052–8.

Gaunt, S. J. 2015. The significance of *Hox* gene collinearity. *Int J Dev Biol,* 59, 159–70.

Gehani, S. S., Agrawal-Singh, S., Dietrich, N., Christophersen, N. S., Helin, K. & Hansen, K. 2010. Polycomb group pro- tein displacement and gene activation through MSK- dependent H3K27me3S28 phosphorylation. *Mol Cell,* 39, 886–900.

Gelbart, W. M. 1982. Synapsis-dependent allelic complemen- tation at the decapentaplegic gene complex in *Drosophila melanogaster. Proc Natl Acad Sci U S A,* 79, 2636–40.

Gelbart, W. M. & Wu, C. T. 1982. Interactions of zeste muta- tions with loci exhibiting transvection effects in *Drosophila melanogaster. Genetics,* 102, 179–89.

Geyer, P. K., Green, M. M. & Corces, V. G. 1990. Tissue- specific transcriptional enhancers may act in trans on the gene located in the homologous chromosome: the molecular basis of transvection in *Drosophila. EMBO J,* 9, 2247–56.

Grau, D. J., Chapman, B. A., Garlick, J. D., Borowsky, M., Francis, N. J. & Kingston, R. E. 2011. Compaction of chro- matin by diverse Polycomb group proteins requires local- ized regions of high charge. *Genes Dev,* 25, 2210–21.

Greaves, I. K., Groszmann, M., Wang, A., Peacock, W. J. & Dennis, E. S. 2014. Inheritance of trans chromosomal methylation patterns from *Arabidopsis* F1 hybrids. *Proc Natl Acad Sci U S A,* 111, 2017–22.

Haering, C. H., Lowe, J., Hochwagen, A. & Nasmyth, K. 2002. Molecular architecture of SMC proteins and the yeast cohesin complex. *Mol Cell, 9*, 773–88.

Haring, M., Bader, R., Louwers, M., Schwabe, A., Van Driel, R. & Stam, M. 2010. The role of DNA methylation, nucleosome occupancy and histone modifications in paramutation. *Plant J, 63*, 366–78.

Henikoff, S. & Shilatifard, A. 2011. Histone modification: cause or cog? *Trends Genet, 27*, 389–96.

Herzog, V. A., Lempradl, A., Trupke, J., Okulski, H., Altmutter, C., Ruge, F., Boidol, B., Kubicek, S., Schmauss, G., Aumayr, K., Ruf, M., Pospisilik, A., Dimond, A., Senergin, H. B., Vargas, M. L., Simon, J. A. & Ringrose, L. 2014. A strand-specific switch in noncoding transcription switches the function of a Polycomb/Trithorax response element. *Nat Genet, 46*, 973–981.

Ho, L. & Crabtree, G. R. 2010. Chromatin remodelling during development. *Nature, 463*, 474–84.

Ingham, P. W. & Whittle, J. R. 1980. Trithorax: a new homeotic mutation of *Drosophila melanogaster* causing transformations of abdominal and thoracic imaginal segments. *Mol Gen Genet, 179*, 607–14.

Juni, N. & Yamamoto, D. 2009. Genetic analysis of chaste, a new mutation of *Drosophila melanogaster* characterized by extremely low female sexual receptivity. *J Neurogenet, 23*, 329–40.

Jurgens, G. 1985. A group of genes controlling the spatial expression of the Bithorax complex in *Drosophila*. *Nature, 316*, 153–4.

Kadoch, C., Williams, R. T., Calarco, J. P., Miller, E. L., Weber, C. M., Braun, S. M., Pulice, J. L., Chory, E. J. & Crabtree, G. R. 2017. Dynamics of BAF-Polycomb complex opposition on heterochromatin in normal and oncogenic states. *Nat Genet, 49*, 213–222.

Kahn, T. G., Dorafshan, E., Schultheis, D., Zare, A., Stenberg, P., Reim, I., Pirrotta, V. & Schwartz, Y. B. 2016. Interdependence of PRC1 and PRC2 for recruitment to Polycomb response elements. *Nucleic Acids Res, 44*, 10132–49.

Kalb, R., Latwiel, S., Baymaz, H. I., Jansen, P. W., Muller, C. W., Vermeulen, M. & Muller, J. 2014. Histone H2A monoubiquitination promotes histone H3 methylation in Polycomb repression. *Nat Struct Mol Biol, 21*, 569–71.

Kassis, J. A. & Brown, J. L. 2013. Polycomb group response elements in *Drosophila* and vertebrates. *Adv Genet, 81*, 83–118.

Kaufman, T. C., Lewis, R. & Wakimoto, B. 1980. Cytogenetic analysis of chromosome 3 in *DROSOPHILA MELANOGASTER*: the homoeotic gene complex in polytene chromosome interval 84a-B. *Genetics, 94*, 115–33.

Kellner, W. A., Van Bortle, K., Li, L., Ramos, E., Takenaka, N. & Corces, V. G. 2013. Distinct isoforms of the *Drosophila* Brd4 homologue are present at enhancers, promoters and insulator sites. *Nucleic Acids Res, 41*, 9274–83.

Kennison, J. A. & Tamkun, J. W. 1988. Dosage-dependent modifiers of polycomb and antennapedia mutations in *Drosophila*. *Proc Natl Acad Sci U S A, 85*, 8136–40.

Kiani, J., Grandjean, V., Liebers, R., Tuorto, F., Ghanbarian, H., Lyko, F., Cuzin, F. & Rassoulzadegan, M. 2013. RNA-mediated epigenetic heredity requires the cytosine methyltransferase Dnmt2. *PLoS Genet, 9*, e1003498.

Kingston, R. E. & Tamkun J. W. 2014. Transcriptional regulation by trithorax-group proteins. *Cold Spring Harb Perspect Biol, 6*. doi: 10.1101/cshperspect.a019349.

Korge, G. 1977. Direct correlation between a chromosome puff and the synthesis of a larval saliva protein in *Drosophila melanogaster*. *Chromosoma, 62*, 155–74.

Ku, M., Koche, R. P., Rheinbay, E., Mendenhall, E. M., Endoh, M., Mikkelsen, T. S., Presser, A., Nusbaum, C., Xie, X., Chi, A. S., Adli, M., Kasif, S., Ptaszek, L. M., Cowan, C. A., Lander, E. S., Koseki, H. & Bernstein, B. E. 2008. Genomewide analysis of PRC1 and PRC2 occupancy identifies two classes of bivalent domains. *PLoS Genet, 4*, e1000242.

Kuzmichev, A., Nishioka, K., Erdjument-Bromage, H., Tempst, P. & Reinberg, D. 2002. Histone methyltransferase activity associated with a human multiprotein complex containing the Enhancer of Zeste protein. *Genes Dev, 16*, 2893–905.

Lanzuolo, C., Roure, V., Dekker, J., Bantignies, F. & Orlando, V. 2007. Polycomb response elements mediate the formation of chromosome higher-order structures in the bithorax complex. *Nat Cell Biol, 9*, 1167–74.

Lau, M. S., Schwartz, M. G., Kundu, S., Savol, A. J., Wang, P. I., Marr, S. K., Grau, D. J., Schorderet, P., Sadreyev, R. I., Tabin, C. J. & Kingston, R. E. 2017. Mutation of a nucleosome compaction region disrupts Polycomb-mediated axial patterning. *Science, 355*, 1081–4.

Lee, A. M. & Wu, C. T. 2006. Enhancer-promoter communication at the yellow gene of *Drosophila melanogaster*: diverse promoters participate in and regulate *trans* interactions. *Genetics, 174*, 1867–80.

Lewis, E. B. 1954. The theory and application of a new method of detecting chromosomal rearrangements in *Drosophila melanogaster*. *Am Nat, 88*, 225–39.

Lewis, E. B. 1978. A gene complex controlling segmentation in *Drosophila*. *Nature, 276*, 565–70.

Li, L., Petsch, K., Shimizu, R., Liu, S., Xu, W. W., Ying, K., Yu, J., Scanlon, M. J., Schnable, P. S., Timmermans, M. C., Springer, N. M. & Muehlbauer, G. J. 2013. Mendelian and non-Mendelian regulation of gene expression in maize. *PLoS Genet, 9*, e1003202.

Liang, Z., Brown, K. E., Carroll, T., Taylor, B., Vidal, I. F., Hendrich, B., Rueda, D., Fisher, A. G. & Merkenschlager, M. 2017. A high-resolution map of transcriptional repression. *Elife, 6*, pii: e22767.

Lynch, M. D., Smith, A. J., De Gobbi, M., Flenley, M., Hughes, J. R., Vernimmen, D., Ayyub, H., Sharpe, J. H., Sloane-Stanley, J. A., Sutherland, L., Meek, S., Burdon, T., Gibbons, R. J., Garrick, D. & Higgs, D. R. 2012. An interspecies analysis reveals a key role for unmethylated CpG dinucleotides in vertebrate Polycomb complex recruitment. *EMBO J*, 31, 317–29.

Messner, P., Christian, R., Kolbe, J., Schulz, G. & Sleytr, U. B. 1992. Analysis of a novel linkage unit of O-linked carbohydrates from the crystalline surface layer glycoprotein of *Clostridium thermohydrosulfuricum* S102-70. *J Bacteriol*, 174, 2236–40.

Milne, T. A., Briggs, S. D., Brock, H. W., Martin, M. E., Gibbs, D., Allis, C. D. & Hess, J. L. 2002. MLL targets SET domain methyltransferase activity to *Hox* gene promoters. *Mol Cell*, 10, 1107–17.

Misulovin, Z., Schwartz, Y. B., Li, X. Y., Kahn, T. G., Gause, M., Macarthur, S., Fay, J. C., Eisen, M. B., Pirrotta, V., Biggin, M. D. & Dorsett, D. 2008. Association of cohesin and Nipped-B with transcriptionally active regions of the *Drosophila melanogaster* genome. *Chromosoma*, 117, 89–102.

Mohan, M., Herz, H. M., Smith, E. R., Zhang, Y., Jackson, J., Washburn, M. P., Florens, L., Eissenberg, J. C. & Shilatifard, A. 2011. The COMPASS family of H3K4 methylases in *Drosophila*. *Mol Cell Biol*, 31, 4310–18.

Morey, L., Aloia, L., Cozzuto, L., Benitah, S. A. & Di Croce, L. 2013. RYBP and Cbx7 define specific biological functions of polycomb complexes in mouse embryonic stem cells. *Cell Rep*, 3, 60–9.

Morey, L., Pascual, G., Cozzuto, L., Roma, G., Wutz, A., Benitah, S. A. & Di Croce, L. 2012. Nonoverlapping functions of the Polycomb group Cbx family of proteins in embryonic stem cells. *Cell Stem Cell*, 10, 47–62.

Moses, A. M., Pollard, D. A., Nix, D. A., Iyer, V. N., Li, X. Y., Biggin, M. D. & Eisen, M. B. 2006. Large-scale turnover of functional transcription factor binding sites in *Drosophila*. *PLoS Comput Biol*, 2, e130.

Muller, J., Hart, C. M., Francis, N. J., Vargas, M. L., Sengupta, A., Wild, B., Miller, E. L., O'connor, M. B., Kingston, R. E. & Simon, J. A. 2002. Histone methyltransferase activity of a *Drosophila* Polycomb group repressor complex. *Cell*, 111, 197–208.

Oktaba, K., Gutierrez, L., Gagneur, J., Girardot, C., Sengupta, A. K., Furlong, E. E. & Muller, J. 2008. Dynamic regulation by polycomb group protein complexes controls pattern formation and the cell cycle in *Drosophila*. *Dev Cell*, 15, 877–89.

Onodera, A., Tumes, D. J., Watanabe, Y., Hirahara, K., Kaneda, A., Sugiyama, F., Suzuki, Y. & Nakayama, T. 2015. Spatial interplay between Polycomb and Trithorax complexes controls transcriptional activity in T lymphocytes. *Mol Cell Biol*, 35, 3841–53.

Oravecz, A., Apostolov, A., Polak, K., Jost, B., Le Gras, S., Chan, S. & Kastner, P. 2015. Ikaros mediates gene silencing in T cells through Polycomb repressive complex 2. *Nat Commun*, 6, 8823.

Pengelly, A. R., Kalb, R., Finkl, K. & Muller, J. 2015. Transcriptional repression by PRC1 in the absence of H2A monoubiquitylation. *Genes Dev*, 29, 1487–92.

Rassoulzadegan, M., Magliano, M. & Cuzin, F. 2002. Transvection effects involving DNA methylation during meiosis in the mouse. *EMBO J*, 21, 440–50.

Regulski, M., Lu, Z., Kendall, J., Donoghue, M. T., Reinders, J., Llaca, V., Deschamps, S., Smith, A., Levy, D., Mccombie, W. R., Tingey, S., Rafalski, A., Hicks, J., Ware, D. & Martienssen, R. A. 2013. The maize methylome influences mRNA splice sites and reveals widespread paramutation-like switches guided by small RNA. *Genome Res*, 23, 1651–62.

Rinn, J. L., Kertesz, M., Wang, J. K., Squazzo, S. L., Xu, X., Brugmann, S. A., Goodnough, L. H.,HELMS, J. A., FARNHAM, P. J., SEGAL, E. & CHANG, H. Y. 2007. Functional demarcation of active and silent chromatin domains in human HOX loci by noncoding RNAs. *Cell*, 129, 1311–23.

Roh, T. Y., Cuddapah, S., Cui, K. & Zhao, K. 2006. The genomic landscape of histone modifications in human T cells. *Proc Natl Acad Sci U S A*, 103, 15782–7.

Ronshaugen, M. & Levine, M. 2004. Visualization of trans-homolog enhancer–promoter interactions at the Abd-B Hox locus in the *Drosophila* embryo. *Dev Cell*, 7, 925–32.

Sauer, B. & Henderson, N. 1989. Cre-stimulated recombination at loxP-containing DNA sequences placed into the mammalian genome. *Nucleic Acids Res*, 17, 147–61.

Schaaf, C. A., Kwak, H., Koenig, A., Misulovin, Z., Gohara, D. W., Watson, A., Zhou, Y., Lis, J. T. & Dorsett, D. 2013. Genome-wide control of RNA polymerase II activity by cohesin. *PLoS Genet*, 9, e1003382.

Scheuermann, J. C., De Ayala Alonso, A. G., Oktaba, K., Ly-Hartig, N., Mcginty, R. K., Fraterman, S., Wilm, M., Muir, T. W. & Muller, J. 2010. Histone H2A deubiquitinase activity of the Polycomb repressive complex PR-DUB. *Nature*, 465, 243–7.

Schiemann, S. M., Martin-Duran, J. M., Borve, A., Vellutini, B. C., Passamaneck, Y. J. & Hejnol, A. 2017. Clustered brachiopod *Hox* genes are not expressed collinearly and are associated with lophotrochozoan novelties. *Proc Natl Acad Sci U S A*, 114, E1913–22.

Schmitges, F. W., Prusty, A. B., Faty, M., Stutzer, A., Lingaraju, G. M., Aiwazian, J., Sack, R., Hess, D., Li, L., Zhou, S., Bunker, R. D., Wirth, U., Bouwmeester, T., Bauer, A., Ly-Hartig, N., Zhao, K., Chan, H., Gu, J., Gut, H., Fischle, W., Muller, J. & Thoma, N. H. 2011. Histone methylation by PRC2 is inhibited by active chromatin marks. *Mol Cell*, 42, 330–41.

Schwartz, Y. B. & Pirrotta, V. 2013. A new world of Polycombs: unexpected partnerships and emerging functions. *Nat Rev Genet*, 14, 853–64.

Shao, Z., Raible, F., Mollaaghababa, R., Guyon, J. R., Wu, C. T., Bender, W. & Kingston, R. E. 1999. Stabilization of chromatin structure by PRC1, a Polycomb complex. *Cell*, 98, 37–46.

Simon, J., Chiang, A., Bender, W., Shimell, M. J. & O'connor, M. 1993. Elements of the *Drosophila* bithorax complex that mediate repression by Polycomb group products. *Dev Biol*, 158, 131–44.

Simon, J. A. & Kingston, R. E. 2013. Occupying chromatin: Polycomb mechanisms for getting to genomic targets, stopping transcriptional traffic, and staying put. *Mol Cell*, 49, 808–24.

Streubel, G., Watson, A., Jammula, S. G., Scelfo, A., Fitzpatrick, D. J., Oliviero, G., Mccole, R., Conway, E., Glancy, E., Negri, G. L., Dillon, E., Wynne, K., Pasini, D., Krogan, N. J., Bracken, A.p. & Cagney, G. 2018. The H3K36me2 methyltransferase Nsd1 demarcates PRC2-mediated H3K27me2 and H3K27me3 domains in embryonic stem cells. *Mol Cell*, 70, 371–9.

Strohner, R., Nemeth, A., Nightingale, K. P., Grummt, I., Becker, P. B. & Langst, G. 2004. Recruitment of the nucleolar remodeling complex NoRC establishes ribosomal DNA silencing in chromatin. *Mol Cell Biol*, 24, 1791–8.

Strubbe, G., Popp, C., Schmidt, A., Pauli, A., Ringrose, L., Beisel, C. & Paro, R. 2011. Polycomb purification by in vivo biotinylation tagging reveals cohesin and Trithorax group proteins as interaction partners. *Proc Natl Acad Sci U S A*, 108, 5572–7.

Styles, E. D. & Brink, R. A. 1969. The metastable nature of paramutable R alleles in maize. IV. Parallel enhancement of R action in heterozygotes with r and in hemizygotes. *Genetics*, 61, 801–11.

Tanaka, Y., Katagiri, Z., Kawahashi, K., Kioussis, D. & Kitajima, S. 2007. Trithorax-group protein ASH1 methylates histone H3 lysine 36. *Gene*, 397, 161–8.

Tavares, L., Dimitrova, E., Oxley, D., Webster, J., Poot, R., Demmers, J., Bezstarosti, K., Taylor, S., Ura, H., Koide, H., Wutz, A., Vidal, M., Elderkin, S. & Brockdorff, N. 2012. RYBP-PRC1 complexes mediate H2A ubiquitylation at polycomb target sites independently of PRC2 and H3K27me3. *Cell*, 148, 664–78.

Tie, F., Banerjee, R., Stratton, C. A., Prasad-Sinha, J., Stepanik, V., Zlobin, A., Diaz, M. O., Scacheri, P. C. & Harte, P. J. 2009. CBP-mediated acetylation of histone H3 lysine 27 antagonizes *Drosophila* Polycomb silencing. *Development*, 136, 3131–41.

Tie, F., Furuyama, T., Prasad-Sinha, J., Jane, E. & Harte, P. J. 2001. The *Drosophila* Polycomb Group proteins ESC and E(Z) are present in a complex containing the histone-binding protein p55 and the histone deacetylase RPD3. *Development*, 128, 275–86.

Treisman, J. 2001. *Drosophila* homologues of the transcriptional coactivation complex subunits TRAP240 and TRAP230 are required for identical processes in eye-antennal disc development. *Development*, 128, 603–15.

Van Bortle, K., Nichols, M. H., Li, L., Ong, C. T., Takenaka, N., Qin, Z. S. & Corces, V. G. 2014. Insulator function and topological domain border strength scale with architectural protein occupancy. *Genome Biol*, 15, R82.

Vieux-Rochas, M., Fabre, P. J., Leleu, M., Duboule, D. & Noordermeer, D. 2015. Clustering of mammalian Hox genes with other H3K27me3 targets within an active nuclear domain. *Proc Natl Acad Sci U S A*, 112, 4672–7.

Walker, E., Chang, W. Y., Hunkapiller, J., Cagney, G., Garcha, K., Torchia, J., Krogan, N. J., Reiter, J. F. & Stanford, W. L. 2010. Polycomb-like 2 associates with PRC2 and regulates transcriptional networks during mouse embryonic stem cell self-renewal and differentiation. *Cell Stem Cell*, 6, 153–66.

Wu, C. T., Jones, R. S., Lasko, P. F. & Gelbart, W. M. 1989. Homeosis and the interaction of zeste and white in *Drosophila*. *Mol Gen Genet*, 218, 559–64.

Yang, Z., Yik, J. H., Chen, R., He, N., Jang, M. K., Ozato, K. & Zhou, Q. 2005. Recruitment of P-TEFb for stimulation of transcriptional elongation by the bromodomain protein Brd4. *Mol Cell*, 19, 535–45.

Yuan, W., Xu, M., Huang, C., Liu, N., Chen, S. & Zhu, B. 2011. H3K36 methylation antagonizes PRC2-mediated H3K27 methylation. *J Biol Chem*, 286, 7983–9.

Zhang, Y., Cao, R., Wang, L. & Jones, R. S. 2004. Mechanism of Polycomb group gene silencing. *Cold Spring Harb Symp Quant Biol*, 69, 309–17.

Zhao, J., Ohsumi, T. K., Kung, J. T., Ogawa, Y., Grau, D. J., Sarma, K., Song, J. J., Kingston, R. E., Borowsky, M. & Lee, J. T. 2010. Genome-wide identification of polycomb-associated RNAs by RIP-seq. *Mol Cell*, 40, 939–53.

CHAPTER 8

DNA methylation and gene expression

Approximately 70–80% of all cytosines in CpG sites in mammalian genomes are methylated (5mC) or hydroxymethylated (5hmC). Regions enriched in CpG dinucleotides, called CpG islands, surround many promoters (Bird et al., 1985). In mammals, 75% of all gene promoter regions are CpG-rich and the rest are CpG-poor. Enhancers are very CpG-poor. CpG-poor regions are usually methylated, but not in all cells within a tissue; CpG-rich promoter regions are largely unmethylated (Weber et al., 2007). Methylation of these regions is prevented by the binding of particular proteins (Blackledge et al., 2010; Thompson et al., 2010). When they are methylated, CpG islands impede transcription either directly by preventing transcription factors from recognizing their binding sites (Watt and Molloy, 1988) or by attracting the members of a family of methyl CpG binding domain (MBD) proteins. The first of these proteins to be described (MeCP1) was found to bind to sequences that contained a minimum of 12 methylated CpGs (Meehan et al., 1989). The second protein to be characterized (MeCP2) can bind single methylated CpGs and is distributed throughout the genome (Lewis et al, 1992). Cytosine methylation that occurs when this base is not associated with guanine (meCpH where H can be either adenine, cytosine or thymine) also serves as a target for MeCP2, albeit with a lower affinity than for meCpG (Chen et al., 2015; Guo et al., 2014). A loss-of-function mutation in the gene that encodes MeCP2 can cause a severe neurodevelopmental disorder (Box 8.1). MBD proteins associate with other proteins to form complexes that often contain histone deacetylases and chromatin remodeling subunits (Wade et al., 1999). The recruitment of deacetylases is consistent with the general observation that histone acetylation is associated with transcriptional activation. A number of additional proteins that bind to meCpGs, but in a sequence-specific context, have been identified (see, for example, Quenneville et al., 2011).

Box 8.1 Rett syndrome (RTT)

In humans, the gene that encodes MeCP2 (methyl-CpG-binding protein 2) is located on the X chromosome (Amir et al., 1999). Loss-of-function mutations cause very early lethality in males; heterozygous females survive because of the presence of a normal allele that is expressed in approximately 50% of their cells (see Chapter 9). These females present the symptoms of Rett syndrome (named after the physician who first described it): autism, seizures, circulatory problems and motor deficiencies. Nevertheless, affected females can survive into adulthood.

The majority of the clinical defects of RTT can be ascribed to structural abnormalities of neuronal development, rather than neuronal loss. In particular regions of the brain, neurons develop with fewer dendrites and spines. Another characteristic of RTT is a decrease in the levels of neurotransmitters, transmitter receptors and other factors involved in normal synapses. MeCP2 is a ubiquitous global repressor of gene activity (Nan et al., 1997) that interacts with numerous proteins (Guy et al., 2011). Not surprisingly, in addition to the neurological symptoms, RTT affects different pathways involving developmental growth factors such as IGF-1 (insulin-like growth factor 1), and metabolic pathways such as cholesterol biosynthesis. With respect to neurogenesis, mutations

continued

Epigenetics, Nuclear Organization and Gene Function: with implications of epigenetic regulation and genetic architecture for human development and health. John C. Lucchesi, Oxford University Press (2019). © John C. Lucchesi 2019.
DOI: 10.1093/oso/9780198831204.001.0001

Box 8.1 *Continued*

in MeCP2 abolish its interaction with NCoR/SMART (nuclear receptor co-repressor, also referred to as silencing mediator of retinoic acid and thyroid), a protein that recruits histone deacetylases (Lyst et al., 2013). The MeCP2–NCoR/SMART– HDAC complex deacetylates the FOXO (Forkhead Box O) transcription factors and enables them to positively regulate a subset of neuronal genes (Nott et al., 2016).

The specific targeting of MeCP2 in the brain has suggested a potential explanation for the relatively late age of onset of RTT. As previously mentioned, DNA methylation occurs on the cytosines of CpG doublets, as well as on CpH doublets (where H stands for adenine, cytosine or thymine). In the normal brain, CpH doublets are not modified during embryonic development; meCpH appears postnatally and its frequency increases progressively, reaching maximal concentration in the mature brain (Chen et al., 2015; Lister et al., 2013). The absence of MeCP2 to bind CpH (a binding that, in normal brains, occurs late during development), rather than CpG, may cause the misregulation (presumably aberrant activation) of CpH-rich genes.

Established silencing by DNA methylation

DNA methylation is clearly associated with instances where gene silencing needs to be permanent and irreversible. Eukaryotic genomes are populated with repetitive DNA sequences of two major types: satellite DNA sequences present in tandem arrays, and transposable elements interspersed throughout the genome. Repetitive sequences are conducive to genome instability and can lead to abnormal cellular differentiation and organismal development. Tandem repeat arrays can undergo lengthening as a result of errors in DNA replication, in DNA damage repair or during homologous recombination (Aguilera and Garcia-Muse, 2013). DNA methylation has been generally associated with the transcription silencing and neutralization of retrotransposons and of tandem arrays of simple repetitive sequences. DNA methylation is also instrumental in the phenomenon of parent-of-origin imprinting (see Genomic imprinting, p. 95) and of X chromosome inactivation in females (see Chapter 9).

The role of DNA methylation in single-copy gene regulation

The methylation of CpG islands usually results in the inactivation of the promoters that they encompass. An example is the DNA methylation of CpG islands of germline genes in the other cell lineages of early mouse embryos (Borgel et al., 2010). As in the case of X chromosome inactivation or imprinting, this methylation represents permanent transcriptional silencing, in spite of the presence of activating transcription factors. Whether DNA methylation is also used as a universal, dynamic switch for the activation and repression of genes has been brought into question (Bestor et al., 2015).

One of the caveats is that, in a population of cells where a particular gene is inactive, some of the inactive gene's promoters are not methylated. Another argument is that certain organisms, such as *Caenorhabditis* and *Drosophila*, lack DNA methylation yet undergo complex developmental pathways. Nevertheless, there are some examples that support a dynamic role for DNA methylation as a regulatory mark for gene activity. Among these is the activation of individual odorant receptor genes.

Olfactory perception occurs in sensory neurons that transmit signals to the olfactory bulbs in the brain. Each olfactory neuron expresses a single odorant receptor by activating a single gene from a family of genes that in rats number around 1200 and in humans around 400 (Niimura 2009). Groups of sensory neurons generate the odorant receptor specific for the detection of the same particular odor; these neurons connect with the neurons of the olfactory bulbs in specialized areas called glomeruli. A group of sensory neurons expressing the odorant receptor specific for the odor of a particular chemical (acetophenone) was characterized; the receptor in question is M71, encoded by the *OLFR151* gene (Bozza et al., 2002; Sullivan et al., 1996). Male mice, conditioned to fear the smell of acetophenone, mated to control females, produced F1 males that had increased sensitivity to this odorant; a similar phenotype was exhibited by F2 males. F1 males had significantly more M71 sensory neurons and larger dedicated glomeruli. In the F0 and F1 males' sperm, the level of DNA methylation of the *OLFR151* gene was significantly less when compared to a control gene encoding another receptor (Dias and Ressler, 2014).

The interpretation that DNA demethylation is the primary cause of the activation of one odorant receptor gene among the many hundreds present in olfactory sensory neurons must be considered in light of the following observations. The lack of methylation in CpG promoter islands depends on the presence of transcription binding sites (Brandeis et al., 1994; Macleod et al., 1994). In genes with unmethylated CpG islands that are not transcribed, just the binding of transcription factors may prevent DNA methylation; in actively transcribed genes, the lack of methylation may be caused by the activity of the transcriptional machinery. Furthermore, the activation of a particular olfactory receptor gene appears to depend on its interaction with a large number of distal enhancers (Markenscoff-Papadimitriou et al., 2014).

The relation of DNA methylation to histone modification

An important and complex correlation exists between DNA methylation and histone methylation (Meissner et al., 2008). Some of the histone modifications in methylated DNA regions of the genome are associated with transcriptional repression and heterochromatin formation. DNA methylation can occur only in the absence of H3K4me2/3 and is positively correlated with the methylation of H3K9. In pericentric regions of human embryonic stem cells, the presence of the histone methyl transferase Suv39h and methylated H3K9 appears to signal the recruitment of the *de novo* DNA methyl transferases DNMT3a and 3b, and the initial establishment of heterochromatin (Muramatsu et al., 2013). In contrast, a number of CpG islands of inactive promoters are not methylated; rather they are silenced by the presence of H3K27me3 laid down by the Polycomb repressive complexes (Lynch et al., 2012). In embryonic stem cells, methylated H3K27, which occurs throughout the genome, is present on unmethylated, but not on methylated, CpG islands (Brinkman et al., 2012); in contrast, H3K27me3 and DNA methylation occur frequently together on CpG islands in somatic cells and especially in cancer cells (Statham et al., 2012).

Active, but also inactive, gene promoters within unmethylated CpG islands exhibit the H3K4me2 isoform. Inactive promoters within high CpG-content islands have elevated levels of H3K4me2, while promoters with few CpGs have lower levels of H3K4me2, suggesting that chromatin structure is involved in preventing promoter DNA methylation (Weber et al., 2007). Histone methylation is carried out by the binding of CFP1 (also known as CXXC1, CXXC finger protein 1) to unmethylated CpGs and the recruitment by this protein of an H3K4 methyl transferase (Thomson et al., 2010). Not surprisingly, unmethylated active promoters are also enriched in the H2A.Z histone variant (Conerly et al., 2010).

Genomic imprinting

In all multicellular organisms, a number of genes are expressed from the copy inherited from one parent, and not the other. This parent of origin-specific expression of some genes has been known since time immemorial to plant and animal breeders. The process that is responsible for the silencing of one of the two alleles of a gene in the germline of one parent is referred to as **genomic imprinting**. Imprinting has been validated, from flowering plants such as *Arabidopsis* to humans. In *Arabidopsis*, where early development is dominated by a preponderance of maternally expressed genes which, following the gradual activation of the paternal alleles, lose their uniparental expression, a few genes remain expressed only from the maternal allele; these genes, together with others that are expressed only from the paternal allele, constitute the imprinted portion of the *Arabidopsis* genome (Autran et al., 2011). In *Drosophila*, there is no evidence for parent-dependent allele-specific expression, and therefore no apparent role of imprinting on development and differentiation. Nevertheless, experimental manipulations have been used to demonstrate that the Y chromosome's normal inheritance through the sperm influences the level of expression of a substantial number of genes (Lemos et al., 2014).

In mammals, the occurrence of sex-specific genomic imprinting was first suggested by Sharat Chandra and Spencer Brown as a model for X chromosome inactivation (Brown and Chandra, 1973). The impetus to study parent-specific gene expression was provided, a decade later, by the observations that embryos produced by nuclear transfer were inviable if the two pronuclei used were both either of maternal or of paternal origin (McGrath and Solter, 1984). The problem was ascribed to a difference in the imprinting of genes in the two parental germlines, leading to a non-equivalence of maternal and paternal pronuclear genomes (Surani et al., 1984). At least 150 imprinted genes have been identified in the mouse genome; many of these genes are imprinted in humans as well. Most imprinted genes occur in clusters that are individually regulated by *cis*-acting **imprinting control regions** (ICRs). The ICRs act in the germline to ensure that the appropriate gene is marked for silencing, and in the developing embryo and beyond to maintain the silenced state. Failure of the imprinting mechanism can lead to severe human pathologies (Box 8.2).

The reason for the existence of parental imprinting, in other words its evolutionary significance, is not clear.

A number of hypotheses have been proposed, including, for example, one suggesting that maternally expressed genes limit fetal growth in order to maximize the chances of development for multiple offspring (Moore and Haig, 1991). A consequence of imprinting, once again with a poorly understood evolutionary advantage, is that it prevents the possibility of parthenogenetic reproduction—the development of progeny from unfertilized eggs.

Epigenetic mechanisms of imprinting

There are several mechanisms that are used to ensure the silencing of imprinted genes. These mechanisms are not mutually exclusive and, in many cases, have DNA methylation at the crux of the silencing process (Li et al., 1993)—all ICRs are differentially methylated in the two parental germlines. Although the recruitment of DNA methyl transferases for the specific methylation of ICRs during parental gametogenesis has not been fully elucidated, it is clear that it involves association with co-factors, and possibly the act of transcription across the ICR (Weaver and Bartolomei, 2014).

One of the best characterized examples of imprinted gene expression is the *H19/Igf2* locus (Bartolomei et al., 1993; Zemel et al., 1992). The *H19* gene expresses a non-coding RNA only from the maternally inherited allele; *Igf2* (insulin-like growth factor 2) is expressed only

---

Box 8.2 Imprinting diseases

Imprinting is a mechanism that ensures the transcription of particular genes from one parental allele in all or particular tissues of the developing embryo and the adult organism. A change in the level of expression of an imprinted gene will lead to pathological consequences. Deviation from monoallelic expression levels can be engendered by cytogenetic mutations such as duplications or deletions. It can also arise from uniparental disomy—as a result of an error during the segregation of homologous chromosomes (during meiosis I) or of sister chromatids (during meiosis II), an offspring receives both copies of a chromosome from the same parent. If the other parent contributes a normal gamete, the fertilized egg will give rise to an individual with three doses of one of the chromosomes; if, by rare chance, the other parent's gamete lacks the chromosome in question, then the offspring will be diploid. More frequently, changes in the level of expression of imprinted genes arise from a failure of methylation at those differentially methylated regions that are responsible for the silencing of one of the two parental alleles. Often, methylation defects occur during early embryonic development, leading to a mosaicism of normal and phenotypically abnormal tissues. Eight imprinting diseases have been characterized to date, among which are the examples discussed below.

Beckwith–Wiedemann syndrome (BWS) is an overgrowth syndrome with a significantly larger-than-average size at birth, enlarged tongue and internal organs and asymmetry in body growth. In addition, affected individuals have a significantly increased chance of developing various types of cancers. Most cases of BWS involve abnormal imprinting in two separate imprinting domains (Fig. 8B.1): the *IGF2/H19* (insulin-like growth factor 2/long intergenic non-protein coding RNA) domain (Bartolomei et al., 1993;

Brandeis et al., 1993; Sasaki et al., 1992) and the *KCNQ1/CDKN1C* (potassium voltage-gated channel subfamily Q member 1/cyclin-dependent kinase inhibitor 1C) domain (Hatada and Mukai, 1995; Lee et al., 1997; Matsuoka et al., 1996). *IGF2* is a maternally imprinted gene expressed

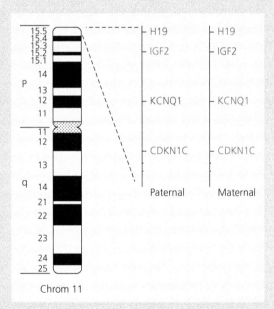

Fig. 8B.1 Diagram of human chromosome 11 showing the location of the imprinted alleles involved in BWS. The genes in red are expressed, and the ones in blue are imprinted and silent.

(Modified from a figure by F.G. Barr (2009) in Atlas of Genetics and Cytogenetics in Oncology and Haematology.)

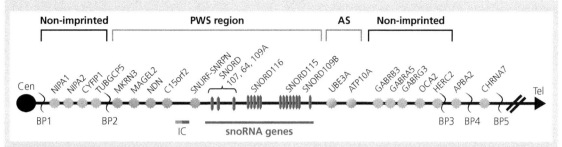

Fig. 8B.2 Diagram of the transcription units of the chromosome 15 region 15q11–13. The PWS region has five protein-coding and several snoRNA genes that are expressed only from the paternal alleles. The AS region contains two genes that have maternal-only expression. The position of the imprinting center (IC) is indicated below the chromosome.

(From Cassidy et al., 2012.)

on the paternal alleles; the reverse is true of the *CDKN1C* gene. Imprinting of the *IGF2/H19* domain is controlled by an intergenic differentially methylated domain (DMD) equivalent to an ICR. The ICR for the *KCNQ1/CDKN1C* domain is in an intron of the *KCNQ1* gene and contains an antisense promoter that allows the transcription of *KCNQ1OT1*, which is an antisense RNA to the *KCNQ1* gene (Mitsuya et al., 1999; Smilinich et al., 1999). The ICR is methylated on the maternal chromosome; therefore, it prevents transcription of the *KCNQ1OT1* RNA and allows activation of the *KCNQ1* and *CDKN1C* genes.

A few cases of BWS involve a gain of methylation of the H19 intergenic DMD of the *IGF2/H19* locus, leading to the biallelic expression of the *IGF2*. Most cases of BWS involve the loss of methylation of the *KCNQ1/CDKN1C* ICR, allowing transcription of the *KCNQ1OT1* antisense RNA and loss of the monoallelic expression of *CDKN1C*—an inhibitor of G1 cyclin complexes that functions as a negative regulator of cellular growth and proliferation. Therefore, both epigenetic changes lead to abnormal overgrowth during development. To date, the factors responsible for the aberrant methylation of the ICRs involved have not been identified. Mention should also be made that BWS can result from classical DNA mutations. Nonsense (leading to a truncate protein), missense (causing the substitution of one amino acid for another) and frame-shift mutations (which change the reading frame) can lead to an inactive *CDKN1C* gene product (Brioude et al., 2015; Hatada et al., 1996).

Prader–Willi (PWS) and Angelman (AS) syndromes are complex neurodevelopmental disorders. PWS is characterized by endocrine problems, poor muscle tone in early infancy, some cognitive deficiency and severely increased appetite that, if left unchecked, can lead to diabetes and cardiovascular problems. AS involves speech deficit, cognitive impairment, motor problems and the occurrence of seizures. Both syndromes are caused by the aberrant function of imprinted genes in a particular region of chromosome 15 (15q11-13)

(Fig. 8B.2). There are five maternally imprinted genes that are expressed only from the paternal chromosome, among which *MKRN3* (makorin ring finger protein 3) and *MAGEL2* (melanoma antigen family member L2) function in protein ubiquitination, *NDN* (necdin melanoma family member) affects the cell cycle, *SNRPN* (small nuclear ribonucleoprotein polypeptide N) plays a role in pre-mRNA processing and the *C15orf2* gene (nuclear pore associated protein 1) has an unknown function. In addition to these five genes, the maternally imprinted region contains a number of small nucleolar RNAs (snoRNAs). The molecular function of a few of these RNAs has been established, and of particular interest is the observation that one of them (SNORD115) is involved in regulating the alternative splicing of one of the serotonin receptors in the brain (Kishore and Stamm, 2006). PWS results from the lack of expression of these genes that can occur by one of several genetic mechanisms: (1) a paternal chromosome with a deletion that encompasses the alleles that should be active, (2) maternal uniparental disomy for chromosome 15 whereby the developing embryo has two maternally imprinted, and therefore silent, sets of alleles and (3) an imprinting defect whereby the paternal alleles are rendered inactive.

Distally located from the maternally imprinted genes, there are two genes that are paternally imprinted and are expressed from the maternal allele: *ATP10A* (ATPase phospholipid transporting 10A) involved in the transport of phospholipids across membranes, and *UBE3A* (ubiquitin protein ligase E3A) involved in protein degradation. A mutation that inactivates the *UBE3A* gene or its erroneous imprinting in the female parent's germline are sufficient to cause AS (Kishino et al., 1997; Matsuura et al., 1997). Interstitial deletions that remove the *UBE3A* gene, as well as adjacent genes, cause a more severe form of the syndrome (Gentile et al., 2010). These deletions are thought to occur because of unequal crossing-over between repeated sequences within the 15q11-13 region (Nicholls, 1994).

from the paternal allele. Both genes are activated by the same endodermal enhancer. Between the two genes, which are approximately 80 kilobases apart, lies a CpG region that is differentially methylated [differentially methylated domain (DMD)]. On the maternal allele, the DMD is not methylated and allows the binding of the insulator protein CTCF (CCCTC-binding factor; see Chapter 10). The enhancer is unable to interact with the *Igf2* gene and activates *H19*. On the paternal allele, the DMD is highly methylated, preventing the binding of CTCF, and thereby allowing the preferred interaction of the enhancer with the *Igf2* gene. It can be said that the DMD acts as an ICR for this locus (Fig. 8.1).

Most, if not all, clusters of imprinted genes contain the transcriptional unit of a long non-coding RNA (lncRNA). These RNAs are transcribed in a parent-specific manner and are required for the silencing of the cluster

(Mancini-DiNardo et al., 2006; Sleutels et al., 2002). The transcription of lncRNAs may overlap the promoter of adjacent genes and interfere with their expression. In other instances, lncRNAs may associate with genes that are located in *cis*, well beyond their transcriptional domain, and initiate a series of silencing epigenetic modifications, for example by recruiting PRCs and the methyl transferases responsible for H3K9 methylation. As expected, imprinted genes replicate asynchronously during development, with the inactivated allele lagging behind the active allele (Simon et al., 1999).

A number of imprinted alleles do not rely on DNA methylation (Okae et al., 2012; Okae et al., 2014). These imprinted genes are paternally expressed and rely on maternal allele-specific silencing, based on the presence of the Polycomb-mediated H3K27me3 (Inoue et al., 2017).

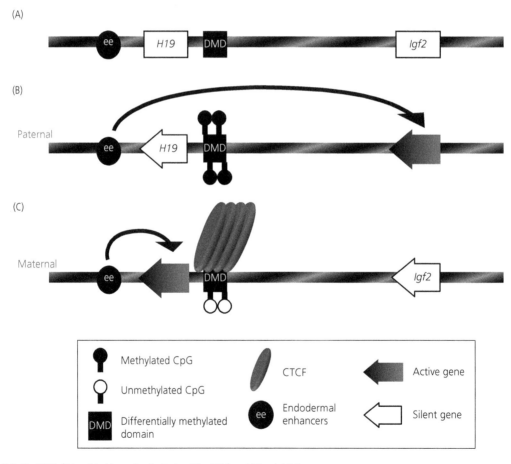

Fig. 8.1 The H19/Igf2 imprinted locus. See the text and Box 8.2 for additional details.

(From Fedoriw et al., 2012.)

Transmission of imprinted alleles

A zygote receives paternally imprinted alleles through the sperm and maternally imprinted alleles through the egg nucleus. In the case of DNA methylation-imprinted genes, silencing must be maintained throughout embryonic and postnatal development by maintenance DNA methyl transferases, in an organism-wide or tissue-specific manner, depending on the particular imprinted gene. The first hurdle encountered by this process is the general demethylation that occurs in mammalian embryos during very early development that is achieved by a failure in DNA methylation maintenance during replication (Arand et al., 2015). Surprisingly, the sex-specific methylation of imprinted regions escapes this early demethylation wave. Of necessity, though, once

the organism reaches sexual maturity, it must transmit to the next generation alleles that are imprinted according to its own sex (Fig. 8.2). This involves erasing the imprinting marks received from its parents in its own primordial germ cells and establishing new marks during gametogenesis (Edwards et al., 2017). In humans, this erasure is achieved by the repression of *de novo* and maintenance DNA methyl transferases and the enrichment of TET hydroxylases (Tang et al., 2015). In these analyses, the downstream products of 5hmC oxidation in the pathway of the eventual conversion of 5mC to thymine were not detected. Nevertheless, in light of earlier observations that high levels of the components of the base excision repair (BER) pathway (see Chapter 15) are present at the time of germ cell genome demethylation (Hajkova et al., 2010), it is

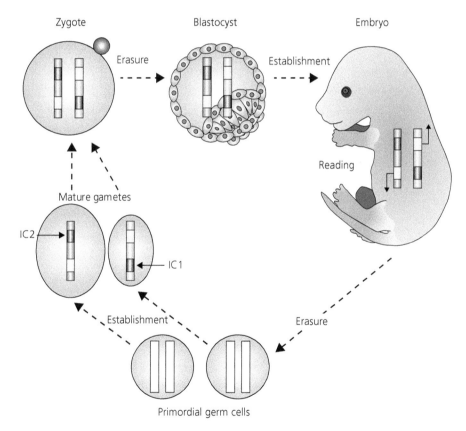

Fig. 8.2 Diagram of the genomic imprinting cycle. Alleles colored gray are modified and inactive; active alleles are colored white. Chromosomes inherited from the male parent are marked in blue, and from the female parent in red. The smaller gamete represents a sperm, and the larger an egg. A general loss of DNA methylation occurs during the development of the zygote into the blastocyst, but imprinted alleles remain methylated. *De novo* methylation of the genome occurs by the gastrula stage. As primordial germ cells are formed in the gonad primordia, the methylation of their genomes, including imprinted genes, is erased. During the early stages of gametogenesis, *de novo* methylation occurs, leading to the levels of methylation found in mature gametes and including the methylation of imprinted genes according to the developing embryo's sex.

(From Reik and Walter, 2001.)

reasonable to conclude that the TET hydroxylases initiated the complete removal of 5mC residues.

The signals that distinguish the genes to be imprinted, or more specifically that mark the specific ICRs for modification during the erasure in the primordial germ cells and the re-establishment of imprinting according to the sex of the individual in whose gonads they are located, are not understood. On the other hand, some experimental evidence suggests aspects of the mechanism involved in maintaining the imprinted loci, following the wave of demethylation that begins in the zygote and terminates in the very early embryo. A protein, first identified in primordial germ cells (PGC7) but also referred to as Stella or Developmental Pluripotency Associated 3 (DPPA3), was found to maintain the methylation of several imprinted genes. This protein is present in mature oocytes, and therefore in zygotes and early embryos, where it prevents the early demethylation of the whole maternal genome and the demethylation of at least two imprinted genes in the paternal genome (Nakamura et al., 2007). It achieves this effect by binding to H3K9me2 and preventing the conversion of methylated cytosine (5mC) to hydroxymethylated cytosine (5hmC) (Nakamura et al., 2012). Another factor involved in maintaining the methylation of imprinted alleles is the transcription repressor TRIM28, which binds to the methylated allele of all known imprinting control regions during and after early embryonic reprogramming (Alexander et al., 2015).

Random monoallelic expression

The term was coined to describe the existence in mammalian genomes of genes that express the maternal allele in some cells and the paternal allele in others. Most of the genes on the X chromosome of placental mammalian females exhibit random monoallelic expression because of dosage compensation and the process of X inactivation (see Chapter 9). A number of autosomal genes, representing a few percent of the human (Gimelbrandt et al., 2007) and of the mouse (Zwemer et al., 2012) genomes, are expressed monoallelically as well. As in the case of the genes present on the X chromosome, monoallelic expression is clonally inherited—once established at random, the allelic choice is transmitted to daughter cells through mitotic divisions. In some cases, a single allele of the gene is expressed in some cells, while other cells, in the same tissue, express both alleles. Genes that exhibit monoallelic expression are marked by the presence of H3K36me3 and H3K27me3 in the gene bodies of active and silent alleles, respectively (Nag et al., 2015).

Chapter summary

DNA methylation is an epigenetic modification that consists of the addition of a methyl, or of a hydroxyl and a methyl group, to the cytosine of CpG dinucleotides. Some gene promoters are rich in CpGs that are predominantly not modified; other promoters and most enhancers are poor in CpGs. These elements, as well as most exons, introns and intergenic regions, tend to be methylated. CpG methylation plays an important role in maintaining transposable elements and tandem arrays of repetitive sequences in a repressed state. CpG methylation is also responsible for the uniparental silencing of imprinted alleles, allowing the monoallelic expression of some genes, and for the silencing and clonal transmission of the inactive X chromosome in mammals. The use of this modification as a means of dynamically turning individual genes on or off, illustrated by the activation of individual odorant receptor genes, is less common.

References

Aguilera, A. & Garcia-Muse, T. 2013. Causes of genome instability. *Annu Rev Genet*, 47, 1–32.

Alexander, K. A., Wang, X., Shibata, M., Clark, A. G. & Garcia-Garcia, M. J. 2015. TRIM28 controls genomic imprinting through distinct mechanisms during and after early genome-wide reprogramming. *Cell Rep*, 13, 1194–205.

Amir, R. E., Van Den Veyver, I. B., Wan, M., Tran, C. Q., Francke, U. & Zoghbi, H. Y. 1999. Rett syndrome is caused by mutations in X-linked *MECP2*, encoding methyl-CpG-binding protein 2. *Nat Genet*, 23, 185–8.

Arand, J., Wossidlo, M., Lepikhov, K., Peat, J. R., Reik, W. & Walter, J. 2015. Selective impairment of methylation maintenance is the major cause of DNA methylation reprogramming in the early embryo. *Epigenetics Chromatin*, 8, 1.

Autran, D., Baroux, C., Raissig, M. T., Lenormand, T., Wittig, M., Grob, S., Steimer, A., Barann, M., Klostermeier, U. C., Leblanc, O., Vielle-Calzada, J. P., Rosenstiel, P., Grimanelli, D. & Grossniklaus, U. 2011. Maternal epigenetic pathways control parental contributions to *Arabidopsis* early embryogenesis. *Cell*, 145, 707–19.

Bartolomei, M. S., Webber, A. L., Brunkow, M. E. & Tilghman, S. M. 1993. Epigenetic mechanisms underlying the imprinting of the mouse *H19* gene. *Genes Dev*, 7, 1663–73.

Bestor, T. H., Edwards, J. R. & Boulard, M. 2015. Notes on the role of dynamic DNA methylation in mammalian development. *Proc Natl Acad Sci U S A*, 112, 6796–9.

Bird, A., Taggart, M., Frommer, M., Miller, O. J. & Macleod, D. 1985. A fraction of the mouse genome that is derived from islands of nonmethylated, CpG-rich DNA. *Cell*, 40, 91–9.

Blackledge, N. P., Zhou, J. C., Tolstorukov, M. Y., Farcas, A. M., Park, P. J. & Klose, R. J. 2010. CpG islands recruit a histone H3 lysine 36 demethylase. *Mol Cell*, 38, 179–90.

Borgel, J., Guibert, S., Li, Y., Chiba, H., Schubeler, D., Sasaki, H., Forne, T. & Weber, M. 2010. Targets and dynamics of promoter DNA methylation during early mouse development. *Nat Genet*, 42, 1093–100.

Bozza, T., Feinstein, P., Zheng, C. & Mombaerts, P. 2002. Odorant receptor expression defines functional units in the mouse olfactory system. *J Neurosci*, 22, 3033–43.

Brandeis, M., Frank, D., Keshet, I., Siegfried, Z., Mendelsohn, M., Nemes, A., Temper, V., Razin, A. & Cedar, H. 1994. Sp1 elements protect a CpG island from *de novo* methylation. *Nature*, 371, 435–8.

Brandeis, M., Kafri, T., Ariel, M., Chaillet, J. R., Mccarrey, J., Razin, A. & Cedar, H. 1993. The ontogeny of allele-specific methylation associated with imprinted genes in the mouse. *EMBO J*, 12, 3669–77.

Brinkman, A. B., Gu, H., Bartels, S. J., Zhang, Y., Matarese, F., Simmer, F., Marks, H., Bock, C., Gnirke, A., Meissner, A. & Stunnenberg, H. G. 2012. Sequential ChIP-bisulfite sequencing enables direct genome-scale investigation of chromatin and DNA methylation cross-talk. *Genome Res*, 22, 1128–38.

Brioude, F., Netchine, I., Praz, F., Le Jule, M., Calmel, C., Lacombe, D., Edery, P., Catala, M., Odent, S., Isidor, B., Lyonnet, S., Sigaudy, S., Leheup, B., Audebert-Bellanger, S., Burglen, L., Giuliano, F., Alessandri, J. L., Cormier-Daire, V., Laffargue, F., Blesson, S., Coupier, I., Lespinasse, J., Blanchet, P., Boute, O., Baumann, C., Polak, M., Doray, B., Verloes, A., Viot, G., Le Bouc, Y. & Rossignol, S. 2015. Mutations of the Imprinted *CDKN1C* gene as a cause of the overgrowth Beckwith–Wiedemann syndrome: clinical spectrum and functional characterization. *Hum Mutat*, 36, 894–902.

Brown, S. W. & Chandra, H. S. 1973. Inactivation system of the mammalian X chromosome. *Proc Natl Acad Sci U S A*, 70, 195–9.

Cassidy, S. B., Schwartz, S., Miller, J. L. & Driscoll, D. J. 2012. Prader–Willi syndrome. *Genet Med*, 14, 10–26.

Chen, L., Chen, K., Lavery, L. A., Baker, S. A., Shaw, C. A., Li, W. & Zoghbi, H. Y. 2015. MeCP2 binds to non-CG methylated DNA as neurons mature, influencing transcription and the timing of onset for Rett syndrome. *Proc Natl Acad Sci U S A*, 112, 5509–14.

Conerly, M. L., Teves, S. S., Diolaiti, D., Ulrich, M., Eisenman, R. N. & Henikoff, S. 2010. Changes in H2A.Z occupancy and DNA methylation during B-cell lymphomagenesis. *Genome Res*, 20, 1383–90.

Dias, B. G. & Ressler, K. J. 2014. Parental olfactory experience influences behavior and neural structure in subsequent generations. *Nat Neurosci*, 17, 89–96.

Edwards, J. R., Yarychkivska, O., Boulard, M. & Bestor, T. H. 2017. DNA methylation and DNA methyltransferases. *Epigenetics Chromatin*, 10, 23.

Fedoriw, A., Mugford, J. & Magnuson, T. 2012. Genomic imprinting and epigenetic control of development. *Cold Spring Harb Perspect Biol*, 4, a008136.

Gentile, J. K., Tan, W. H., Horowitz, L. T., Bacino, C. A., Skinner, S. A., Barbieri-Welge, R., Bauer-Carlin, A., Beaudet, A. L., Bichell, T. J., Lee, H. S., Sahoo, T., Waisbren, S. E., Bird, L. M. & Peters, S. U. 2010. A neurodevelopmental survey of Angelman syndrome with genotype–phenotype correlations. *J Dev Behav Pediatr*, 31, 592–601.

Gimelbrant, A., Hutchinson, J. N., Thompson, B. R. & Chess, A. 2007. Widespread monoallelic expression on human autosomes. *Science*, 318, 1136–40.

Guo, J. U., Su, Y., Shin, J. H., Shin, J., Li, H., Xie, B., Zhong, C., Hu, S., Le, T., Fan, G., Zhu, H., Chang, Q., Gao, Y., Ming, G. L. & Song, H. 2014. Distribution, recognition and regulation of non-CpG methylation in the adult mammalian brain. *Nat Neurosci*, 17, 215–22.

Guy, J., Cheval, H., Selfridge, J. & Bird, A. 2011. The role of MeCP2 in the brain. *Annu Rev Cell Dev Biol*, 27, 631–52.

Hajkova, P., Jeffries, S. J., Lee, C., Miller, N., Jackson, S. P. & Surani, M. A. 2010. Genome-wide reprogramming in the mouse germ line entails the base excision repair pathway. *Science*, 329, 78–82.

Hatada, I. & Mukai, T. 1995. Genomic imprinting of p57KIP2, a cyclin-dependent kinase inhibitor in mouse. *Nat Genet*, 11, 204–6.

Hatada, I., Ohashi, H., Fukushima, Y., Kaneko, Y., Inoue, M., Komoto, Y., Okada, A., Ohishi, S., Nabetani, A., Morisaki, H., Nakayama, M., Niikawa, N. & Mukai, T. 1996. An imprinted gene *p57KIP2* is mutated in Beckwith–Wiedemann syndrome. *Nat Genet*, 14, 171–3.

Inoue, A., Jiang, L., Lu, F., Suzuki, T. & Zhang, Y. 2017. Maternal H3K27me3 controls DNA methylation-independent imprinting. *Nature*, 547, 419–24.

Kishino, T., Lalande, M. & Wagstaff, J. 1997. UBE#3A/E6-AP mutations cause Angelman syndrome. *Nat Genet*, 15, 70–3.

Kishore, S. & Stamm, S. 2006. Regulation of alternative splicing by snoRNAs. *Cold Spring Harb Symp Quant Biol*, 71, 329–34.

Lee, M. P., Hu, R. J., Johnson, L. A. & Feinberg, A. P. 1997. Human *KVLQT1* gene shows tissue-specific imprinting and encompasses Beckwith–Wiedemann syndrome chromosomal rearrangements. *Nat Genet*, 15, 181–5.

Lemos, B., Branco, A. T., Jiang, P. P., Hartl, D. L. & Meiklejohn, C. D. 2014. Genome-wide gene expression effects of sex chromosome imprinting in *Drosophila*. *G3 (Bethesda)*, 4, 1–10.

Lewis, J. D., Meehan, R. R., Henzel, W. J., Maurer-Fogy, I., Jeppesen, P., Klein, F. & Bird, A. 1992. Purification, sequence, and cellular localization of a novel chromosomal protein that binds to methylated DNA. *Cell*, 69, 905–14.

Li, E., Beard, C. & Jaenisch, R. 1993. Role for DNA methylation in genomic imprinting. *Nature*, 366, 362–5.

Lister, R., Mukamel, E. A., Nery, J. R., Urich, M., Puddifoot, C. A., Johnson, N. D., Lucero, J., Huang, Y., Dwork, A. J., Schultz, M. D., Yu, M., Tonti-Filippini, J., Heyn, H., Hu, S., Wu, J. C., Rao, A., Esteller, M., He, C., Haghighi, F. G., Sejnowski, T. J., Behrens, M. M. & Ecker, J. R. 2013. Global epigenomic reconfiguration during mammalian brain development. *Science*, 341, 1237905.

Lynch, M. D., Smith, A. J., De Gobbi, M., Flenley, M., Hughes, J. R., Vernimmen, D., Ayyub, H., Sharpe, J. A., Sloane-Stanley, J. A., Sutherland, L., Meek, S., Burdon, T., Gibbons, R. J., Garrick, D. & Higgs, D. R. 2012. An interspecies analysis reveals a key role for unmethylated CpG dinucleotides in vertebrate Polycomb complex recruitment. *EMBO J*, 31, 317–29.

Lyst, M. J., Ekiert, R., Ebert, D. H., Merusi, C., Nowak, J., Selfridge, J., Guy, J., Kastan, N. R., Robinson, N. D., De Lima Alves, F., Rappsilber, J., Greenberg, M. E. & Bird, A. 2013. Rett syndrome mutations abolish the interaction of MeCP2 with the NCoR/SMRT co-repressor. *Nat Neurosci*, 16, 898–902.

Macleod, D., Charlton, J., Mullins, J. & Bird, A. P. 1994. Sp1 sites in the mouse aprt gene promoter are required to prevent methylation of the CpG island. *Genes Dev*, 8, 2282–92.

Mancini-Dinardo, D., Steele, S. J., Levorse, J. M., Ingram, R. S. & Tilghman, S. M. 2006. Elongation of the Kcnq1ot1 transcript is required for genomic imprinting of neighboring genes. *Genes Dev*, 20, 1268–82.

Markenscoff-Papadimitriou, E., Allen, W. E., Colquitt, B. M., Goh, T., Murphy, K. K., Monahan, K., Mosley, C. P., Ahituv, N. & Lomvardas, S. 2014. Enhancer interaction networks as a means for singular olfactory receptor expression. *Cell*, 159, 543–57.

Matsuoka, S., Thompson, J. S., Edwards, M. C., Bartletta, J. M., Grundy, P., Kalikin, L. M., Harper, J. W., Elledge, S. J. & Feinberg, A. P. 1996. Imprinting of the gene encoding a human cyclin-dependent kinase inhibitor, p57KIP2, on chromosome 11p15. *Proc Natl Acad Sci U S A*, 93, 3026–30.

Matsuura, T., Sutcliffe, J. S., Fang, P., Galjaard, R. J., Jiang, Y. H., Benton, C. S., Rommens, J. M. & Beaudet, A. L. 1997. De novo truncating mutations in E6-AP ubiquitin-protein ligase gene (UBE3A) in Angelman syndrome. *Nat Genet*, 15, 74–7.

Mcgrath, J. & Solter, D. 1984. Completion of mouse embryogenesis requires both the maternal and paternal genomes. *Cell*, 37, 179–83.

Meehan, R. R., Lewis, J. D., Mckay, S., Kleiner, E. L. & Bird, A. P. 1989. Identification of a mammalian protein that binds specifically to DNA containing methylated CpGs. *Cell*, 58, 499–507.

Meissner, A., Mikkelsen, T. S., Gu, H., Wernig, M., Hanna, J., Sivachenko, A., Zhang, X., Bernstein, B. E., Nusbaum, C., Jaffe, D. B., Gnirke, A., Jaenisch, R. & Lander, E. S. 2008. Genome-scale DNA methylation maps of pluripotent and differentiated cells. *Nature*, 454, 766–70.

Mitsuya, K., Meguro, M., Lee, M. P., Katoh, M., Schulz, T. C., Kugoh, H., Yoshida, M. A., Niikawa, N., Feinberg, A. P. & Oshimura, M. 1999. LIT1, an imprinted antisense RNA in the human KvLQT1 locus identified by screening for differentially expressed transcripts using monochromosomal hybrids. *Hum Mol Genet*, 8, 1209–17.

Moore, T. & Haig, D. 1991. Genomic imprinting in mammalian development: a parental tug-of-war. *Trends Genet*, 7, 45–9.

Muramatsu, D., Singh, P. B., Kimura, H., Tachibana, M. & Shinkai, Y. 2013. Pericentric heterochromatin generated by HP1 protein interaction-defective histone methyltransferase Suv39h1. *J Biol Chem*, 288, 25285–96.

Nag, A., Vigneau, S., Savova, V., Zwemer, L. M. & Gimelbrant, A. A. 2015. Chromatin signature identifies monoallelic gene expression across mammalian cell types. *G3 (Bethesda)*, 5, 1713–20.

Nakamura, T., Arai, Y., Umehara, H., Masuhara, M., Kimura, T., Taniguchi, H., Sekimoto, T., Ikawa, M., Yoneda, Y., Okabe, M., Tanaka, S., Shiota, K. & Nakano, T. 2007. PGC7/Stella protects against DNA demethylation in early embryogenesis. *Nat Cell Biol*, 9, 64–71.

Nakamura, T., Liu, Y. J., Nakashima, H., Umehara, H., Inoue, K., Matoba, S., Tachibana, M., Ogura, A., Shinkai, Y. & Nakano, T. 2012. PGC7 binds histone H3K9me2 to protect against conversion of 5mC to 5hmC in early embryos. *Nature*, 486, 415–19.

Nan, X., Campoy, F. J. & Bird, A. 1997. MeCP2 is a transcriptional repressor with abundant binding sites in genomic chromatin. *Cell*, 88, 471–81.

Nicholls, R. D. 1994. Recombination model for generation of a submicroscopic deletion in familial Angelman syndrome. *Hum Mol Genet*, 3, 9–11.

Niimura, Y. 2009. Evolutionary dynamics of olfactory receptor genes in chordates: interaction between environments and genomic contents. *Hum Genomics*, 4, 107–18.

Nott, A., Cheng, J., Gao, F., Lin, Y. T., Gjoneska, E., Ko, T., Minhas, P., Zamudio, A. V., Meng, J., Zhang, F., Jin, P. & Tsai, L. H. 2016. Histone deacetylase 3 associates with MeCP2 to regulate FOXO and social behavior. *Nat Neurosci*, 19, 1497–505.

Okae, H., Hiura, H., Nishida, Y., Funayama, R., Tanaka, S., Chiba, H., Yaegashi, N., Nakayama, K., Sasaki, H. & Arima, T. 2012. Re-investigation and RNA sequencing-based identification of genes with placenta-specific imprinted expression. *Hum Mol Genet*, 21, 548–58.

Okae, H., Matoba, S., Nagashima, T., Mizutani, E., Inoue, K., Ogonuki, N., Chiba, H., Funayama, R., Tanaka, S., Yaegashi, N., Nakayama, K., Sasaki, H., Ogura, A. & Arima, T. 2014. RNA sequencing-based identification of aberrant imprinting in cloned mice. *Hum Mol Genet*, 23, 992–1001.

Quenneville, S., Verde, G., Corsinotti, A., Kapopoulou, A., Jakobsson, J., Offner, S., Baglivo, I., Pedone, P. V., Grimaldi, G., Riccio, A. & Trono, D. 2011. In embryonic stem cells, ZFP57/KAP1 recognize a methylated hexanucleotide to affect chromatin and DNA methylation of imprinting control regions. *Mol Cell*, 44, 361–72.

Reik, W. & Walter, J. 2001. Genomic imprinting: parental influence on the genome. *Nat Rev Genet*, 2, 21–32.

Sasaki, H., Jones, P. A., Chaillet, J. R., Ferguson-Smith, A. C., Barton, S. C., Reik, W. & Surani, M. A. 1992. Parental imprinting: potentially active chromatin of the repressed maternal allele of the mouse insulin-like growth factor II (*Igf2*) gene. *Genes Dev,* 6, 1843–56.

Simon, I., Tenzen, T., Reubinoff, B. E., Hillman, D., Mccarrey, J. R. & Cedar, H. 1999. Asynchronous replication of imprinted genes is established in the gametes and maintained during development. *Nature,* 401, 929–32.

Sleutels, F., Zwart, R. & Barlow, D. P. 2002. The non-coding Air RNA is required for silencing autosomal imprinted genes. *Nature,* 415, 810–13.

Smilinich, N. J., Day, C. D., Fitzpatrick, G. V., Caldwell, G. M., Lossie, A. C., Cooper, P. R., Smallwood, A. C., Joyce, J. A., Schofield, P. N., Reik, W., Nicholls, R. D., Weksberg, R., Driscoll, D. J., Maher, E. R., Shows, T. B. & Higgins, M. J. 1999. A maternally methylated CpG island in KvLQT1 is associated with an antisense paternal transcript and loss of imprinting in Beckwith–Wiedemann syndrome. *Proc Natl Acad Sci U S A,* 96, 8064–9.

Statham, A. L., Robinson, M. D., Song, J. Z., Coolen, M. W., Stirzaker, C. & Clark, S. J. 2012. Bisulfite sequencing of chromatin immunoprecipitated DNA (BisChIP-seq) directly informs methylation status of histone-modified DNA. *Genome Res,* 22, 1120–7.

Sullivan, S. L., Adamson, M. C., Ressler, K. J., Kozak, C. A. & Buck, L. B. 1996. The chromosomal distribution of mouse odorant receptor genes. *Proc Natl Acad Sci U S A,* 93, 884–8.

Surani, M. A., Barton, S. C. & Norris, M. L. 1984. Development of reconstituted mouse eggs suggests imprinting of the genome during gametogenesis. *Nature,* 308, 548–50.

Tang, W. W., Dietmann, S., Irie, N., Leitch, H. G., Floros, V. I., Bradshaw, C. R., Hackett, J. A., Chinnery, P. F. & Surani, M. A. 2015. A unique gene regulatory network resets the human germline epigenome for development. *Cell,* 161, 1453–67.

Thomson, J. P., Skene, P. J., Selfridge, J., Clouaire, T., Guy, J., Webb, S., Kerr, A. R., Deaton, A., ANDAndrews, R., James, K. D., Turner, D. J., Illingworth, R. & Bird, A. 2010. CpG islands influence chromatin structure via the CpG-binding protein Cfp1. *Nature,* 464, 1082–6.

Wade, P. A., Gegonne, A., Jones, P. L., Ballestar, E., Aubry, F. & Wolffe, A. P. 1999. Mi-2 complex couples DNA methylation to chromatin remodelling and histone deacetylation. *Nat Genet,* 23, 62–6.

Watt, F. & Molloy, P. L. 1988. Cytosine methylation prevents binding to DNA of a HeLa cell transcription factor required for optimal expression of the adenovirus major late promoter. *Genes Dev,* 2, 1136–43.

Weaver, J. R. & Bartolomei, M. S. 2014. Chromatin regulators of genomic imprinting. *Biochim Biophys Acta,* 1839, 169–77.

Weber, M., Hellmann, I., Stadler, M. B., Ramos, L., Paabo, S., Rebhan, M. & Schubeler, D. 2007. Distribution, silencing potential and evolutionary impact of promoter DNA methylation in the human genome. *Nat Genet,* 39, 457–66.

Zemel, S., Bartolomei, M. S. & Tilghman, S. M. 1992. Physical linkage of two mammalian imprinted genes, H19 and insulin-like growth factor 2. *Nat Genet,* 2, 61–5.

Zwemer, L. M., Zak, A., Thompson, B. R., Kirby, A., Daly, M. J., Chess, A. & Gimelbrant, A. A. 2012. Autosomal monoallelic expression in the mouse. *Genome Biol,* 13, R10.

CHAPTER 9

Regulation of domains and whole chromosomes

Transcriptional regulation of gene clusters

Differential gene activity during cellular maintenance and differentiation requires the simultaneous or sequential activation of sets of genes that function in particular biochemical pathways. This can be accomplished by transcription factors whose cognate binding sequence is present in the promoter of different genes. Another strategy used by eukaryotes is the grouping of genes into clusters located in particular chromosomal locations that can be regulated in a coordinated fashion. It is generally accepted that gene clusters have arisen by duplication of initially single genes, followed by diversification of function of the original gene and the copies. The continued presence over evolutionary time of these duplicated genes as clusters indicates that positive selective pressure has maintained their physical linkage. Although diversified, all the genes often perform related functions in a particular metabolic pathway, in a specific tissue type, and at least some of the parameters regulating the transcription of the initial gene are maintained for the regulation of the copies. The following three examples of gene clusters—the β-globin genes, the ribosomal RNA genes and the canonical histone genes—are regulated by distinct parameters.

The β-globin genes

In vertebrates, hemoglobin is a multimeric protein consisting of α-globin and β-globin chains. During the course of evolution, the ancestral globin gene was duplicated and the two copies were physically separated by a chromosome rearrangement. Further duplication of each copy gave rise to two separate gene clusters that encode the α-globin and β-globin chains, and led to the functional diversification of the members of each cluster.

The transcriptional regulation of vertebrate β-globin genes has been studied in detail in chickens (McGhee et al.,1983), in mice and in humans (Curtis, 1980; Proudfoot et al., 1980). The human β-globin cluster (Fig. 9.1) includes a gene for embryonic (ε), two genes for fetal (^{G}γ, ^{A}γ) and two genes for adult hemoglobins (δ, β). The different globin chains impart specific oxygen-carrying capacities to the hemoglobin in accordance to embryonic and postnatal requirements. During development, the embryonic ε-globin gene is expressed in the yolk sac; later, it is silenced, and the fetal ^{G}γ and ^{A}γ genes are activated in the liver. Nearing birth, the fetal genes are progressively silenced and replaced by the expression of the adult genes in the bone marrow, with β-globin being the major component. The sequential activation of the different genes during development and adult life is the responsibility of regulatory sequences located a few kilobases upstream of the gene cluster, and collectively referred to as the locus control region (LCR). Discovery of the LCR occurred when a chromosomal deletion that included a DNA segment upstream of the β-globin genes resulted in the absence of transcription in differentiating erythroid cells; furthermore, the genes lost their sensitivity to DNase digestion, a mark of open and transcribing chromatin, and replicated late in the S phase of the cell cycle, a mark of repressed and heterochromatic regions of the genome (Forrester et al., 1990). Activation of the individual β-globin genes occurs via their physical interaction with the LCR (Tolhuis et al., 2002) (see Chapter 10).

Epigenetics, Nuclear Organization and Gene Function: with implications of epigenetic regulation and genetic architecture for human development and health. John C. Lucchesi, Oxford University Press (2019). © John C. Lucchesi 2019.
DOI: 10.1093/oso/9780198831204.001.0001

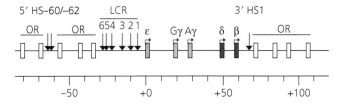

Fig. 9.1 The human β-globin locus consists of a gene for embryonic (ε), two genes for fetal (^Gγ, ^Aγ) and two genes for adult hemoglobins (δ, β). The LCR is defined by six distinct hypersensitive sites. In addition, there are 5′ and 3′ hypersensitive sites (HS). The locus is embedded in a chromosomal region rich in olfactory response genes (OR).

(Adapted from Katsumura et al., 2013; Palstra et al., 2003.)

The principal transcription factor involved in the expression of the globin genes is GATA-1, the first member of a family of transcription factors that were found to bind to A/TGATAA/G sequences (Evans et al., 1988; Ko and Engel, 1993). During hemoglobin synthesis, GATA-1 associates with components of the Mediator complex and recruits several transcriptional co-regulators such as FOG-1 (friend of Gata-1) and KLF1 (Kruppel-like factor 1), acetyl transferases (CBP/p300) and remodeling complexes [SWI/SNF and NURD (nucleosome remodeling and deacetylase)]. An important aspect of the function of GATA-1 is its role in moving the β-globin locus away from the nuclear periphery, a zone of inactive chromatin (see Chapter 11); this subnuclear relocalization requires FOG-1 and the NURD complex (Lee et al., 2011).

As previously mentioned, the sequential activation of the β-globin genes during development requires the silencing of the early-acting embryonic and fetal genes. A number of factors and chromatin modifications, some of which are listed here, have been associated with this effect. The ε and γ genes are methylated by the DNA methyl transferase DNMT1 and are bound by the methyl-binding domain protein MBD2 that, in turn,

recruits the NURD remodeling complex. The Mi2b subunit of NURD binds to, and activates, the expression of BCL11A (B-cell lymphoma-leukemia A) and KLF1 (Kruppel-like factor). BCL11A is stimulated by KLF1 and has a dominant effect on ε and γ genes silencing (Ginder, 2015; Suzuki et al., 2014).

rRNA genes and the nucleolus

In eukaryotes, the rRNAs are encoded in tandemly arrayed genes, clustered on one or more chromosomes (Fig. 9.2). The clusters constitute the nucleolus organizer region (NOR). The rRNA genes (Froberg et al., 2013) are transcribed by RNA polymerase I (RNAPI) into large 45S transcripts (S, or Svedberg, is a unit of sedimentation rate that is used to indicate the size of a molecule) that are processed into smaller RNA species (Scherrer et al., 1963). The resulting RNAs are found in the two ribosomal subunits—18S RNAs in the small subunit, and 5.8S and 28S RNAs in the large subunit (Pontvianne et al., 2013). In addition, the large subunit contains a 5S RNA encoded by genes located elsewhere. Transcription and processing of the 45S precursor RNAs, as well as ribosome biogenesis, occur in the

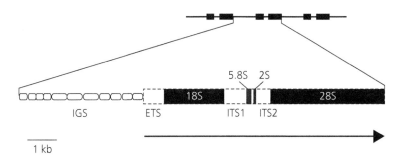

Fig. 9.2 Diagram of a *Drosophila* rRNA gene showing the 18S, 5.8S and 28S genes, and a 2S gene found in insects. The arrow indicates the extent of the primary transcript. Intergenic spacers consist of a number of internal subrepeats.

(From Stage and Eickbush, 2007.)

nucleolus (see Chapter 12). The 45S rRNA genes can be subdivided into active and silent genes (Conconi et al., 1989). The ratio of active to inactive genes differs in particular tissues, depending on the requirement for protein synthesis; it also varies as normal differentiation and development proceed, or upon the onset of oncogenesis.

Transcription of rRNA genes depends on the recruitment of RNAPI by the upstream binding transcription factor (UBTF), first discovered in *Xenopus* (Pikaard et al., 1989), by the promoter selectivity factor (SL1 in humans, TIF-IB in mice), which includes the TATA-binding protein (TBP) and several TBP-associated factors (TAFs) (Heix et al., 1997; Learned et al., 1985), and by nucleolin (Lapeyre et al., 1987; Rickards et al., 2007). Transcription is terminated by the binding of transcription termination factor 1 (TTF-1) to the 3′ end of active rRNA genes (Grummt et al., 1985). This factor also binds at the 5′ end of rRNA genes and mediates the formation of loops that bring the promoter and terminator into contact (Nemeth et al., 2008). Such loops are thought to enhance the rate of transcription by facilitating reinitiation. Other positive transcriptional regulators associated with active rRNA genes are an ATPase (Cockayne syndrome protein B, CSB), the elongation factors PAF (polymerase-associated factor) and FACT (facilitates chromatin transcription) and DNA topoisomerase I (Birch et al., 2009; Grierson et al., 2013; Yuan et al., 2007; Zhang et al., 2009). Transcribing rRNA genes exhibit all of the epigenetic marks of active chromatin (H3K4 methylation, H4K16 acetylation and, in mammals, DNA hypomethylation).

Some rRNA genes are not active, lack RNAPI and yet exhibit many of the characteristics of active transcription units. These genes are poised for activation; their status is established by the association of the nucleosome remodeling and deacetylation (NuRD) complex with the CSB protein and TTF-1 (Xie et al., 2012).

The repression of rRNA genes involves the recruitment of the nucleolar remodeling complex (NoRC). This complex is a member of the SWI/SNF family of ATP-dependent remodeling complexes (Strohner et al., 2001). NoRC is attracted to the TTF-1 factor bound to the 5′ region of rRNA genes and targets histone deacetylases and DNA methyl transferases to the promoter (Strohner et al., 2004). The function of the latter enzymes depends on the association with the promoter of a non-coding RNA, originating from the intergenic spacer region that lies between rRNA genes (Santoro et al., 2010; Schmitz et al., 2010). Additional proteins associated with silenced rRNA genes are PARP [poly (ADP ribose) polymerase] (Guetg et al., 2012) and histone macro-H2A. Silent rRNA

Fig. 9.3 Diagram of one set of *Drosophila* linker and core histone genes. The H1, H2A/B and H3/H4 promoters are indicated by red bars. (From Isogai et al., 2007.)

genes are marked by histone H4 hypoacetylation, methylation of H3K9, H4K20 and H3K27 and, in mammals, by DNA hypermethylation.

In *Drosophila*, approximately 50% of the rRNA genes are inactivated by insertions of transposable elements belonging to two conserved families R1 and R2 (Jakubczak et al., 1990). In fact, R1 and R2 transposons are present in subsets of 28S genes in many insects (Jakubczak et al., 1991).

During mitosis, all synthesis of RNA transcripts, including rRNA, is repressed. The Cdk1/cyclin B activated kinase, which triggers the nuclear envelope breakdown (see Chapter 11), phosphorylates one of the subunits of the promoter selectivity factor SL1, preventing its interaction with UBTF and the formation of the RNAPI pre-initiation complex (Heix et al., 1997). An additional SL1 subunit involved in DNA binding is deacetylated (Voit et al., 2015). UBF, as well as some transcription factors such as TTF-1, remain associated with rRNA genes through mitosis in order to mark them for reactivation in the daughter cells (Sirri et al.,1999), a process that involves reversing the silencing modifications of SL1.

Histone genes

In mice and humans, the canonical histones and the linker histone (H2A, H2B, H3, H4 and H1) are encoded by multiple gene copies present in three clusters (Fig. 9.3) located in two different chromosomes (Albig and Doenecke, 1997; Clark et al., 1981; Marzluff et al., 2002). These histones, synthesized in a replication-coupled manner (see Chapters 5 and 14), are translated from transcripts that are not polyadenylated and that lack introns. In contrast, the genes that give rise to the variants of the core histones are present in single copies dispersed throughout the genome and generate polyadenylated and processed transcripts.

In *Drosophila*, sets of the five canonical histones and the linker histone genes (*His2A*, *His2B*, *His3*, *His4* and *His1*) are tandemly repeated, more than 100 times, in a single cluster. Histone synthesis begins at the time of the cell cycle transition between G1 and S phases (when DNA replication is initiated) and ceases at the end of the S phase when histone gene transcripts are rapidly degraded. The cyclin/cyclin-dependent kinase combination CyclinE/Cdk2 that mediates the G1 to S phase transition is responsible for this regulation by phosphorylating a transcriptional activator Mxc (multi sex combs), an ortholog of human NPAT (nuclear protein of ataxia telangiectasia). CyclinE/Cdk2 phosphorylates other epitopes within a spherical nuclear body associated with the histone gene cluster, called the histone locus body (HLB; see Chapter 13). The HLB is rich in histone transcript processing factors that play a key role in the production of functional mRNAs (White et al., 2007). The transcription of the different genes within a set, in a particular cell type, has been studied by single-cell imaging techniques (Guglielmi et al., 2013). The two genes that encode the proteins of each of the heterodimers in the histone octamers (H2A-H2B and H3-H4) share promoter sequences. These promoters use TBP andTAFs for the formation of the pre-initiation complex. The *His1* promoter lacks a TATA box and uses a TBP-related factor (TRF2) that recognizes the DRE (DNA replication-related) core promoter element. The pre-initiation complex assembled at the promoters of the two pairs of core histone genes is unusual in that some of the general transcription factors (such as TFIIB) or some of their components (the TAFs of TFIID) appear to be dispensable.

Transcriptional regulation of whole chromosomes

In many animal species, differentiation into a male or a female is based on the presence of particular sex chromosomes. In mammals and in flies, females have two X chromosomes, and males have one X and a chromosome that is different from the X in shape and genetic content and is referred to as the Y chromosome. In some insects and in roundworms, the only sex chromosome present in males is the X. In other insects and in birds, the sex chromosome situation is reversed—males have two identical sex chromosomes (Z chromosomes) and females have one Z and another sex chromosome that is different from the Z and is called the W chromosome. Sex chromosomes were derived from autosomes, and the morphological and functional differentiation between the X and the Y chromosomes (or between the Z and the W chromosomes) was triggered by the presence of a pair of sex-determining alleles, one member of which induced the differentiation of one sex and, therefore, was limited to that sex. Absence of this allele in the genome would lead, by default, to differentiation of the other sex. A reduced level of genetic recombination in the individuals that received the sex-determining allele would favor the accumulation of random mutations and of transposable elements on the allele-bearing chromosome; the result was the eventual partial degeneration of that ancestral autosome into a Y chromosome or to its loss from the genome (Lucchesi, 1978).

In some animal groups, the Y (or W) chromosome contains the genetic directives for differentiation into a particular sex. This is true in mammals where the Y chromosome is male-determining and in birds where female differentiation is determined by the presence of the W chromosome. In some other species, the presence of two X chromosomes vs. only one X leads to sexual differentiation. This is the case in flies and round worms. A common problem presented by all of these systems of sex determination is that they would lead to an inequality between males and females in the dosage of genes on the X or Z chromosomes. Many of these genes, which were present in the ancestral autosome and have been retained on the X or Z chromosome, are equally important to the development and maintenance of both sexes, and a difference in the level of product of these genes could lead to an unequal selection in males and females. In order to avoid this problem, mechanisms have evolved to compensate for the difference in the dosage of X-linked (or Z-linked) genes between the sexes. These mechanisms extend to the entire chromosome or are limited to those genes for which the organism is particularly dose-sensitive. They have been studied at the molecular level in distant taxa: round worms, flies (where the first mechanism was discovered), birds and mammals. Although they ultimately accomplish the same goal, they were evolved independently from established regulatory pathways. In the case of chromosome-wide regulation, whether the regulatory process targets individual genes or whether genes respond to global chromosomal-level epigenetic modifications are distinctions that have been challenging to make.

Dosage compensation in *Drosophila melanogaster*

The original observations that genes present on the X chromosome lead to the same phenotype in males and females were made in *Drosophila*, using partial loss-of-function mutations. Although these mutant alleles lead to a dose-dependent phenotype within either sex, males with a single X chromosome, and therefore a single dose of the alleles, have a phenotype similar to females with two doses. These observations led to the conclusion that a regulatory mechanism must have evolved to compensate for the difference in dosage of these alleles in the two sexes in order to obtain phenotypic equality (Muller, 1932). Using transcription autoradiography with larval salivary gland chromosomes, a technique consisting of measuring the incorporation of radioactive (H3-labelled) uridine into chromosomal RNA following a brief exposure, A. S. Mukherjee demonstrated that the level of RNA synthesis on the single male X chromosome was equivalent to that of both X chromosomes in females and that dosage compensation operates at the level of transcription of the whole sex chromosome (Mukherjee and Beermann, 1965). Rather than reducing its level in females, equalization is achieved by enhancing the transcriptional rate of X-linked genes in males (Belote and Lucchesi, 1980). This regulatory mechanism is the responsibility of the MSL (male-specific lethal) complex that associates with the X chromosome in male somatic cells. The MSL complex consists of five core subunits that include an ATPase/helicase that preferentially unwinds short RNA:DNA hybrid substrates *in vitro* (Lee et al., 1997), an E3 ubiquitin ligase (Wu et al., 2011) and a histone acetyl transferase that acetylates histone H4 at lysine 16 (H4K16ac) (Hilfiker et al., 1997). As a member of the MSL complex, this subunit globally acetylates the X chromosome genes in males (Smith et al., 2000). In addition to its protein subunits, the MSL complex contains one of two long non-coding RNAs that are transcribed from the two *roX* (RNA on the X) genes *roX1* and *roX2*, located on the X chromosome. Although very different in size and having limited sequence similarity, either one of these RNAs can provide the initial scaffold for the formation of the complex (Ilik et al., 2013; Maenner et al., 2013). In females, the complex is prevented from forming by the presence of the SXL (sex lethal) female-determining protein (see

Chapter 3, Box 3.3), which interferes with the translation of one of the protein subunits (MSL2, male-specific lethal 2), as well as by the absence of the roX RNAs that are expressed only in males.

Several factors ensure that, in males, the MSL complex is found only on the X chromosome. Because the roX RNAs are degraded, unless they are associated with the MSL proteins, the complex initiates its assembly at the site of roX RNA synthesis (Fig. 9.4). Following assembly, the complex binds a series of sites along the X chromosome for which it has a high degree of affinity, termed chromatin entry sites (CES) or high-affinity sites (HAS) that share a common degenerate DNA sequence, and then spreads from these sites to numerous additional locations on the X (Alekseyenko et al., 2008; Straub et al., 2008). The first of these sites to be bound by the MSL complex have a slightly different sequence signature and are referred to as PionX (pioneering sites on the X) (Villa et al., 2016). As the DNA sequence present at the CES is only slightly enriched on the X chromosome, a number of additional factors must participate in strictly limiting the presence of the MSL complex to this chromosome. One of these factors is a chromatin-linked adaptor (CLAMP) that binds to the DNA sequence common to CES and is necessary for the recruitment of the MSL complex to these sites (Larschan et al., 2012; Soruco et al., 2013). Absence of the CLAMP protein is lethal in both sexes and, therefore, appears to be required in both sexes (Urban et al., 2017a). This requirement is explained by the association of CLAMP with many other proteins, including the negative elongation factor NELF (Urban et al., 2017b).

In vitro structural studies, complemented by *in vivo* mutagenesis, have identified two domains in MSL2 that bind to the GA-rich sequence common to the HAS

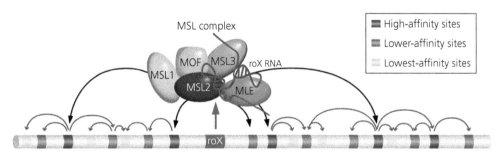

Fig. 9.4 The MSL complex assembles at the site of transcription of the *roX* genes that are both present on the X chromosome. Following assembly, the complexes target a number of high-affinity sites (HAS) and spread to numerous other sites for which they have varying levels of affinity.

(From Lucchesi and Kuroda, 2015.)

(Villa et al., 2016; Zheng et al., 2014). One of the two domains had been shown to bind to the roX RNA, thereby modifying the conformation of MSL2 to allow its association with the GA-rich sequence (Li et al., 2008). The ultimate targets of the MSL complex are the active genes along the X chromosome (Sass et al., 2003) with which it associates, in a sequence-independent manner, via the affinity of one of its subunits (MSL3, male-specific lethal 3) for the H3K36me3 mark of active genes (Larschan et al., 2007).

An alternate approach to uncover the molecular basis of the specific targeting of the X chromosome by the MSL complex has involved the analysis of repetitive sequences restricted to this chromosome. Original observations were obtained by determining the distribution of synthetic CA/GT polymers on polytene chromosomes (Pardue et al., 1987). Although hybridization of this polymer occurred throughout the *Drosophila* genome, its presence along the X chromosome was substantially greater than on the autosomes. The enrichment of the repetitive sequence recognized by the polymer probe suggested a connection with the unique functional characteristic of the X chromosome, namely dosage compensation. This possibility was further reinforced by the identical results that were obtained in *Drosophila* species where a second compensated X chromosome had evolved. Similar results, i.e. enrichment on the X chromosomes, were obtained with CT/GA and C/G polymers, but not with any of the remaining possible polymers (Lowenhaupt et al., 1989). Although it is impossible to compare frequencies, it is interesting to note that the major component of the consensus sequence of the HAS consists of three GA nucleotide pairs.

More recently, another family of repetitive sequences has been implicated in the specific X chromosome targeting by the MSL complex. This experimental approach was spurred by the observation that a reduction of the small interfering RNA (siRNA) synthetic pathway had a male-specific deleterious effect, suggesting that siRNA was involved in some aspect of dosage compensation (Menon and Meller, 2012). A specific repetitive DNA sequence that is unique to the X chromosome (DiBartolomeis et al., 1992; Waring and Pollack, 1987) is the source of the siRNA in question. The absence of this RNA results in failure of the MSL complex to associate with the X chromosome (Menon et al., 2014). The repetitive DNA sequences ectopically relocated to autosomal sites can recruit the MSL complex; this recruitment and the spreading of the complex from the ectopic site are greatly increased if a roX-encoding gene is present at the autosomal site as well, indicating that the repeated sequences and roX RNA act synergistically (Joshi and Meller, 2017).

Most of the experimental evidence supports the notion that the enhanced level of transcription of X-linked genes in males is achieved by increasing the rate of elongation of RNAPII complexes (Ferrari et al., 2013; Larschan et al., 2011). Several parameters contribute to this increase. The acetylation of H4K16 mediated by MOF, the acetyl transferase subunit of the complex occurs throughout the length of the genes but is particularly enhanced towards the 3′ end (Smith et al., 2001). This specific histone mark weakens nucleosome packing in reconstituted chromatin fibers and results in a more disordered architecture (Dunlap et al., 2012), rendering it more accessible to factors or complexes (Bell et al., 2010) such as the elongation factor SPT5 (Prabhakaran and Kelley, 2012). Further, the MSL complex targets the DNA-unwinding topoisomerase II to active X-linked genes, resulting in reduced negative supercoiling, once again facilitating RNAPII elongation (Cugusi et al., 2013).

Following a characteristic of eukaryotes, the X chromosome occupies its own intra-nuclear territory (Strukov et al., 2011). Within its territory, the male X chromosome exhibits a unique higher-order topology manifested by the physical clustering of the CES. This clustering depends on the presence in the nucleus of those MSL complex subunits (MSL1 and MSL2) that bind to the CES, even in the absence of the other components; in the absence of MSL1 or MSL2, the distance between the CES becomes similar to that in the non-compensated female cells (Grimaud and Becker, 2009). It is reasonable to assume that the function of this male-specific X chromosome topology is to facilitate the spreading of the MSL complex to all active genes. More recently, using chromosome conformation capture techniques (see Chapter 10), the fine-grained topological organization of the X chromosome was seen to be identical in males and female cell lines (Ramirez et al., 2015; Ulianov et al., 2016). In males, though the presence of the MSL complex results in an enhanced level of interaction among active X chromosome loci, the overall structure of this chromosome is very similar in both sexes (Schauer et al., 2017). As discussed in Chapter 10, chromosomes are organized in a series of domains whose boundaries are defined by the presence of insulator molecules. The clustering of CES reflects this type of topological arrangement (Ramirez et al., 2015), a conclusion that is supported by the association of CES with a known type of boundary factor complex (LBC) that, incidentally, contains the CLAMP adaptor protein (Kaye et al., 2017).

Dosage compensation in *Caenorhabditis elegans*

Individuals of this model organism are either males or hermaphrodites (somatically equivalent to females, although they do produce sperm during a brief period of their life cycle). Males have a single X chromosome and hermaphrodites have two Xs, leading to the assumption that some form of compensation for the difference in the dosage of X-linked genes between the sexes must occur (Hodgkin, 1983; Meneely and Wood, 1984). In contrast to the mechanism that operates in *Drosophila*, dosage compensation in *Caenorhabditis* is the responsibility of a multi-subunit dosage compensation complex (DCC) that decreases the level of transcription of both X chromosomes in hermaphrodites (Meyer and Casson, 1986). The DCC contains five proteins that are present only in this complex and an additional five proteins that are found also in condensin I, a ubiquitous complex which functions to condense chromosomes in preparation for mitosis or meiosis (Fig. 9.5).

The DCC does not form in males because one of the components that triggers its assembly and targets it to the X chromosome [SDC-2 (sex and dosage compensation 2)] is synthesized only in hermaphrodites (Dawes et al., 1999). In this latter sex, the complex associates exclusively with the two X chromosomes, with a single exception—an autosomal sex determination gene (*her-1*) that must be repressed in order to allow hermaphrodite development. Targeting of the DCC to the X chromosome shares some common features with the targeting of the *Drosophila* MSL complex. A number of recruitment elements on the *Caenorhabditis* X chromosome (referred to as *rex*) that share a DNA sequence motif to which the complex binds have been identified (Csankovszki et al., 2004; Ercan et al., 2007; McDonel et al., 2006). As in the case of the CES of *Drosophila*, although the rex consensus sequence is slightly enriched on the X chromosome, it is also found in the autosomes, albeit not targeted by the complex. This specific recruitment to the X chromosome is initiated at a small number of high-affinity rex sites that contain an additional sequence identification mark (Albritton et al., 2017). From these rex elements, the DCC spreads along the X chromosome to other rex sites in a sequence-independent manner, landing eventually near the promoter regions of active genes. Attraction to these regions appears to be facilitated by the presence of the histone variant H2A.Z (Petty et al., 2009).

The DCC alters the compaction of X chromosome chromatin in interphase (Lau et al., 2014). Although none of its components have any demonstrable histone-modifying activity, some epigenetic characteristics are correlated with the down-regulation of X-linked genes—the level of histone H4 acetylated at lysine 16 (H4K16ac), a canonical mark of active genes, is reduced, and the presence of histone H4 monomethylated at lysine 20 (H4K20me1) is increased on the hermaphrodite's X chromosomes (Vielle et al., 2012; Wells et al., 2012). This last modification has been correlated with condensed chromatin (Beck et al., 2012; Yang and Mizzen, 2009). Enrichment of H4K20me1 is achieved by one of the DCC proteins (DPY-21) that specifically demethylates H4K20me2; inactivation of DPY-21 reduces the levels of H4K20me1 and results in decreased X chromosome compaction and loss of dosage compensation function (Brejc et al., 2017).

The DCC also alters the topology of the X chromosomes. In *Caenorhabditis*, the genome is subdivided into functional units that are equivalent to the topologically associating domains (TADs) of flies and mammals (see Chapter 10). Binding of the DCC creates a unique distribution of TADs that function to repress gene expression along the entire chromosome (Crane et al., 2015). Here again, the histone H4 isoform H4K20me1 is responsible for diminishing the formation of TADs, resulting in greater long-range interactions across the X chromosome (Bian et al., 2018; Brejc et al., 2017).

In hermaphrodites, the two X chromosomes appear to be randomly localized within the nucleus, while in males, the X chromosome is preferentially located at the nuclear periphery where it interacts with a nuclear pore protein (Sharma et al., 2014). The nuclear periphery contains the lamina, to which silent heterochromatic regions of the genome are anchored, and nuclear

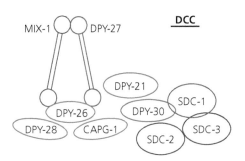

Fig. 9.5 The DCC complex of *Caenorhabditis*. Some of the proteins (SDC1-3, DPY-21 and DPY-30) are found exclusively on the X chromosomes of hermaphrodites. They target five other proteins (MIX-1, DPY26-28 and CAPG-1) that are found in the canonical condensing 1 complex involved in the condensation of the genome for cell division. The MIX-1 and DPY-27 proteins are drawn to indicate their structure and the fact that they form a dimer.

(From Ercan, 2015.)

pores with which active regions of the genome are associated (see Chapter 11). The presence of the male X chromosome near nuclear pores would be consistent with its increased level of transcriptional activity.

Dosage compensation in mammals

The study of dosage compensation in this group was launched by the observation that the heterochromatic body seen for the first time in the interphase nuclei of female cat neurons (Barr and Bertram, 1949) is formed by the condensation of one of the two X chromosomes (Ohno et al., 1959). These results led to the formulation of the hypothesis that the condensed X chromosome is inactive (Lyon, 1961). By allowing only one copy of each gene to be active in females, X inactivation represents the mechanism used by mammals to equalize the output of X-linked genes between the sexes (Davidson et al., 1963). In eutherian mammals, X inactivation is initiated by the transcription of a gene (*Xist*, X inactive specific transcript) located in a region of the X that is designated as the X chromosome inactivation center (XIC). The transcription of *Xist* is induced by the RLIM (RING finger LIM-domain-interacting, formerly known as RNF12) protein (Jonkers et al., 2009). RLIM is a ubiquitin ligase that regulates the activity of different transcription factors and co-factors. The gene that encoded RLIM is present on the X chromosome; therefore, the level of RLIM in females is twice that in males and is sufficient to activate Xist in one of the two X chromosomes. The eutherian XIC contains additional genes that regulate the transcription of *Xist*. An example is *Jpx* (just proximal to *Xist*) (Tian et al., 2010). Only one of the two *Xist* alleles present in females is transcribed, and the long non-coding RNA that it generates induces silencing of the X chromosome (Xi) from which it is expressed (Brown et al., 1991). On the other X chromosome, a gene that is also present in the XIC—*Tsix* (the name spells *Xist* backwards)—produces a non-coding antisense RNA that inhibits the transcription of *Xist*, allowing this chromosome (Xa) to remain active (Lee and Lu, 1999). *Tsix* transcription may physically interfere with the transcription of *Xist* or it may stimulate the DNA methylation of the Xist promoter (Sun et al., 2006). Several elements (*Xite*, X inactivation intergenic transcription elements), located in close proximity of the *Tsix* gene, function as enhancers of *Tsix* (Ogawa and Lee, 2003). In very early embryos, *Xist* is repressed on both X chromosomes by a number of factors that include *Tsix* RNA and the major pluripotency factors (see Chapter 17). In methatheria, which include marsupials, dosage compensation is also achieved via X

inactivation, although different molecules and factors are involved (Gendrel and Heard, 2014).

X chromosome inactivation occurs at the onset of cellular differentiation; it is random with an equal probability of inactivation of either of the two chromosomes, and clonally inherited—once the decision to inactivate either the maternal or paternal chromosome is made in a particular cell, the same chromosome will be inactivated in all of its descendants. Exceptions to the random nature of X inactivation are found in marsupials where the paternal X chromosome is always inactivated (Sharman, 1971), and in the extra-embryonic tissues of some eutherian mammals (Takagi and Sasaki, 1975). These instances are clearly cases of parent of origin imprinting (discussed in Chapter 8). In eutherian mammals, deviation from random X inactivation can occur as well and have been ascribed to the presence of alleles at the XIC locus that either enhance the transcription of the *Xist* or lower the level of transcription of the *Tsix* gene (Lee and Lu, 1999; Nesterova et al., 2003).

Not all genes on the X chromosome are inactivated. In humans, for example, around 15% of X-linked genes are expressed from both X chromosomes in females (Balaton et al., 2015). Some of these are active in some tissues, but not in others. Among the genes that escape inactivation are those that are present in the **pseudoautosomal region**. This region has avoided degeneration and heterochromatization during the evolution of the Y chromosome; therefore, the active genes that it includes are present in two active doses in males. Dosage compensation of these genes requires that both doses remain active in females.

Clearly, a mechanism exists that allows an early embryo to count the number of X chromosomes in its genome and ensures that females inactivate one and only one of their two X chromosomes and that the single X chromosome in male embryos remains active. This surveillance mechanism is extended to aneuploid embryos with supernumerary X chromosomes in their genome (Box 9.1). In these individuals, a single X chromosome remains active, while all of the others are inactivated. A number of hypotheses have been advanced to explain the molecular basis of the counting process, and only recently, the existence of a regulatory mechanism has been reported in humans (Migeon et al., 2017). Through studying partial autosomal aneuploids (trisomics) in fetuses that survive gestation, the dosage of a region on chromosome 19 was discovered to regulate X inactivation. Importantly, the copy number of the region determined that a single X chromosome remains active in diploid females and

Box 9.1 Human aneuploidy

Deviations from the precise diploid constitution of the genome (22 pairs of autosomes plus one pair of sex chromosomes), whether in the form of loss or duplication of genetic material, are almost invariably discernible at the molecular level and may lead to phenotypic alterations. Depending on the particular transcription units involved, if the extent of the duplicated or deleted material is limited, and if an individual is heterozygous for the deletion, the consequences may be negligible. In some cases, though, even small differences may have a very significant impact on development and function, for example the effect of deletions of imprinted regions (see Chapter 8, Box 8.2). As expected, the presence of an extra chromosome (**trisomy**) or the absence of a chromosome (**monosomy**) lead to severe clinical disorders. Similar disorders can also be engendered by the absence or addition of a substantial portion of a chromosome.

Different events can cause variation in the chromosomal constitution (karyotype) of an individual. Chromosome breakage, usually induced by external factors, followed by the re-associations of the broken fragments, can lead to the loss of some genetic material. Slippage during recombination and the particular distribution of the components of reciprocal translocations during meiosis can generate both duplications and deletions. But the most common causes of aneuploidy are errors in the separation (non-disjunction) of homologous chromosomes or of sister chromatids during meiosis (Fig. 9B.1), or the failure of replicated chromosomes to separate during cell division in the early embryo. This last occurrence would give rise to individuals who are mosaic for normal (diploid) and aneuploidy cells.

Autosomal aneuploidy for chromosome 21 of the human karyotype was first reported as the cause of Down syndrome (Jacobs et al., 1959; Lejeune et al., 1959). Affected individuals carry an additional copy of the whole chromosome or, in some cases, have inherited translocation components that include a large segment of the chromosome. For this reason, the syndrome is also referred to as "**trisomy 21**." During the past decades, numerous cases of full or partial trisomy and monosomy for different autosomes have been reported (Kunze, 1980; Witters et al., 2011).

Sex chromosome aneuploidy consists of an abnormal number of X or Y chromosome in the karyotype. The most familiar of these aneuploidies are the presence of two X chromosomes and a Y chromosome (47,XXY, Klinefelter syndrome) (Ford et al., 1959a), the presence of a single X (45,X, Turner syndrome) (Ford et al., 1959b), the presence of one X and two Y chromosomes (47,XYY syndrome) (Jacobs et al., 1965) and the presence of three X chromosomes (47,XXX) (Jacobs et al., 1959). Other sex chromosome aneuploidies that involve the occurrence of more than three sex chromosomes in the karyotype do occur but are very rare and have not been sufficiently studied. Almost invariably, individuals

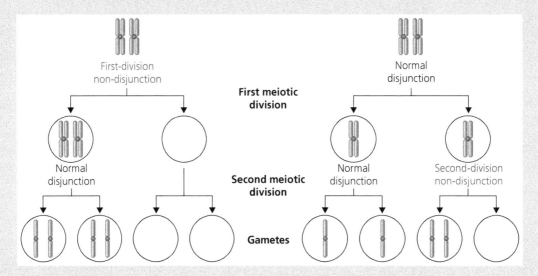

Fig. 9B.1 Non-disjunction affecting one chromosome pair in the human genome. The example could represent non-disjunction of the X chromosomes in a human female.

(From *Essentials of Genetics*, Sixth Edition)

affected by the more common syndromes live to adulthood. This is due to the counting mechanism that ensures that only one X chromosome in the karyotype remains active while all the others are inactivated.

Klinefelter syndrome (47,XXY) individuals are males whose development and physical adult characteristics are so similar to those of normal males that an accurate diagnosis is often delayed. A defining symptom of the syndrome is the presence of an inactive X chromosome, identifiable as a cytologically visible Barr body, similar to the one seen in the somatic cell nuclei of normal females. Nevertheless, the presence of genes that normally escape X chromosome inactivation in females, added to the activity of genes in the pseudoautosomal region that normally occurs in males, is probably responsible for the phenotypic differences (neurological and cognitive) between Klinefelter syndrome and normal males. There are a number of reported variants of the Klinefelter type of aneuploidy: individuals with the karyotypes 48,XXYY, 48,XXXY and 49,XXXXY. In these much rarer cases, although all X chromosomes but one are inactivated, the residual genetic imbalance that gives rise to the abnormal Klinefelter syndrome phenotype is exacerbated and leads to increased levels of abnormality.

Turner syndrome (45,X) is the only chromosomal monosomy that is viable in humans. Individuals with this syndrome lack an entire X chromosome or are heterozygous for deletions of the long or short arm of the X chromosome. Due to the absence of a Y chromosome, Turner syndrome individuals develop as females. The genes in the pseudoautosomal region that are normally present in two expressed doses in normal females are present in a single dose in Turner syndrome females, most probably leading to the developmental abnormalities characteristic of the syndrome: short stature, physiological abnormalities and webbing of the neck.

47,XYY syndrome individuals represent the most common form of sex chromosome aneuploidy in males. The physical characteristics of these individuals—increased average stature, some verbal impairments and cognitive problems—are difficult to distinguish from those of the normal diploid male population. This consideration may explain that fewer studies have been carried out on this syndrome than on the Klinefelter and Turner syndromes.

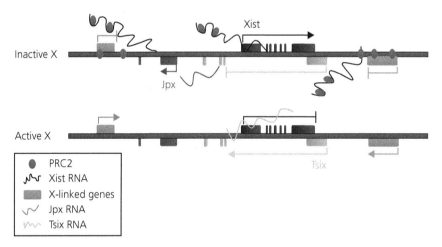

Fig. 9.6 Diagram representing the transcription events responsible for inactivating one of the two X chromosomes of females and maintaining the other one in the active state. The transcription and spreading of the Xist RNA prevents the activation of most genes along the X chromosome. On the other X chromosome, transcription of the *Tsix* gene interferes with *Xist* function and prevents inactivation.

(Graphical abstract from Froberg et al., 2013.)

two X chromosomes are active in triploid females (XXX; AAA where A represents one set of autosomes) or triploid males (XXY; AAA). Presumably, the dosage of one or more genes in this region inhibits directly or indirectly the synthesis of Xist RNA.

X chromosome inactivation involves the interaction of different non-coding regulatory RNAs, with the ultimate purpose of regulating the transcription of the *Xist* and *Tsix* genes (Lee and Lu, 1999) (Fig. 9.6). The Xist RNA spreads in *cis*, first to various entry points brought close to the XIC by the folding of the X chromosome in interphase nuclei; this is followed by a general association to gene-rich, and then to gene-poor, regions (Engreitz et al. 2013; Simon et al., 2013). The early entry sites do not seem to share a common sequence that would attract Xist RNA; although the mechanism that achieves subsequent coating of the entire X chromosome is still not fully understood, repetitive elements that are enriched on this chromosome (such as the long interspersed element Line 1 or L1 predominantly found in the proximity of inactivated genes) have been thought to facilitate Xist RNA spreading (Cotton et al., 2014; Bala Tannan et al., 2014). Other studies have demonstrated a negative correlation between the target sites of Xist RNA and Line 1 elements (Engreitz et al., 2013; Simon et al., 2013).

A number of histone modifications, as well as DNA methylation, are correlated with X inactivation (Csankovszki et al., 2001; Gendrel and Heard, 2014). Following the spread of Xist RNA, a number of interacting proteins associate with the RNA, including histone deacetylases and the Polycomb repressive complexes (Chu et al., 2015; McHugh et al., 2015). Recent observations in mouse embryonic stem cells have uncovered an unusual pattern of recruitment of the latter complexes. A non-canonical PRC1 complex associates with Xist RNA and ubiquitinates histone H2A (H2AK119ub); this modification mediates the binding of PRC2 that is responsible for the presence H3K27me3 along the inactive X chromosome (Almeida et al., 2017). Loading of PRC2 on the RNA appears to involve the ATRX (alpha thalassemia/mental retardation X-linked), a protein that has an ATPase and a helicase domain (Sarma et al., 2014), and the co-factor JARID2 (da Rocha et al., 2014).

The histone variant macroH2A is also incorporated into the Xi chromosome (Costanzi and Pehrson, 1998). Histone H4 acetylation of lysines 5, 12, 8 and 16 is substantially abrogated (Jeppesen and Turner, 1993), and the Xi exhibits other epigenetic marks associated with gene silencing: reduced H3K4me and H3K9ac with a corresponding increase of H3K9me3; a high level of DNA methylation in the CpG islands of silenced genes (Norris et al., 1991); and a high level of H4K20me1. Xist RNA is edited and is marked by the presence of N6-methyladenosine (Patil et al., 2016).

The nuclear localization of the inactive X chromosome is mediated by different factors. Among the many proteins that interact with Xist is the lamin B receptor (LBR), suggesting that it plays a role in the frequent presence of Xi at the nuclear periphery (Chen et al., 2016). Occasionally, Xi is found at the periphery of the nucleolus. A lncRNA (Firre), encoded by a gene on the X chromosome that escapes inactivation, is responsible for this association (Yang et al., 2015; see Chapter 6).

A reversal of X chromosome inactivation takes place in primordial germ cells. This event is necessary to ensure the transmission of an active X chromosome to male progeny and the random inactivation of the maternal or paternal X chromosomes in female progeny. The molecular signals that initiate and result in reactivation are not fully understood. Xist RNA transcription is abrogated, but that may not be the key factor since the presence of an active *Xist* gene is not required for the maintenance of an inactive X chromosome in somatic cells (Brown and Willard, 1994). A particular feature of the inactive X chromosome during the early development of primordial germ cells is the absence of methylated CpG islands (Grant et al., 1992). This distinction between germ and somatic cells may trigger and facilitate the reversal of the chromatin modifications associated with inactivation.

Dosage compensation in birds

The avian W chromosome is a degenerated form of the Z. Like all other heteromorphic sex chromosomes, the Z and W are thought to have evolved from a pair of autosomes that had acquired a gene with a sex-specific determining allele. Yet, having arisen from a different autosomal pair than the X and Y of eutherian or metatherian mammals, the two sets of sex chromosomes are not homologous (Graves, 2014). The W chromosome has lost most of the genes that are present on the Z, leading to the presumption that a robust system of dosage compensation must have evolved to remediate the difference in gene product levels between males and females. This is in fact not the case, since only a few Z-linked genes are fully compensated while a very large number of genes are over-expressed to a variable extent in males (Itoh et al., 2007; Zhang et al., 2010).

In the chicken, a number of the more dosage-compensated genes are clustered in a small region of the Z chromosome around a gene called *MHM* (male hypermethylated region). This gene is hypermethylated and silent in males; it expresses a non-coding RNA in females where it is flanked by regions of open chromatin, rich in H4K16ac (Bisoni et al., 2005; Teranishi et al., 2001). These features suggest that *MHM* is a regulator for the dosage compensation of neighboring genes.

Gene duplication is a corollary to dosage compensation

In *Drosophila* females, both X chromosomes are active, ensuring that the level of products generated by X-linked and autosomal genes is generally concordant.

Drosophila males reach the same level of X-linked gene products and the same general concordance with autosomal genes by doubling, on average the transcriptional level of their single X chromosome. In contrast, mammals and *Caenorhabditis* achieve dosage compensation by inactivating an entire X chromosome in females or by reducing the expression of both X chromosomes in hermaphrodites. In these organisms, males with their single X chromosome should have a deficiency in the level of X-linked gene products in relation to autosomal gene products; the same would be true in females and hermaphrodites where dosage compensation reduces the level of X-linked gene products to that of the male. This realization led to the early hypothesis that the single X chromosome in mammalian males must have doubled its transcriptional level during the evolution of the sex chromosomes, bringing its output to the same level as that of autosomal genes. The presence of two such hyperactive X chromosomes in females, though, would lead to an unacceptable overexpression of X-linked genes and necessitated the concomitant development of a dosage compensation mechanism (Ohno, 1967). This is, in fact, the case in both *Caenorhabditis* and mammals (Gupta et al., 2006; Nguyen and Disteche, 2006). In mammalian cells, X chromosome up-regulation involves an increased level of MOF-mediated H4K16 acetylation at X-linked gene promoters in comparison to a comparable set of autosomal genes; it also involves a greater frequency of paused RNAPII and a greater stability of transcripts, although no evidence of differences at the 5′ or 3′ ends of these RNA molecules has been obtained (Deng et al., 2013).

Chapter summary

In eukaryotic genomes, there exist clusters of genes that encode similar, but not identical, products and that are regulated in a coordinated fashion. Examples of such gene clusters are the β-globin genes that include five genes that are sequentially activated during development by their individual association with a single regulatory region (the locus control region or LCR). In other instances, clustering provides access to a localized concentration of specific transcription factors. Such is likely the case for the genes that encode the ribosomal RNAs that are activated by a promoter region present at each gene copy, or the core histone genes that include one individual and two divergent promoters per set of five genes.

An extreme case of coordinated regulation is the modulation or abrogation of most or all of the genes present on one chromosome. This type of regulation is best exemplified by dosage compensation, a regulatory system designed to equalize the transcription output of the genes present on the sex chromosome that is common to both sexes (X or Z chromosome, depending on the animal species involved). Different mechanisms of dosage compensation have evolved in different animal groups. In *Drosophila* males, a complex (MSL) associates with the X chromosome where it carries out the global acetylation of H4K16 and the recruitment of topoisomerase II and of the elongation factor SPT5; these modifications enhance the activity of most X-linked genes by increasing the rate of RNAPII elongation. In *Caenorhabditis*, a complex (DCC) that contains a number of proteins involved in condensing chromosomes decreases the level of transcription of both X chromosomes in the XX hermaphrodite. In mammals, dosage compensation is achieved by the inactivation, early during development, of most X-linked genes on one of the two X chromosomes in females. While in marsupials and in the extraembryonic membranes of some eutherian mammals, the paternal X chromosome is inactivated, in the mammalian embryo proper X inactivation of either X chromosome is random and clonally inherited. The mechanism involves the synthesis of an RNA (Tsix) that protects one of the two Xs from inactivation, and of another RNA (Xist) that coats the other X chromosome and recruits histone- and DNA-modifying enzymes.

References

Albig, W. & Doenecke, D. 1997. The human histone gene cluster at the D6S105 locus. *Hum Genet*, 101, 284–94.

Albritton, S. E., Kranz, A. L., Winterkorn, L. H., Street, L. A. & Ercan, S. 2017. Cooperation between a hierarchical set of recruitment sites targets the X chromosome for dosage compensation. *Elife*, 6, pii: e23645.

Alekseyenko, A. A., Peng, S., Larschan, E., Gorchakov, A. A., Lee, O. K., Kharchenko, P., Mcgrath, S. D., Wang, C. I., Mardis, E. R., Park, P. J. & Kuroda, M. I. 2008. A sequence motif within chromatin entry sites directs MSL establishment on the *Drosophila* X chromosome. *Cell*, 134, 599–609.

Almeida, M., Pintacuda, G., Masui, O., Koseki, Y., Gdula, M., Cerase, A., Brown, D., Mould, A., Innocent, C., Nakayama, M., Schermelleh, L., Nesterova, T. B., Koseki, H. & Brockdorff, N. 2017. PCGF3/5-PRC1 initiates Polycomb recruitment in X chromosome inactivation. *Science,* 356, 1081–4.

Bala Tannan, N., Brahmachary, M., Garg, P., Borel, C., Alnefaie, R., Watson, C. T., Thomas, N. S. & Sharp, A. J. 2014. DNA methylation profiling in X;autosome translocations supports a role for L1 repeats in the spread of X chromosome inactivation. *Hum Mol Genet,* 23, 1224–36.

Balaton, B. P., Cotton, A. M. & Brown, C. J. 2015. Derivation of consensus inactivation status for X-linked genes from genome-wide studies. *Biol Sex Differ,* 6, 35.

Barr, M. L. & Bertram, E. G. 1949. A morphological distinction between neurones of the male and female, and the behaviour of the nucleolar satellite during accelerated nucleoprotein synthesis. *Nature,* 163, 676.

Beck, D. B., Oda, H., Shen, S. S. & Reinberg, D. 2012. PR-Set7 and H4K20me1: at the crossroads of genome integrity, cell cycle, chromosome condensation, and transcription. *Genes Dev,* 26, 325–37.

Bell, O., Schwaiger, M., Oakeley, E. J., Lienert, F., Beisel, C., Stadler, M. B. & Schubeler, D. 2010. Accessibility of the *Drosophila* genome discriminates PcG repression, H4K16 acetylation and replication timing. *Nat Struct Mol Biol,* 17, 894–900.

Belote, J. M. & Lucchesi, J. C. 1980. Control of X chromosome transcription by the maleless gene in *Drosophila. Nature,* 285, 573–5.

Bian, Q., Anderson, E. C., Brejc, K. & Meyer, B. J. 2018. Dynamic control of chromosome topology and gene expression by a chromatin modification. *Cold Spring Harb Symp Quant Biol.* doi: 10.1101/sqb.2017.82.034439.

Birch, J. L., Tan, B. C., Panov, K. I., Panova, T. B., Andersen, J. S., OWEN-HUGHES, T. A., RUSSELL J., LEE, S. C. & ZOMERDIJK, J. C. 2009. FACT facilitates chromatin transcription by RNA polymerases I and III. *EMBO J,* 28, 854–65.

Bisoni, L., Batlle-Morera, L., Bird, A. P., Suzuki, M. & Mcqueen, H. A. 2005. Female-specific hyperacetylation of histone H4 in the chicken Z chromosome. *Chromosome Res,* 13, 205–14.

Brejc, K., Bian, Q., Uzawa, S., Wheeler, B. S., Anderson, E. C., King, D. S., Kranzusch, P. J., Preston, C. G. & Meyer, B. J. 2017. Dynamic control of X chromosome conformation and repression by a histone H4K20 demethylase. *Cell,* 171, 85–102 e23.

Brown, C. J., Ballabio, A., Rupert, J. L., Lafreniere, R. G., Grompe, M., Tonlorenzi, R. & Willard, H. F. 1991. A gene from the region of the human X inactivation centre is expressed exclusively from the inactive X chromosome. *Nature,* 349, 38–44.

Brown, C. J. & Willard, H. F. 1994. The human X-inactivation centre is not required for maintenance of X-chromosome inactivation. *Nature,* 368, 154–6.

Chen, C. K., Blanco, M., Jackson, C., Aznauryan, E., Ollikainen, N., Surka, C., Chow, A., Cerase, A., Mcdonel, P. & Guttman, M. 2016. Xist recruits the X chromosome to the nuclear lamina to enable chromosome-wide silencing. *Science,* 354, 468–72.

Chu, C., Zhang, Q. C., Da Rocha, S. T., Flynn, R. A., Bharadwaj, M., Calabrese, J. M., Magnuson, T., Heard, E.

& Chang, H. Y. 2015. Systematic discovery of Xist RNA binding proteins. *Cell,* 161, 404–16.

Clark, S. J., Krieg, P. A. & Wells, J. R. 1981. Isolation of a clone containing human histone genes. *Nucleic Acids Res,* 9, 1583–90.

Conconi, A., Widmer, R. M., Koller, T. & Sogo, J. M. 1989. Two different chromatin structures coexist in ribosomal RNA genes throughout the cell cycle. *Cell,* 57, 753–61.

Costanzi, C. & Pehrson, J. R. 1998. Histone macroH2A1 is concentrated in the inactive X chromosome of female mammals. *Nature,* 393, 599–601.

Cotton, A. M., Chen, C. Y., Lam, L. L., Wasserman, W. W., Kobor, M. S. & Brown, C. J. 2014. Spread of X-chromosome inactivation into autosomal sequences: role for DNA elements, chromatin features and chromosomal domains. *Hum Mol Genet,* 23, 1211–23.

Crane, E., Bian, Q., Mccord, R. P., Lajoie, B. R., Wheeler, B. S., Ralston, E. J., Uzawa, S., Dekker, J. & Meyer, B. J. 2015. Condensin-driven remodelling of X chromosome topology during dosage compensation. *Nature,* 523, 240–4.

Csankovszki, G., Mcdonel, P. & Meyer, B. J. 2004. Recruitment and spreading of the *C. elegans* dosage compensation complex along X chromosomes. *Science,* 303, 1182–5.

Csankovszki, G., Nagy, A. & Jaenisch, R. 2001. Synergism of Xist RNA, DNA methylation, and histone hypoacetylation in maintaining X chromosome inactivation. *J Cell Biol,* 153, 773–84.

Cugusi, S., Ramos, E., Ling, H., Yokoyama, R., Luk, K. M. & Lucchesi, J. C. 2013. Topoisomerase II plays a role in dosage compensation in *Drosophila. Transcription,* 4, 238–50.

Curtis, P. J. 1980. Globin mRNA in Friend cells: its structure, function and synthesis. *Biochim Biophys Acta,* 605, 347–64.

Da Rocha, S. T., Boeva, V., Escamilla-Del-Arenal, M., Ancelin, K., Granier, C., Matias, N. R., Sanulli, S., Chow, J., Schulz, E., Picard, C., Kaneko, S., Helin, K., Reinberg, D., Stewart, A. F., Wutz, A., Margueron, R. & Heard, E. 2014. Jarid2 is implicated in the initial Xist-induced targeting of PRC2 to the inactive X chromosome. *Mol Cell,* 53, 301–16.

Davidson, R. G., Nitowsky, H. M. & Childs, B. 1963. Demonstration of two populations of cells in the human female heterozygous for glucose-6-phosphate dehydrogenase variants. *Proc Natl Acad Sci U S A,* 50, 481–5.

Dawes, H. E., Berlin, D. S., Lapidus, D. M., Nusbaum, C., Davis, T. L. & Meyer, B. J. 1999. Dosage compensation proteins targeted to X chromosomes by a determinant of hermaphrodite fate. *Science,* 284, 1800–4.

Deng, X., Berletch, J. B., Ma, W., Nguyen, D. K., Hiatt, J. B., Noble, W. S., Shendure, J. & Disteche, C. M. 2013. Mammalian X upregulation is associated with enhanced transcription initiation, RNA half-life, and MOF-mediated H4K16 acetylation. *Dev Cell,* 25, 55–68.

Dibartolomeis, S. M., Tartof, K. D. & Jackson, F. R. 1992. A superfamily of *Drosophila* satellite related (SR) DNA repeats restricted to the X chromosome euchromatin. *Nucleic Acids Res,* 20, 1113–16.

Dunlap, D., Yokoyama, R., Ling, H., Sun, H. Y., Mcgill, K., Cugusi, S. & Lucchesi, J. C. 2012. Distinct contributions of MSL complex subunits to the transcriptional enhancement responsible for dosage compensation in *Drosophila. Nucleic Acids Res,* 40, 11281–91.

Engreitz, J. M., Pandya-Jones, A., Mcdonel, P., Shishkin, A., Sirokman, K., Surka, C., Kadri, S., Xing, J., Goren, A., Lander, E. S., Plath, K. & Guttman, M. 2013. The Xist lncRNA exploits three-dimensional genome architecture to spread across the X chromosome. *Science,* 341, 1237973.

Ercan, S. 2015. Mechanisms of x chromosome dosage compensation. *J Genomics,* 3, 1–19.

Ercan, S., Giresi, P. G., Whittle, C. M., Zhang, X., Green, R. D. & Lieb, J. D. 2007. X chromosome repression by localization of the *C. elegans* dosage compensation machinery to sites of transcription initiation. *Nat Genet,* 39, 403–8.

Evans, T., Reitman, M. & Felsenfeld, G. 1988. An erythrocyte-specific DNA-binding factor recognizes a regulatory sequence common to all chicken globin genes. *Proc Natl Acad Sci U S A,* 85, 5976–80.

Ferrari, F., Plachetka, A., Alekseyenko, A. A., Jung, Y. L., Ozsolak, F., Kharchenko, P. V., Park, P. J. & Kuroda, M. I. 2013. "Jump start and gain" model for dosage compensation in *Drosophila* based on direct sequencing of nascent transcripts. *Cell Rep,* 5, 629–36.

Ford, C. E., Jones, K. W., Miller, O. J., Mittwoch, U., Penrose, L. S., Ridler, M. & Shapiro, A. 1959a. The chromosomes in a patient showing both mongolism and the Klinefelter syndrome. *Lancet,* 1, 709–10.

Ford, C. E., Jones, K. W., Polani, P. E., De Almeida, J. C. & Briggs, J. H. 1959b. A sex-chromosome anomaly in a case of gonadal dysgenesis (Turner's syndrome). *Lancet,* 1, 711–13.

Forrester, W. C., Epner, E., Driscoll, M. C., Enver, T., Brice, M., Papayannopoulou, T. & Groudine, M. 1990. A deletion of the human beta-globin locus activation region causes a major alteration in chromatin structure and replication across the entire beta-globin locus. *Genes Dev,* 4, 1637–49.

Froberg, J. E., Yang, L. & Lee, J. T. 2013. Guided by RNAs: X-inactivation as a model for lncRNA function. *J Mol Biol,* 425, 3698–706.

Gendrel, A. V. & Heard, E. 2014. Noncoding RNAs and epigenetic mechanisms during X-chromosome inactivation. *Annu Rev Cell Dev Biol,* 30, 561–80.

Ginder, G. D. 2015. Epigenetic regulation of fetal globin gene expression in adult erythroid cells. *Transl Res,* 165, 115–25.

Grant, M., Zuccotti, M. & Monk, M. 1992. Methylation of CpG sites of two X-linked genes coincides with X-inactivation in the female mouse embryo but not in the germ line. *Nat Genet,* 2, 161–6.

Graves, J. A. 2014. The epigenetic sole of sex and dosage compensation. *Nat Genet,* 46, 215–17.

Grierson, P. M., Acharya, S. & Groden, J. 2013. Collaborating functions of BLM and DNA topoisomerase I in regulating human rDNA transcription. 2013. *Mutat Res,* 743–4, 89–96.

Grimaud, C. & Becker, P. B. 2009. The dosage compensation complex shapes the conformation of the X chromosome in *Drosophila. Genes Dev,* 23, 2490–5.

Grummt, I., Maier, U., Ohrlein, A., Hassouna, N. & Bacellerie, J. P. 1985. Transcription of mouse rDNA terminates downstream of the 3′ end of 28S RNA and involves interaction of factors with repeated sequences in the 3′ spacer. *Cell,* 43, 801–10.

Guetg, C., Scheifele, F., Rosenthal, F., Hottiger, M. O. & Santoro, R. 2012. Inheritance of silent rDNA chromatin is mediated by PARP1 via noncoding RNA. *Mol Cell,* 45, 790–800.

Guglielmi, B., La Rochelle, N. & Tjian, R. 2013. Gene-specific transcriptional mechanisms at the histone gene cluster revealed by single-cell imaging. *Mol Cell,* 51, 480–92.

Gupta, V., Parisi, M., Sturgill, D., Nuttall, R., Doctolero, M., Dudko, O. K., Malley, J. D., Eastman, P. S. & Oliver, B. 2006. Global analysis of X-chromosome dosage compensation. *J Biol,* 5, 3.

Heix, J., Zomerdijk, J. C., Ravanpay, A., Tjian, R. & Grummt, I. 1997. Cloning of murine RNA polymerase I-specific TAF factors: conserved interactions between the subunits of the species-specific transcription factor TIF-IB/SL1. *Proc Natl Acad Sci U S A,* 94, 1733–8.

Hilfiker, A., Hilfiker-Kleiner, D., Pannuti, A. & Lucchesi, J. C. 1997. *mof,* a putative acetyl transferase gene related to the *Tip60* and *MOZ* human genes and to the *SAS* genes of yeast, is required for dosage compensation in *Drosophila. EMBO J,* 16, 2054–60.

Hodgkin, J. 1983. X chromosome dosage gene expression in *C. elegans*: two unusual dumpy genes. *Mol Gen Genet,* 192, 452–8.

Ilik, I. A., Quinn, J. J., Georgiev, P., Tavares-Cadete, F., Maticzka, D., Toscano, S., Wan, Y., Spitale, R. C., Luscombe, N., Backofen, R., Chang, H. Y. & Akhtar, A. 2013. Tandem stem-loops in roX RNAs act together to mediate X chromosome dosage compensation in *Drosophila. Mol Cell,* 51, 156–73.

Isogai, Y., Keles, S., Prestel, M., Hochheimer, A. & Tjian, R. 2007. Transcription of histone gene cluster by differential core-promoter factors. *Genes Dev,* 21, 2936–49.

Itoh, Y., Melamed, E., Yang, X., Kampf, K., Wang, S., Yehya, N., Van Nas, A., Replogle, K., Band, M. R., Clayton, D. F., Schadt, E. E., Lusis, A. J. & Arnold, A. P. 2007. Dosage compensation is less effective in birds than in mammals. *J Biol,* 6, 2.

Jacobs, P. A., Baikie, A. G., Brown, W. M., Macgregor, T. N., Maclean, N. & Harnden, D. G. 1959. Evidence for the existence of the human "super female." *Lancet,* 2, 423–5.

Jacobs, P. A., Brunton, M., Melville, M. M., Brittain, R. P. & Mcclemont, W. F. 1965. Aggressive behavior, mental subnormality and the XYY male. *Nature*, 208, 1351–2.

Jakubczak, J. L., Burke, W. D. & Eickbush, T. H. 1991. Retrotransposable elements R1 and R2 interrupt the rRNA genes of most insects. *Proc Natl Acad Sci U S A*, 88, 3295–9.

Jakubczak, J. L., Xiong, Y. & Eickbush, T. H. 1990. Type I (R1) and type II (R2) ribosomal DNA insertions of *Drosophila melanogaster* are retrotransposable elements closely related to those of *Bombyx mori*. *J Mol Biol*, 212, 37–52.

Jeppesen, P. & Turner, B. M. 1993. The inactive X chromosome in female mammals is distinguished by a lack of histone H4 acetylation, a cytogenetic marker for gene expression. *Cell*, 74, 281–9.

Jonkers, I., Barakat, T. S., Achame, E. M., Monkhorst, K., Kenter, A., Rentmeester, E., Grosveld, F., Grootegoed, J. A. & Gribnau, J. 2009. RNF12 is an X-encoded dose-dependent activator of X chromosome inactivation. *Cell*, 139, 999–1011.

Joshi, S. S. & Meller, V. H. 2017. Satellite repeats identify X chromatin for dosage compensation in *Drosophila melanogaster* males. *Curr Biol*, 27, 1393–1402 e2.

Katsumura, K. R., Devilbiss, A. W., Pope, N. J., Johnson, K. D. & Bresnick, E. H. 2013. Transcriptional mechanisms underlying hemoglobin synthesis. *Cold Spring Harb Perspect Med*, 3, a015412.

Kaye, E. G., Kurbidaeva, A., Wolle, D., Aoki, T., Schedl, P. & Larschan, E. 2017. *Drosophila* dosage compensation loci associate with a boundary-forming insulator complex. *Mol Cell Biol*, 37, pii: e00253-17.

Ko, L. J. & Engel, J. D. 1993. DNA-binding specificities of the GATA transcription factor family. *Mol Cell Biol*, 13, 4011–22.

Kunze, J. 1980. Neurological disorders in patients with chromosomal anomalies. *Neuropediatrics*, 11, 203–49.

Lapeyre, B., Bourbon, H. & Amalric, F. 1987. Nucleolin, the major nucleolar protein of growing eukaryotic celle: an unusual protein structure revealed by the nucleotide sequence. *Proc Natl Acad Sci U S A*, 84, 1472–6.

Larschan, E., Alekseyenko, A. A., Gortchakov, A. A., Peng, S., Li, B., Yang, P., Workman, J. L., Park, P. J. & Kuroda, M. I. 2007. MSL complex is attracted to genes marked by H3K36 trimethylation using a sequence-independent mechanism. *Mol Cell*, 28, 121–33.

Larschan, E., Bishop, E. P., Kharchenko, P. V., Core, L. J., Lis, J. T., Park, P. J. & Kuroda, M. I. 2011. X chromosome dosage compensation via enhanced transcriptional elongation in *Drosophila*. *Nature*, 471, 115–18.

Larschan, E., Soruco, M. M., Lee, O. K., Peng, S., Bishop, E., Chery, J., Goebel, K., Feng, J., Park, P. J. & Kuroda, M. I. 2012. Identification of chromatin-associated regulators of MSL complex targeting in *Drosophila* dosage compensation. *PLoS Genet*, 8, e1002830.

Lau, A. C., Nabeshima, K. & Csankovszki, G. 2014. The *C. elegans* dosage compensation complex mediates interphase X chromosome compaction. *Epigenetics Chromatin*, 7, 31.

Learned, R. M., Cordes, S. & Tjian, R. 1985. Purification and characterization of a transcription factor that confers promoter specificity to human RNA polymerase I. *Mol Cell Biol*, 5, 1358–69.

Lee, C. G., Chang, K. A., Kuroda, M. I. & Hurwitz, J. 1997. The NTPase/helicase activities of *Drosophila* maleless, an essential factor in dosage compensation. *EMBO J*, 16, 2671–81.

Lee, H. Y., Johnson, K. D., Boyer, M. E. & Bresnick, E. H. 2011. Relocalizing genetic loci into specific subnuclear neighborhoods. *J Biol Chem*, 286, 18834–44.

Lee, J. T. & Lu, N. 1999. Targeted mutagenesis of *Tsix* leads to nonrandom X inactivation. *Cell*, 99, 47–57.

Lejeune, J., Turpin, R. & Gautier, M. 1959. Mongolism: a chromosomal disease (trisomy). *Bull Acad Natl Med*, 143, 256–65.

Li, F., Schiemann, A. H. & Scott, M. J. 2008. Incorporation of the noncoding roX RNAs alters the chromatin-binding specificity of the *Drosophila* MSL1/MSL2 complex. *Mol Cell Biol*, 28, 1252–64.

Lowenhaupt, K., Rich, A. & Pardue, M. L. 1989. Nonrandom distribution of long mono- and dinucleotide repeats in *Drosophila* chromosomes: correlations with dosage compensation, heterochromatin, and recombination. *Mol Cell Biol*, 9, 1173–82.

Lucchesi, J. C. 1978. Gene dosage compensation and the evolution of sex chromosomes. *Science*, 202, 711–16.

Lucchesi, J. C. & Kuroda, M. I. 2015. Dosage compensation in *Drosophila*. *Cold Spring Harb Perspect Biol*, 7, pii: a019398.

Lyon, M. F. 1961. Gene action in the X-chromosome of the mouse (*Mus musculus L.*). *Nature*, 190, 372–3.

Maenner, S., Muller, M., Frohlich, J., Langer, D. & Becker, P. B. 2013. ATP-dependent roX RNA remodeling by the helicase maleless enables specific association of MSL proteins. *Mol Cell*, 51, 174–84.

Marzluff, W. F., Gongidi, P., Woods, K. R., Jin, J. & Maltais, L. J. 2002. The human and mouse replication-dependent histone genes. *Genomics*, 80, 487–98.

Mcdonel, P., Jans, J., Peterson, B. K. & Meyer, B. J. 2006. Clustered DNA motifs mark X chromosomes for repression by a dosage compensation complex. *Nature*, 444, 614–18.

Mcghee, J. D., Rau, D. C. & Felsenfeld, G. 1983. Chromatin structure of the chicken adult beta-globin gene: is gene activity associated with local or long-range perturbations in chromatin structure? *Prog Clin Biol Res*, 134, 143–57.

Mchugh, C. A., Chen, C. K., Chow, A., Surka, C. F., Tran, C., Mcdonel, P., Pandya-Jones, A., Blanco, M., Burghard, C., Moradian, A., Sweredoski, M. J., Shishkin, A. A., Su, J., Lander, E. S., Hess, S., Plath, K. & Guttman, M. 2015. The Xist lncRNA interacts directly with SHARP to silence transcription through HDAC3. *Nature*, 521, 232–6.

Meneely, P. M. & Wood, W. B. 1984. An autosomal gene that affects X chromosome expression and sex determination in *Caenorhabditis elegans. Genetics,* 106, 29–44.

Menon, D. U., Coarfa, C., Xiao, W., Gunaratne, P. H. & Meller, V. H. 2014. siRNAs from an X-linked satellite repeat promote X-chromosome recognition in *Drosophila melanogaster. Proc Natl Acad Sci U S A,* 111, 16460–5.

Menon, D. U. & Meller, V. H. 2012. A role for siRNA in X-chromosome dosage compensation in *Drosophila melanogaster. Genetics,* 191, 1023–8.

Meyer, B. J. & Casson, L. P. 1986. *Caenorhabditis elegans* compensates for the difference in X chromosome dosage between the sexes by regulating transcript levels. *Cell,* 47, 871–81.

Migeon, B. R., Beer, M. A. & Bjornsson, H. T. 2017. Embryonic loss of human females with partial trisomy 19 identifies region critical for the single active X. *PLoS One,* 12, e0170403.

Mukherjee, A. S. & Beermann, W. 1965. Synthesis of ribonucleic acid by the X-chromosomes of *Drosophila melanogaster* and the problem of dosage compensation. *Nature,* 207, 785–6.

Muller, H. J. 1932. Further studies on the nature and causes of gene mutations. *Proc 6th Int Congr Genet,* 1, 213–55.

Nemeth, A., Guibert, S., Tiwari, V. K., Ohlsson, R. & Langst, G. 2008. Epigenetic regulation of TTF-I-mediated promoter-terminator interactions of rRNA genes. *EMBO J,* 27, 1255–65.

Nesterova, T. B., Johnston, C. M., Appanah, R., Newall, A. E., Godwin, J., Alexiou, M. & Brockdorff, N. 2003. Skewing X chromosome choice by modulating sense transcription across the Xist locus. *Genes Dev,* 17, 2177–90.

Nguyen, D. K. & Disteche, C. M. 2006. Dosage compensation of the active X chromosome in mammals. *Nat Genet,* 38, 47–53.

Norris, D. P., Brockdorff, N. & Rastan, S. 1991. Methylation status of CpG-rich islands on active and inactive mouse X chromosomes. *Mamm Genome,* 1, 78–83.

Ogawa, Y. & Lee, J. T. 2003. Xite, X-inactivation intergenic transcription elements that regulate the probability of choice. *Mol Cell,* 11, 731–43.

Ohno, S. 1967. *Sex Chromosomes and Sex-linked Genes.* Berlin: Springer-Verlag.

Ohno, S., Kaplan, W. D. & Kinosita, R. 1959. Formation of the sex chromatin by a single X-chromosome in liver cells of *Rattus norvegicus. Exp Cell Res,* 18, 415–18.

Palstra, R. J., Tolhuis, B., Splinter, E., Nijmeijer, R., Grosveld, F. & De Laat, W. 2003. The beta-globin nuclear compartment in development and erythroid differentiation. *Nat Genet,* 35, 190–4.

Pardue, M. L., Lowenhaupt, K., Rich, A. & Nordheim, A. 1987. (dC-dA)n.(dG-dT)n sequences have evolutionarily conserved chromosomal locations in *Drosophila* with implications for roles in chromosome structure and function. *EMBO J,* 6, 1781–9.

Patil, D. P., Chen, C. K., Pickering, B. F., Chow, A., Jackson, C., Guttman, M. & Jaffrey, S. R. 2016. m(6)A RNA methylation promotes XIST-mediated transcriptional repression. *Nature,* 537, 369–73.

Petty, E. L., Collette, K. S., Cohen, A. J., Snyder, M. J. & Csankovszki, G. 2009. Restricting dosage compensation complex binding to the X chromosomes by H2A.Z/HTZ-1. *PLoS Genet,* 5, e1000699.

Pikaard, C. S., Mcstay, B., Schultz, M. C., Bell, S. P. & Reeder, R. H. 1989. The *Xenopus* ribosomal gene enhancers bind an essential polymerase I transcription factor, xUBF. *Genes Dev,* 3, 1779–88.

Pontvianne, F., Blevins, T., Chandrasekhara, C., Mozgova, I., Hassel, C., Pontes, O. M., Tucker, S., Mokros, P., Muchova, V., Fajkus, J. & Pikaard, C. S. 2013. Subnuclear partitioning of rRNA genes between the nucleolus and nucleoplasm reflects alternative epiallelic states. *Genes Dev,* 27, 1545–50.

Prabhakaran, M. & Kelley, R. L. 2012. Mutations in the transcription elongation factor SPT5 disrupt a reporter for dosage compensation in *Drosophila. PLoS Genet,* 8, e1003073.

Proudfoot, N. J., Shander, M. H., Manley, J. L., Gefter, M. L. & Maniatis, T. 1980. Structure and *in vitro* transcription of human globin genes. *Science,* 209, 1329–36.

Ramirez, F., Lingg, T., Toscano, S., Lam, K. C., Georgiev, P., Chung, H. R., Lajoie, B. R., De Wit, E., Zhan, Y., De Laat, W., Dekker, J., Manke, T. & Akhtar, A. 2015. High-affinity sites form an interaction network to facilitate spreading of the MSL complex across the X chromosome in *Drosophila. Mol Cell,* 60, 146–62.

Rickards, B., Flint, S. J., Cole, M. D. & Leroy, G. 2007. Nucleolin is required for RNA polymerase I transcription in vivo. *Mol Cell Biol,* 27, 937–4.

Santoro, R., Schmitz, K. M., Sandoval, J. & Grummt, I. 2010. Intergenic transcripts originating from a subclass of ribosomal DNA repeats silence ribosomal RNA genes in trans. *EMBO Rep,* 11, 52–8.

Sarma, K., Cifuentes-Rojas, C., Ergun, A., Del Rosario, A., Jeon, Y., White, F., Sadreyev, R. & Lee, J. T. 2014. ATRX directs binding of PRC2 to Xist RNA and Polycomb targets. *Cell,* 159, 869–83.

Sass, G. L., Pannuti, A. & Lucchesi, J. C. 2003. Male-specific lethal complex of *Drosophila* targets activated regions of the X chromosome for chromatin remodeling. *Proc Natl Acad Sci U S A,* 100, 8287–91.

Schauer, T., Ghavi-Helm, Y., Sexton, T., Albig, C., Regnard, C., Cavalli, G., Furlong, E. E. & Becker, P. B. 2017. Chromosome topology guides the *Drosophila* dosage compensation complex for target gene activation. *EMBO Rep.* doi: 10.15252/embr.201744292.

Scherrer, K., Latham, H. & Darnell, J. E. 1963. Demonstration of an unstable RNA and of a precursor to ribosomal RNA in HeLa cells. *Proc Natl Acad Sci U S A,* 49, 240–8.

Schmitz, K. M., Mayer, C., Postepska, A. & Grummt, I. 2010. Interaction of noncoding RNA with the rDNA promoter mediates recruitment of DNMT3b and silencing of rRNA genes. *Genes Dev,* 24, 2264–9.

Sharma, R., Jost, D., Kind, J., Gomez-Saldivar, G., Van Steensel, B., Askjaer, P., Vaillant, C. & Meister, P. 2014. Differential spatial and structural organization of the X chromosome underlies dosage compensation in *C. elegans. Genes Dev,* 28, 2591–6.

Sharman, G. B. 1971. Late DNA replication in the paternally derived X chromosome of female kangaroos. *Nature,* 230, 231–2.

Simon, M. D., Pinter, S. F., Fang, R., Sarma, K., Rutenberg-Schoenberg, M., Bowman, S. K., Kesner, B. A., Maier, V. K., Kingston, R. E. & Lee, J. T. 2013. High-resolution Xist binding maps reveal two-step spreading during X-chromosome inactivation. *Nature,* 504, 465–9.

Smith, E. R., Allis, C. D. & Lucchesi, J. C. 2001. Linking global histone acetylation to the transcription enhancement of X-chromosomal genes in *Drosophila* males. *J Biol Chem,* 276, 31483–6.

Smith, E. R., Pannuti, A., Gu, W., Steurnagel, A., Cook, R. G., Allis, C. D. & Lucchesi, J. C. 2000. The *Drosophila* MSL complex acetylates histone H4 at lysine 16, a chromatin modification linked to dosage compensation. *Mol Cell Biol,* 20, 312–18.

Sirri, V., Roussel, P. & Hernandez-Verdun, D. 2000. *In vivo* release of mitotic silencing of ribosomal gene transcription does not give rise to precursor ribosomal RNA processing. *J Cell Biol,* 148, 259–70.

Soruco, M. M., Chery, J., Bishop, E. P., Siggers, T., Tolstorukov, M. Y., Leydon, A. R., Sugden, A. U., Goebel, K., Feng, J., Xia, P., Vedenko, A., Bulyk, M. L., Park, P. J. & Larschan, E. 2013. The CLAMP protein links the MSL complex to the X chromosome during *Drosophila* dosage compensation. *Genes Dev,* 27, 1551–6.

Stage, D. E. & Eickbush, T. H. 2007. Sequence variation within the rRNA gene loci of 12 *Drosophila* species. *Genome Res,* 17, 1888–97.

Straub, T., Grimaud, C., Gilfillan, G. D., Mitterweger, A. & Becker, P. B. 2008. The chromosomal high-affinity binding sites for the *Drosophila* dosage compensation complex. *PLoS Genet,* 4, e1000302.

Strohner, R., Nemeth, A., Jansa, P., Hofmann-Rohrer, U., Santoro, R., Langst, G. & Grummt, I. 2001. NoRC—a novel member of mammalian ISWI-containing chromatin remodeling machines. *EMBO J,* 20, 4892–900.

Strohner, R., Nemeth, A., Nightingale, K. P., Grummt, I., Becker, P. B. & Langst, G. 2004. Recruitment of the nucleolar remodeling complex NoRC establishes ribosomal DNA silencing in chromatin. *Mol Cell Biol,* 24, 1791–8.

Strukov, Y. G., Sural, T. H., Kuroda, M. I. & Sedat, J. W. 2011. Evidence of activity-specific, radial organization of mitotic chromosomes in *Drosophila. PLoS Biol,* 9, e1000574.

Sun, B. K., Deaton, A. M. & Lee, J. T. 2006. A transient heterochromatic state in Xist preempts X inactivation choice without RNA stabilization. *Mol Cell,* 21, 617–28.

Suzuki, M., Yamamoto, M. & Engel, J. D. 2014. Fetal globin gene repressors as drug targets for molecular therapies to treat the beta-globinopathies. *Mol Cell Biol,* 34, 3560–9.

Takagi, N. & Sasaki, M. 1975. Preferential inactivation of the paternally derived X chromosome in the extraembryonic membranes of the mouse. *Nature,* 256, 640–2.

Teranishi, M., Shimada, Y., Hori, T., Nakabayashi, O., Kikuchi, T., Macleod, T., Pym, R., Sheldon, B., Solovei, I., Macgregor, H. & Mizuno, S. 2001. Transcripts of the MHM region on the chicken Z chromosome accumulate as noncoding RNA in the nucleus of female cells adjacent to the DMRT1 locus. *Chromosome Res,* 9, 147–65.

Tian, D., Sun, S. & Lee, J. T. 2010. The long noncoding RNA, Jpx, is a molecular switch for X chromosome inactivation. *Cell,* 143, 390–403.

Tolhuis, B., Palstra, R. J., Splinter, E., Grosveld, F. & De Laat, W. 2002. Looping and interaction between hypersensitive sites in the active beta-globin locus. *Mol Cell,* 10, 1453–65.

Ulianov, S. V., Khrameeva, E. E., Gavrilov, A. A., Flyamer, I. M., Kos, P., Mikhaleva, E. A., Penin, A. A., Logacheva, M. D., Imakaev, M. V., Chertovich, A., Gelfand, M. S., Shevelyov, Y. Y. & Razin, S. V. 2016. Active chromatin and transcription play a key role in chromosome partitioning into topologically associating domains. *Genome Res,* 26, 70–84.

Urban, J. A., Doherty, C. A., Jordan, W. T., 3RD, Bliss, J. E., Feng, J., Soruco, M. M., Rieder, L. E., Tsiarli, M. A. & Larschan, E. N. 2017. The essential *Drosophila* CLAMP protein differentially regulates non-coding roX RNAs in male and females. *Chromosome Res,* 25, 101–13.

Urban, J. A., Urban, J. M., Kuzu, G. & Larschan, E. N. 2017. The *Drosophila* CLAMP protein associates with diverse proteins on chromatin. *PLoS One,* 12, e0189772.

Vielle, A., Lang, J., Dong, Y., Ercan, S., Kotwaliwale, C., Rechtsteiner, A., Appert, A., Chen, Q. B., Dose, A., Egelhofer, T., Kimura, H., Stempor, P., Dernberg, A., Lieb, J. D., Strome, S. & Ahringer, J. 2012. H4K20me1 contributes to downregulation of X-linked genes for *C. elegans* dosage compensation. *PLoS Genet,* 8. doi: 10.1371/journal.pgen.1002933.

Villa, R., Schauer, T., Smialowski, P., Straub, T. & Becker, P. B. 2016. PionX sites mark the X chromosome for dosage compensation. *Nature,* 537, 244–8.

Voit, R., Seiler, J. & Grummt, I. 2015. Cooperative action of Cdk1/cyclin B and SIRT1 is required for mitotic repression of rRNA synthesis. *PLoS Genet,* 11. doi: 10.1371/journal.pgen.1005246.

Waring, G. L. & Pollack, J. C. 1987. Cloning and characterization of a dispersed, multicopy, X chromosome sequence in *Drosophila melanogaster. Proc Natl Acad Sci U S A,* 84, 2843–7.

Wells, M.b., Snyder, M. J., Custer, L. M. & Csankovski, G. 2012. *Caenorhabditis elegans* dosage compensation regulates histone H4 chromatin state on X chromosomes. *Mol Cell Biol*, 32, 1710–19.

White, A. E., Leslie, M. E., Calvi, B. R., Marzluff, W. F. & Duronio, R. J. 2007. Developmental and cell cycle regulation of the *Drosophila* histone locus body. *Mol Biol Cell*, 18, 2491–502.

Witters, G., Van Robays, J., Willekes, C., Coumans, A., Peeters, H., Gyselaers, W. & Fryns, J. P. 2011. Trisomy 13, 18, 21, triploidy and Turner syndrome: the 5T's. Look at the hands. *Facts Views Vis Obgyn*, 3, 15–21.

Wu, L., Zee, B. M., Wang, Y., Garcia, B. A. & Dou, Y. 2011. The RING finger protein MSL2 in the MOF complex is an E3 ubiquitin ligase for H2B K34 and is involved in crosstalk with H3 K4 and K79 methylation. *Mol Cell*, 43, 132–44.

Xie, W., Ling, T., Zhou, Y., Feng, W., Zhu, Q., Stunnenberg, H. G., Grummt, I. & Tao, W. 2012. The chromatin remodeling complex NuRD establishes the poised state of rRNA genes characterized by bivalent histone modifications and altered nucleosome positions. *Proc Natl Acad Sci U S A*, 109, 8161–6.

Yang, F., Deng, X., Ma, W., Berletch, J. B., Rabaia, N., Wei, G., Moore, J. M., Filippova, G. N., Xu, J., Liu, Y., Noble, W. S., Shendure, J. & Disteche, C. M. 2015. The lncRNA Firre anchors the inactive X chromosome to the nucleolus by binding CTCF and maintains H3K27me3 methylation. *Genome Biol*, 16, 52.

Yang, H. & Mizzen, C. A. 2009. The multiple facets of histone H4-lysine 20 methylation. *Biochem Cell Biol*, 87, 151–61.

Yuan, X., Feng, W., Imhof, A., Grummt, I. & Zhou, Y. 2007. Activation of polymerase I transcription by Cockayne syndrome group B protein and histone methyltransferase G9a. *Mol Cell*, 27, 585–95.

Zhang, S. O., Mathur, S., Hattem, G., Tassy, O. & Pourquie, O. 2010. Sex-dimorphic gene expression and ineffective dosage compensation of Z-linked genes in gastrulating chicken embryos. *BMC Genomics*, 11, 13.

Zhang, Y., Sikes, M. L., Beyer, A. L. & Schneider D. A. 2009. The Paf1 complex is required for efficient transcriptioelongation by RNA polymerase I. *Proc Natl Acad Sci U S A*, 106, 2153–8.

Zheng, S., Villa, R., Wang, J., Feng, Y., Wang, J., Becker, P. B. & Ye, K. 2014. Structural basis of X chromosome DNA recognition by the MSL2 CXC domain during *Drosophila* dosage compensation. *Genes Dev*, 28, 2652–62.

PART III

The Interplay Between Transcription and Nuclear Structures

The process of genomic transcription is controlled by three general mechanisms that work in unison. The first mechanism is mediated by transcription factors that bind to specific DNA motifs and initiate the sequence of events that lead to active transcription or that prepare and poise a particular gene for eventual transcription. The second mechanism consists of post-translational covalent modifications of histones, the methylation of DNA and the physical rearrangement of the nucleosomal organization of chromatin. The third mechanism of transcriptional regulation involves the dynamic localization of genomic regions in specific nuclear compartments and sub-compartments.

The cell nucleus is the structure that distinguishes eukaryotes from all other organisms. Its most obvious function is to separate the genetic material and the molecular processes that retrieve genetic information from the portion of the cell that is in direct contact with the environment and that contains the translation mechanism needed to convert a subset of the transcribed information into proteins. In fact, the nucleus functions critically in regulating how the genetic information is retrieved and used. In eukaryotes, the genome is organized into chromosomes that occupy specific positions in the nucleus called **chromosome territories**. The relative position of these territories is influenced by a variety of parameters that include gene density and chromosome size, and can vary during differentiation. Each chromosome contains active regions that tend to localize towards the middle portion of the nucleus, and inactive regions that tend to associate with the inner region of the nuclear membrane—the **lamina**. Individual active and inactive regions within a chromosome form **topologically associating domains** (TADs); inactive regions that localize at the nuclear membrane are referred to as **lamin-associated domains** (LADS).

A substantial portion of the regulation of gene expression mediated by the nucleus involves a group of specialized, structured regions referred to as **nuclear bodies** that include **nucleoli**, **PML bodies**, **Cajal bodies** and several others. Each type of nuclear bodies is composed of a specific assemblage of proteins and RNA and appears to be involved in an amazing variety of cellular pathways. What is currently known of the form and functions of these specialized nuclear regions is discussed in the chapters of Part III.

CHAPTER 10

Architectural organization of the genome

Following cell division, as the newly generated nuclei of most eukaryotic cells enter interphase, their chromosomes lose their condensed appearance and become indistinguishable. These observations had led to the generally accepted belief that, in this phase of the cell cycle, chromosomes constitute an unorganized mass of intermingled chromatin fibers. Yet a number of exceptions to this view, some of them quite ancient (Boveri, 1909; Rabl, 1885), had begun to accumulate suggesting that, in fact, chromosomes occupy individual regions of the interphase nucleus. Using a term introduced by Boveri, these regions were named **chromosome territories** (CTs).

Evidence for the existence of chromosome territories

Occasionally, interphase nuclei, treated with cytological stains used to visualize chromosomes during the various stages of mitosis, would exhibit a series of condensed regions, leading to the conclusion that these regions represented individual, distinct chromosomes (Stack et al., 1977). This interpretation was eventually tested experimentally by determining that ultraviolet (UV) microbeams aimed at a specific area of a nucleus resulted in damage that was not randomly distributed among all of the chromosomes; rather, the effect of the microbeam was limited to one member or to a small subset of the chromosome complement (Cremer et al., 1982). The most direct and, therefore compelling, evidence for the existence of CTs was obtained using the technique of DNA–DNA *in situ* hybridization—human DNA probes that specifically recognized a human

chromosome in mitotic metaphases of hamster/human or mouse/human hybrid cells, hybridized to a very localized and limited region in interphase nuclei (Manuelidis, 1985; Schardin et al., 1985). More recently, the spatial arrangement of CTs in the nuclear space was obtained by combining chromosome-specific *in situ* hybridization with confocal microscopy, an optical technique that allows the three-dimensional (3D) reconstruction of nuclei (Bolzer et al., 2005).

The arrangement of CTs within a nucleus

In the nuclei of most eukaryotes, the arrangement of chromosomes is not random; rather, CTs appear to occupy particular positions, with some chromosomes preferring to be near the outer edge, some near the interior and others in intermediate locations (Fig. 10.1). This radial arrangement is consistent with the distribution of translocations—rearrangements resulting when two chromosomes break and exchange segments with one another. These rearrangements occur more frequently between neighboring chromosomes than between chromosomes in more distally located territories (Branco and Pombo, 2006; Roukos and Misteli, 2014). Factors thought to influence the non-random positioning of chromosomes are size and gene density, with smaller and more gene-rich chromosomes in closer proximity to one another and occupying more central locations (Boyle et al., 2001; Lieberman-Aiden et al., 2009). In a given tissue, the arrangement of CTs is not invariant; rather, it is probabilistic in that a particular configuration is preferred and occurs more frequently than any other. The predominant configuration is not

Epigenetics, Nuclear Organization and Gene Function: with implications of epigenetic regulation and genetic architecture for human development and health. John C. Lucchesi, Oxford University Press (2019). © John C. Lucchesi 2019. DOI: 10.1093/oso/9780198831204.001.0001

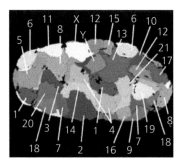

Fig. 10.1 False color representation of all CTs visible in a confocal microscope mid-section of a human fibroblast cell nucleus.

(From Bolzer et al., 2005.)

altered during the cell cycle. In contrast, the relative position of some CTs is different in different cell types (Fritz et al., 2014a) and is altered very substantially when a tissue undergoes malignant transformation (Fritz et al., 2014b).

Chromatin domains

Within each CT, the chromosome is subdivided into active chromatin domains that include promoters, enhancers, insulators and transcribed regions, interspersed with inactive domains that include constitutively repressed and Polycomb-repressed regions (discussed in Chapters 5 and 7). Summing these two types of domains across all chromosomes, the genome is divided into active (A) and inactive (B) compartments (Lieberman-Aiden et al., 2009). While the protein composition of all of the chromatin domains and their functional characteristics have been described by classical biochemical and molecular genetic approaches, it is only more recently that their physical associations have come to light. The discovery of these interactions was made possible by the development of the chromosome conformation capture (3C) and related techniques based on chemical crosslinking of genomic regions that are physically associated within the cell's nucleus. The detection of these associations and the characterization of the genomic segments involved are accomplished by polymerase chain reaction (PCR) amplification, by hybridization to DNA microarrays or by high-throughput sequencing (Dekker et al., 2002; Dekker et al., 2013) (Box 10.1). These techniques revealed that the architecture of active and inactive chromatin domains consists of loops that are formed and organized to accommodate all aspects of transcriptional regulation.

Promoters physically associate with distant enhancers

The initial interactions that were detected by the chromosome conformation capture method were the physical association of specific promoters with distant enhancers. These studies established the validity of the looping model of gene regulation. Proposed on the basis of results obtained with viruses and bacteria, the model posited that enhancers and promoters interact physically and the DNA stretch that separates them forms a loop (Ptashne, 1986). Although some early evidence for such an interaction had been obtained (Cullen et al., 1993), it was the study of the sequential activation of the β-globin gene cluster (discussed in Chapter 9) that provided fundamental information on the interaction of enhancers and insulators with gene promoters (Fig. 10.2). Activation of the β-globin genes requires the association of the hypersensitive sites 5' and 3' from the cluster, as well as those found in the locus control region (LCR), to form an active chromatin hub (ACH) (Tolhuis et al., 2002). Within this hub, the different β-globin genes are activated in a development-dependent succession by interacting with the LCR enhancers; inactive globin genes loop out.

Following those studies, numerous examples of the physical association between enhancers and promoters in a variety of organisms have demonstrated the commonality of this mechanism for transcription activation. The physical and functional connection between enhancers and promoters involves the Mediator complex (see Chapter 3) and cohesin (Kagey et al., 2010). Recently, another bridging protein, Yin Yang 1 (YY1), was reported to participate in enhancer promoter interactions (Weintraub et al., 2017). YY1 proteins bind to enhancers and promoter proximal sequences and bring these elements together by forming homodimers.

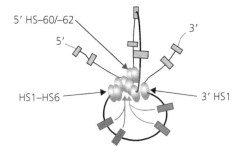

Fig. 10.2 Diagram representing the loop formation that results from the sequential activation of the β-globin genes (colored rectangles) by their interaction with the regulatory elements of the LCR (orange ovals).

(From Palstra et al., 2003.)

Box 10.1 Chromosome conformation capture techniques

The chromosome conformation capture (3C) technique (Dekker et al., 2002) was developed to determine whether two particular regions of the genome associate physically. Its principal usefulness was the demonstration that enhancers interact directly with target genes, causing a looping out of the intervening chromatin. This technique is valuable for detecting interactions between two sites that are relatively close but is less reliable for sites that are separated by several hundreds of kilobases. A number of related techniques have been developed that achieve greater scale or specificity (Fig. 10B.1).

4C (circularized chromosome conformation capture): the 3C ligation products are circularized, following digestion with a second restriction enzyme. Inverse PCR is used to identify the ligation partners. This procedure allows the detection of genome-wide interactions with a specific locus.

5C (carbon copy chromosome conformation capture): this procedure is used to establish all of the interactions between different types of loci, for example between enhancers and promoters. Because of the current high cost of using unique primers, the application of 5C is usually limited to a few megabases of the genome. The 3C ligation products are hybridized to a mixture of oligonucleotides that extend into the restriction sites of a particular restriction enzyme that occur in a region of interest. Two adjacent oligonucleotides spanning the site are ligated, amplified and sequenced.

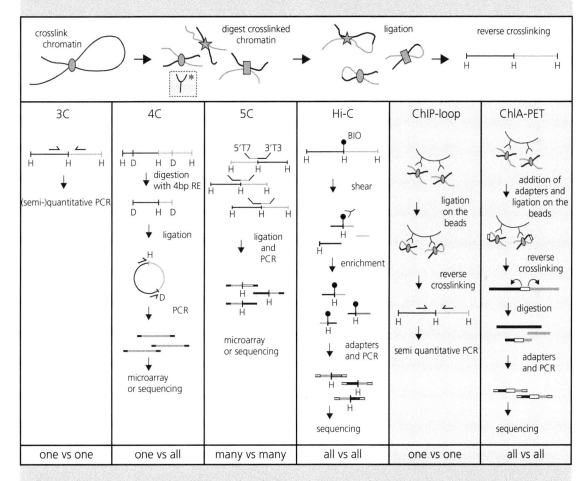

Fig. 10B.1 Diagrams summarizing the different variations of the 3C method. The top panel represents the cross-linking, the digestion with a restriction enzyme, and the ligation of the digested fragments.

(From de Wit and de Laat, 2012.)

continued

Box 10.1 *Continued*

Hi-C: this uses a modification in the establishment of the 3C ligation products. Following restriction enzyme digestion, the staggered ends are filled in with biotinylated nucleotides prior to blunt-end ligation. Biotinylation is used to select by immunoprecipitation those fragments that represent joined fragments.

ChIA-PET (Chromatin Interaction Analysis by Paired-End Tag Sequencing): this procedure is designed to detect the association of genomic regions that are brought together by the binding of a specific protein, for example a hormone receptor. These fragments are preselected from 3C ligation products by immunoprecipitation.

Insulators circumscribe the activity of enhancers

Since enhancers and genes are interspersed along chromosome arms, how do given enhancers interact only with specific genes and do not inappropriately activate other genes? This specificity is achieved through the presence of **insulators** that divide the genome into regions where enhancers can interact with given promoters but are separated from other genes.

The expectation that individual genes and their associated regulatory elements should be isolated into individual domains led to the search for special genomic sequences that would represent domain boundaries, insulating genetic units from one another. The first to be identified were the sequences flanking a pair of heat-shock genes of *Drosophila* (Kellum and Schedl, 1992; Udvardy et al., 1985). Since that time, analogous elements have been found to exist in large numbers in the genomes of metazoans and have been most extensively characterized in flies and mammals. Most insulators have been identified by the use of two types of functional assay: an enhancer blocking assay that tests the ability of a DNA segment interposed between an enhancer and a target gene promoter to block activation of the gene, and a barrier assay that prevents the inactivation of the target gene when it is placed adjacent to heterochromatin (Fig. 10.3). Insulators range from several dozen base-pairs to more than a kilobase in length and harbor one to several DNase hypersensitive sites, as well as the binding sites for insulator-specific proteins. In mammals, the first dedicated insulator-binding protein discovered is CTCF (CCCTC-binding factor); in fact, this protein is conserved in most metazoan phyla. In *Drosophila*, in addition to a CTCF homolog, there are a number of additional proteins that associate directly or indirectly with insulator sequences and contribute to their function: Su(Hw) (Suppressor of Hairy-wing), BEAF32 (boundary element associated factor 32 kilodaltons), Zw5 (Zeste-white 5), GAF (GAGA factor), CP190 (centrosomal protein 190) and Mod(mdg4) (modifier of transposable element

mdg4). Other proteins that are found in insulator protein clusters are the transcription factor DREF (DNA replication element binding factor), the mitotic spindle protein Chromator, and the PITA, ZIPIC and M1BP proteins.

CTCF was discovered as a protein that bound repeats of the core sequence CCCTC located a few hundred base-pairs upstream of the *c-myc* gene in chickens and that appeared to regulate the transcription of this gene (Lobanenkov et al., 1990). Further studies have suggested that, in many vertebrate species, CTCF is associated with insulators located between independently regulated genes and that it carries out enhancer blocking activity (Bell et al, 1999). CTCF sites were long thought to be the only insulator elements in mammalian genomes. This notion was challenged by the discovery that transfer RNA (tRNA) genes mediate a robust insulator function that is conserved from yeast to mammals (Donze et al., 1999; Ebersole et al., 2011; Raab et al., 2012). This function of tRNA genes requires the binding of TFIIIC, a transcription factor that recruits RNA polymerase III, and the presence of the cohesin

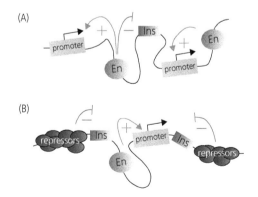

Fig. 10.3 (A) Enhancer-blocking insulator assay: an insulator placed between an enhancer and its cognate gene promoter fails to activate the gene. (B) Barrier insulator assay: an insulator can shield a gene from inactivation by neighboring repressors.

(From Chetverina et al., 2014.)

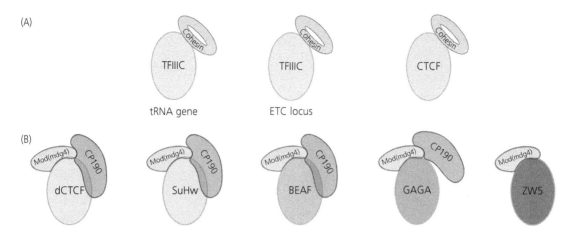

Fig. 10.4 Insulator complexes. The insulators represented in (A) are present in mammals; those represented in both (A) and (B) are found in *Drosophila*. ETC is an extra TFIIIC-binding locus. See the text for a description of the various components.

(From Van Bortle and Corces, 2012.)

(Carriere et al., 2012; Donze and Kamakaka, 2001) or condensin complexes (Van Bortle et al., 2014). TFIIIC is found at numerous additional genomic sites named ETC (extra TFIIIC) sites where it binds a DNA sequence that is not associated with RNAPIII promoters and recruits the cohesin complex (Moqtaderi and Struhl, 2004; Moqtaderi et al., 2010). A SET motif-containing protein PRDM5 is usually associated with both CTCF and TFIIIC insulator elements (Galli et al., 2013).

In *Drosophila*, in addition to the TFIIIC–cohesin/ condensin combinations, a number of DNA-binding proteins with different associated factors constitute a variety of insulator complexes (Schwartz et al., 2012). Two associated proteins that are present in a majority of insulator elements are Mod(mdg4) and CP190 (Fig. 10.4). The different insulator complexes are present at thousands of binding sites in the genome where most of them do not act as insulators or barrier sites but perform other functions, depending on the combination of proteins with which they associate and on the covalent modifications [sumoylation, ubiquitination or poly(ADP ribosyl)ation] of these proteins (Capelson and Corces, 2005; Capelson and Corces, 2006; Golovnin et al., 2012; Ong et al., 2013). For these reasons, insulator proteins are more appropriately called **architectural proteins**, and their genomic locations architectural protein binding sites (APBSs), unless they have been experimentally demonstrated to have enhancer-blocking activity (Van Bortle et al., 2014). This is particularly useful in the case of the newly identified proteins PITA, ZIPIC, IBF1 and 2 and M1BP. PITA and ZIPIC (zinc-finger protein interacting with CP190), discovered through their association with CP190, function as insulators (Maksimenko et al., 2015); these proteins are often located at the boundaries of TADs and therefore qualify as architectural proteins (Zolotarev et al., 2016). IBF1 and IBF2 (insulator-binding factor 1 and 2) co-localize with CP190 throughout the genome and, as heterodimers, mediate the binding of a subclass of CP190 proteins to DNA (Cuartero et al., 2014); their greater overlap with CTCF, rather than with Su(Hw) and BEAF, and their function as insulators in an experimental assay suggest that this may be their biological role, rather than that of architectural proteins. M1BP (motif 1 binding protein) is enriched at TAD boundaries and qualifies as an architectural protein (Ramirez et al., 2018).

Insulators subdivide the genome into functional units

CTCF exhibits a number of different functions, all derived from its ability to self-aggregate and, thereby, to subdivide the genome into circumscribed regions. This property is facilitated by the usual association of CTCF with cohesin complexes (Parelho et al., 2008; Rubio et al., 2008; Wendt et al., 2008). A plethora of recent experimental observations have led to the conclusion that, depending on its genomic location and context, CTCF can act as a barrier element or as an enhancer blocker, or can facilitate contact between regulatory elements and gene promoters. Using ChIA-PET (chromatin interaction analysis by paired-end-tag sequencing), a variation of the 3C method (Box 10.1), the loops formed by the association of CTCF-binding sites throughout the genome of embryonic cells were analyzed with respect to their genetic content and epigenetic characteristics (Handoko et al., 2011). In addition to the loops caused by enhancer–promoter associations, three additional categories were discovered: (1) loops consisting of a number of active genes and characterized by the epigenetic marks of active chromatin, (2) loops that include mostly inactive genes and (3) loops with no specific chromatin marks, indicating activity or repression within the loop, but with such marks present on either side of the boundary, with active marks on one side and inactive marks on the other. These loops more than likely serve as separators between active and repressed chromatin domains.

Topologically associating domains (TADs)

Loops that include active genes and loops that include repressed chromatin associate, respectively, into domains called TADs. In mammalian cells, TADs range in size from several hundred kilobases to several megabases; in *Drosophila*, they are much smaller, averaging around 60 kilobases. The positioning of TADs is conserved among different cell types, but numerous changes occur within these domains, suggesting that they are not involved in the process of cellular differentiation; rather, they provide the framework for cell-specific, differential gene activity to occur (Dixon et al., 2012; Wendt and Grosveld, 2014). During differentiation, some TADs shift from the more active (A) to the less active (B) compartment of the nucleus, with a

concomitant decrease in many of the interactions that occur within them; in TADs that remain in the A compartment, many of these interactions may increase (Dixon et al., 2015). At the species level, while the larger loops are conserved in vertebrates, the smaller ones, located within the large loops, tend to be species-specific (Vietri Rudan et al., 2015).

TADs are defined by binding sites that are present at their borders and that are occupied by architectural proteins. Housekeeping genes are particularly prevalent at TAD boundaries; in *Drosophila*, these genes are also abundant in inter-TAD segments (Ulianov et al., 2016). The upstream and downstream borders associate in order to establish the TADs. In mammals, TAD boundaries are marked by the presence of CTCF, cohesin complexes and TFIIIC, the DNA-binding complex that serves as a transcription factor for RNA polymerase III (Van Bortle et al., 2014). Topoisomerase IIβ has recently been found associated with CTCF-cohesin at approximately half of the TAD boundaries in mouse cells (Uuskula-Reimand et al., 2016). The association of the two boundaries of a TAD may require the presence of this enzyme activity to release the supercoiling generated during replication. In *Drosophila*, these boundaries are defined by combinations of architectural proteins (Fig. 10.5) that include CTCF, Su(Hw), BEAF32, CP190, Mod(mdg4), GAGA-binding factor and TFIIIC. The level of TAD boundary occupancy by these proteins determines the extent to which genic interactions occur between different TADs (Van Bortle et al., 2014).

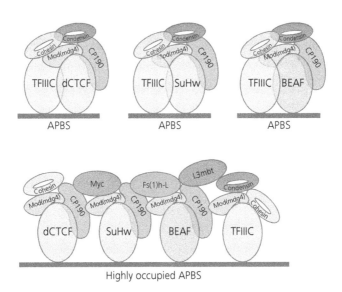

Fig. 10.5 Examples of *Drosophila* APBSs exhibiting low or high levels of architectural protein occupancy. See the text for a description of the various components.

(From Gomez-Diaz and Corces, 2014.)

Box 10.2 Multiplexed reporter assay

The purpose of the assay is to determine the level of variation in gene expression as a function of genomic position. Each copy of a reporter gene is tagged at one end with one of a very large number of randomly generated short DNA sequences. This process, referred to as "barcoding," distinguishes each copy of the reporter gene from all the others. The reporter genes are integrated randomly in the genome and excised by treatment with restriction enzymes. Each fragment that includes one of the reporter genes also includes the unique barcode that identifies that particular gene, as well as some of the DNA flanking the

site of integration. Therefore, the particular genomic location of all of the integrated reporter genes can be identified by high-throughput sequencing. Individual barcodes and the flanking genomic DNA can also be used to determine the level of expression of each reporter gene (Fig. 10B.2).

An alternative method consists of introducing into the genome a single copy of a reporter gene carried by a transposable element. Induced transposition leads to the almost random integration of multiple copies of the reporter gene throughout the genome (Symmons et al., 2014).

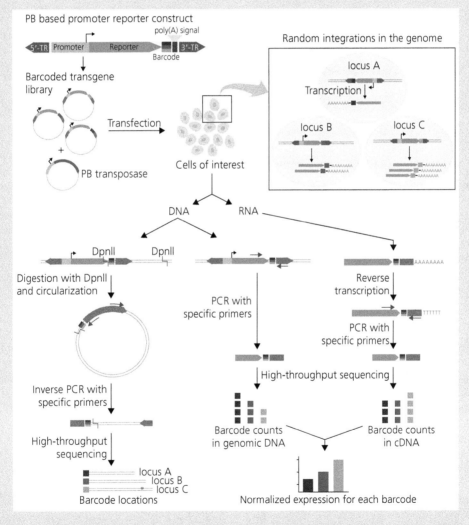

Fig. 10B.2 A large number of transcription reporters, individually marked with a unique DNA sequence, are integrated randomly in the genome. Following digestion with a restriction enzyme and circularization, the location of each reporter is determined by sequencing. For each inserted reporter, the number of copies present in the genome and the level of expression are determined.

(From Akhtar et al., 2013.)

Interactions between regulatory elements that are located within the same TAD are facilitated; interactions between similar elements residing in different TADs are generally inhibited (Dixon et al., 2012; Nora et al., 2012; Sexton et al., 2012).

In mammalian cells, TAD boundaries can be categorized by their level of strength, which reflects the level of CTCF binding, suggesting that stronger CTCF binding may confer stronger insulation. Superenhancers are found significantly more frequently within TADs that are marked by strong boundaries (Gong et al., 2018). Depleting the level of CTCF in mammalian cells leads to the merging of neighboring TADs and to a significant level of ectopic contacts across TAD boundaries; these effects disappear after CTCF levels are restored (Nora et al., 2017).

Enhancers and the promoters that they target, regardless of the distance that separates them, are more frequently located in the same TAD (Jin et al., 2013). Using an experimental procedure consisting of a regulatory sensor (a reporter gene with a weak promoter inserted at numerous random sites in the mouse genome) (Box 10.2), several properties of enhancers were uncovered. Enhancers distribute their activity within relatively broad regions that surround their cognate target genes; within regulatory domains, there exist occasional sites where no enhancer activity is manifest, suggesting the presence of factors (CTCF and cohesin?) that shield genes from activation (Symmons et al., 2014). When TAD boundaries are disrupted, the normal topological relationship between genes and their enhancers are altered, resulting in misexpression and phenotypic defects (Lupiáñez et al., 2015).

Loop extrusion as a basis of TAD formation

While TADs represent a major level of organization, using modified Hi-C methods (Box 10.1), a very large number of contacts were seen to occur within the genome of human and mouse cell lines, many of which were contacts between distant sites generating loops (Rao et al., 2014). The ends of these loops displayed the architectural proteins CTCF and cohesin. The CTCF binding sites at both ends of the loops are almost exclusively in a convergent orientation (Mourad et al., 2017; Rao et al., 2014; Vietri Rudan et al., 2015). Other loops are attached to tandemly oriented CTCF sites (Fig. 10.6). Changing the orientation of one CTCF binding site causes a reconfiguration of the chromatin loop topology (Guo et al., 2015). These observations led to a loop extrusion model (Nichols and Corces, 2015). The model proposes that the association of two convergent CTCF

Fig. 10.6 Diagram illustrating the convergent position of the CTCF binding sites in evolutionarily conserved genomic loops (A), and the two possible divergent positions of those binding sites in less conserved and species-specific loops (B).

(From Ghirlando and Felsenfeld, 2016.)

sites may be the result of the ability of CTCF to bend DNA, forming the beginning of a loop that would increase in size until a CTCF site of the appropriate orientation is met. Using only the genomic localization of CTCF and the orientation of the CTCF binding sites, a loop extrusion model can be used to simulate the topological data obtained by chromosome conformation capture techniques (Sanborn et al., 2015). Further modeling analyses have suggested that cohesin is responsible for loop extrusion and that this process is halted when CTCF binding sites in the appropriate orientation are encountered (Busslinger et al., 2017; Fudenberg et al., 2016). These contentions have been validated by: (1) the demonstration that cohesin subunits promote loop extension and that WAPL, a factor that causes cohesin to release DNA, stops this process (Haarhuis et al., 2017) and (2) the depletion of CTCF leads to the disappearance of TAD boundaries and the loss of insulation between neighboring TADs (Nora et al., 2017).

The complete elimination of cohesin binding can be obtained by inducing the loss-of-function of *Nipb*l (nipped-like protein) that is necessary for loading cohesin onto chromatin. The absence of cohesin in mouse cells has no effect on CTCF binding but results in the disappearance of TADs; it reveals the existence of fine-scale compartments that reflect the chromatin states, i.e. the epigenetic marks, of the active (A) and inactive (B) chromatin compartments (Schwarzer et al., 2017). When TADs form, they can include both types of these finer compartments.

TADs can be visualized in *Drosophila* polytene chromosomes

Polytene chromosomes occur in certain tissues of Dipteran insects. They are produced by a process of endoreplication whereby chromosomes replicate successively without undergoing the subsequent condensation characteristic of mitotic chromosomes or the separation of the daughter chromatids. The many resulting chromatids remain associated in perfect register throughout their length. Each chromatid exhibits regions of localized coiling separated by uncoiled segments that are reflected in the polytene chromosomes' banded appearance (Fig. 10.7). Polytene chromosomes have two types of bands: very dark bands reflecting a high degree of condensation and gray bands where the DNA fibers are less compact. While interbands consist of fully extended nucleosomal arrays, gray bands and dark bands are ten times and 30 times more compacted, respectively. Previous experiments had determined that the regulatory regions of genes are present in interbands and the coding regions of the majority of active genes, most of which are housekeeping genes, are in the gray bands; tissue-specific genes are present in the dark, highly compacted bands (Zhimulev et al., 2014). To allow the transcription of these genes, the bands may decondense, leading to the formation of structures descriptively referred to as puffs.

Using one of the conformation capture techniques on salivary gland chromatin (Hi-C) (Box 10.1), a correspondence between bands and TADs was revealed across all of the chromosomes, with the majority of TAD and band boundaries very close to overlapping (Eagen et al., 2015; Ulianov et al., 2016). Interbands correspond to the chromatid segments that separate TADs. The distribution of TADs was similar to the one

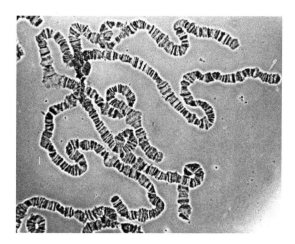

Fig. 10.7 Polytene chromosomes from the salivary gland of a *Drosophila* third instar female larva. All of the chromosome arms extend from the chromocenter, a region of the nucleus where the blocks of constitutive heterochromatin that flank the centromeres aggregate.

(From *Molecular Cell Biology*, 7th Edition.)

previously established in diploid cells (Hou et al., 2012) and in embryos (Sexton et al., 2012). Superimposing Hi-C data on a polytene chromosome region that had been very finely mapped by electron microscopy, a precise correspondence was found between faint bands and TADs, and between interbands and the boundary elements that separate the TADs (Stadler et al., 2017).

TAD formation during early development

In *Drosophila*, following fertilization, very rapid, synchronous mitotic divisions produce a syncytial mass of nuclei. These nuclei eventually migrate to the surface of the egg and, following cell membrane formation, give rise to a cellular blastoderm, the equivalent of the blastula in vertebrate development. Prior to nuclear migration, a small number of genes are activated. Immediately following the formation of the blastoderm, global genome activation occurs with the recruitment of RNAPII. Before global activation, the genome is largely unorganized; following genome activation, TADs are formed, housekeeping genes are enriched at their boundaries and the level of intra-TAD contacts increases (Hug et al., 2017). TAD formation is independent of transcription and RNAPII localization, although the presence of Zelda, a transcription factor involved in early embryonic transcription, is required for the formation of TAD boundaries at Zelda-rich loci.

The genome of mouse mature oocytes is devoid of TADs; that of mature sperm does contain TADs. In sperm, the interactions within TADs are extra long-range and there are frequent interactions between TADs. In early embryos, the high-order chromatin structures are gradually re-established (Du et al., 2017; Ke et al., 2017). As in *Drosophila*, TAD formation is independent from the onset of genomic transcription.

Insulator bodies and PcG bodies

In *Drosophila* and perhaps in other organisms, including mammals, groups of border APBS associate at specific nuclear locations to form *insulator bodies* (Bushey et al., 2009; Gerasimova et al., 2000). This arrangement may facilitate the occurrence of long-range inter-TAD interactions in order to bring into proximity genes that should be co-regulated. These interactions can occur in *cis*, between TADs located on the same chromosome (Fig. 10.8) or, less frequently, in *trans*, between TADs in different chromosomes. Some of the parameters that underlie this type of coordinated regulation were established experimentally by relocating the β-globin LCR to an ectopic location in the mouse genome

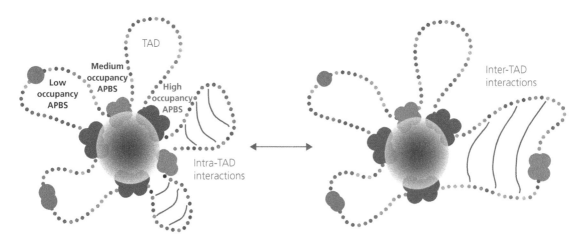

Fig. 10.8 Differential role of APBSs in facilitating intergenic interactions and in limiting or permitting these interactions to occur between TADs. Low-occupancy APBSs occur almost exclusively within active TADs where they regulate individual gene function.

(From Van Bortle et al., 2014.)

(Noordermeer et al., 2011). In its new location, the LCR searches for genes with which it can interact on the basis of shared transcription factors, but it only enhances the activity of endogenous β-globin genes. Furthermore, enhancement occurs only in a subset of cells, indicating that the distance that the LCR can cover is finite and that contact between enhancer and target genes located on different chromosomes is limited to cells where these elements happen to be in the appropriate spatial proximity. Inter-TAD chromatin interactions have also been documented in stem cells (see Chapter 16).

Intra- and inter-TAD associations also occur among regions of the chromosomes that are enriched for transcriptionally repressive PcG protein complexes. In *Drosophila*, PcG proteins are attracted to PRE elements dispersed throughout the genome, from whence they spread to silence neighboring genes (see Chapter 7). Yet, early cytological examination had revealed that PcG proteins are concentrated in a relatively small number of spots that were named **PcG bodies** (Saurin et al., 1998). A number of these bodies may represent the general tendency for PcG targets to be close to one another in chromosome domains of approximately 200 kilobases. But many PcG bodies involve the co-localization of PcG target genes that are far from one another on a given chromosome or are on different chromosomes. The first example of these interactions was provided by the observation that two genes belonging, respectively, to the HOX complexes ANT-C and BX-C (discussed in Chapter 7) are frequently physically associated when they are repressed by PcG proteins, even though the two complexes are 10 megabases distant from each other. This association is disrupted when

either of the genes is active (Bantignies et al., 2011). At least in *Drosophila*, the interactions of PcG-repressed genes are generally limited to their presence on the same chromosomal arms (Tolhuis et al., 2011).

The aggregation of PcG protein domains is not a consequence of some affinity that these proteins have for one another; rather, it is mediated by the association of neighboring insulators. This somewhat surprising conclusion was suggested by the observations that two remotely located PRE insertions, introduced into the *Drosophila* genome by transgenesis, interacted only if the insulator activities associated with these elements were present in the constructs (Li et al., 2011). Insulators were also found to target PREs to gene promoters for repression, in a similar manner that they target enhancers to promoters for activation (Comet et al., 2011).

In addition to the insulator-mediated associations just described, PRC-silenced domains exhibit a much more fine-grained architecture (Kundu et al., 2017). In mouse embryonic stem cells (ESCs), the chromatin of individual or small groups of PRC-targeted genes is compacted, presumably preventing their interaction with enhancers. These domains are much smaller than TADs, are not bound by architectural proteins and disappear as ESCs begin differentiation.

Transcription and DNA replication can be regulated by relocating genes to different nuclear sub-compartments

In addition to the established role of transcription factors, of DNA methylation and of the post-translational modification of histones in regulating gene activity, the location of genes in different nuclear sub-compartments can determine whether transcription is allowed to occur or not. This regulatory mechanism relies on the ability of chromosomes to form large extraterritorial loops that allow specific genes, which are distant from one another, to be co-transcribed in transcription factories (see Chapter 13); it is also mediated by the presence of genomic sequences that direct the interaction of particular chromatin regions with silencing environments such as the nuclear lamina and the nucleolus (see Chapters 11 and 12).

DNA replication originates at specific sites (Cairns, 1966; Taylor, 1963), termed origins of replication, which occur in clusters throughout the genome (see Chapter 14). The different origins in a cluster initiate the process of replication synchronously. Early studies established that actively transcribed genes appeared to replicate earlier than many inactive genes during the S phase, and that the chromosomal location of genes reflected the timing of their replication (O'Keefe et al., 1992). This assessment was confirmed by the results of 3C-based experiments—the genome consists of segments that differ in the timing of their replication during the S phase (Hiratani et al., 2008; Ryba et al., 2010). Generally, early replicating domains contain active genes, while late-replicating domains are silent. These observations suggest a correlation between TADs and replication segments.

Chapter summary

Contrary to early belief, chromatin is not randomly organized in the interphase nucleus. The spatial arrangement of the genome contributes importantly to the regulation of gene function.

To begin, chromosomes occupy particular positions (territories) based on their size and gene content. Within each territory, chromosomes are subdivided into active and inactive regions, consisting of loops containing a number of active genes and characterized by the epigenetic marks of active chromatin, loops that include mostly inactive genes and loops with no specific chromatin marks, indicating activity or repression within the loop. Loops that include active genes and loops that include repressed chromatin associate, respectively, into domains called topologically associating domains (TADs). These domains are defined by binding sites present at their borders and occupied by architectural proteins that, in mammals, include CTCF (CCCTC-binding factor), cohesin complexes and TFIIIC (a transcription factor for RNA polymerase III). Enhancers and their target promoters are located within the same TAD where they are prevented from promiscuous interactions with other promoters by the presence of insulators. Interactions between similar elements located in different TADs are generally suppressed.

In *Drosophila*, architectural protein binding sites associate to group TADs together and form insulator bodies. These bodies are thought to facilitate the rare inter-TAD interactions that may bring into proximity genes that should be co-regulated. A similar situation appears to exist for regions within and between TADs that are enriched for PCG repressive complexes; grouping of these regions forms PcG bodies.

An important additional aspect of the non-random organization of the genetic material is its plastic and dynamic functional interaction with the different nuclear compartments and sub-compartments discussed in the remaining chapters of Part III.

References

Akhtar, W., De Jong, J., Pindyurin, A. V., Pagie, L., Meuleman, W., De Ridder, J., Berns, A., Wessels, L. F., Van Lohuizen, M. & Van Steensel, B. 2013. Chromatin position effects assayed by thousands of reporters integrated in parallel. *Cell*, 154, 914–27.

Bantignies, F., Roure, V., Comet, I., Leblanc, B., Schuettengruber, B., Bonnet, J., Tixier, V., Mas, A. & Cavalli, G. 2011. Polycomb-dependent regulatory contacts between distant Hox loci in *Drosophila*. *Cell*, 144, 214–26.

Bell, A. C., West, A. G. & Felsenfeld, G. 1999. The protein CTCF is required for the enhancer blocking activity of vertebrate insulators. *Cell*, 98, 387–96.

Bolzer, A., Kreth, G., Solovei, I., Koehler, D., Saracoglu, K., Fauth, C., Muller, S., Eils, R., Cremer, C., Speicher, M. R. & Cremer, T. 2005. Three-dimensional maps of all chromosomes in human male fibroblast nuclei and prometaphase rosettes. *PLoS Biol*, 3, e157.

Boveri, T. 1909. [Die blastomerenkerne von Ascaris megalocephala und die theorie der chromosomenindividualitat]. *Arch Zellforscung*, 3, 181–286.

Boyle, S., Gilchrist, S., Bridger, J. M., Mahy, N. L., Ellis, J.a. & Bickmore, W. A. 2001. The spatial organization of human chromosomes within the nuclei of normal and emerin-mutated cells. *Hum Mol Genet*, 10, 211–19.

Branco, M. R. & Pombo, A. 2006. Intermingling of chromosome territories in interphase suggests role in translocations and transcription-dependent associations. *PLoS Biol*, 4, e138.

Bushey, A. M., Ramos, E. & Corces, V. G. 2009. Three subclasses of a *Drosophila* insulator show distinct and cell type-specific genomic distributions. *Genes Dev*, 23, 1338–50.

Busslinger, G. A., Stocsits, R. R., Van Der Lelij, P., Axelsson, E., Tedeschi, A., Galjart, N. & Peters, J. M. 2017. Cohesin is positioned in mammalian genomes by transcription, CTCF and Wapl. *Nature*, 544, 503–7.

Cairns, J. 1966. Autoradiography of HeLa cell DNA. *J Mol Biol*, 15, 372–3.

Capelson, M. & Corces, V. G. 2005. The ubiquitin ligase dTopors directs the nuclear organization of a chromatin insulator. *Mol Cell*, 20, 105–16.

Capelson, M. & Corces, V. G. 2006. SUMO conjugation attenuates the activity of the gypsy chromatin insulator. *EMBO J*, 25, 1906–14.

Carriere, L., Graziani, S., Alibert, O., Ghavi-Helm, Y., Boussouar, F., Humbertclaude, H., Jounier, S., Aude, J. C., Keime, C., Murvai, J., Foglio, M., Gut, M., Gut, I., Lathrop, M., Soutourina, J., Gerard, M. & Werner, M. 2012. Genomic binding of Pol III transcription machinery and relationship with TFIIS transcription factor distribution in mouse embryonic stem cells. *Nucleic Acids Res*, 40, 270–83.

Chetverina, D., Aoki, T., Erokhin, M., Georgiev, P. & Schedl, P. 2014. Making connections: insulators organize eukaryotic chromosomes into independent *cis*-regulatory networks. *Bioessays*, 36, 163–72.

Comet, I., Schuettengruber, B., Sexton, T. & Cavalli, G. 2011. A chromatin insulator driving three-dimensional Polycomb response element (PRE) contacts and Polycomb association with the chromatin fiber. *Proc Natl Acad Sci U S A*, 108, 2294–9.

Cremer, T., Cremer, C., Baumann, H., Luedtke, E. K., Sperling, K., Teuber, V. & Zorn, C. 1982. Rabl's model of the interphase chromosome arrangement tested in Chinese hamster cells by premature chromosome condensation and laser-UV-microbeam experiments. *Hum Genet*, 60, 46–56.

Cuartero, S., Fresan, U., Reina, O., Planet, E. & Espinas, M. L. 2014. Ibf1 and Ibf2 are novel CP190-interacting proteins required for insulator function. *EMBO J*, 33, 637–47.

Cullen, K. E., Kladde, M. P. & Seyfred, M. A. 1993. Interaction between transcription regulatory regions of prolactin chromatin. *Science*, 261, 203–6.

Dekker, J., Marti-Renom, M. A. & Mirny, L. A. 2013. Exploring the three-dimensional organization of genomes: interpreting chromatin interaction data. *Nat Rev Genet*, 14, 390–403.

Dekker, J., Rippe, K., Dekker, M. & Kleckner, N. 2002. Capturing chromosome conformation. *Science*, 295, 1306–11.

De Wit, E. & De Laat, W. 2012. A decade of 3C technologies: insights into nuclear organization. *Genes Dev*, 26, 11–24.

Dixon, J. R., Jung, I., Selvaraj, S., Shen, Y., Antosiewicz-Bourget, J. E., Lee, A. Y., Ye, Z., Kim, A., Rajagopal, N., Xie, W., Diao, Y., Liang, J., Zhao, H., Lobanenkov, V. V., Ecker, J. R., Thomson, J. A. & Ren, B. 2015. Chromatin architecture reorganization during stem cell differentiation. *Nature*, 518, 331–6.

Dixon, J. R., Selvaraj, S., Yue, F., Kim, A., Li, Y., Shen, Y., Hu, M., Liu, J. S. & Ren, B. 2012. Topological domains in

mammalian genomes identified by analysis of chromatin interactions. *Nature*, 485, 376–80.

Donze, D., Adams, C. R., Rine, J. & Kamakaka, R. T. 1999. The boundaries of the silenced HMR domain in *Saccharomyces cerevisiae*. *Genes Dev*, 13, 698–708.

Donze, D. & Kamakaka, R. T. 2001. RNA polymerase III and RNA polymerase II promoter complexes are heterochromatin barriers in *Saccharomyces cerevisiae*. *EMBO J*, 20, 520–31.

Du, Z., Zheng, H., Huang, B., Ma, R., Wu, J., Zhang, X., He, J., Xiang, Y., Wang, Q., Li, Y., Ma, J., Zhang, X., Zhang, K., Wang, Y., Zhang, M. Q., Gao, J., Dixon, J. R., Wang, X., Zeng, J. & Xie, W. 2017. Allelic reprogramming of 3D chromatin architecture during early mammalian development. *Nature*, 547, 232–5.

Eagen, K. P., Hartl, T. A. & Kornberg, R. D. 2015. Stable chromosome condensation revealed by chromosome conformation capture. *Cell*, 163, 934–46.

Ebersole, T., Kim, J. H., Samoshkin, A., Kouprina, N., Pavlicek, A., White, R. J. & Larionov, V. 2011. tRNA genes protect a reporter gene from epigenetic silencing in mouse cells. *Cell Cycle*, 10, 2779–91.

Fritz, A. J., Stojkovic, B., Ding, H., Xu, J., Bhattacharya, S. & Berezney, R. 2014. Cell type specific alterations in interchromosomal networks across the cell cycle. *PLoS Comput Biol*, 10, e1003857.

Fritz, A. J., Stojkovic, B., Ding, H., Xu, J., Bhattacharya, S., Gaile, D. & Berezney, R. 2014. Wide-scale alterations in interchromosomal organization in breast cancer cells: defining a network of interacting chromosomes. *Hum Mol Genet*, 23, 5133–46.

Fudenberg, G., Imakaev, M., Lu, C., Goloborodko, A., Abdennur, N. & Mirny, L. A. 2016. Formation of chromosomal domains by loop extrusion. *Cell Rep*, 15, 2038–49.

Galli, G. G., Carrara, M., Francavilla, C., De Lichtenberg, K. H., Olsen, J. V., Calogero, R. A. & Lund, A. H. 2013. Genomic and proteomic analyses of Prdm5 reveal interactions with insulator binding proteins in embryonic stem cells. *Mol Cell Biol*, 33, 4504–16.

Gerasimova, T. I., Byrd, K. & Corces, V. G. 2000. A chromatin insulator determines the nuclear localization of DNA. *Mol Cell*, 6, 1025–35.

Ghirlando, R. & Felsenfeld, G. 2016. CTCF: making the right connections. *Genes Dev*, 30, 881–91.

Golovnin, A., Volkov, I. & Georgiev, P. 2012. SUMO conjugation is required for the assembly of *Drosophila* Su(Hw) and Mod(mdg4) into insulator bodies that facilitate insulator complex formation. *J Cell Sci*, 125, 2064–74.

Gomez-Diaz, E. & Corces, V. G. 2014. Architectural proteins: regulators of 3D genome organization in cell fate. *Trends Cell Biol*, 24, 703–11.

Gong, Y., Lazaris, C., Sakellaropoulos, T., Lozano, A., Kambadur, P., Ntziachristos, P., Aifantis, I. & Tsirigos, A. 2018. Stratification of TAD boundaries reveals preferential insulation of super-enhancers by strong boundaries. *Nat Commun, 9,* 542.

Guo, Y., Xu, Q., Canzio, D., Shou, J., Li, J., Gorkin, D. U., Jung, I., Wu, H., Zhai, Y., Tang, Y., Lu, Y., Wu, Y., Jia, Z., Li, W., Zhang, M. Q., Ren, B., Krainer, A. R., Maniatis, T. & Wu, Q. 2015. CRISPR inversion of CTCF sites alters genome topology and enhancer/promoter function. *Cell, 162,* 900–10.

Haarhuis, J. H. I., Van Der Weide, R. H., Blomen, V. A., Yanez-Cuna, J. O., Amendola, M., Van Ruiten, M. S., Krijger, P. H. L., Teunissen, H., Medema, R. H., Van Steensel, B., Brummelkamp, T. R., De Wit, E. & Rowland, B. D. 2017. The cohesin release factor WAPL restricts chromatin loop extension. *Cell, 169,* 693–707 e14.

Handoko, L., Xu, H., Li, G., Ngan, C. Y., Chew, E., Schnapp, M., Lee, C. W., Ye, C., Ping, J. L., Mulawadi, F., Wong, E., Sheng, J., Zhang, Y., Poh, T., Chan, C. S., Kunarso, G., Shahab, A., Bourque, G., Cacheux-Rataboul, V., Sung, W. K., Ruan, Y. & Wei, C. L. 2011. CTCF-mediated functional chromatin interactome in pluripotent cells. *Nat Genet, 43,* 630–8.

Hiratani, I., Ryba, T., Itoh, M., Yokochi, T., Schwaiger, M., Chang, C. W., Lyou, Y., Townes, T. M., Schubeler, D. & Gilbert, D. M. 2008. Global reorganization of replication domains during embryonic stem cell differentiation. *PLoS Biol, 6,* e245.

Hou, C., Li, L., Qin, Z. S. & Corces, V. G. 2012. Gene density, transcription, and insulators contribute to the partition of the Drosophila genome into physical domains. *Mol Cell, 48,* 471–84.

Hug, C. B., Grimaldi, A. G., Kruse, K. & Vaquerizas, J. M. 2017. Chromatin architecture emerges during zygotic genome activation independent of transcription. *Cell, 169,* 216–28 e19.

Jin, F., Li, Y., Dixon, J. R., Selvaraj, S., Ye, Z., Lee, A. Y., Yen, C. A., Schmitt, A. D., Espinoza, C. A. & Ren, B. 2013. A high-resolution map of the three-dimensional chromatin interactome in human cells. *Nature, 503,* 290–4.

Kagey, M. H., Newman, J. J., Bilodeau, S., Zhan, Y., Orlando, D. A., Van Berkum, N. L., Ebmeier, C. C., Goossens, J., Rahl, P. B., Levine, S. S., Taatjes, D. J., Dekker, J. & Young, R. A. 2010. Mediator and cohesin connect gene expression and chromatin architecture. *Nature, 467,* 430–5.

Ke, Y., Xu, Y., Chen, X., Feng, S., Liu, Z., Sun, Y., Yao, X., Li, F., Zhu, W., Gao, L., Chen, H., Du, Z., Xie, W., Xu, X., Huang, X. & Liu, J. 2017. 3D chromatin structures of mature gametes and structural reprogramming during mammalian embryogenesis. *Cell, 170,* 367–81 e20.

Kellum, R. & Schedl, P. 1992. A group of scs elements function as domain boundaries in an enhancer-blocking assay. *Mol Cell Biol, 12,* 2424–31.

Kundu, S., Ji, F., Sunwoo, H., Jain, G., Lee, J. T., Sadreyev, R. I., Dekker, J. & Kingston, R. E. 2017. Polycomb repressive complex 1 generates discreet compacted domains that change during differentiation. *Mol Cell, 65,* 432–46.

Li, H. B., Muller, M., Bahechar, I. A., Kyrchanova, O., Ohno, K., Georgiev, P. & Pirrotta, V. 2011. Insulators, not Polycomb response elements, are required for long-range interactions between Polycomb targets in *Drosophila melanogaster. Mol Cell Biol, 31,* 616–25.

Lieberman-Aiden, E., Van Berkum, N. L., Williams, L., Imakaev, M., Ragoczy, T., Telling, A., Amit, I., Lajoie, B. R., Sabo, P. J., Dorschner, M. O., Sandstrom, R., Bernstein, B., Bender, M. A., Groudine, M., Gnirke, A., Stamatoyannopoulos, J., Mirny, L. A., Lander, E. S. & Dekker, J. 2009. Comprehensive mapping of long-range interactions reveals folding principles of the human genome. *Science, 326,* 289–93.

Lovanenkov, V. V., Nicolas, R. H., Adler, V. V., Paterson, H., Klenova, E. M., Polotskata, A. V. & Goodwin, G. H. 1990. A novel sequence-specific DNA binding protein which interacts with three regularly spaced direct repeats of the CCCTC-motif in the 5′-flanking sequence of the chicken *c-myc* gene. *Oncogene, 5,* 1743–53.

Lupianez, D. G., Kraft, K., Heinrich, V., Krawitz, P., Brancati, F., Klopocki, E., Horn, D., Kayserili, H., Opitz, J. M., Laxova, R., SANTOS-Simarro, F., Gilbert-Dussardier, B., Wittler, L., Borschiwer, M., Haas, S. A., Osterwalder, M., Franke, M., Timmermann, B., Hecht, J., Spielmann, M., Visel, A. & Mundlos, S. 2015. Disruptions of topological chromatin domains cause pathogenic rewiring of gene-enhancer interactions. *Cell, 161,* 1012–25.

Maksimenko, O., Bartkuhn, M., Stakhov, V., Herold, M., Zolotarev, N., Jox, T., Buxa, M. K., Kirsch, R., Bonchuk, A., Fedotova, A., Kyrchanova, O., Renkawitz, R. & Georgiev, P. 2015. Two new insulator proteins, Pita and ZIPIC, target CP190 to chromatin. *Genome Res, 25,* 89–99.

Manuelidis, L. 1985. Individual interphase chromosome domains revealed by *in situ* hybridization. *Hum Genet, 71,* 288–93.

Moqtaderi, Z. & Struhl, K. 2004. Genome-wide occupancy profile of the RNA polymerase III machinery in *Saccharomyces cerevisiae* reveals loci with incomplete transcription complexes. *Mol Cell Biol, 24,* 4118–27.

Moqtaderi, Z., Wang, J., Raha, D., White, R. J., Snyder, M., Weng, Z. & Struhl, K. 2010. Genomic binding profiles of functionally distinct RNA polymerase III transcription complexes in human cells. *Nat Struct Mol Biol, 17,* 635–40.

Mourad, R., Li, L. & Cuvier, O. 2017. Uncovering direct and indirect molecular determinants of chromatin loops using a computational integrative approach. *PLoS Comput Biol, 13,* e1005538.

Nichols, M. H. & Corces, V. G. 2015. A CTCF code for 3D genome architecture. *Cell, 162,* 703–5.

Noordermeer, D., De Wit, E., Klous, P., Van De Werken, H., Simonis, M., Lopez-Jones, M., Eussen, B., De Klein, A., Singer, R. H. & De Laat, W. 2011. Variegated gene expression caused by cell-specific long-range DNA interactions. *Nat Cell Biol, 13,* 944–51.

Nora, E. P., Goloborodko, A., Valton, A. L., Gibcus, J. H., Uebersohn, A., Abdennur, N., Dekker, J., Mirny, L. A. & Bruneau, B. G. 2017. Targeted degradation of CTCF decouples local insulation of chromosome domains from genomic compartmentalization. *Cell*, 169, 930–44 e22.

Nora, E. P., Lajoie, B. R., Schulz, E. G., Giorgetti, L., Okamoto, I., Servant, N., Piolot, T., Van Berkum, N. L., Meisig, J., Sedat, J., Gribnau, J., Barillot, E., Bluthgen, N., Dekker, J. & Heard, E. 2012. Spatial partitioning of the regulatory land-scape of the X-inactivation centre. *Nature*, 485, 381–5.

O'keefe, R. T., Henderson, S. C. & Spector, D. L. 1992. Dynamic organization of DNA replication in mammalian cell nuclei: spatially and temporally defined replication of chromosome-specific alpha-satellite DNA sequences. *J Cell Biol*, 116, 1095–110.

Ong, C. T., Van Bortle, K., Ramos, E. & Corces, V. G. 2013. Poly(ADP-ribosyl)ation regulates insulator function and intrachromosomal interactions in *Drosophila*. *Cell*, 155, 148–59.

Palstra, R. J., Tolhuis, B., Splinter, E., Nijmeijer, R., Grosveld, F. & De Laat, W. 2003. The beta-globin nuclear compart-ment in development and erythroid differentiation. *Nat Genet*, 35, 190–4.

Parelho, V., Hadjur, S., Spivakov, M., Leleu, M., Sauer, S., Gregson, H. C., Jarmuz, A., Canzonetta, C., Webster, Z., Nesterova, T., Cobb, B. S., Yokomori, K., Dillon, N., Aragon, L., Fisher, A. G. & Merkenschlager, M. 2008. Cohesins functionally associate with CTCF on mamma-lian chromosome arms. *Cell*, 132, 422–33.

Ptashne, M. 1986. Gene regulation by proteins acting nearby and at a distance. *Nature*, 322, 697–701.

Raab, J. R., Chiu, J., Zhu, J., Katzman, S., Kurukuti, S., Wade, P. A., Haussler, D. & Kamakaka, R. T. 2012. Human tRNA genes function as chromatin insulators. *EMBO J*, 31, 330–50.

Rabl, C. 1885. Uber zelltheilung. *Morphologisches jahrbuch.* Gegenbaur C (ed) 10, 214–330.

Ramirez, F., Bhardwaj, V., Arrigoni, L., Lam, K. C., Gruning, B. A., Villaveces, J., Habermann, B., Akhtar, A. & Manke, T. 2018. High-resolution TADs reveal DNA sequences underlying genome organization in flies. *Nat Commun*, 9, 189.

Rao, S. S., Huntley, M. H., Durand, N. C., Stamenova, E. K., Bochkov, I. D., Robinson, J. T., Sanborn, A. L., Machol, I., Omer, A. D., Lander, E. S. & Aiden, E. L. 2014. A 3D map of the human genome at kilobase resolution reveals prin-ciples of chromatin looping. *Cell*, 159, 1665–80.

Roukos, V. & Misteli, T. 2014. The biogenesis of chromosome translocations. *Nat Cell Biol*, 16, 293–300.

Rubio, E. D., Reiss, D. J., Welcsh, P. L., Disteche, C. M., Filippova, G. N., Baliga, N. S., Aebersold, R., Ranish, J. A. & Krumm, A. 2008. CTCF physically links cohesin to chromatin. *Proc Natl Acad Sci U S A*, 105, 8309–14.

Ryba, T., Hiratani, I., Lu, J., Itoh, M., Kulik, M., Zhang, J., Schulz, T. C., Robins, A. J., Dalton, S. & Gilbert, D. M. 2010. Evolutionarily conserved replication timing profiles predict long-range chromatin interactions and distinguish closely related cell types. *Genome Res*, 20, 761–70.

Sanborn, A. L., Rao, S. S., Huang, S. C., Durand, N. C., Huntley, M. H., Jewett, A. I., Bochkov, I. D., Chinnappan, D., Cutkosky, A., Li, J., Geeting, K. P., Gnirke, A., Melnikov, A., Mckenna, D., Stamenova, E. K., Lander, E. S. & Aiden, E. L. 2015. Chromatin extrusion explains key features of loop and domain formation in wild-type and engineered genomes. *Proc Natl Acad Sci U S A*, 112, E6456–65.

Saurin, A. J., Shiels, C., Williamson, J., Satijn, D. P., Otte, A. P., Sheer, D. & Freemont, P. S. 1998. The human poly-comb group complex associates with pericentromeric het-erochromatin to form a novel nuclear domain. *J Cell Biol*, 142, 887–98.

Schardin, M., Cremer, T., Hager, H. D. & Lang, M. 1985. Specific staining of human chromosomes in Chinese hamster × man hybrid cell lines demonstrates interphase chromosome territories. *Hum Genet*, 71, 281–7.

Schwartz, Y. B., Linder-Basso, D., Kharchenko, P. V., Tolstorukov, M. Y., Kim, M., Li, H. B., Gorchakov, A. A., Minoda, A., Shanower, G., Alekseyenko, A. A., Riddle, N. C., Jung, Y. L., Gu, T., Plachetka, A., Elgin, S. C., Kuroda, M. I., Park, P. J., Savitsky, M., Karpen, G. H. & Pirrotta, V. 2012. Nature and function of insulator protein binding sites in the *Drosophila* genome. *Genome Res*, 22, 2188–98.

Schwarzer, W., Abdennur, N., Goloborodko, A., Pekowska, A., Fudenberg, G., Loe-Mie, Y., Fonseca, N. A., Huber, W., C, H. H., Mirny, L. & Spitz, F. 2017. Two independent modes of chromatin organization revealed by cohesin removal. *Nature*, 551, 51–6.

Sexton, T., Yaffe, E., Kenigsberg, E., Bantignies, F., Leblanc, B., Hoichman, M., Parrinello, H., Tanay, A. & Cavalli, G. 2012. Three-dimensional folding and functional organiza-tion principles of the *Drosophila* genome. *Cell*, 148, 458–72.

Stack, S. M., Brown, D. B. & Dewey, W. C. 1977. Visualization of interphase chromosomes. *J Cell Sci*, 26, 281–99.

Stadler, M. R., Haines, J. E. & Eisen, M. B. 2017. Convergence of topological domain boundaries, insulators, and poly-tene interbands revealed by high-resolution mapping of chromatin contacts in the early *Drosophila melanogaster* embryo. *Elife*, 6, pii: e29550.

Symmons, O., Uslu, V. V., Tsujimura, T., Ruf, S., Nassari, S., Schwarzer, W., Ettwiller, L. & Spitz, F. 2014. Functional and topological characteristics of mammalian regulatory domains. *Genome Res*, 24, 390–400.

Talamas, J. A. & Capelson, M. 2015. Nuclear envelope and genome interactions in cell fate. *Front Genet*, 6, 95.

Taylor, J. H. 1963. DNA synthesis in relation to chromosome reproduction and the reunion of breaks. *J Cell Comp Physiol*, 62, SUPPL1: 73–86.

Tolhuis, B., Blom, M., Kerkhoven, R. M., Pagie, L., Teunissen, H., Nieuwland, M., Simonis, M., De Laat, W., Van Lohuizen, M. & Van Steensel, B. 2011. Interactions among Polycomb domains are guided by chromosome architecture. *PLoS Genet, 7*, e1001343.

Tolhuis, B., Palstra, R. J., Splinter, E., Grosveld, F. & De Laat W. 2002. Looping abd interactions between hypersensitove sitesin the active beta-globin locus. *Mol Cell, 10*, 1453–65.

Udvardy, A., Maine, E. & Schedl, P. 1985. The 87A7 chromomere. Identification of novel chromatin structures flanking the heat shock locus that may define the boundaries of higher order domains. *J Mol Biol, 185*, 341–58.

Ulianov, S. V., Khrameeva, E. E., Gavrilov, A. A., Flyamer, I. M., Kos, P., Mikhaleva, E. A., Penin, A. A., Logacheva, M. D., Imakaev, M. V., Chertovich, A., Gelfand, M. S., Shevelyov, Y. Y. & Razin, S. V. 2016. Active chromatin and transcription play a key role in chromosome partitioning into topologically associating domains. *Genome Res, 26*, 70–84.

Uuskula-Reimand, L., Hou, H., Samavarchi-Tehrani, P., Rudan, M. V., Liang, M., Medina-Rivera, A., Mohammed, H., Schmidt, D., Schwalie, P., Young, E. J., Reimand, J., Hadjur, S., Gingras, A. C. & Wilson, M. D. 2016. Topoisomerase II beta interacts with cohesin and CTCF at topological domain borders. *Genome Biol, 17*, 182.

Van Bortle, K. & Corces, V. G. 2012. tDNA insulators and the emerging role of TFIIIC in genome organization. *Transcription, 3*, 277–84.

Van Bortle, K., Nichols, M. H., Li, L., Ong, C. T., Takenaka, N., Qin, Z. S. & Corces, V. G. 2014. Insulator function and topological domain border strength scale with architectural protein occupancy. *Genome Biol, 15*, R82.

Vietri Rudan, M., Barrington, C., Henderson, S., Ernst, C., Odom, D. T., Tanay, A. & Hadjur, S. 2015. Comparative Hi-C reveals that CTCF underlies evolution of chromosomal domain architecture. *Cell Rep, 10*, 1297–309.

Weintraub, A. S., Li, C. H., Zamudio, A. V., Sigove, A. A., Hannett, N. M., Day, D. S., Abraham, B. J., Cohen, M. A., Nabet, B., Buckley, D. L., Guo, Y. E., Hnisz, D., Jaenisch, R., Bradner, J. E., Gray, N. S. & Young, R. A. 2017. YY1 is a structural regulator of enhancer promoter loops. Cell, 172, 1573-88.

Wendt, K. S. & Grosveld, F. G. 2014. Transcription in the context of the 3D nucleus. *Curr Opin Genet Dev,* 25, 62–7.

Wendt, K. S., Yoshida, K., Itoh, T., Bando, M., Koch, B., Schirghuber, E., Tsutsumi, S., Nagae, G., Ishihara, K., Mishiro, T., Yahata, K., Imamoto, F., Aburatani, H., Nakao, M., Imamoto, N., Maeshima, K., Shirahige, K. & Peters, J. M. 2008. Cohesin mediates transcriptional insulation by CCCTC-binding factor. *Nature,* 451, 796–801.

Zhao, H., Sifakis, E. G., Sumida, N., Millan-Arino, L., Scholz, B. A., Svensson, J. P., Chen, X., Ronnegren, A. L., Mallet De Lima, C. D., Varnoosfaderani, F. S., Shi, C., Loseva, O., Yammine, S., Israelsson, M., Rathje, L. S., Nemeti, B., Fredlund, E., Helleday, T., Imreh, M. P. & Gondor, A. 2015. PARP1- and CTCF-mediated interactions between active and repressed chromatin at the lamina promote oscillating transcription. *Mol Cell,* 59, 984–97.

Zhimulev, I. F., Zykova, T. Y., Goncharov, F. P., Khoroshko, V. A., Demakova, O. V., Semeshin, V. F., Pokholkova, G. V., Boldyreva, L. V., Demidova, D. S., Babenko, V. N., Demakov, S. A. & Belyaeva, E. S. 2014. Genetic organization of interphase chromosome bands and interbands in *Drosophila melanogaster. PLoS One,* 9, e101631.

Zolotarev, N., Fedotova, A., Kyrchanova, O., Bonchuk, A., Penin, A. A., Lando, A. S., Eliseeva, I. A., Kulakovskiy, I. V., Maksimenko, O. & Georgiev, P. 2016. Architectural proteins Pita, Zw5,and ZIPIC contain homodimerization domain and support specific long-range interactions in *Drosophila. Nucleic Acids Res,* 44, 7228–41.

CHAPTER 11

The nuclear envelope

The nuclear envelope is a multifunctional structure that serves as a dynamic interface between the nucleus and the cytoplasm. It provides protection for the genome and plays an important role in gene regulation. It is perforated by numerous nuclear pores that allow the selective passage of molecules into and out of the nucleus.

The nuclear envelope is a double membrane, consisting of two lipid bilayers. The outer membrane is continuous with the endoplasmic reticulum—an extensive network of intercommunicating flattened sacs and tubules that occupies most of the extra-nuclear cellular space. The inner membrane is associated with a layer of enmeshed protein intermediate filaments called the nuclear lamina. The two membranes are separated by the perinuclear space and are traversed by nuclear pores that allow the passage of molecules or molecular complexes between the nucleus and the cytoplasm. Both layers of the nuclear envelope are studded with transmembrane proteins that vary in different cells (Korfali et al., 2012; Schirmer et al., 2003) or in different developmental lineages (Chen et al., 2006) and have been implicated in cell type-specific biological characteristics.

Inner and outer membranes

Some of the proteins of the inner nuclear membrane interact with the nuclear lamina and are referred to as lamin-associated proteins (LAPs). LAPs bind directly, or via intermediate proteins, to chromatin, anchoring it to the nuclear envelope. Some LAPs interact with outer membrane proteins that, themselves, connect to cytoskeletal components such as actin and intermediate filaments. These LAPs have SUN domains (120-amino acid residue domains present in the lumen of the nuclear membrane) that bind the KASH domains of outer membrane proteins (regions of approximately 60 amino acids,

comprising a transmembrane segment and a C-terminal region that lies in the lumen). The resulting protein complexes are termed LINC because they link the nucleoskeleton with the cytoskeleton. An example of the functional significance of LINCs is the transmission of low-magnitude mechanical signals from the cellular environment to the nucleus (Uzer et al., 2015). Such signals have been shown to affect cell differentiation in a variety of cell types such as myocytes, endothelial cells, chondrocytes and fibroblasts by altering transcription (Tajik et al., 2016); the molecular mechanisms that translate extra-cellular mechanical input into parameters that control gene activity are not well understood.

The nuclear envelope through the cell cycle

During mitosis, the nuclear envelope disintegrates in order to allow attachment of the newly replicated chromosomes to the mitotic spindle and to ensure their distribution to the two daughter cells. This process is initiated by the phosphorylation of the lamins and some LAPs, causing the nuclear lamina to disintegrate; in addition, certain components of the nuclear pores are phosphorylated, leading to the disassembly of nuclear pore complexes into subcomplexes. Although several kinases act during mitosis, CDK1 (cyclin-dependent kinase 1), activated by cyclin B, is the major kinase involved in the breakdown of the nuclear envelope (Santamaria et al., 2007). MASTL (microtubule-associated serine/threonine kinase-like protein, a.k.a. GREATWALL) contributes to the process by activating various phosphatase inhibitors (Castilho et al., 2009).

During telophase, the CDK1 substrates are dephosphorylated in order for the nuclear membrane to reform. Regions of contact between the endoplasmic

Epigenetics, Nuclear Organization and Gene Function: with implications of epigenetic regulation and genetic architecture for human development and health. John C. Lucchesi, Oxford University Press (2019). © John C. Lucchesi 2019.
DOI: 10.1093/oso/9780198831204.001.0001

reticulum and the decondensing chromosomes are established via inner membrane proteins, directly or through chromatin-binding adaptor proteins; these regions eventually coalesce to generate the nuclear envelope. Nuclear pore complexes form on the newly assembled nuclear membrane from pre-existing components. Dephosphorylated lamins reassemble into the lamina and interact with inner membrane proteins, as well as with the chromatin. In order to be enclosed in a functional nuclear envelope, the genome must be in its natural state of nucleosome-based chromatin; naked DNA can lead to the formation of a membrane that lacks certain components and is devoid of nuclear pores (Inoue and Zhang, 2014; Zierhut et al., 2014).

The nuclear lamina

The lamina is a mesh of protein fibers consisting mostly of lamins

Lamins have a short amino-terminal globular domain followed by a central alpha-helical rod domain, a nuclear localization signal and a carboxy-terminal domain containing an immunoglobulin-like fold and a C-terminal CAAX motif (C is cysteine, A is an aliphatic amino acid and X is any amino acid). Four different lamins are present in mammals (two A-type: A and C; and two B-type: B1 and B2) and are expressed at particular levels in different cell types. A- and B-type lamins form separate filamentous networks that overlap and interact to some extent.

Lamins A and B undergo extensive modifications during their maturation. The cysteine residue of their CAAX motif is modified by the addition of a farnesyl group, the -AAX amino acid residues are excised and the cysteine is methylated. Lamin A is further modified by removal of 15 residues from the carboxy terminus that eliminates the farnesylated and methylated cysteine (Davies et al., 2009). Lamin C lacks the CAAX motif and does not undergo these modifications. The presence of the farnesylated cysteine in B-type, but not A-type, lamins may be responsible for their different behavior when the lamina disassembles during mitosis—A-type lamins diffuse through the cytoplasm, while B-type lamins remain associated with the endoplasmic reticulum. This differential distribution may explain the distinct behaviors of the two types of lamins during telophase when the nuclear envelope reforms. As endoplasmic reticulum membranes associate with the decondensing chromatin, B-type lamins assemble into filaments, while A-type lamins enter the nucleus at a later time, after a functional nuclear envelope has formed. The behavior of lamins during these processes appears to be regulated by phosphorylation and dephosphorylation of serine and threonine residues by specific kinases and phosphatases. Although most of the lamins form the lamina, some fraction remains in the nucleoplasm (Goldman et al., 1992; Lutz et al., 1992).

A number of additional proteins associate with the lamins and participate in a variety of regulatory functions. The lamina is responsible for much of the mechanical stability of the nucleus and plays important roles in DNA replication, in cell proliferation and in the organization of chromatin and gene expression during interphase. Mutations in lamin proteins are responsible for a series of human diseases (laminopathies) and for aging, giving this structure special importance in the understanding of epigenetic processes related to human health.

Association of the genome with the nuclear lamina

A substantial portion of the genome of eukaryotes contacts the nuclear lamina (Belmont et al., 1993). The first methodical mapping of such contacts was performed in *Drosophila* (Pickersgill et al., 2006) and revealed the presence of approximately 500 genes that interact with lamins. In mammals, the number of **lamina-associated domains** (LADs) is greater than 1000 and involves approximately 40% of the genome (Guelen et al., 2008). Transcriptionally inactive genes associate with heterochromatin foci (Brown et al., 1997). Consistent with the long-standing observation that heterochromatin is localized at the nuclear periphery, LADs usually position silent or gene-poor regions of the genome near the lamina and nuclear membrane. Attachment of the chromatin to these nuclear compartments is mediated by a number of different proteins (Kubben et al., 2010) (Fig. 11.1). Some inner membrane proteins, such as LAP2β (lamin-associated protein 2β), EMD (Emerin) and MAN1 [a.k.a. LEMD3 (LEM domain containing 3)], attach to chromatin through the chromatin-binding adaptor protein BAF (barrier-to-autointegration factor) (Montes de Oca et al., 2009) or through association with transcription factors (for example ThPOK, the vertebrate homolog of the *Drosophila* GAGA factor) (Zullo et al., 2012). Proteins such as the inner membrane-associated LBR (lamin B receptor) or the lamina protein PRR14 (proline-rich 14) bind to heterochromatin protein 1 (HP1) (Poleshko et al., 2013). Consistent with the observation that LADs represent transcriptionally inactive genomic regions is their enrichment in heterochromatin histone marks H3K9me2/3, H3K27me3 and H3K20me2 (Guelen et al., 2008; Pickersgill et al., 2006).

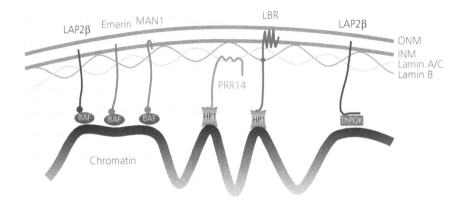

Fig. 11.1 Diagram illustrating the different molecular systems that anchor chromatin to the lamina and the nuclear membrane during interphase. (From Poleshko and Katz, 2014.)

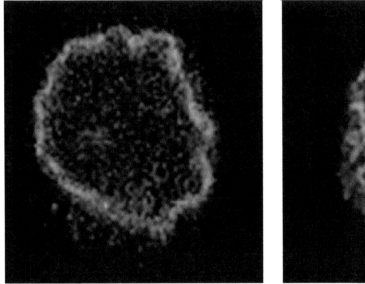
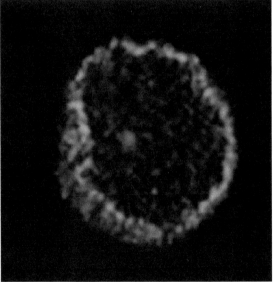

Fig. 11.2 The IgH locus (immunoglobulin heavy locus) is associated with the lamina in hematopoietic cells where it is inactive but moves to the more transcriptionally active center of the nucleus in pro-B lymphocytes. (From Kosak et al., 2002.)

These epigenetic features may be the key to tethering inactive regions to the lamina.

Transcriptional and epigenetic features of genes associated with the lamina

Although, in general, genes that are associated with the lamina are silent, some are active. These genes are characterized by a low level of transcription and by the position of their promoters that are not associated with the lamina, suggesting that lamina–promoter interactions repress transcription. Although the repressive H3K27me3 and H3K9me3 modified histones are enriched in LADs and mark a number of lamina-associated silent genes, a significant group of silent genes lack these modifications (Wu and Yao, 2017).

Facultative LADs

Not surprisingly, the position of some LADs (referred to as facultative LADs) changes during differentiation

(Fig. 11.2). Genes that are responsible for the maintenance of pluripotency and the execution of the cell cycle become associated with the lamina and are repressed, while genes that are specific to different cell types leave the lamina and are actively transcribed (Peric-Hupkes et al., 2010). Repositioning of LADs during differentiation may be achieved by the differential, tissue-specific expression of particular inner membrane or lamina-associated proteins (Korfali et al., 2010). Surprisingly, following each cell division in a given cell type, different sets of LADs appear to associate with the lamina (Kind et al., 2013). In light of this observation, the conclusion that differentiation is achieved by orchestrated changes in the tissue-specific association of LADs to the lamina will require the identification of the subset of these associations that is differentiation-specific. A number of LADs are conserved in different cell types. These so-called constitutive LADs are characterized by long stretches of A/T-rich DNA, suggesting a potential role for a sequence-related determinant of interaction with the nuclear lamina (Meuleman et al., 2013).

Perhaps surprisingly, after mitosis, the LADs of each daughter cell are different from one another and different from the LADs of the mother cell (Kind et al., 2013). This observation indicates that the positioning of genomic loci at the nuclear lamina is not transmitted and that, although some regions interact more frequently with the lamina, in any given cell, only a subset of LADs contact the lamina. In daughter cells, some of the maternal LADs associate with the perinucleolar region, the other gene-silenced compartment in the nucleus. Whether LADs do associate with these compartments depends on their level of H3K9 methylation. These observations also indicate that in some cells of the same tissue, some genes are active, while in other cells, these genes are repressed (Kind et al., 2015).

Nuclear pore complexes

Nuclear pore complexes (NPCs) are large multiprotein structures

Nuclear pores allow the passage of molecules between the nucleus and the cytoplasm. In vertebrates, each NPC is made up of approximately 30 different proteins called nucleoporins (Nups) assembled in repetitive subunits arranged around a central channel (Fig. 11.3). The Nups form a series of distinct components that include a core scaffold surrounded by two rings (cytoplasmic and nuclear). The core scaffold is lined by a set of Nups that are responsible for the regulation of transport through the central channel. Three different Nups

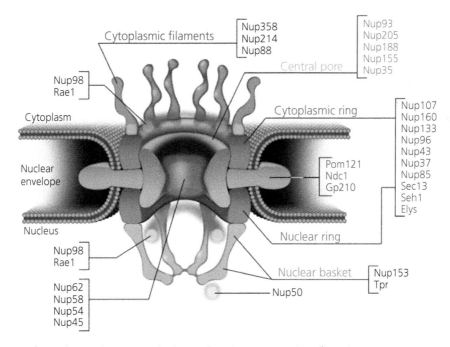

Fig. 11.3 Structure of a vertebrate nuclear pore complex showing the nucleoporin composition of its various components.

(From Ibarra and Hertzer, 2015.)

anchor the complex to the cell membrane (Fig. 11.4). Several Nups form filaments that extend into the cytoplasm or that form a basket on the nuclear side (Ibarra and Hetzer, 2015). Small molecules can pass, unassisted, through the NPCs. Larger molecules, such as proteins, require association with receptors that bind to specific amino acid sequences: nuclear localization sequences for import into the nucleus and nuclear export sequences for movement in the opposite direction. Once into the nucleoplasm, receptors are released by the action of a GTP hydrolase; in contrast, within the nucleus, the same GTP hydrolase facilitates the association of receptors with the proteins to be exported to the cytoplasm. In both cases, the receptor–protein complexes react with Nups for translocation through the NPCs. Non-coding RNAs, such as tRNAs or miRNAs, can utilize the same receptor-mediated mechanisms as proteins. Messenger RNAs use a different receptor and receptor release system.

Nuclear pores participate in the regulation of gene expression

The first indication that NPCs participate in gene regulation was obtained in yeast where the association of active genes with nuclear pores was thought to facilitate

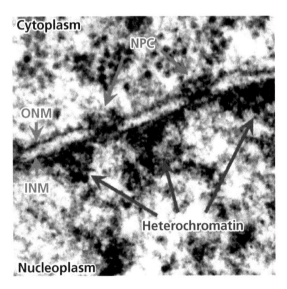

Fig. 11.4 Electron micrograph of a mammalian cell showing the association of heterochromatic regions with the inner nuclear membrane (INM) and the regions maintained free of heterochromatin by the presence of NPCs that traverse both the outer nuclear membrane (ONM) and the INM.

(From Ptak et al., 2014.)

the export of transcripts to the cytoplasm (Blobel, 1985). Subsequently, numerous examples of genes that require association with the NPC for maximal induction (Brickner and Walter, 2004; Casolari et al., 2004), or genes that are constitutively expressed (Casolari et al., 2004), were reported. The promoter regions of these genes harbor specific recruiting sequences (GRS) (Ahmed et al., 2010; Light et al., 2010). In contrast to their role in gene activation, contacts with NPCs can, in some cases, lead to repression (Galy et al., 2000; Van de Vosse et al., 2013). In multicellular organisms, although transcriptional regulation by association with the NPCs occurs as it does in yeast, some Nups associate with transcribing genes away from the nuclear membrane, within the nuclear interior. In *Drosophila*, nucleoporins stimulate the expression of developmental and cell cycle genes in the nucleoplasm and at the nuclear periphery (Capelson et al., 2010; Kalverda et al., 2010). A genome-wide study in the fruit fly revealed that Nups binding occurs along chromatin domains containing up to 500 kbp of DNA (Vaquerizas et al., 2010). Once again, as in yeast, some NPC-free Nups associate with repressed genes (Capelson et al., 2010). In mammalian cells, some Nups bind to superenhancers, regulatory sequences that activate groups of genes involved in cellular identity (Ibarra et al., 2016). Absence of these Nups has a significant effect on the expression of these genes.

Although a number of observations suggest that Nups exert their influence on transcription by facilitating chromatin remodeling activities or gene looping, the molecular bases of these mechanisms are still largely uncharacterized (Ibarra and Hetzer, 2015). An ancillary aspect of the action of Nups is the transcriptional memory that they confer to the genes with which they associate. In yeast, the association of inducible genes with the NPC leads to faster reactivation of these genes in subsequent cell generations (Brickner et al., 2007; Tan-Wong et al., 2009). A similar phenomenon is manifested in human cells (Light et al., 2013).

Chapter summary

The nuclear envelope is a double membrane sheath made up of two lipid bilayers—an outer and an inner membrane. The inner surface of the inner membrane is associated with a meshwork of filaments made up of lamins and of lamin-associated proteins that constitute the lamina. The lamina is responsible for much of the mechanical stability of the nucleus and plays important roles in the organization of chromatin and gene expression during interphase, and in DNA replication and cell proliferation. A substantial portion of the genome contacts the lamina through lamina-associated domains (LADs). LADs usually position silent or gene-poor regions of the genome near the lamina and nuclear membrane. Nevertheless, some lamina-associated genes whose promoters escape this association exhibit a low level of transcription. The position of some LADs is different in some cells of the same tissue, reflecting the stochastic nature of gene activity; it can also change during differentiation, allowing the necessary activation of particular genes.

Nuclear pores are large multiprotein structures that allow the passage of molecules between the nucleus and the cytoplasm. Contact of transcription units with nuclear pores can result in activation or, sometimes, repression. Some of the proteins that contribute to the structure of the pores can activate transcription by associating with genes or with super-enhancers away from the nuclear membrane.

References

Ahmed, S., Brickner, D. G., Light, W. H., Cajigas, I., Mcdonough, M., Froyshteter, A. B., Volpe, T. & Brickner, J. H. 2010. DNA zip codes control an ancient mechanism for gene targeting to the nuclear periphery. *Nat Cell Biol*, 12, 111–18.

Belmont, A. S., Zhai, Y. & Thilenius, A. 1993. Lamin B distribution and association with peripheral chromatin revealed by optical sectioning and electron microscopy tomography. *J Cell Biol*, 123, 1671–85.

Blobel, G. 1985. Gene gating: a hypothesis. *Proc Natl Acad Sci U S A*, 82, 8527–9.

Brickner, D. G., Cajigas, I., Fondufe-Mittendorf, Y., Ahmed, S., Lee, P. C., Widom, J. & Brickner, J. H. 2007. H2A.Z-mediated localization of genes at the nuclear periphery confers epigenetic memory of previous transcriptional state. *PLoS Biol*, 5, e81.

Brickner, J. H. & Walter, P. 2004. Gene recruitment of the activated INO1 locus to the nuclear membrane. *PLoS Biol*, 2, e342.

Brown, K. E., Guest, S. S., Smale, S. T., Hahm, K., Merkenschlager, M. & Fisher, A. G. 1997. Association of transcriptionally silent genes with Ikaros complexes at centromeric heterochromatin. *Cell*, 91, 845–54.

Capelson, M., Liang, Y., Schulte, R., Mair, W., Wagner, U. & Hetzer, M. W. 2010. Chromatin-bound nuclear pore components regulate gene expression in higher eukaryotes. *Cell*, 140, 372–83.

Casolari, J. M., Brown, C. R., Komili, S., West, J., Hieronymus, H. & Silver, P. A. 2004. Genome-wide localization of the nuclear transport machinery couples transcriptional status and nuclear organization. *Cell*, 117, 427–39.

Castilho, P. V., Williams, B. C., Mochida, S., Zhao, Y. & Goldberg, M. L. 2009. The M phase kinase Greatwall (Gwl) promotes inactivation of PP2A/B55delta, a phosphatase directed against CDK phosphosites. *Mol Biol Cell*, 20, 4777–89.

Chen, I. H., Huber, M., Guan, T., Bubeck, A. & Gerace, L. 2006. Nuclear envelope transmembrane proteins (NETs) that are up-regulated during myogenesis. *BMC Cell Biol*, 7, 38.

Davies, B. S., Fong, L. G., Yang, S. H., Coffinier, C. & Young, S. G. 2009. The posttranslational processing of prelamin A and disease. *Annu Rev Genomics Hum Genet*, 10, 153–74.

Galy, V., Olivo-Marin, J. C., Scherthan, H., Doye, V., Rascalou, N. & Nehrbass, U. 2000. Nuclear pore complexes in the organization of silent telomeric chromatin. *Nature*, 403, 108–12.

Goldman, A. E., Moir, R. D., Montag-Lowy, M., Stewart, M. & Goldman, R. D. 1992. Pathway of incorporation of microinjected lamin A into the nuclear envelope. *J Cell Biol*, 119, 725–35.

Guelen, L., Pagie, L., Brasset, E., Meuleman, W., Faza, M. B., Talhout, W., Eussen, B. H., De Klein, A., Wessels, L., De Laat, W. & Van Steensel, B. 2008. Domain organization of human chromosomes revealed by mapping of nuclear lamina interactions. *Nature*, 453, 948–51.

Ibarra, A., Benner, C., Tyagi, S., Cool, J. & Hetzer, M. W. 2016. Nucleoporin-mediated regulation of cell identity genes. *Genes Dev*, 30, 2253–8.

Ibarra, A. & Hetzer, M. W. 2015. Nuclear pore proteins and the control of genome functions. *Genes Dev*, 29, 337–49.

Inoue, A. & Zhang, Y. 2014. Nucleosome assembly is required for nuclear pore complex assembly in mouse zygotes. *Nat Struct Mol Biol*, 21, 609–16.

Kalverda, B., Pickersgill, H., Shloma, V. V. & Fornerod, M. 2010. Nucleoporins directly stimulate expression of developmental and cell-cycle genes inside the nucleoplasm. *Cell*, 140, 360–71.

Kind, J., Pagie, L., De Vries, S. S., Nahidiazar, L., Dey, S. S., Bienko, M., Zhan, Y., Lajoie, B., De Graaf, C. A., Amendola, M., Fudenberg, G., Imakaev, M., Mirny, L. A., Jalink, K., Dekker, J., Van Oudenaarden, A. & Van Steensel, B. 2015. Genome-wide maps of nuclear lamina interactions in single human cells. *Cell*, 163, 134–47.

Kind, J., Pagie, L., Ortabozkoyun, H., Boyle, S., De Vries, S. S., Janssen, H., Amendola, M., Nolen, L. D., Bickmore, W. A. & Van Steensel, B. 2013. Single-cell dynamics of genome-nuclear lamina interactions. *Cell*, 153, 178–92.

Korfali, N., Wilkie, G. S., Swanson, S. K., Srsen, V., Batrakou, D. G., Fairley, E. A., Malik, P., Zuleger, N., Goncharevich, A., De Las Heras, J., Kelly, D. A., Kerr, A. R., Florens, L. & Schirmer, E. C. 2010. The leukocyte nuclear envelope proteome varies with cell activation and contains novel transmembrane proteins that affect genome architecture. *Mol Cell Proteomics*, 9, 2571–85.

Korfali, N., Wilkie, G. S., Swanson, S. K., Srsen, V., De Las Heras, J., Batrakou, D. G., Malik, P., Zuleger, N., Kerr, A. R., Florens, L. & Schirmer, E. C. 2012. The nuclear envelope proteome differs notably between tissues. *Nucleus,* 3, 552–64.

Kosak, S. T., Skok, J. A., Medina, K. L., Riblet, R., Le Beau, M. M., Fisher, A. G. & Singh, H. 2002. Subnuclear compartmentalization of immunoglobulin loci during lymphocyte development. *Science,* 296, 158–62.

Kubben, N., Voncken, J. W., Demmers, J., Calis, C., Van Almen, G., Pinto, Y. & Misteli, T. 2010. Identification of differential protein interactors of lamin A and progerin. *Nucleus,* 1, 513–25.

Light, W. H., Brickner, D. G., Brand, V. R. & Brickner, J. H. 2010. Interaction of a DNA zip code with the nuclear pore complex promotes H2A.Z incorporation and INO1 transcriptional memory. *Mol Cell,* 40, 112–25.

Light, W. H., Freaney, J., Sood, V., Thompson, A., D'urso, A., Horvath, C. M. & Brickner, J. H. 2013. A conserved role for human Nup98 in altering chromatin structure and promoting epigenetic transcriptional memory. *PLoS Biol,* 11, e1001524.

Lutz, R. J., Trujillo, M. A., Denham, K. S., Wenger, L. & Sinensky, M. 1992. Nucleoplasmic localization of prelamin A: implications for prenylation-dependent lamin A assembly into the nuclear lamina. *Proc Natl Acad Sci U S A,* 89, 3000–4.

Meuleman, W., Peric-Hupkes, D., Kind, J., Beaudry, J. B., Pagie, L., Kellis, M., Reinders, M., Wessels, L. & Van Steensel, B. 2013. Constitutive nuclear lamina-genome interactions are highly conserved and associated with A/T-rich sequence. *Genome Res,* 23, 270–80.

Montes De Oca, R., Shoemaker, C. J., Gucek, M., Cole, R. N. & Wilson, K. L. 2009. Barrier-to-autointegration factor proteome reveals chromatin-regulatory partners. *PLoS One,* 4, e7050.

Peric-Hupkes, D., Meuleman, W., Pagie, L., Bruggeman, S. W., Solovei, I., Brugman, W., Graf, S., Flicek, P., Kerkhoven, R. M., Van Lohuizen, M., Reinders, M., Wessels, L. & Van Steensel, B. 2010. Molecular maps of the reorganization of genome-nuclear lamina interactions during differentiation. *Mol Cell,* 38, 603–13.

Pickersgill, H., Kalverda, B., De Wit, E., Talhout, W., Fornerod, M. & Van Steensel, B. 2006. Characterization of the *Drosophila melanogaster* genome at the nuclear lamina. *Nat Genet,* 38, 1005–14.

Poleshko, A. & Katz, R. A. 2014. Specifying peripheral heterochromatin during nuclear lamina reassembly. *Nucleus,* 5, 32–9.

Poleshko, A., Mansfield, K. M., Burlingame, C. C., Andrake, M. D., Shah, N. R. & Katz, R. A. 2013. The human protein PRR14 tethers heterochromatin to the nuclear lamina during interphase and mitotic exit. *Cell Rep,* 5, 292–301.

Ptak, C., Aitchison, J. D. & Wozniak, R. W. 2014. The multifunctional nuclear pore complex: a platform for controlling gene expression. *Curr Opin Cell Biol,* 28, 46–53.

Santamaria, D., Barriere, C., Cerqueira, A., Hunt, S., Tardy, C., Newton, K., Caceres, J. F., Dubus, P., Malumbres, M. & Barbacid, M. 2007. Cdk1 is sufficient to drive the mammalian cell cycle. *Nature,* 448, 811–15.

Schirmer, E. C., Florens, L., Guan, T., Yates, J. R., 3RD & Gerace, L. 2003. Nuclear membrane proteins with potential disease links found by subtractive proteomics. *Science,* 301, 1380–2.

Tajik, A., Zhang, Y., Wei, F., Sun, J., Jia, Q., Zhou, W., Singh, R., Kjanna, N., Belmont, A. S. & Wang, N. 2016. Transcription upregulation via force-induced direct stretching of chromatin. *Nat Mater,* 15, 1287–96.

Tan-Wong, S. M., Wijayatilake, H. D. & Proudfoot, N. J. 2009. Gene loops function to maintain transcriptional memory through interaction with the nuclear pore complex. *Genes Dev,* 23, 2610–24.

Uzer, G., Thompson, W. R., Sen, B., Xie, Z., Yen, S. S., Miller, S., Bas, G., Styner, M., Rubin, C. T., Judex, S., Burridge, K. & Rubin, J. 2015. Cell mechanosensitivity to extremely low-magnitude signals is enabled by a LINCed nucleus. *Stem Cells,* 33, 2063–76.

Van De Vosse, D. W., Wan, Y., Lapetina, D. L., Chen, W. M., Chiang, J. H., Aitchison, J. D. & Wozniak, R. W. 2013. A role for the nucleoporin Nup170p in chromatin structure and gene silencing. *Cell,* 152, 969–83.

Vaquerizas, J. M., Suyama, R., Kind, J., Miura, K., Luscombe, N. M. & Akhtar, A. 2010. Nuclear pore proteins nup153 and megator define transcriptionally active regions in the *Drosophila* genome. *PLoS Genet,* 6, e1000846.

Wu, F. & Yao, J. 2017. Identifying novel transcriptional and epigenetic features of nuclear lamina-associated genes. *Sci Rep,* 7, 100.

Zierhut, C., Jenness, C., Kimura, H. & Funabiki, H. 2014. Nucleosomal regulation of chromatin composition and nuclear assembly revealed by histone depletion. *Nat Struct Mol Biol,* 21, 617–25.

Zullo, J. M., Demarco, I. A., Pique-Regi, R., Gaffney, D. J., Epstein, C. B., Spooner, C. J., Luperchio, T. R., Bernstein, B. E., Pritchard, J. K., Reddy, K. L. & Singh, H. 2012. DNA sequence-dependent compartmentalization and silencing of chromatin at the nuclear lamina. *Cell,* 149, 1474–87.

CHAPTER 12

The nucleolus

The existence of the nucleolus as a visible inclusion within the nucleus of most cells during interphase was first reported in the 1830s. Approximately 100 years later, Barbara McClintock discovered that the nucleolus was associated with a specific chromosomal region termed the **nucleolus organizer**. The first indication of a function that could be ascribed to the nucleolus was that its absence in anucleolate mutant frog embryos led to a complete absence of ribosomal RNA (rRNA) synthesis (Brown and Gurdon, 1964). This correlation was validated at the molecular level by the demonstration that the DNA in the nucleolus organizer region of *Drosophila* is complementary to rRNA (Ritossa and Spiegelman, 1965). Soon thereafter, transcription of the genes that encoded rRNA was illustrated in one of the most iconic photomicrographs ever published (Miller and Beatty, 1969). These seminal investigations led to a complete understanding of the nucleolus's role in rRNA synthesis and the biogenesis of ribosomal subunits. Concomitantly with some of these studies, a number of additional nucleolar functions have been discovered, establishing this organelle as a key hub in epigenetic regulation.

Ribosome biogenesis

The nucleolus forms at specific chromosomal locations called nucleolus organizer regions (NORs) that consist of clusters of repeated rRNA genes. The number of NORs and of the rRNA genes they contain varies among different organisms. In *Drosophila melanogaster*, there are two NORs, present on the X and Y chromosomes, harboring approximately 100 and 300 rRNA gene copies, respectively, arranged in a head-to-tail orientation. In both males and females, the two NORs are associated in the chromocenter (formed by the aggregation of centro-meric heterochromatin regions) and give rise to a single nucleolus. A peculiar and unexplained case of epigenetic regulation involves the exclusive expression of the Y chromosome rRNA gene array in males (Greil and Ahmad, 2012; Zhou et al., 2012). This phenomenon, termed nucleolar dominance, was originally discovered in interspecific hybrids in which only the rRNA genes inherited from one parent give rise to a functioning nucleolus (Pikaard, 2000). In *Drosophila* males, nucleolar dominance is not due to gametic imprinting and suggests the existence of a specific chromatin regulatory system. In human cells, there are approximately 400 RNA gene repeats, each 43 kbp long, present in groups of approximately 80 copies on chromosomes 13, 14, 15, 21 and 22. The NORs on different chromosomes aggregate so that, in general, only two or three nucleoli are formed in a cell.

Early ultrastructural studies revealed that nucleoli consist of morphologically distinct sub-compartments: a fibrillar center, a dense fibrillar component and a granular component (Fig. 12.1). The fibrillar center contains the rRNA genes that are transcribed therein or at the border between this sub-compartment and the dense fibrillar component by RNA polymerase I (Koberna et al., 2002). The dense fibrillar component contains the rRNA gene transcripts and is the site of their processing and modification. In this sub-compartment, the primary 45S transcript is cleaved to yield the mature 28S, 18S and 5.8S rRNAs (S, or Svedberg, is a unit of sedimentation rate that is used to indicate the size of a molecule). The granular component is the region where ribosomal proteins, which have been synthesized in the cytoplasm and have re-entered the nucleus, associate with the rRNAs to produce the two ribosome subunits. In mammals, the large subunit includes approximately 60 proteins plus 28S and 5.8S

Epigenetics, Nuclear Organization and Gene Function: with implications of epigenetic regulation and genetic architecture for human development and health. John C. Lucchesi, Oxford University Press (2019). © John C. Lucchesi 2019.
DOI: 10.1093/oso/9780198831204.001.0001

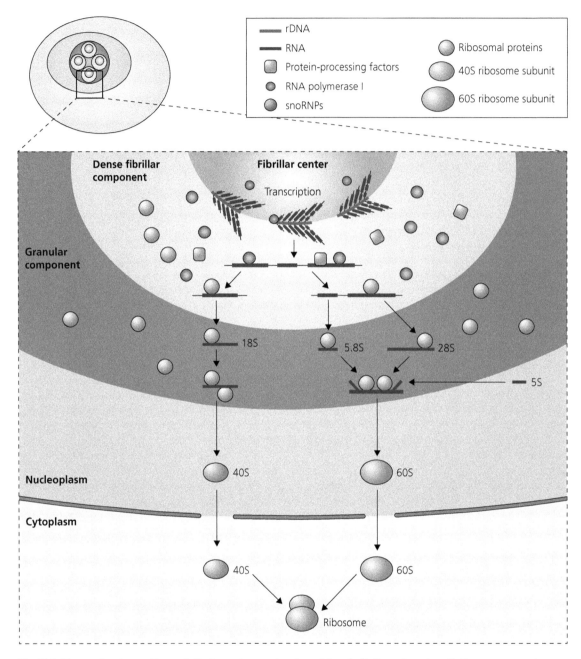

Fig. 12.1 Diagram of nucleosome biogenesis. The three sub-compartments consisting of a fibrillar center, a dense fibrillar component and a granular component, and the sequence of events that occur in each of them are illustrated.

(From Boisvert et al., 2007.)

RNAs, as well as a 5S RNA; the latter is encoded by clusters of genes that are usually located in regions other than the NORs and that are transcribed by RNA polymerase III. In *Drosophila melanogaster*, the 5S RNA genes are present as a cluster of 100–200 repeats on chromosome 2; in humans, the repeated genes are significantly more numerous, with one large cluster present on chromosome 1 and smaller clusters on other chromosomes. The small ribosomal subunit includes approximately 20 proteins and the 18S RNA.

During processing, rRNAs undergo extensive modifications consisting of ribose 2′-O-methylation and pseudouridine formation. These steps are guided by the action of numerous small nucleolar RNAs (snoRNAs) that associate with a set of core proteins, including either the 2′-O-methyl transferase fibrillarin or the pseudouridylase dyskerin. The resulting snoRNPs are targeted to the appropriate regions of the rRNAs by means of sequence complementarity (Li et al., 1990; Tollervey et al., 1993).

Only subsets of the rRNA genes present in cells are transcribed (Conconi et al., 1989). The ratio of active to inactive genes varies with cell differentiation and is greater in cells that are involved in extensive protein synthesis. This ratio is also affected by aging and by the onset of oncogenesis. Active rRNA gene arrays are present within the nucleolus, whereas silent gene arrays form loops that protrude into the nucleoplasm. A detailed description of this distribution was obtained in the model plant *Arabidopsis thaliana*—only active rRNA genes are included in isolated nucleoli, while silenced genes are excluded (Pontvianne et al., 2013). The epigenetic control of rRNA gene transcription has been described in Chapter 9.

The role that the linker histone H1 plays in the condensation of inactive chromatin and the formation of heterochromatin led to the investigation of its specific role in the repression of inactive rRNA genes. Using purified nucleoli from human cells with normal H1 expression, as well as from cells where H1 had been abrogated, a large number of nucleolar proteins were found to associate directly or indirectly with H1 or to

depend on H1 for their nucleolar concentration. Prevalent among them were proteins involved in both positive and negative RNAPI regulation, rRNA processing and mRNA splicing (Szerlong et al., 2015). H1 and many of the interacting proteins form large multimolecular complexes, leading to the conclusion that this linker histone is a major protein hub responsible for maintaining many of the important functions of the nucleolus.

The nucleolus as a gene-silencing environment

The nucleolus is surrounded by a zone of heterochromatin consisting of silenced rRNA gene arrays, DNA repeats that flank the centromeres (pericentric heterochromatin) and chromatin domains that include gene-poor, as well as silent, regions of the genome (Fig. 12.2). A particularly informative example of the latter case is provided by the large group of genes that encode the olfactory receptors in the airway epithelium of mice (introduced in Chapter 8). Each cell in the epithelium expresses a single receptor gene (Chess et al., 1994) that associates with an enhancer located in a euchromatic region of chromosome 12. All of the other genes converge in a small number of clusters enriched in H3K9me3, H4K20me3 and HP1 that are localized to the nuclear lamina (Clowney et al., 2012), but more extensively to the nucleolar periphery (Nemeth et al., 2010). Although the molecular mechanisms responsible for targeting genomic regions to the nucleolar periphery are still largely elusive, long non-coding

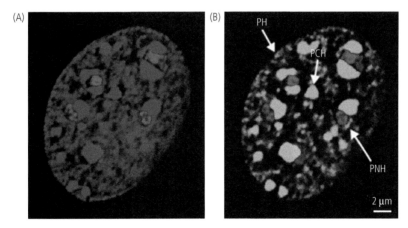

Fig. 12.2 Heterochromatin is associated with the perinucleolar region. (A) Nucleoli are stained red with an antibody that recognizes fibrillarin; the genomic DNA is stained blue. (B) Same nucleus with the nucleoli stained as before, but the genome is stained with an antibody that recognizes H3K9me3. PH indicates peripheral heterochromatin; PNH indicates perinucleolar heterochromatin.

(From Politz et al., 2016.)

RNAs (lncRNAs) appear to be involved in some cases. A lncRNA that causes the silencing in *cis* of a cluster of imprinted paternal alleles targets the silenced alleles to the nucleolar periphery (Mohammad et al., 2008). Another lncRNA, transcribed from a gene that escapes X chromosome inactivation, mediates the association of the inactive X to the nucleolar periphery (Yang et al., 2015; Zhang et al., 2007).

The nucleolus is the site of many diverse molecular processes

Nucleoli from human cells contain around 700–900 different proteins (Jarboui et al., 2011; Leung et al., 2006), only some of which are involved in ribosome biogenesis-related processes. Other proteins are known to function in chromatin remodeling, gene expression, DNA replication and repair and chromosome segregation.

As mentioned in Ribosome biogenesis, p. 147, the NORs present on different chromosomes aggregate so that, in general, a smaller number of nucleoli are formed in a cell. The consequence of this aggregation is that genes adjacent to different NORs are brought into close vicinity (Strongin et al., 2014). A number of other genomic regions that are distant from the NORs are targeted to the nucleolus by mechanisms that are poorly understood. Although these regions, termed nucleolus-associated domains (NADs), generally include genes that are repressed, some of the genes that they contain are actively transcribed (Nemeth et al., 2010; van Koningsbruggen et al., 2010). In addition, under conditions of stress (change in pH, impaired ribosome biosynthesis, DNA damage and other oncogenic alterations), a number of proteins travel to the nucleolus where they are retained. Targeting of these proteins depends on the interaction between a specific domain in their amino acid sequence and non-coding RNAs produced by intergenic rRNA spacers (Audas et al., 2012). One should keep in mind that the nucleolar proteome also changes in response to transcriptional inhibition or senescence.

Some of the nucleolar proteins and the processes with which they are involved are discussed in the following paragraphs in order to illustrate the multifunctional nature of the nucleolus.

Telomerase assembly

Telomerase is required for the maintenance of telomere length during DNA replication (see Chapter 20); it consists of a complex that includes a reverse transcriptase enzyme, an RNA component that serves as a template for the synthesis of the DNA repeats at the end of chromosomes and several additional proteins that are involved in the assembly and function of the complex. Although telomerase accumulates in nuclear Cajal bodies (see Chapter 13) from where it accesses telomeres (Cristofari et al., 2007), it is assembled in the dense fibrillar component of the nucleolus (Lee et al., 2014). By associating with the nucleolar phosphoprotein nucleolin (see Chapter 9), telomerase is retained in the nucleolus until late S phase when the telomeres are replicated and is transported to the Cajal bodies by the telomerase Cajal body protein 1 (TCAB1). Telomerase is recruited to telomeres by one of the proteins that protect telomeres from nucleolytic and other attacks.

Regulation of p53 stability

The tumor suppressor gene *p53* produces a transcription factor that induces the expression of a number of genes, leading to cell cycle arrest, cellular senescence or apoptosis. Under normal cellular conditions, p53 levels are kept low by the presence of MDM2 (mouse double minute 2), a ubiquitin ligase that binds to p53, ubiquitinates it and targets it for degradation. Different proteins can bind to MDM2 and alter its p53-regulatory activity. If ribosome biogenesis is impaired, free ribosomal proteins inhibit MDM2 and lead to the activation and stabilization of p53 (Lohrum et al., 2003; Pestov et al., 2001; Zhang et al., 2003). Another tumor suppressor ARF (alternate reading frame) is absent in normal cells through the repression of its locus by PcG complexes (Bracken et al., 2007). ARF regulates ribosome biogenesis by modulating rRNA synthesis and processing. In addition, oncogenic signals and other cellular stress conditions cause ARF to contribute to an increase in active p53 levels by binding and sequestering MDM2 in the granular component of the nucleolus (Pomerantz et al., 1998; Stott et al., 1998). Another factor involved in p53 activation is a ribonucleoprotein particle (RNP) that contains 5S RNA.

Modifications and processing of non-ribosomal RNAs

A number of micro RNAs (miRNAs) are found predominantly in the nucleolus. Cellular stress, in the form of foreign RNA or DNA insertions, causes a shift of some of these miRNAs to the cytoplasm, suggesting that they are stored in the nucleolus and are released when their function is needed (Li et al., 2013). Some miRNAs may participate in the regulation of ribosomal

RNA maturation and modification. Other RNAs, such as 5S RNA and tRNAs, are generated in the nucleolus. In addition, RNP complexes, including complexes involved in the splicing of mRNAs, are also assembled in the nucleolus.

Eukaryotic tRNAs are encoded by genes that are dispersed throughout the genome. Active genes form clusters in the nucleolus where they are transcribed by RNA polymerase III. The primary transcripts are extensively processed to produce the mature tRNAs, which are then issued into the cytoplasm by export receptors (Hopper et al., 2010). RNAPIII also transcribes the 5S RNA genes; this RNA and two specific ribosomal proteins (RPL5 and RPL11) form the 5S RNP complex that is an essential component of the large ribosomal subunit. The two ribosomal proteins had been previously shown to inhibit MDM2 and, thereby, to activate p53. Recent evidence suggests that the 5S RNA itself is required for p53 induction and stabilization following cellular stress, and that the 5S RNP is required for p53 activation by ARF (Sloan et al., 2013).

The nucleolus has not been shown directly to be a center for primary transcript processing. Nevertheless, a very substantial number of pre-mRNA processing proteins have been found in various nucleolar proteomic studies (Jarboui et al., 2011; Leung et al., 2006; Szerlong et al., 2015). In the plant model organism *Arabidopsis*, the nucleolus is enriched in aberrantly spliced mRNAs and in the proteins of the nonsense-mediated decay pathway, indicating that it is involved in mRNA quality control (Kim et al., 2009).

Chapter summary

The nucleolus forms at specific chromosomal locations called nucleolus organizer regions (NORs) that consist of clusters of repeated rRNA genes. The number of NORs and of the rRNA genes they contain varies among different organisms. Transcription of the rRNA genes and processing of the transcripts yield the three types of RNAs necessary for the biogenesis of ribosomes. Only subsets of the rRNA genes present in cells are transcribed. The ratio of active to inactive genes varies with cell differentiation and is greater in cells that are involved in extensive protein synthesis. This ratio is also affected by aging and by the onset of oncogenesis. The linker histone H1 plays a specific role in the repression of inactive rRNA genes and in many of the other functions of the nucleolus. One of these functions is gene silencing—the nucleolus is surrounded by a zone of heterochromatin consisting of silenced rRNA gene arrays, DNA repeats that flank the centromeres and

chromatin domains that include gene-poor, as well as silent, regions of the genome; any gene associating with this zone is subjected to repression. Other functions include the assembly of telomerase, the regulation of p53 stability and the synthesis of 5S and tRNAs whose genes form clusters in the nucleolus.

References

Audas, T. E., Jacob, M. D. & Lee, S. 2012. Immobilization of proteins in the nucleolus by ribosomal intergenic spacer noncoding RNA. *Mol Cell*, 45, 147–57.

Boisvert, F. M., Van Koningsbruggen, S., Navascues, J. & Lamond, A. I. 2007. The multifunctional nucleolus. *Nat Rev Mol Cell Biol*, 8, 574–85.

Bracken, A. P., Kleine-Kohlbrecher, D., Dietrich, N., Pasini, D., Gargiulo, G., Beekman, C., Theilgaard-Monch, K., Minucci, S., Porse, B. T., Marine, J. C., Hansen, K. H. & Helin, K. 2007. The Polycomb group proteins bind throughout the INK4A-ARF locus and are disassociated in senescent cells. *Genes Dev*, 21, 525–30.

Brown, D. D. & Gurdon, J. B. 1964. Absence of ribosomal RNA synthesis in the anucleolate mutant of *Xenopus laevis*. *Proc Natl Acad Sci U S A*, 51, 139–46.

Chess, A., Simon, I., Cedar, H. & Axel, R. 1994. Allelic inactivation regulates olfactory receptor gene expression. *Cell*, 78, 823–34.

Clowney, E. J., Legros, M. A., Mosley, C. P., Clowney, F. G., Markenskoff-Papadimitriou, E. C., Myllys, M., Barnea, G., Larabell, C. A. & Lomvardas, S. 2012. Nuclear aggregation of olfactory receptor genes governs their monogenic expression. *Cell*, 151, 724–37.

Conconi, A., Widmer, R. M., Koller, T. & Sogo, J. M. 1989. Two different chromatin structures coexist in ribosomal RNA genes throughout the cell cycle. *Cell*, 57, 753–61.

Cristofari, G., Adolf, E., Reichenbach, P., Sikora, K., Terns, R. M., Terns, M. P. & Lingner, J. 2007. Human telomerase RNA accumulation in Cajal bodies facilitates telomerase recruitment to telomeres and telomere elongation. *Mol Cell*, 27, 882–9.

Greil, F. & Ahmad, K. 2012. Nucleolar dominance of the Y chromosome in *Drosophila melanogaster*. *Genetics*, 191, 1119–28.

Hopper, A. K., Pai, D. A. & Engelke, D. R. 2010. Cellular dynamics of tRNAs and their genes. *FEBS Lett*, 584, 310–17.

Jarboui, M. A., Wynne, K., Elia, G., Hall, W. W. & Gautier, V. W. 2011. Proteomic profiling of the human T-cell nucleolus. *Mol Immunol*, 49, 441–52.

Kim, S. H., Koroleva, O. A., Lewandowska, D., Pendle, A. F., Clark, G. P., Simpson, C. G., Shaw, P. J. & Brown, J. W. 2009. Aberrant mRNA transcripts and the nonsense-mediated decay proteins UPF2 and UPF3 are enriched in the *Arabidopsis* nucleolus. *Plant Cell*, 21, 2045–57.

Koberna, K., Malinsky, J., Pliss, A., Masata, M., Vecerova, J., Fialova, M., Bednar, J. & Raska, I. 2002. Ribosomal genes in focus: new transcripts label the dense fibrillar components and form clusters indicative of "Christmas trees" *in situ*. *J Cell Biol*, 157, 743–8.

Lee, J. H., Lee, Y. S., Jeong, S. A., Khadka, P., Roth, J. & Chung, I. K. 2014. Catalytically active telomerase holoenzyme is assembled in the dense fibrillar component of the nucleolus during S phase. *Histochem Cell Biol*, 141, 137–52.

Leung, A. K., Trinkle-Mulcahy, L., Lam, Y. W., Andersen, J. S., Mann, M. & Lamond, A. I. 2006. NOPdb: nucleolar proteome database. *Nucleic Acids Res*, 34, D218–20.

Li, Z. F., Liang, Y. M., Lau, P. N., Shen, W., Wang, D. K., Cheung, W. T., Xue, C. J., Poon, L. M. & Lam, Y. W. 2013. Dynamic localisation of mature microRNAs in human nucleoli is influenced by exogenous genetic materials. *PLoS One*, 8, e70869.

Lohrum, M. A., Ludwig, R. L., Kubbutat, M. H., Hanlon, M. & Vousden, K. H. 2003. Regulation of HDM2 activity by the ribosomal protein L11. *Cancer Cell*, 3, 577–87.

Miller, O. L.,JR. & Beatty, B. R. 1969. Visualization of nucleolar genes. *Science*, 164, 955–7.

Mohammad, F., Pandey, R. R., Nagano, T., Chakalova, L., Mondal, T., Fraser, P. & Kanduri, C. 2008. Kcnq1ot1/Lit1 noncoding RNA mediates transcriptional silencing by targeting to the perinucleolar region. *Mol Cell Biol*, 28, 3713–28.

Nemeth, A., Conesa, A., Santoyo-Lopez, J., Medina, I., Montaner, D., Peterfia, B., Solovei, I., Cremer, T., Dopazo, J. & Langst, G. 2010. Initial genomics of the human nucleolus. *PLoS Genet*, 6, e1000889.

Pestov, D. G., Strezoska, Z. & Lau, L. F. 2001. Evidence of p53-dependent cross-talk between ribosome biogenesis and the cell cycle: effects of nucleolar protein Bop1 on G(1)/S transition. *Mol Cell Biol*, 21, 4246–55.

Pikaard, C. S. 2000. The epigenetics of nucleolar dominance. *Trends Genet*, 16, 495–500.

Politz, J. C. R., Scalzo, D. & Groudine, M. 2016. The redundancy of the mammalian heterochromatic compartment. *Curr Opin Genet Dev*, 37, 1–8.

Pomerantz, J., Schreiber-Agus, N., Liegeois, N. J., Silverman, A., Alland, L., Chin, L., Potes, J., Chen, K., Orlow, I., Lee, W., Cordon-Cardo, C. & Depinho, R. A. 1998. The Ink4a tumor suppressor gene product, p19Arf, interacts with MDM2 and neutralizes MDM2's inhibition of p53. *Cell*, 92, 713–23.

Pontvianne, F., Blevins, T., Chandrasekhara, C., Mozgova, I., Hassel, C., Pontes, O. M., Tucker, S., Mokros, P., Muchova, V., Fajkus, J. & Pikaard, C. S. 2013. Subnuclear partitioning of rRNA genes between the nucleolus and nucleoplasm reflects alternative epiallelic states. *Genes Dev*, 27, 1545–50.

Ritossa, F. M. & Spiegelman, S. 1965. Localization of DNA complementary to ribosomal RNA in the nucleolus organizer region of *Drosophila melanogaster*. *Proc Natl Acad Sci U S A*, 53, 737–45.

Sloan, K. E., Bohnsack, M. T. & Watkins, N. J. 2013. The 5S RNP couples p53 homeostasis to ribosome biogenesis and nucleolar stress. *Cell Rep*, 5, 237–47.

Stott, F. J., Bates, S., James, M. C., Mcconnell, B. B., Starborg, M., Brookes, S., Palmero, I., Ryan, K., Hara, E., Vousden, K. H. & Peters, G. 1998. The alternative product from the human CDKN2A locus, p14(ARF), participates in a regulatory feedback loop with p53 and MDM2. *EMBO J*, 17, 5001–14.

Strongin, D. E., Groudine, M. & Politz, J. C. 2014. Nucleolar tethering mediates pairing between the IgH and Myc loci. *Nucleus*, 5, 474–81.

Szerlong, H. J., Herman, J. A., Krause, C. M., Deluca, J. G., Skoultchi, A., Winger, Q. A., Prenni, J. E. & Hansen, J. C. 2015. Proteomic characterization of the nucleolar linker histone H1 interaction network. *J Mol Biol*, 427, 2056–71.

Tollervey, D., Lehtonen, H., Jansen, R., Kern, H. & Hurt, E. C. 1993. Temperature-sensitive mutations demonstrate roles for yeast fibrillarin in pre-rRNA processing, pre-rRNA methylation, and ribosome assembly. *Cell*, 72, 443–57.

Van Koningsbruggen, S., Gierlinski, M., Schofield, P., Martin, D., Barton, G. J., Ariyurek, Y., Den Dunnen, J. T. & Lamond, A. I. 2010. High-resolution whole-genome sequencing reveals that specific chromatin domains from most human chromosomes associate with nucleoli. *Mol Biol Cell*, 21, 3735–48.

Yang, F., Deng, X., Ma, W., Berletch, J. B., Rabaia, N., Wei, G., Moore, J. M., Filippova, G. N., Xu, J., Liu, Y., Noble, W. S., Shendure, J. & Disteche, C. M. 2015. The lncRNA Firre anchors the inactive X chromosome to the nucleolus by binding CTCF and maintains H3K27me3 methylation. *Genome Biol*, 16. doi: 10.1186/s13059-015-0618-0.

Zhang, L. F., Huynh, K. D. & Lee, J. T. 2007. Perinucleolar targeting of the inactive X during S phase: evidence for a role in the maintenance of silencing. *Cell*, 129, 693–706.

Zhang, Y., Wolf, G. W., Bhat, K., Jin, A., Allio, T., Burkhart, W. A. & Xiong, Y. 2003. Ribosomal protein L11 negatively regulates oncoprotein MDM2 and mediates a p53-dependent ribosomal-stress checkpoint pathway. *Mol Cell Biol*, 23, 8902–12.

Zhou, J., Sackton, T. B., Martinsen, L., Lemos, B., Eickbush, T. H. & Hartl, D. L. 2012. Y chromosome mediates ribosomal DNA silencing and modulates the chromatin state in *Drosophila*. *Proc Natl Acad Sci U S A*, 109, 9941–6.

CHAPTER 13

Nuclear bodies

In addition to the nuclear lamina and the nucleoli, the nuclei of eukaryotic cells contain a variety of specialized regions or nuclear bodies, sometimes referred to as organelles, which include perinucleolar compartments, nuclear speckles, paraspeckles, promyelocytic leukemia (PML) bodies, Cajal bodies (CBs), histone locus bodies (HLBs), nuclear stress bodies (nSBs) and orphan nuclear bodies. The function of many of these nuclear regions is not well understood. They are not bound by membranes and seem to represent areas of concentration of proteins, often seeded by RNA transcripts (Shevtsov and Dundr, 2011), for the purpose of enhancing the occurrence of various gene regulatory functions, RNA processing in particular. How some of these nuclear bodies are formed remains a matter of debate. Of the many hypotheses advanced on this subject, one of the simplest and increasingly favored is based on the principle of phase transition (Weber, 2017; Zhu and Brangwynne, 2015). Phase transition represents the tendency of systems to change into equilibrium configurations. In nuclei, phases would depend on the concentration of particular molecules, the salt concentration of the nucleoplasm, the temperature and the greater attraction of these particular molecules for one another than for other dissolved molecules. A similar model relies on the tendency of macromolecules to phase-separate when their attraction for one another is sufficiently stronger than their interaction with a solvent, water in the case of biological entities (Banani et al., 2017).

An important complement to these cytologically visible regions are the transcription factories, dedicated nuclear sites where gene transcription occurs.

The perinucleolar compartment

This compartment was first described during a study of the cytological localization of the polypyrimidine tract binding (PTB) protein in HeLa cells (Ghetti et al., 1992). Polypyrimidine tracts are present towards the 3′ ends of introns and promote the assembly of the splicing complex. To date, the function of the perinucleolar compartment (PNC) has not been elucidated. As its name indicates, the PNC is present at the periphery of the nucleolus. In contrast to the zone of heterochromatin that surrounds the nucleolus and that is characterized by the presence of silenced regions of the genome (discussed in Chapter 12), the PNC is rich in RNAPII transcripts processing proteins and in RNAPIII-produced small non-coding RNAs; the presence of these RNAs, which are transcribed elsewhere and accumulate in the PNC, is necessary for the PNC's integrity and maintenance (Pollock and Huang, 2010). Another difference between the zone of perinucleolar heterochromatin and the PNC is that the former is found in all eukaryotic cells, while the PNC appears following the onset of oncogenesis and is a hallmark of a number of different cancers (Norton and Huang, 2013; Wen et al., 2013).

The Cajal bodies

These nuclear structures were initially called coiled bodies and later were renamed after their discoverer Santiago Ramon y Cajal who, in 1910, described a number of different nuclear sub-regions (Fig. 13.1). The major structural component of CBs is coilin, a protein

Epigenetics, Nuclear Organization and Gene Function: with implications of epigenetic regulation and genetic architecture for human development and health. John C. Lucchesi, Oxford University Press (2019). © John C. Lucchesi 2019.
DOI: 10.1093/oso/9780198831204.001.0001

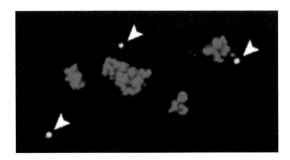

Fig. 13.1 Cajal bodies in a HeLa cell nucleus. The nucleus is immunostained with antibodies against coilin (green) and fibrillarin (red). The three Cajal bodies appear yellow because they stain with both antibodies. The nucleoli stain only for fibrillarin.

(From Gall, 2003.)

that self-associates and is essential for CB assembly (Hebert and Matera, 2000). Another component of CBs is the survival motor neuron (SMN) protein. SMN and other proteins form complexes that function in the assembly of various ribonucleoprotein particles (RNPs) involved in different RNA processing pathways. Some RNPs contain small nuclear RNAs (snRNAs) involved in primary RNA transcript splicing. In a manner similar to the modification of ribosomal RNAs (rRNAs) by small nucleolar RNPs (snoRNPs) (see Chapter 12), snRNAs are ribose 2′-O-methylated and pseudouridylated by Cajal body-specific RNPs (scaRNPs) that contain fibrillarin. The scaRNAs that they contain have a sequence that interacts with telomerase Cajal body protein 1 (TCAB1), a protein responsible for their localization to the CBs. Following these aspects of their biogenesis, small nuclear ribonucleic particles (snRNPs) move to nuclear speckles and finally to the sites of active transcription where the actual splicing process takes place. A number of snoRNAs that lack the CB localization sequence nevertheless pass through the CBs on their way to the nucleoli. This reinforces the notion that CBs are centers of RNP assembly (Machyna et al., 2014).

Consistent with this conclusion, CBs are the site of biogenesis of telomerase. Telomerase is an RNP composed of an RNA and several proteins that include a reverse transcriptase, an enzyme that can synthesize a DNA strand using an RNA as template (see Chapter 20). The telomerase RNA belongs to the scaRNA family and contains the characteristic sequence necessary for CB localization. The protein involved in this localization (TCAB1) is part of the telomerase holoenzyme and is required for the synthesis of telomeres (Venteicher et al., 2009).

CBs are also involved in the 3′ end processing of histone pre-mRNAs by their close association with HLBs (see below).

Gemini of coiled bodies and histone locus bodies

Gemini of coiled bodies (Gems) are nuclear structures that are found near, or co-localized with, CBs (Liu and Dreyfuss, 1996). They do not contain snRNPs or coilin but are highly enriched in the SMN protein. This protein serves as a platform for the nucleation of a multi-protein complex that plays a key role in the assembly of the common core of the snRNPs involved in RNA splicing (Fischer et al., 1997; Meister et al., 2001). Gems are absent in a number of cell types but appear when CBs, due to some epigenetic modification, fail to retain the SMN protein (Boisvert et al., 2002).

The histone locus bodies (HLBs) are subnuclear regions that form at the sites of the histone gene clusters and are often in close proximity to the CBs, with which they share a number of factors. HLBs are implicated in the coordinated expression of the five core histones (see Chapter 9). In human cells, the major histone gene cluster, located on chromosome 6, is subdivided into three sub-clusters that are located at the base of small chromatin loops within a larger loop (Fritz et al., 2018). A major HLB forms around these sub-clusters. HLBs are enriched in the transcription factor HINFP (histone nuclear factor P), the co-factor nuclear protein of ataxia telangiectasia (NPAT) and specific snRNPs (Liu et al., 2006), as well as FLASH, a protein involved in the 3′ end processing of histone mRNA and YARP, a protein likely to act as a repressor of histone gene transcription (Yang et al., 2014). Targeting of the HLB components to the histone gene clusters has been investigated recently in *Drosophila*. The CLAMP protein, discovered for its role in the association of the male-specific lethal (MSL) dosage compensation complex with the X chromosome in males (see Chapter 9), binds to the histone H3 and H4 common promoter region and promotes HLB formation (Rieder et al., 2017).

Redundancy or collaboration? Storage or active processing?

Active histone genes and processing factors of rRNA transcripts have been identified in the fibrillar components of the nucleoli, in the HLBs and in association with CBs. Fibrillarin, the ribose 2′-O-methyl transferase, named after its localization in the fibrillar part of the nucleolus (Raska et al., 1991), is also found in

CBs. The U7 snRNP, involved in the 3′ end processing of the rRNA primary transcript (Mowry and Steitz, 1987), is present in CBs of vertebrate cells; in *Drosophila* cells, it is found in the HLBs. Clearly, these subnuclear regions are functionally redundant or collaborate in rRNA biogenesis. A related question is whether CBs and HLBs are independent structures where specific proteins and factors are simply stored or whether they are regions of the genome that are brought into association by the presence of specific DNA-bound or RNA-bound proteins related to specific processes. Studies using fluorescence recovery after photobleaching (FRAP) have shown that proteins exchange rapidly between CBs, HLBs and the surrounding nucleoplasm, suggesting a more active role in rRNA metabolism than simply storage (Phair and Misteli, 2000). Although the function of SMN complexes has been defined in molecular terms, the actual purpose of their concentration in Gems is not yet understood.

The same considerations obtain with respect to nuclear speckles, paraspeckles and PML bodies (see the following sections).

Nuclear speckles

Nuclear speckles, also called interchromatin granule clusters (IGCs), were first observed by Ramon y Cajal. They are characterized by the presence of various proteins involved in pre-mRNA splicing, including snRNPs, other spliceosome subunits and other non-snRNP splicing factors (Fig. 13.2). In addition, speckles contain proteins involved in binding, packaging and transporting mRNA (Saitoh et al., 2004).

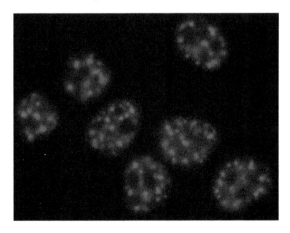

Fig. 13.2 Nuclear speckles in HeLa cells stained by indirect immunofluorescence using a splicing factor primary antiserum.

(From Spector and Lamond, 2011.)

The basis for the organization of speckles is not fully understood. Actin and one of the proteins previously identified—the RNA-binding protein SON—may provide the scaffolding necessary for their assembly. SON binds RNA and promotes splicing, particularly of transcripts with poor splice sites. A role for SON in speckle assembly is supported by the observation that small interfering RNA (siRNA) inhibition of its expression leads to the mislocalization of nuclear speckle components (Sharma et al., 2010; Sharma et al., 2011).

Speckles are often observed closely associated with sites of very active transcription, presumably in order to deliver the factors necessary for co-transcriptional splicing and other steps. Speckles change in number, size and shape in response to changes in gene expression. If transcription is inhibited, they become round and enlarge due to the accumulation of splicing factors, suggesting that they may function in the modification, assembly or storage of these factors, rather than being sites where splicing occurs (O'Keefe et al., 1994). On the other hand, when transcription increases, for example during viral infection, splicing factors redistribute from the speckles to sites of transcription (Jimenez-Garcia and Spector, 1993). The accumulation or release of splicing factors from speckles is regulated by phosphorylation or de-phosphorylation of these factors by kinases and phosphatases present in the speckles.

In addition to pre-mRNA splicing factors, proteomic analyses of purified nuclear speckles revealed the presence of proteins involved in every aspect of gene regulation such as transcription activation, elongation, transcript processing and repression (Saitoh et al., 2004). Other proteins are also found in different nuclear bodies, highlighting once again the dynamic exchange of components that occurs among the different nuclear compartments, a process whose function is not yet fully understood (Galganski et al., 2017).

Paraspeckles

Paraspeckles have been observed only in mammalian nuclei (Fig. 13.3). These relatively new nuclear organelles were discovered in the process of characterizing the proteome of nucleoli from human cells. One of the newly discovered proteins, now called paraspeckle protein 1 (PSPC1), was found to associate with 5–20 subnuclear foci distinct from any of those previously characterized (Fox et al., 2002). Although PSPC1 is normally present in paraspeckles, it traffics through nucleoli and, in transcriptionally inactive cells, it can be found associated with nucleoli in so-called perinucleolar

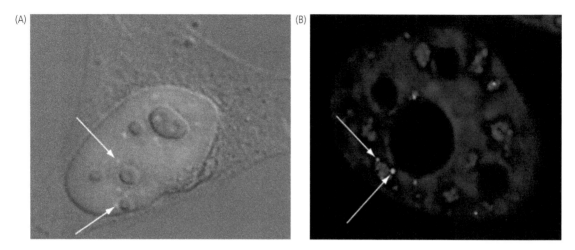

Fig. 13.3 Paraspeckles in HeLa cells. (A) Differential interference contrast (DIC) microscopy of cells immunostained with antisera to show the nucleoli (red) and the paraspeckles (green). (B) Immunostaining distinguishes paraspeckles (green) from nuclear speckles (red).

(From Fox and Lamond, 2010.)

caps, thus explaining its presence among nucleolar proteins. PSPC1 is a member of the *Drosophila* behavior human splicing (DBHS) protein family, which includes several other paraspeckle components. These proteins contain RNA-binding and protein–protein interaction domains that are essential for their localization to paraspeckles, and they appear to be necessary for their structural integrity. Members of this protein family are involved in various nuclear processes, including transcription initiation, termination, splicing and RNA surveillance.

Two classes of RNA molecules are associated with paraspeckles. The first consists of a group of mRNAs that are retained in the paraspeckles, rather than being exported to the cytoplasm. These RNAs are characterized by long 3′ untranslated regions (UTRs) that are repetitive and highly edited (editing is the enzymatic conversion of cytosine to uracil, or of adenine to inosine, that occurs presumably to expand the variety of protein isoforms that can be produced by particular genes). Retention is mediated by the affinity of certain paraspeckle proteins for the edited sequences (Zhang and Carmichael, 2001), and the release of these RNAs by cleavage of the 3′ UTRs constitutes a form of gene function regulation. The other class of RNA molecules present in paraspeckles consists of the isoforms of the long intergenic non-coding RNA NEAT1 that serves as the scaffolding for the assembly of the many proteins of these granules (Chen and Carmichael, 2009; Clemson et al., 2009; Sasaki et al., 2009; Sunwoo et al., 2009). Recently, NEAT1, in association

with two other proteins [NONO (non-POU domain-containing octamer-binding) and PSF (splicing factor proline- and glutamine-rich)], was shown to anchor the Microprocessor complex involved in the generation of miRNA, expanding the role of paraspeckles in the post-transcriptional processing of pri-miRNAs (Jiang et al., 2017). An unexpected player in the synthesis of paraspeckles—the remodeling SWI/SNF complex—was discovered in a search for proteins that interact with paraspeckle proteins. In contrast to its function in chromatin remodeling, the ATPase activity of the complex is not involved in its role in paraspeckle biogenesis (Kawaguchi et al., 2015).

PML nuclear bodies

PML nuclear bodies (PML-NBs), originally referred to as nuclear domain-10 (ND-10s), were discovered by their appearance under the electron microscope as dense spherical objects and later visualized by immunofluorescence microscopy (Szostecki et al., 1987). Their current name comes from the presence of the acute PML protein that, in normal cells, is present in ND-10s. In patients with this disease, fusion to the retinoic acid receptor alpha protein (PML/RARα) causes the relocation of the PML protein from the usual 5–15 PML-NBs to hundreds of very small intra-nuclear dots and to the cytoplasm (Daniel et al., 1993). The PML protein is responsible for recruiting the other proteins that are found in PML-NBs, and it must be sumoylated in order to perform this function (Ishov et al., 1999; Zhong et al.,

2000). Many of the PML protein partners are sumoylated as well and, like PML, contain SUMO interacting motifs, suggesting a mechanism for recruitment and retention.

Because of the large size and functional variety of the proteome of PML-NBs, these bodies have been implicated in a multitude of nuclear and cellular functions. In reality, the significance of PML-NBs in cellular biology is still unclear. PML-NBs are present in most, but not all, cell lines. Mice deficient for the *PML* gene exhibit no phenotypic defects. Yet, the PML protein clearly has tumor suppressor activity through its positive regulation of p53 (a protein that responds to cellular stress by inducing cell cycle arrest and apoptosis), which the PML protein achieves by sequestering the p53 inhibitor Mdm2. PML-NBs preferentially associate with chromatin regions that are transcriptionally active, yet they do not directly affect the level of transcription of these genes (Wang et al., 2004). Under conditions of stress (oxidative and other types such as DNA damage, exposure to heavy metals, viral infection or treatment with interferon—a family of cytokines that trigger the immune system in the presence of pathogens), the number and size of PML-NBs increase (Sahin et al., 2014). PML has a significant impact on epigenetic nucleosome modifications. Absence of PML reduces the level of H3.3 deposited by the DAXX/ATRX chaperones in the heterochromatic regions associated with the NBs; it also reduces the level of H3K9me3 while increasing the level of H3K27me3 in these regions (Delbarre et al., 2017).

Nuclear stress bodies

nSBs are subnuclear organelles that form in response to stress agents, such as heat shock, hypoxia and certain toxins, and are especially evident following heat shock as ribonucleoprotein complexes called perichromatin granules (Nover et al., 1989; Sarge et al., 1993). They form as a result of stalled transcription and are not detectable in cells under normal growth conditions. nSBs seem to form by recruitment of the heat shock transcription factor 1 (HSF1) to pericentric tandem repeats of satellite III sequences (Jolly et al., 2002). These sequences are transcribed during the heat shock response and remain associated with nSBs in close proximity to their origin. In addition to HSF1, nSBs contain a heterogeneous nuclear ribonucleoprotein (hnRNP) named HAP/Saf-B (Weighardt et al., 1999). It is not currently clear whether nSBs have a function in the response of the cell to stress or whether their formation is a byproduct of the response without functional consequences.

Transcription factories

Clues to the possible existence of specific, dedicated nuclear sites where gene transcription may occur were provided by the discovery of nuclear areas where pre-mRNA splicing factors are concentrated (Fu and Maniatis, 1990; Spector, 1984). Using a modified deoxyribonucleotide bromouridine triphosphate (BrUTP) that is incorporated into nascent RNA and can be detected with a specific antiserum, Peter Cook and Luitzen de Jong, independently, were the first to report the presence of distinct visible sites of RNA synthesis that Cook termed transcription factories (Jackson et al., 1993; Wansink et al., 1993).

Formally, a transcription factory is a nuclear region where several RNA polymerase complexes are transcribing two or more genes (Papantonis and Cook, 2013). Transcription factories are so numerous in cells that their actual number is difficult to determine by any form of light microscopy. A successful experimental approach to this problem has been to label all of the nascent transcripts in a cell with gold particles that can then be visualized in thin sections using an electron microscope (Iborra et al., 1996). Clusters of particles visible in a section represent transcription factories and, knowing the thickness of the section, their total number in the cell can be extrapolated. This methodology has led to the conclusion that most genes are transcribed in transcription factories. These earlier studies were conducted on fixed cells and open to the criticism that fixation can cause the artifactual aggregation of molecules. Evidence that transcription factories occur in living cells was obtained by using super-resolution microscopy (Cisse et al., 2013) or *in vivo* fluorescence labeling (Ghamari et al., 2013).

Dynamic association of polymerases and transcription units

A conceptual change brought about by the existence of transcription factories was the realization that the vast majority of RNA polymerase complexes do not associate with stationary templates and generate transcripts by tracking along a stationary coding strand; rather, it is the genes that manage to access stable regions of polymerase concentration in the nucleus, and transcription is achieved by threading the DNA template through fixed polymerase complexes. One of the more elegant experiments supporting this contention made use of two genes that located at some distance from each other on a chromosome and that can be induced by a common activator; one gene is quite long and

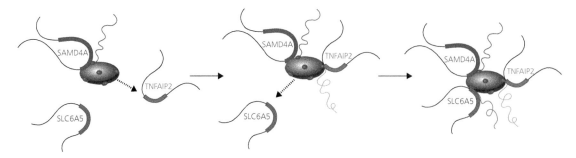

Fig. 13.4 Formation of a transcription factory following TNFα stimulation of human cells. The *SAMD4A* and *TNFAIP2* genes are on chromosome 14, and the *SLC6A5* gene is located on chromosome 11. The gray sphere represents the localized concentration of transcription and co-transcription processing factors. The pink spheres are RNA polymerases.

(From Fanucchi et al., 2013.)

requires over an hour to produce a full transcript, while the other gene is much shorter and is fully transcribed in a few minutes. Following induction, chromosome conformation capture (3C) was used to determine the relative position of the two genes. If the genes were transcribed at their respective positions on the chromosome, they would not be found to associate. In fact, the two genes do associate and, as time passes, the short gene associates with successive regions of the long gene. These results indicate that both genes are present in the same transcription factory, that the short gene is repeatedly pulled in by a polymerase complex for transcription and that 3C captures its proximity to the longer gene as the latter is reeled in by a neighboring polymerase (Papantonis et al., 2010).

Co-localization of genes in transcription factories may lead to their co-regulation

The experiment just discussed illustrates an important feature of transcription factories—genes that participate in a common biochemical pathway can be subjected to the same regulatory signals if they preferentially associate in the same transcription factory. The first genome-wide analysis of functionally related genes uncovered a high level of physical associations between the mouse globin genes Hbb and Hba and a large number of genes expressed in erythroid cells. These genes were present in over 200 clusters that contained elongating RNA polymerases. Different groups of genes were associated in these transcription factories, reflecting either a physical limitation imposed by the genomic locations of the genes or some intergenic regulatory interactions (Schoenfelder et al., 2010).

Another example of this specific type of association between co-regulated transcriptional units is provided by a group of genes induced by a cytokine [tumor necrosis factor alpha (TNFα)] that signals through the transcription activator nuclear factor kappa B (NFκB). Upon activation, genes that are normally distant from one another on the same chromosome or present on different chromosomes are seen to assemble into discrete transcription factories (Papantonis et al., 2012). In a particular factory, three of these genes—*SAMD4A* and *TNFAIP2* present on the same chromosome and *SLC6A5* located on a different chromosome—are co-activated (Fig. 13.4). Surprisingly, their respective transcriptional states are interdependent—activation of *SAMD4A* is required for the activation of *TNFAIP2*, which, in turn, is necessary for the transcription of *SLC6A5* (Fanucchi et al., 2013).

An example of the correlation between co-localization and co-regulation is found in mouse erythroid cells where active genes that are up to 40 megabases away from one another on the same chromosome, and genes such as those of the β-globin and α-globin loci located on different chromosomes, often co-localize in transcription foci, i.e. transcription factories (Osborne et al., 2004). Another striking example is provided by the association of immunoglobulin genes during mouse B-cell development. The *Igk, IgH* and *IgJ* genes, located on three different chromosomes, as well as their respective enhancers, are found associated in the same transcription factory (Park et al., 2014).

How do transcription factories originate?

Chromosome conformation capture techniques have established that the most frequent interactions occur among active genes that are located in close proximity to one another on chromosomes. These associations would lead to localized concentrations of polymerases and general transcription factors that would be shared by these genes as they undergo rounds of reinitiation. These types of transcription factories would be the result of the activation and transcription of individual genes that may be co-regulated (Buckley and Lis, 2014). This model for the origin of transcription factories is not easily extended to cases where distant genes, sometimes on different chromosomes, are seen to associate during transcription. To explain these instances, the most widely accepted model maintains that there are pre-existing concentrations of polymerases and general transcription factors to which activated genes are attracted. The numerous examples supporting this model include the early observation that during erythroid cell differentiation, the mouse β-globin gene relocates from the nuclear periphery to foci of highly active polymerases; this relocation requires the presence of the locus control region (Ragoczy et al., 2006). Another example is provided by the proto-oncogene *Myc* and the immunoglobulin heavy chain gene *Igh* that are on separate chromosomes and are found to associate physically only when they are induced to transcribe in B lymphocytes (Osborne et al., 2007).

To date, the signals and mechanisms responsible for the targeting to transcription factories of genes that are to be activated are poorly understood. A preliminary insight into one of the topological parameters that may be involved was provided by the observation that chromatin decondensation, in the absence of transcriptional induction, resulted in the relocation of genes from the nuclear periphery to a more central location. Several genes, individually induced by targeted engineered transcription factors, were seen to move away from the nuclear lamina towards the center of the nucleus; the same movement was reproduced by targeting a peptide that decondensed chromatin without inducing transcription (Therizols et al., 2014).

Chapter summary

In addition to the nuclear lamina and the nucleoli, the nucleus is subdivided into a number of functional compartments or nuclear bodies. These compartments are not enclosed by membranes and consist of local concentrations of specific proteins and nucleic acids. Their main functions are transcriptional regulation and RNA processing. An important consideration is that many of the same proteins and snRNPs are found in different nuclear bodies, highlighting the dynamic

exchange of components that occurs among the different nuclear compartments, a process whose function is not yet fully understood.

The **perinucleolar compartment** (PNC), present at the periphery of the nucleolus, is rich in RNAPII transcript processing proteins and in RNAPIII-produced small non-coding RNAs. PNCs appear following the onset of oncogenesis and are a hallmark of a number of different cancers. **Cajal bodies** (CBs) are sites of processing and assembly of small nuclear ribonucleic particles (snRNPs) that function in messenger RNA or ribosomal RNA processing and in the biogenesis of telomerase. In addition, and through their close association with **histone locus bodies**, they are involved in the 3′ end processing of the five core histone gene transcripts. **Nuclear speckles** are another type of nuclear bodies that contain pre-mRNA splicing components; in addition, nuclear speckles contain proteins involved in every aspect of gene regulation. **Paraspeckles** are involved in the retention of hairpin-containing RNAs, in the sequestration of various RNA-binding proteins and in the processing and maturation of miRNAs. **PML nuclear bodies** (PML-NBs) contain the promyelocytic leukemia (PML) protein that, in turn, recruits a large number of other proteins. Under conditions of stress, such as oxidative damage, viral infection, etc., the number and size of PML-NBs increases. The PML protein also has tumor suppressor activity by preventing the inactivation of the tumor suppressor p53. **Nuclear stress bodies** only form in response to stress agents, either as a defense mechanism against stress or as a byproduct of the cellular reaction to stress. **Transcription factories** are nuclear regions where several RNAPII complexes are transcribing several genes. The co-localization of genes in transcription factories may lead to their co-regulation.

References

Banani, S. F., Lee, H. O., Hyman, A. A. & Rosen, M. K. 2017. Biomolecular condensates: organizers of cellular biochemistry. *Nat Rev Mol Cell Biol*, 18, 285–98.

Boisvert, F. M., Cote, J., Boulanger, M. C., Cleroux, P., Bachand, F., Autexier, C. & Richard, S. 2002. Symmetrical dimethylarginine methylation is required for the localization of SMN in Cajal bodies and pre-mRNA splicing. *J Cell Biol*, 159, 957–69.

Buckley, M. S. & Lis, J. T. 2014. Imaging RNA polymerase II transcription sites in living cells. *Curr Opin Genet Dev*, 25, 126–30.

Chen, L. L. & Carmichael, G. G. 2009. Altered nuclear retention of mRNAs containing inverted repeats in human embryonic stem cells: functional role of a nuclear noncoding RNA. *Mol Cell,* 35, 467–78.

Cisse, Ii, Izeddin, I., Causse, S. Z., Boudarene, L., Senecal, A., Muresan, L., Dugast-Darzacq, C., Hajj, B., Dahan, M. & Darzacq, X. 2013. Real-time dynamics of RNA polymerase II clustering in live human cells. *Science,* 341, 664–7.

Clemson, C. M., Hutchinson, J. N., Sara, S. A., Ensminger, A. W., Fox, A. H., Chess, A. & Lawrence, J. B. 2009. An architectural role for a nuclear noncoding RNA: NEAT1 RNA is essential for the structure of paraspeckles. *Mol Cell,* 33, 717–26.

Daniel, M. T., Koken, M., Romagne, O., Barbey, S., Bazarbachi, A., Stadler, M., Guillemin, M. C., Degos, L., Chomienne, C. & De The, H. 1993. PML protein expression in hematopoietic and acute promyelocytic leukemia cells. *Blood,* 82, 1858–67.

Delbarre, E., Ivanauskiene, K., Spirkoski, J., Shah, A., Vekterud, K., Moskaug, J. O., Boe, S. O., Wong, L. H., Kuntziger, T. & Collas, P. 2017. PML protein organizes heterochromatin domains where it regulates H3.3 deposition by ATRX/DAXX. *Genome Res,* 27, 913-21

Fanucchi, S., Shibayama, Y., Burd, S., Weinberg, M. S. & Mhlanga, M. M. 2013. Chromosomal contact permits transcription between coregulated genes. *Cell,* 155, 606–20.

Fischer, U., Liu, Q. & Dreyfuss, G. 1997. The SMN-SIP1 complex has an essential role in spliceosomal snRNP biogenesis. *Cell,* 90, 1023–9.

Fox, A. H., Lam, Y. W., Leung, A. K., Lyon, C. E., Andersen, J., Mann, M. & Lamond, A. I. 2002. Paraspeckles: a novel nuclear domain. *Curr Biol,* 12, 13–25.

Fox, A. H. & Lamond, A. I. 2010. Paraspeckles. *Cold Spring Harb Perspect Biol,* 2, a000687.

Fritz, A. J., Ghule, P. N., Boyd, J. R., Tye, C. E., Page, N. A., Hong, D., Shirley, D. J., Weinheimer, A. S., Barutcu, A. R., Gerrard, D. L., Frietze, S., van Wijnen, A. J., Zaidi, S. K., Imbalzano, A. N., Lian, J. B., Stein, J. L. & Stein, G. S. 2018. Intranuclear and higher-order chromatin organization of the major histone gene cluster in breast cancer. *J Cell Physiol,* 233,1278-90.

Fu, X. D. & Maniatis, T. 1990. Factor required for mammalian spliceosome assembly is localized to discrete regions in the nucleus. *Nature,* 343, 437–41.

Galganski, L., Urbanek, M. O. & Krzyzosiak, W. J. 2017. Nuclear speckles: molecular organization, biological function and role in disease. *Nucleic Acids Res,* 45, 10350–68.

Gall, J. G. 2003. The centennial of the Cajal body. *Nat Rev Mol Cell Biol,* 4, 975–80.

Ghamari, A., Van De Corput, M. P., Thongjuea, S., Van Cappellen, W. A., Van Ijcken, W., Van Haren, J., Soler, E., Eick, D., Lenhard, B. & Grosveld, F. G. 2013. *In vivo* live imaging of RNA polymerase II transcription factories in primary cells. *Genes Dev,* 27, 767–77.

Ghetti, A., Pinol-roma, S., Michael, W. M., Morandi, C. & Dreyfuss, G. 1992. hnRNP I, the polypyrimidine tract-binding protein: distinct nuclear localization and association with hnRNAs. *Nucleic Acids Res,* 20, 3671–8.

Hebert, M. D. & Matera, A. G. 2000. Self-association of coilin reveals a common theme in nuclear body localization. *Mol Biol Cell,* 11, 4159–71.

Iborra, F. J., Pombo, A., Jackson, D. A. & Cook, P. R. 1996. Active RNA polymerases are localized within discrete transcription "factories" in human nuclei. *J Cell Sci,* 109 (Pt 6), 1427–36.

Ishov, A. M., Sotnikov, A. G., Negorev, D., Vladimirova, O. V., Neff, N., Kamitani, T., Yeh, E. T., Strauss, J. F., 3RD & Maul, G. G. 1999. PML is critical for ND10 formation and recruits the PML-interacting protein daxx to this nuclear structure when modified by SUMO-1. *J Cell Biol,* 147, 221–34.

Jackson, D. A., Hassan, A. B., Errington, R. J. & Cook, P. R. 1993. Visualization of focal sites of transcription within human nuclei. *EMBO J,* 12, 1059–65.

Jiang, L., Shao, C., Wu, Q. J., Chen, G., Zhou, J., Yang, B., Li, H., Gou, L. T., Zhang, Y., Wang, Y., Yeo, G. W., Zhou, Y. & Fu, X. D. 2017. NEAT1 scaffolds RNA-binding proteins and the Microprocessor to globally enhance pri-miRNA processing. *Nat Struct Mol Biol,* 24, 816–24.

Jimenez-Garcia, L. F. & Spector, D. L. 1993. *In vivo* evidence that transcription and splicing are coordinated by a recruiting mechanism. *Cell,* 73, 47–59.

Jolly, C., Konecny, L., Grady, D. L., Kutskova, Y. A., Cotto, J. J., Morimoto, R. I. & Vourc'h, C. 2002. *In vivo* binding of active heat shock transcription factor 1 to human chromosome 9 heterochromatin during stress. *J Cell Biol,* 156, 775–81.

Kawaguchi, T., Tanigawa, A., Naganuma, T., Ohkawa, Y., Souquere, S., Pierron, G. & Hirose, T. 2015. SWI/SNF chromatin-remodeling complexes function in noncoding RNA-dependent assembly of nuclear bodies. *Proc Natl Acad Sci U S A,* 112, 4304–9.

Liu, J. L., Murphy, C., Buszczak, M., Clatterbuck, S., Goodman, R. & Gall, J. G. 2006. The *Drosophila melanogaster* Cajal body. *J Cell Biol,* 172, 875–84.

Liu, Q. & Dreyfuss, G. 1996. A novel nuclear structure containing the survival of motor neurons protein. *EMBO J,* 15, 3555–65.

Machyna, M., Kehr, S., Straube, K., Kappei, D., Buchholz, F., Butter, F., Ule, J., Hertel, J., Stadler, P. F. & Neugebauer, K. M. 2014. The coilin interactome identifies hundreds of small noncoding RNAs that traffic through Cajal bodies. *Mol Cell,* 56, 389–99.

Meister, G., Buhler, D., Pillai, R., Lottspeich, F. & Fischer, U. 2001. A multiprotein complex mediates the ATP-dependent assembly of spliceosomal U snRNPs. *Nat Cell Biol,* 3, 945–9.

Mowry, K. L. & Steitz, J. A. 1987. Identification of the human U7 snRNP as one of several factors involved in the 3′ end maturation of histone premessenger RNAs. *Science,* 238, 1682–7.

Norton, J. T. & Huang, S. 2013. The perinucleolar compartment: RNA metabolism and cancer. *Cancer Treat Res*, 158, 139–52.

Nover, L., Scharf, K. D. & Neumann, D. 1989. Cytoplasmic heat shock granules are formed from precursor particles and are associated with a specific set of mRNAs. *Mol Cell Biol*, 9, 1298–308.

O'keefe, R. T., Mayeda, A., Sadowski, C. L., Krainer, A. R. & Spector, D. L. 1994. Disruption of pre-mRNA splicing *in vivo* results in reorganization of splicing factors. *J Cell Biol*, 124, 249–60.

Osborne, C. S., Chakalova, L., Brown, K. E., Carter, D., Horton, A., Debrand, E., Goyenechea, B., Mitchell, J. A., Lopes, S., Reik, W. & Fraser, P. 2004. Active genes dynamically colocalize to shared sites of ongoing transcription. *Nat Genet*, 36, 1065–71.

Osborne, C. S., Chakalova, L., Mitchell, J. A., Horton, A., Wood, A. L., Bolland, D. J., Corcoran, A. E. & Fraser, P. 2007. Myc dynamically and preferentially relocates to a transcription factory occupied by Igh. *PLoS Biol*, 5. doi: 10.1371/journal.pbio.0050192.

Papantonis, A. & Cook, P. R. 2013. Transcription factories: genome organization and gene regulation. *Chem Rev*, 113, 8683–705.

Papantonis, A., Kohro, T., Baboo, S., Larkin, J. D., Deng, B., Short, P., Tsutsumi, S., Taylor, S., Kanki, Y., Kobayashi, M., Li, G., Poh, H. M., Ruan, X., Aburatani, H., Ruan, Y., Kodama, T., Wada, Y. & Cook, P. R. 2012. TNFalpha signals through specialized factories where responsive coding and miRNA genes are transcribed. *EMBO J*, 31, 4404–14.

Papantonis, A., Larkin, J. D., Wada, Y., Ohta, Y., Ihara, S., Kodama, T. & Cook, P. R. 2010. Active RNA polymerases: mobile or immobile molecular machines? *PLoS Biol*, 8, e1000419.

Park, S. K., Xiang, Y., Feng, X. & Garrard, W. T. 2014. Pronounced cohabitation of active immunoglobulin genes from three different chromosomes in transcription factories during maximal antibody synthesis. *Genes Dev*, 28, 1159–64.

Phair, R. D. & Misteli, T. 2000. High mobility of proteins in the mammalian cell nucleus. *Nature*, 404, 604–9.

Pollock, C. & Huang, S. 2010. The perinucleolar compartment. *Cold Spring Harb Perspect Biol*, 2, a000679.

Ragoczy, T., Bender, M. A., Telling, A., Byron, R. & Groudine, M. 2006. The locus control region is required for association of the murine beta-globin locus with engaged transcription factories during erythroid maturation. *Genes Dev*, 20, 1447–57.

Raska, I., Andrade, L. E., Ochs, R. L., Chan, E. K., Chang, C. M., Roos, G. & Tan, E. M. 1991. Immunological and ultrastructural studies of the nuclear coiled body with autoimmune antibodies. *Exp Cell Res*, 195, 27–37.

Rieder, L. E., Koreski, K. P., Boltz, K. A., Kuzu, G., Urban, J. A., Bowman, S. K., Zeidman, A., Jordan, W. T., 3RD, Tolstorukov, M. Y., Marzluff, W. F., Duronio, R. J. & Larschan, E. N. 2017. Histone locus regulation by the *Drosophila* dosage compensation adaptor protein CLAMP. *Genes Dev*, 31, 1494–508.

Sahin, U., Lallemand-Breitenbach, V. & De The, H. 2014. PML nuclear bodies: regulation, function and therapeutic perspectives. *J Pathol*, 234, 289–91.

Saitoh, N., Spahr, C. S., Patterson, S. D., Bubulya, P., Neuwald, A. F. & Spector, D. L. 2004. Proteomic analysis of interchromatin granule clusters. *Mol Biol Cell*, 15, 3876–90.

Sarge, K. D., Murphy, S. P. & Morimoto, R. I. 1993. Activation of heat shock gene transcription by heat shock factor 1 involves oligomerization, acquisition of DNA-binding activity, and nuclear localization and can occur in the absence of stress. 1993. *Mol Cell Biol*, 13, 1392–407.

Sasaki, Y. T., Ideue, T., Sano, M., Mituyama, T. & Hirose, T. 2009. MENepsilon/beta noncoding RNAs are essential for structural integrity of nuclear paraspeckles. *Proc Natl Acad Sci U S A*, 106, 2525–30.

Schoenfelder, S., Sexton, T., Chakalova, L., Cope, N. F., Horton, A., Andrews, S., Kurukuti, S., Mitchell, J. A., Umlauf, D., Dimitrova, D. S., Eskiw, C. H., Luo, Y., Wei, C. L., Ruan, Y., Bieker, J. J. & Fraser, P. 2010. Preferential associations between co-regulated genes reveal a transcriptional interactome in erythroid cells. *Nat Genet*, 42, 53–61.

Sharma, A., Markey, M., Torres-Munoz, K., Varia, S., Kadakia, M., Bubulya, A. & Bubulya, P. A. 2011. Son maintains accurate splicing for a subset of human pre-mRNAs. *J Cell Sci*, 124, 4286–98.

Sharma, A., Takata, H., Shibahara, K., Bubulya, A. & Bubulya, P. A. 2010. Son is essential for nuclear speckle organization and cell cycle progression. *Mol Biol Cell*, 21, 650–63.

Shevtsov, S. P. & Dundr, M. 2011. Nucleation of nuclear bodies by RNA. *Nat Cell Biol*, 13, 167–73.

Spector, D. L. 1984. Colocalization of U1 and U2 small nuclear RNPs by immunocytochemistry. *Biol Cell*, 51, 109–12.

Spector, D. L. & Lamond, A. I. 2011. Nuclear speckles. *Cold Spring Harb Perspect Biol*, 3, pii: a000646.

Sunwoo, H., Dinger, M. E., Wilusz, J. E., Amaral, P. P., Mattick, J. S. & Spector, D. L. 2009. MEN epsilon/beta nuclear-retained non-coding RNAs are up-regulated upon muscle differentiation and are essential components of paraspeckles. *Genome Res*, 19, 347–59.

Szostecki, C., Krippner, H., Penner, E. & Bautz, F. A. 1987. Autoimmune sera recognize a 100 kD nuclear protein antigen (sp-100). *Clin Exp Immunol*, 68, 108–16.

Therizols, P., Illingworth, R. S., Courilleau, C., Boyle, S., Wood, A. J. & Bickmore, W. A. 2014. Chromatin decondensation is sufficient to alter nuclear organization in embryonic stem cells. *Science*, 346, 1238–42.

Venteicher, A. S., Abreu, E. B., Meng, Z., Mccann, K. E., Terns, R. M., Veenstra, T. D., Terns, M. P. & Artandi, S. E. 2009. A human telomerase holoenzyme protein required for Cajal body localization and telomere synthesis. *Science*, 323, 644–8.

Wang, J., Shiels, C., Sasieni, P., Wu, P. J., Islam, S. A., Freemont, P. S. & Sheer, D. 2004. Promyelocytic leukemia nuclear bodies associate with transcriptionally active genomic regions. *J Cell Biol,* 164, 515–26.

Wansink, D. G., Schul, W., Van Der Kraan, I., Van Steensel, B., Van Driel, R. & De Jong, L. 1993. Fluorescent labeling of nascent RNA reveals transcription by RNA polymerase II in domains scattered throughout the nucleus. *J Cell Biol,* 122, 283–93.

Weber, S. C. 2017. Sequence-encoded material properties dictate the structure and function of nuclear bodies. *Curr Opin Cell Biol,* 46, 62–71.

Weighardt, F., Cobianchi, F., Cartegni, L., Chiodi, I., Villa, A., Riva, S. & Biamonyi, G. 1999. A novel hnRNP protein (HAP/SAF-B) enters a subset of hnRNP complexes and relocates in nuclear granules in response to heat shock. *J Cell Sci,* 112, 1465–76.

Wen, Y., Wang, C. & Huang, S. 2013. The perinucleolar compartment associates with malignancy. *Front Biol (Beijing),* 8, 10.1007/s11515-013-1265-z.

Yang, X. C., Sabath, I., Kunduru, L., Van Wijnen, A. J., Marzluff, W. F. & Dominski, Z. 2014. A conserved interaction that is essential for the biogenesis of histone locus bodies. *J Biol Chem,* 289, 33767–82.

Zhang, Z. & Carmichael, G. G. 2001. The fate of dsRNA in the nucleus: a p54(nrb)-containing complex mediates the nuclear retention of promiscuously A-to-I edited RNAs. *Cell,* 106, 465–75.

Zhong, S., Muller, S., Ronchetti, S., Freemont, P. S., Dejean, A. & Pandolfi, P. P. 2000. Role of SUMO-1-modified PML in nuclear body formation. *Blood,* 95, 2748–52.

Zhu, L. & Brangwynne, C. P. 2015. Nuclear bodies: the emerging biophysics of nucleoplasmic phase. *Curr Opin Cell Biol,* 34, 23–30.

PART IV

Inheritance of Chromatin Structure and Functional States

CHAPTER 14

Chromatin replication

Cell division and the faithful distribution of the genetic material from parent to daughter cells are fundamental characteristics common to all living forms. These events involve replicating each parental chromosome during the S (synthesis) phase of the cell cycle and allotting the resulting sister chromatids to the two daughter cells during mitosis. Chromosome replication includes the synthesis of DNA molecules identical in sequence to the DNA in the parent cell by a process termed **semiconservative replication** whereby each strand of the DNA serves as a template for the synthesis of a new strand. This universal mode of DNA replication was first demonstrated by Matt Meselson and Frank Stahl in one of the most elegant and fundamental experiments in molecular biology (Meselson and Stahl, 1958). As expected, DNA replication must be coordinated with histone deposition and the other steps of chromatin assembly in order to reconstitute, in daughter cells, the parental form of the genetic material.

An overview of the process of DNA replication

Given the staggering length of DNA molecules in eukaryotic genomes, it is not surprising that DNA replication originates at multiple sites, termed **origins of replication** (ORIs), established during the G1 phase of the cell cycle by the binding of **pre-replication complex** (MCM2–7, minichromosome maintenance complex 2–7), a multi-subunit helicase required for unwinding the DNA region to be replicated. This step is known as "licensing" (Fig. 14.1). In most eukaryotes that have been studied, with the exception of budding yeast, these sites appear to be determined by chromatin structural features such as the presence of nucleosome-free regions, rather than by a specific DNA sequence. A subset of sites will become active replication origins by the addition of components to form **pre-initiation complexes**. Actual DNA synthesis is triggered in the S phase of the cell cycle by the further addition of factors forming a **replisome**. Present in these large complexes are cyclin-dependent kinases that activate the helicase. This step is referred to as origin "firing." Operationally, it is useful to consider the length of DNA replicated by the two replication forks that diverge from an active replication origin as a unit of replication, or **replicon**. Within the replicon, several origins of replication are found, yet only one is activated and the choice of the origin that is activated varies in the different cells of a given population (Cayrou et al., 2011). An important question, yet to be answered, is how particular origins of replication are selected for activation. A stringent control mechanism exists to ensure that DNA is replicated only once during the cell cycle. The major factors involved in this control are the cyclin-dependent kinases that activate the replicative helicase complex at the ORI destined to fire; simultaneously, these kinases inhibit the loading of the complex onto neighboring ORIs, thereby ensuring that the latter are not activated. If, for some reason, the activated replication fork stalls or collapses, some of these flexible ORIs are activated in order to continue the replication process.

The antiparallel structure of the two strands in a DNA molecule and the need for DNA polymerase to have a 3′-OH residue on which to add nucleotides allow the replication of one strand (the **leading strand**) to proceed continuously in the 5′ to 3′ direction and force the other strand (the **lagging strand**) to be replicated

Epigenetics, Nuclear Organization and Gene Function: with implications of epigenetic regulation and genetic architecture for human development and health. John C. Lucchesi, Oxford University Press (2019). © John C. Lucchesi 2019. DOI: 10.1093/oso/9780198831204.001.0001

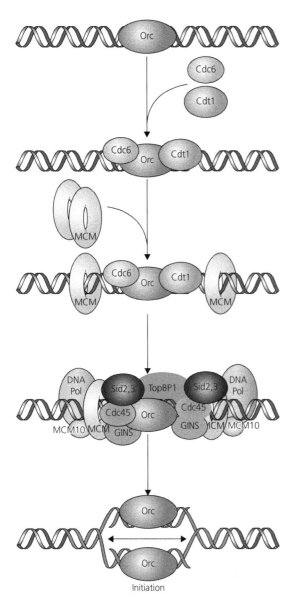

Fig. 14.1 Successive assembly of the pre-replication and pre-initiation complexes. ORC is the origin recognition complex made up of six subunits; Cdc are cell division cycle proteins; Cdt1 is a chromatin licensing protein; MCM (minichromosome maintenance) proteins form a complex that exhibits helicase activity.

(From Conner and Aladjen, 2012.)

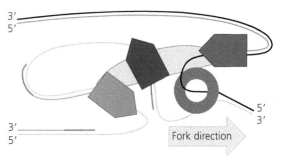

Fig. 14.2 Synthesis of the leading and lagging strands at the replication fork. The leading strand (black line) is replicated continuously. Replication of the lagging strand (gray line) requires the successive synthesis of RNA primers (red). The pentagons represent the different DNA polymerases: Polε (blue), Polα/primase (brown) and Polδ (green). The gray circle represents the DNA helicase.

(From Lujan et al., 2016.)

synthesis of the Okazaki fragments (Sugino et al., 1972). The leading strand requires only one primer at the origin, while the lagging strand requires a primer for the synthesis of each Okazaki fragment. DNA synthesis involves the action of three major polymerases, each consisting of a multi-subunit aggregate—on the lagging strand, Polα, a four-subunit complex which contains the RNA primase activity and synthesizes the beginning of the Okazaki fragments, and Polδ which completes the fragments; Polε is the DNA replicase for the leading strand. As replication of the lagging strand proceeds, the Okazaki fragments are processed into a continuous DNA strand by clipping off or digesting away the RNA primer, allowing the 3′ end of the previous fragment to be extended in order to fill the gap, and ligating the newly exposed 5′ end of one Okazaki segment and the 3′ end of the adjacent segment. There are two intrinsic problems incurred by this process. The first is that some of the DNA synthesized by Polα remains in the genome after replication is finished; since this polymerase lacks proofreading activity (see Errors can occur during DNA replication, p. 168), this DNA is frequently associated with the presence of mutations at the junction between adjacent Okazaki fragments (Reijns et al., 2015). The second problem is that, when the last Okazaki fragment is synthesized and the last RNA primer of the lagging strand is digested, there is no 3′-OH residue present to allow DNA polymerases to fill in this gap. In the terminal replicons, the forks that replicate the left and right tips of the chromatid DNA move in opposite directions (Fig. 14.3). This causes a terminal gap to occur at the 5′ end of both newly synthesized DNA strands. If this

one short segment at a time (Fig. 14.2); these segments of replicated DNA are termed **Okazaki fragments** after their discoverer Reiji Okazaki (Okazaki et al., 1967). Short RNA primers synthesized by a **primase** provide the 3′-OH residues required to initiate the continuous replication of the leading strand and the

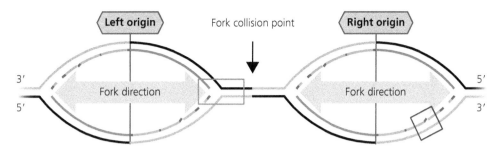

Fig. 14.3 Diagram representing the replication of a chromatid consisting of only two replicons. Note that the opposite direction of the two most distal replication forks causes the two new DNA strands to be shorter than the template strands. See Fig. 14.2 for a key to the strand colors.
(From Lujan et al., 2016.)

were allowed to proceed unattended, the chromatid DNA would shorten further during each replication cycle. In fact, this problem is resolved by the presence of telomeres (see Chapter 20).

A key participant in replication is the proliferating cell nuclear antigen (PCNA) discovered as the target of an autoimmune antiserum. Three PCNA molecules form a sliding clamp that associates with Polε and Polδ and increases their processivity by preventing their dissociation from the DNA template. In addition, PCNA serves as a binding platform for many other proteins involved in DNA replication proper, as well as in the correction of errors made during the replicative process. Another important feature of the replication process is the binding of specialized proteins to the single-stranded DNA, as it is generated by the action of the helicase. These proteins protect the singe-stranded DNA from degradation and prevent the formation of secondary structures; in addition, they participate directly in the execution of replication by recruiting Polα/primase to the single-stranded DNA and, after the Okazaki fragments have been initiated, by recruiting the nucleases that remove the RNA primers.

In human cells, activated replication initiation sites are more numerous in uncondensed chromatin where they flank active genes and, therefore, are enriched in the expected epigenetic marks such as H3K4me3, H3K9ac and H3K36me2 (Petryk et al., 2016); initiation sites are also present at enhancers and at topologically associating domain (TAD) borders (Pope et al., 2014). Actively transcribing regions of the genome, as well as untranscribed regions of open chromatin, replicate early during the cell cycle, while genetically silent and more condensed regions have fewer activated initiation sites and are replicated later (Goldman et al., 1984; Hansen et al., 2010). In general, replication origins in

hyperacetylated chromatin fire earlier and more efficiently. This effect nay be due to the uncondensing effect of acetylation (Gindin et al., 2014); it also appears to be mediated by bromodomain-containing proteins that recognize the modification and that interact and target essential replication factors to the ORIs (Sansam et al., 2018).

Coordination of DNA replication with histone synthesis and nucleosome assembly

As mentioned in previous chapters, the establishment of the varied cell types that constitute a multicellular organism relies on the differential expression of genes initiated by the action of pioneer transcription factors and established and stabilized by post-translational modifications of histones, the introduction of histone variants and the repositioning of nucleosomes by remodeling complexes. Of equal importance to differential gene expression is the repression of particular portions of the genome that is generally achieved by mechanisms that are analogous to those involved in activation. The particular epigenetic landscape characteristic of a cell type is disrupted during DNA replication by the dismantling and eviction of nucleosomes and must be reconstituted on the newly synthesized DNA molecules. The transmission of epigenetic modifications from mother to daughter cells is discussed in Chapter 16. Here, the general processes responsible for the reconstitution of nucleosomes will be considered.

Unwinding the DNA at the replication fork requires the disassembly of the nucleosomes that lie in its path. This process, which involves chaperones and chaperone-like complexes, is initiated by the relative weakness of the association between the H2A–H2B dimers with the H3–H4 octamer (Böhm et al., 2011). The ASF1 (anti-silencing function 1) histone chaperone and the FACT (facilitates chromatin transcription) complex can associate with histones at the replication fork and may help in the disassembly of nucleosomes (Groth et al., 2007; Jasencakova et al., 2010). The FACT complex is also involved in re-establishing nucleosomes behind the replication fork (VanDemark et al., 2006). More recently, the MCM2 subunit of the ATP-dependent replicative helicase was shown to have chaperone activity and to bind H3–H4 dimers from evicted nucleosomes (Huang et al., 2015). Following DNA replication, the number of nucleosomes that existed in chromatin prior to replication must be supplemented by a proportional amount of newly synthesized nucleosomes to re-establish the appropriate nucleosome density on the two "daughter" genomes. The replication of the core portion of nucleosomes is conservative—old H3–H4 tetramers and new H3–H4 tetramers associate with the newly replicated DNA strands with no evidence of mixed tetramers made up of new and old H3–H4 dimers (Leffak, 1983; Leffak et al., 1977; Yamasu and Senshu, 1990). In contrast, new and old H2A–H2B dimers seem to associate at random (Jackson, 1987). A possible explanation for the absence of hybrid tetramers made up of new and old H3–H4 dimers may be that new dimers contain H4 molecules that are acetylated at lysines 5 and 12 (Sobel et al., 1995), making it difficult for such dimers to associate with old dimers containing H4 that is differently modified. A series of chaperones—ASF1, CAF-1 (chromatin assembly factor 1) and Nap1 (nucleosome assembly protein 1)—as well as the FACT remodeling complex, are clearly involved in replication-related nucleosome deposition, although their interactions remain to be fully understood. Given the role that chromatin organization plays in gene function, the nucleosome pattern on the newly replicated DNA is of special importance to the maintenance of established cellular differentiation (see Chapter 16).

Errors can occur during DNA replication

A major source of these errors is the incorporation of the wrong DNA nucleotide; other errors consist of the addition of too many nucleotides or the failure to add nucleotides on the newly synthesized strand. The last two types of error, referred to as **replication slippage**, are most frequent during the replication of repetitive sequences when some of the repeats of the newly synthesized DNA strand pair with upstream or downstream repeats on the template strand. The major DNA

replicative polymerases Polε and Polδ have the ability to recognize a misincorporated nucleotide, which causes them to stall and to remove the aberrant nucleotide using an exonuclease activity present in their multi-protein complexes. If this **proofreading** property of the polymerases fails to recognize the presence of an error, a different repair system, called **mismatch repair**, comes into play (see Chapter 15). Relatively frequently, a ribonucleotide is incorporated during DNA synthesis. In general, DNA polymerases are not very efficient in detecting this type of error that, therefore, is eliminated by the action of a particular RNase that cleaves the DNA strand at the site of incorporation and allows removal of the region of the error.

The simultaneous occurrence of DNA replication and transcription can lead to potential encounters between replisomes and transcriptional complexes. Cells use two strategies to solve this type of conflict—the temporal separation of the two processes whereby early replicating genes are transcribed later during the S phase of the cell cycle, and vice versa (Meryet-Figuiere et al., 2014), or in the case of genes that are transcribed throughout the S phase, such as the repeated ribosomal genes, individual genes or group of genes alternating between being either transcribed or replicated (Smirnov et al., 2014).

Another impediment to replication is the occasional formation of **R-loops** during transcription. R-loop formation can occur when a nascent transcript anneals to the template DNA strand or when it fails to denature from that strand due to the high thermodynamic stability of the hybrid (Fig. 14.4). Genes with non-methylated promoter CpG islands contain stretches of GC-rich sequences (i.e. exhibit so-called GC skewing) and can give rise to R-loops that are generated by pairing of G-rich nascent transcripts with the corresponding C-rich region of the template strand that has just been copied. R-loops can be the source of genomic instability because the single-stranded DNA region within the loop is more sensitive to the action of chemicals and nucleases, resulting in mutations or nicks and gaps that can lead to a stalling and eventual collapse of replication forks (Gan et al., 2011). The effect of R-loops on DNA replication may not be direct—there appears to be a strong correlation between the presence of these loops and the presence of nucleosomes bearing histone H3 phosphorylated at serine 10 (Castellano-Pozo et al., 2013). H3S10P is a mark of chromosome condensation, suggesting that a change in chromatin architecture associated with R-loop formation may be responsible for interfering with replication. This suggestion may also explain the role of the FACT reorganizing complex

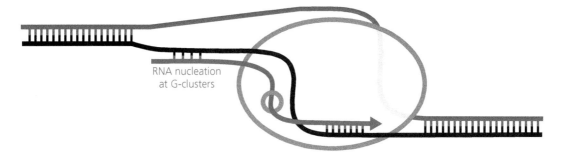

Fig. 14.4 Diagram illustrating R-loop formation when a G-rich region is present in the nascent transcript. The large oval represents the processive RNA polymerase complex.

(From Roy and Lieber, 2009.)

in DNA replication. The FACT complex's well-established role in transcription involves destabilizing nucleosomes, to allow the passage of the transcription machinery, and acting as a chaperone to help reconstitute nucleosomes following transcription (Belotserko-vskaya et al., 2003). Depletion of the FACT complex in human cells leads to an accumulation of chromosomal defects, characteristic of R-loop interference with replication; these observations indicate that the complex functions in resolving the problem generated by the conflict between replication and transcription (Herrera-Moyano et al., 2014).

An additional problem for DNA replication is the occurrence of G-quadruplexes, structures that are formed in DNA regions that possess at least four tracks of 3–4 guanines (Sen and Gilbert, 1988). During replication, these structures form more frequently on the lagging strand since it is replicated discontinuously, one segment at a time, and tends to remain single-stranded, longer than the leading strand. There are several thousands of G-quadruplexes in the human genome usually found in regulatory regions depleted of nucleosomes (Hansel-Hertsch et al., 2017). These structures must be resolved before DNA replication can proceed, a task that is performed by several helicases (Bochman et al., 2012).

The DNA damage response

Many of the problems just discussed that can occur during the course of DNA replication will cause the fork to stall, resulting in the presence of single-stranded DNA breaks (SSBs) or double-stranded breaks (DSBs), with potentially severe consequences such as mitotic arrest, the occurrence of complex chromosome rearrangements or cell death. DNA breaks initiate damage response pathways that include arresting the cell cycle at the replication checkpoint, and repair. SSBs are repaired by a specific pathway, while DSBs are repaired by two major pathways: the **non-homologous end-joining (NHEJ) pathway** that operates throughout the cell cycle, and the **homologous recombination (HR) pathway** that is commonly used during the S (synthesis) and G2 (gap 2) phases. The NHEJ process relies on the direct ligation of the two broken ends. In the HR pathway, information that is missing at the break is copied by using the homologous chromosome or the sister chromatid as a template. These pathways and the parameters that determine which one is chosen for DSB repair are discussed in greater detail in Chapter 15.

Chromosome replication in the context of nuclear organization

Early attempts to visualize DNA replication revealed the appearance of a number of fine granules that grew larger and eventually fused as the S phase proceeded (Nakamura et al., 1986). These granules were shown to contain both Polα and PCNA and were much fewer in number than the estimated number of replicons involved in the DNA synthesis of the entire genome; they were named **replication factories** (Hozak et al., 1993). The "factory" concept implied that the different DNA polymerases and the other factors necessary for replication are concentrated in fixed positions in the nucleus into which the parental DNA molecules slide and the daughter molecules are pushed out as loops (Cook, 1999) (Fig. 14.5). Using three-dimensional super-resolution microscopy, a form of light microscopy that circumvents the limitations of light diffraction (Huang et al., 2008), the replication foci (Fig. 14.6) that gave rise to the factory model were shown to consist of clusters of replicons firing individually (Chagin et al., 2016).

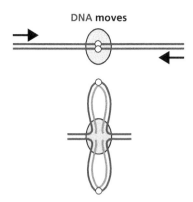

DNA moves

Fig. 14.5 Diagram illustrating the concept underlying the DNA replication mechanism of replication factories. Small white circles represent replication origins; blue lines represent parental DNA strands, and green lines are the newly synthesized strands.

(From Cook, 1999.)

Template DNA
Nascent DNA
Replisome

Fig. 14.6 Diagram illustrating the organization of replicons on extended chromatin fibers and their clustering to form replication foci.

(From Chagin et al., 2016.)

An early topological correlation was made between the timing of replication of particular genomic regions and the position of these regions within the nucleus— late-replicating regions are associated with the nuclear lamina or the perinucleolar region (Hansen et al., 2010; Ragoczy et al., 2014). Consistent with this observation, the replication timing of specific genomic sites differs as cellular differentiation proceeds and the sites switch between activation and repression (Shufaro et al., 2010). Yet, the regulation timing profile of particular tissues appear to be similar between mouse and human cell types, suggesting that tissue-specific replication programs are conserved, at least among mammals (Ryba et al., 2010).

Following the realization that genomes are subdivided into TADs and given the close relationship between replication timing and open chromatin or transcription, it is not surprising that clusters of replicons firing individually, but in unison, form large **replication domains** (RDs) whose boundaries correspond to TAD boundaries (Pope et al., 2014). In a given cell type, the different RDs of the genome do not replicate simultaneously; rather they follow a particular timing pattern that is characteristic for a given cell type. Reflecting the changes that occur in the activity of TADs as a result of cellular differentiation, the replication timing pattern of stem cells changes during differentiation and is different in different cell types (Hiratani et al., 2008; Rivera-Mulia et al., 2015).

Chapter summary

Chromosome replication includes the synthesis of DNA molecules identical in sequence to the DNA in the parent cell by a process termed **semiconservative replication** whereby each strand of the DNA serves as a template for the synthesis of a new strand. DNA replication originates with the binding of pre-replication complexes (MCM2–7) at multiple sites, termed **origins of replication** (ORIs). Additional factors are added, forming **replisomes** that become activated and initiate replication. Although numerous possible ORIs exist throughout the length of the chromosomes, only specific ones are chosen for replication by a mechanism that is still largely elusive. One strand of the parent DNA molecule (the leading strand) is replicated continuously in the 5′ to 3′ direction; the other, or lagging strand, is replicated one short segment at a time. These segments are referred to as Okazaki fragments, and their generation requires the synthesis of short complementary RNA primers.

In order for replication to proceed, DNA must be unwound and, therefore, must be freed from its normal association with nucleosomes. Soon after replication, the nucleosomal structure is re-established using a mixture of old and newly synthesized histones. Areas of open, active chromatin tend to replicate early, and areas of condensed, inactive chromatin tend to replicate later during the cell cycle. For this reason, it is not surprising that large replication domains correspond to topologically associating domains, or TADs.

DNA replication can encounter problematic conditions such as nucleotide depletion, DNA damage or topologically unfavorable DNA structures that generate so-called **replication stress** and replication fork stalling. The DNA replication process itself can also be subjected to errors. Cells have evolved a battery of mechanisms designed to bypass or repair damaged DNA strands. These mechanisms are discussed in Chapter 15.

References

Belotserkovskaya, R., Oh, S., Bondarenko, V. A., Orphanides, G., Studitsky, V. M. & Reinberg, D. 2003. FACT facilitates transcription-dependent nucleosome alteration. *Science,* 301, 1090–3.

Bochman, M. L., Paeschke, K. & Zakian, V. A. 2012. DNA secondary structures: stability and function of G-quadruplex structures. *Nat Rev Genet,* 13, 770–80.

Bohm, V., Hieb, A. R., Andrews, A. J., Gansen, A., Rocker, A., Toth, K., Luger, K. & Langowski, J. 2011. Nucleosome accessibility governed by the dimer/tetramer interface. *Nucleic Acids Res,* 39, 3093–102.

Castellano-Pozo, M., Santos-Pereira, J. M., Rondon, A. G., Barroso, S., Andujar, E., Perez-Alegre, M., Garcia-Muse, T. & Aguilera, A. 2013. R loops are linked to histone H3 S10 phosphorylation and chromatin condensation. *Mol Cell,* 52, 583–90.

Cayrou, C., Coulombe, P., Vigneron, A., Stanojcic, S., Ganier, O., Peiffer, I., Rivals, E., Puy, A., Laurent-Chabalier, S., Desprat, R. & Mechali, M. 2011. Genome-scale analysis of metazoan replication origins reveals their organization in specific but flexible sites defined by conserved features. *Genome Res,* 21, 1438–49.

Chagin, V. O., Casas-Delucchi, C. S., Reinhart, M., Schermelleh, L., Markaki, Y., Maiser, A., Bolius, J. J., Bensimon, A., Fillies, M., Domaing, P., Rozanov, Y. M., Leonhardt, H. & Cardoso, M. C. 2016. 4D visualization of replication foci in mammalian cells corresponding to individual replicons. *Nat Commun,* 7, 11231.

Conner, A. L. & Aladjem, M. I. 2012. The chromatin backdrop of DNA replication: lessons from genetics and genome-scale analyses. *Biochim Biophys Acta,* 1819, 794–801.

Cook, P. R. 1999. The organization of replication and transcription. *Science,* 284, 1790–5.

Gan, W., Guan, Z., Liu, J., Gui, T., Shen, K., Manley, J. L. & Li, X. 2011. R-loop-mediated genomic instability is caused by impairment of replication fork progression. *Genes Dev,* 25, 2041–56.

Gindin, Y., Valenzuela, M. S., Aladjem, M. I., Meltzer, P. S. & Bilke, S. 2014. A chromatin structure-based model accurately predicts DNA replication timing in human cells. *Mol Syst Biol,* 10, 722.

Goldman, M. A., Holmquist, G. P., Gray, M. C., Caston, L. A. & Nag, A. 1984. Replication timing of genes and middle repetitive sequences. *Science,* 224, 686–92.

Groth, A., Corpet, A., Cook, A. J., Roche, D., Bartek, J., Lukas, J. & Almouzni, G. 2007. Regulation of replication fork progression through histone supply and demand. *Science,* 318, 1928–31.

Hansen, R. S., Thomas, S., Sandstrom, R., Canfield, T. K., Thurman, R. E., Weaver, M., Dorschner, M. O., Gartler, S. M. & Stamatoyannopoulos, J. A. 2010. Sequencing newly replicated DNA reveals widespread plasticity in human replication timing. *Proc Natl Acad Sci U S A,* 107, 139–44.

Hansel-Hertsch, R., Di Antonio, M. & Balasubramanian, S.2017. DNA G-quadruplexes in the huuman genome: detection, functions and therapeutic potential. *Nat Rev Mol Cell Biol,* 18, 279–84.

Herrera-Moyano, E., Mergui, X., Garcia-Rubio, M. L., Barroso, S. & Aguilera, A. 2014. The yeast and human FACT chromatin-reorganizing complexes solve R-loop-mediated transcription-replication conflicts. *Genes Dev,* 28, 735–48.

Hiratani, I., Ryba, T., Itoh, M., Yokochi, T., Schwaiger, M., Chang, C. W., Lyou, Y., Townes, T. M., Schubeler, D. & Gilbert, D. M. 2008. Global reorganization of replication domains during embryonic stem cell differentiation. *PLoS Biol,* 6, e245.

Hozak, P., Hassan, A. B., Jackson, D. A. & Cook, P. R. 1993. Visualization of replication factories attached to nucleoskeleton. *Cell,* 73, 361–73.

Huang, B., Wang, W., Bates, M. & Zhuang, X. 2008. Three-dimensional super-resolution imaging by stochastic optical reconstruction microscopy. *Science,* 319, 810–13.

Huang, H., Stromme, C. B., Saredi, G., Hodl, M., Strandsby, A., Gonzalez-Aguilera, C., Chen, S., Groth, A. & Patel, D. J. 2015. A unique binding mode enables MCM2 to chaperone histones H3–H4 at replication forks. *Nat Struct Mol Biol,* 22, 618–26.

Jackson, V. 1987. Deposition of newly synthesized histones: new histones H2A and H2B do not deposit in the same nucleosome with new histones H3 and H4. *Biochemistry,* 26, 2315–25.

Jasencakova, Z., Scharf, A. N., Ask, K., Corpet, A., Imhof, A., Almouzni, G. & Groth, A. 2010. Replication stress interferes with histone recycling and predeposition marking of new histones. *Mol Cell,* 37, 736–43.

Leffak, I. M. 1983. Stability of the conservative mode of nucleosome assembly. *Nucleic Acids Res*, 11, 2717–32.

Leffak, I. M., Grainger, R. & Weintraub, H. 1977. Conservative assembly and segregation of nucleosomal histones. *Cell*, 12, 837–45.

Lujan, S. A., Williams, J. S. & Kunkel, T. A. 2016. DNA polymerases divide the labor of genome replication. *Trends Cell Biol*, 26, 640–54.

Meryet-Figuiere, M., Alaei-Mahabadi, B., Ali, M. M., Mitra, S., Subhash, S., Pandey, G. K., Larsson, E. & Kanduri, C. 2014. Temporal separation of replication and transcription during S-phase progression. *Cell Cycle*, 13, 3241–8.

Meselson, M. & Stahl, F. W. 1958. The replication of DNA in *Escherichia coli. Proc Natl Acad Sci U S A*, 44, 671–82.

Nakamura, H., Morita, T. & Sato, C. 1986. Structural organizations of replicon domains during DNA synthetic phase in the mammalian nucleus. *Exp Cell Res*, 165, 291–7.

Okazaki, R., Okazaki, T., Sakabe, K. & Sugimoto, K. 1967. Mechanism of DNA replication possible discontinuity of DNA chain growth. *Jpn J Med Sci Biol*, 20, 255–60.

Petryk, N., Kahli, M., D'aubenton-Carafa, Y., Jaszczyszyn, Y., Shen, Y., Silvain, M., Thermes, C., Chen, C. L. & Hyrien, O. 2016. Replication landscape of the human genome. *Nat Commun*, 7, 10208.

Pope, B. D., Ryba, T., Dileep, V., Yue, F., Wu, W., Denas, O., Vera, D. L., Wang, Y., Hansen, R. S., Canfield, T. K., Thurman, R. E., Cheng, Y., Gulsoy, G., Dennis, J. H., Snyder, M. P., Stamatoyannopoulos, J. A., Taylor, J., Hardison, R. C., Kahveci, T., Ren, B. & Gilbert, D. M. 2014. Topologically associating domains are stable units of replication-timing regulation. *Nature*, 515, 402–5.

Ragoczy, T., Telling, A., Scalzo, D., Kooperberg, C. & Groudine, M. 2014. Functional redundancy in the nuclear compartmentalization of the late-replicating genome. *Nucleus*, 5, 626–35.

Reijns, M. A. M., Kemp, H., Ding, J., De Proce, S. M., Jackson, A. P. & Taylor, M. S. 2015. Lagging-strand replication shapes the mutational landscape of the genome. *Nature*, 518, 502–5.

Rivera-Mulia, J. C., Buckley, Q., Sasaki, T., Zimmerman, J., Didier, R. A., Nazor, K., Loring, J. F., Lian, Z., Weissman, S., Robins, A. J., Schulz, T. C., Menendez, L., Kulik, M. J., Dalton, S., Gabr, H., Kahveci, T. & Gilbert, D. M. 2015. Dynamic changes in replication timing and gene expression during lineage specification of human pluripotent stem cells. *Genome Res*, 25, 1091–103.

Roy, D. & Lieber, M. R. 2009. G clustering is important for the initiation of transcription-induced R-loops *in vitro*, whereas high G density without clustering is sufficient thereafter. *Mol Cell Biol*, 29, 3124–33.

Ryba, T., Hiratani, I., Lu, J., Itoh, M., Kulik, M., Zhang, J., Schulz, T. C., Robins, A. J., Dalton, S. & Gilbert, D. M. 2010. Evolutionarily conserved replication timing profiles predict long-range chromatin interactions and distinguish closely related cell types. *Genome Res*, 20, 761–70.

Sansam, C. G., Pietrzak, K., Majchrzycka, B., Kerlin, M. A., Chen, J., Rankin, S. & Sansam, C. L. 2018. A mechanism for epigenetic control of DNA replication. *Genes Dev*, 32, 224–9.

Sen, D. & Gilbert, W. 1988. Formation of parallel four-stranded complexes by guanine-rich motifs in DNA and its implications for meiosis. *Nature*, 334, 364–6.

Shufaro, Y., Lacham-Kaplan, O., Tzuberi, B. Z., Mclaughlin, J., Trounson, A., Cedar, H. & Reubinoff, B. E. 2010. Reprogramming of DNA replication timing. *Stem Cells*, 28, 443–9.

Smirnov, E., Borkovec, J., Kovacik, L., Svidenska, S., Schrofel, A., Skalnikova, M., Svindrych, Z., Krizek, P., Ovesny, M., Hagen, G. M., Juda, P., Michalova, K., Cardoso, M. C., Cmarko, D. & Raska, I. 2014. Separation of replication and transcription domains in nucleoli. *J Struct Biol*, 188, 259–66.

Sobel, R. E., Cook, R. G., Perry, C. A., Annunziato, A. T. & Allis, C. D. 1995. Conservation of deposition-related acetylation sites in newly synthesized histones H3 and H4. *Proc Natl Acad Sci U S A*, 92, 1237–41.

Sugino, A., Hirose, S. & Okazaki, R. 1972. RNA-linked nascent DNA fragments in *Escherichia coli. Proc Natl Acad Sci U S A*, 69, 1863–7.

Vandemark, A. P., Blanksma, M., Ferris, E., Heroux, A., Hill, C. P. & Formosa, T. 2006. The structure of the yFACT Pob3-M domain, its interaction with the DNA replication factor RPA, and a potential role in nucleosome deposition. *Mol Cell*, 22, 363–74.

Yamasu, K. & Senshu, T. 1990. Conservative segregation of tetrameric units of H3 and H4 histones during nucleosome replication. *J Biochem*, 107, 15–20.

CHAPTER 15

DNA repair and genomic stability

In addition to the mistakes that can occur during DNA replication, the genome is under constant attack by a variety of external and internal agents such as ultraviolet (UV) or ionizing radiation, chemical carcinogens and oxidative radicals. All of these agents can lead to DNA instability, in the form of mutations, single (SSBs) or double-strand DNA breaks (DSBs) or chromosomal rearrangements.

A number of DNA repair pathways have evolved in order to maintain the integrity of the genetic material. These include **mismatch repair** (MMR), **base excision repair** (BER), **nucleotide excision repair** (NER), **single-strand break repair** (SSBR) and **double-strand break repair** (DSBR). MMR operates when an improper nucleotide is used or when an insertion or deletion occurs during DNA replication. BER deletes and replaces a base that has been chemically modified. NER repairs damage that distorts the DNA helix such as the presence of pyrimidine dimers induced by UV light or the presence of DNA adducts. Repair is achieved by removing a stretch of several nucleotides that includes the damaged area and resynthesizing the removed DNA segment. SSBR uses the same enzymatic steps as BER, while DSBR is achieved either by **non-homologous end-joining** (NHEJ) or by **homologous recombination** (HR). HR uses the same enzymes that are responsible for crossing-over during meiosis and depends on the presence of the sister chromatid or a chromatid from the homologous chromosome to repair the break.

While some of the factors involved in DNA repair were first detected in bacteria, most of the steps involved in carrying out the different repair pathways in eukaryotes were first identified in budding yeast.

Mismatch repair

The proteins of the MMR system recognize the replication error and specifically cleave the new DNA strand, to allow an exonuclease to remove the region of the error; the appropriate DNA polymerase then fills the gap, and a ligase seals the strand break (Fig. 15.1). MMR is also the primary repair system for replication slippage errors.

Mispaired bases are recognized by MutS homolog (MSH) complexes that are thought to associate with proliferating cell nuclear antigen (PCNA) and travel with the DNA replication components (Kolodner and Marsischky, 1999). MSH recruits a second complex MutL homolog (MHL), which has endonuclease activity (cleaves the phosphodiester bond within a polynucleotide chain) and nicks the newly synthesized DNA. The region containing the mismatch is excised by exonucleases (enzymes that cleave nucleotides successively from the end of a polynucleotide chain). The basis of strand discrimination, i.e. the recognition of the newly synthesized DNA strand, and the identity of the different endonucleases involved in the excision reaction are not yet fully understood in eukaryotes. Once excision has occurred, the gap is filled by Polδ using the parental DNA strand as a template, and the nicks are sealed after synthesis by a ligase.

As expected, epigenetic marks on histones and chromatin remodeling factors influence the initiation of the MMR pathway. One of the histone marks involved is methylated lysine 36 on histone H3 (H3K36me3), a mark that is generally associated with active chromatin (see Chapter 5). In MMR, H3K36me3 recruits the most common of the two MSH complexes (MutSα) to

Epigenetics, Nuclear Organization and Gene Function: with implications of epigenetic regulation and genetic architecture for human development and health. John C. Lucchesi, Oxford University Press (2019). © John C. Lucchesi 2019.
DOI: 10.1093/oso/9780198831204.001.0001

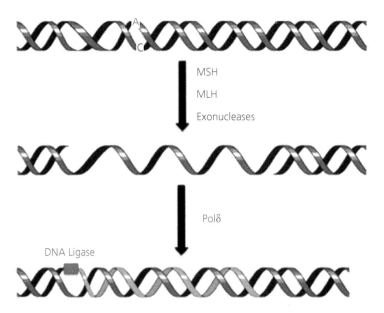

Fig. 15.1 Diagram illustrating the major steps in the mismatch repair pathway. The mismatch is recognized by the MSH complex that, in turn, recruits the MLH complex. The segment containing the mismatch is excised by endonucleases; the gap is filled by a replicative polymerase, and the nicks are sealed by a DNA ligase.

(Modified from Fishel et al., 2015.)

chromatin before the beginning of DNA replication (Li et al., 2013). As the replication machinery disrupts nucleosomes, the MSH complex associates with it and scans the newly replicated DNA for errors (Gorman et al., 2007).

Nucleotide excision and base excision repair pathways

NER deals with a number of different DNA base lesions and bulky adducts caused by various chemical carcinogens or exposure to UV irradiation. There are two NER pathways in eukaryotic cells: one termed transcription-coupled repair (TC-NER) that functions to remove blocks that would stop RNA polymerase elongation, and the other termed global genome repair (GG-NER) that checks the entire genome and targets damage in expressed, as well as silent, regions. The first factors responsible for NER were discovered as mutations in patients with **xeroderma pigmentosum**, a disease that is characterized by acute sensitivity to UV rays from sunlight and a high risk of developing skin cancer (Hwang et al., 1999; Shivji et al., 1994; Sugasawa et al., 1998), and in patients with **Cockayne syndrome** characterized by growth retardation, skeletal abnormalities and severe UV sensitivity (van Hoffen et al., 1993).

Detection of the DNA lesion involves different complexes in the two NER pathways (Sugasawa et al., 1998; van Hoffen et al., 1993), but the steps that mediate the repair are common to both and employ a number of similar proteins (Fig. 15.2). These steps consist of recruitment of the TFIIH complex and its associated XPA factor to verify the occurrence of a problem, unwinding the DNA segment that contains the lesion, recruitment of two different endonucleases that nick the DNA on either side of the lesion and DNA synthesis by one of the two major DNA polymerases Polε and Polδ, followed by ligation of the newly synthesized segment. Surprisingly, Dicer, the enzyme involved in short non-coding RNA (sncRNA) synthesis, is responsible for the chromatin decondensation necessary to enable the repair factors to better access the site of the lesion (Chitale and Richly, 2018).

Ubiquitin ligases are recruited to UV-damaged chromatin (Bergink et al., 2006; Kapetanaki et al., 2006; Wang et al., 2006). In GG-NER, H2A is mono-ubiquitinated and attracts ZRF1 (zuotin-related factor 1), leading to the remodeling of the multiprotein complexes that recognize the presence of DNA lesions (Gracheva et al., 2016). In TC-NER, ubiquitination is less well understood; it consists mostly of poly-ubiquitination of repair complexes to regulate their

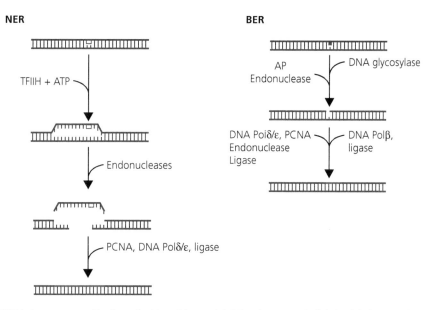

Fig. 15.2 Bulky DNA lesions are removed by the nucleotide excision repair (NER) pathway. Steps include local duplex unwinding by the TFIIH complex, DNA 5′ and 3′ strand incision by different endonucleases, DNA synthesis and ligation. Base excision repair (BER) corrects non-bulky damage to bases. An apyrimidinic/apurinic (AP) or abasic site is created and is processed by an AP endonuclease. Either a single nucleotide is added, followed by ligation (right side of the diagram), or DNA synthesis of multiple nucleotides occurs, leading to an overhang of displaced nucleotides that are removed by an endonuclease (left side).

(Modified from Peterson and Cote, 2004.)

presence by targeting them for degradation (Stadler and Richly, 2017). An additional histone modification associated with NER is the dimethylation of histone 4 at lysine 20 (H4K20me2) by a methyl transferase recruited to the lesion in a Dicer-dependent manner (Chitale and Richly, 2018).

BER is a repair system devoted to the removal of damaged or altered DNA bases that are not bulky, i.e. that do not result in a conformational change in the chromatin (Fig. 15.2). Bases can be altered by oxidation, methylation or deamination, or they can be excised by hydrolysis of the bond that connects them to the sugar–phosphate backbone. BER is initiated by the excision of a damaged base by a glycosylase (glycosylases were first discovered in bacteria), followed by an endonuclease that nicks the DNA at the abasic site. The next steps depend on the extent of the region to be repaired. If a single nucleotide is removed, a single nucleotide is added by the DNA polymerase (Polβ), followed by ligation. Polε and Polδ are the polymerases involved if a longer segment of DNA is resynthesized; in this case, the overhanging segment of DNA on the corrected strand is removed by endonuclease digestion.

DNA breaks

Breaks in one of the two strands of a DNA molecule arise as a consequence of inherent errors in the processes of replication and transcription. As seen earlier, they can also occur as transient intermediates during other DNA repair pathways. Unless they are repaired, SSBs will most often give rise to DSBs, with highly deleterious consequences. DNA breaks initiate damage response pathways that include arresting the cell cycle at the replication checkpoint, DNA replication and repair. These pathways are coordinated and regulated by three special protein kinases: ATM (ataxia telangiectasia mutated), ATR (ATM- and Rad3-related) or DNA-PK (DNA-dependent protein kinase).

Single-strand break repair

Errors in DNA replication can stall the replication fork and result in the formation of a SSB. Other prevalent sources of SSBs are the action of endogenous reactive oxygen species that can interrupt the sugar–phosphate DNA backbone (Pogozelski and Tullius, 1998), and the direct effects of ionizing radiation. If left unrepaired,

SSBs can lead to DSBs in dividing cells and to RNA polymerase stalling and transcription defects in non-dividing cells.

Regions of single-stranded DNA (SS-DNA) are coated by RPA (replication protein A) that recruits the ATR complex. ATR activates the checkpoint protein kinase CHK1 that arrests the cell cycle (Zou and Elledge, 2003). ATR functions also to activate proteins that stabilize the stalled fork and restart DNA replication. SSBs are detected by poly(ADP ribose) polymerase 1 (PARP-1), which binds to the DNA strand broken ends and recruits a "scaffold" protein [X-ray cross complementing group 1 (XRCC1)] that interacts with the various proteins that effect the repair pathway (Whitehouse et al., 2001). The subsequent steps use the enzyme complexes and factors of the BER pathway.

Given the importance of resolving SSBs before they become DSBs, cells appear to have developed a mechanism that prevents the immediate replication of DNA and allows more time for SSBR to function. While a small number of cells are not able to perform SSBR before the start or continuation of replication, in some cells, following the occurrence of a SSB, ATM is activated, leading to delayed replication and repair (Khoronenkova and Dianov, 2015).

Double-strand break repair

The presence of a DSB initiates a set of responses that arrest the cell cycle, prepare the chromatin at the site of the break and repair it. Repair is achieved either by **NHEJ** or by **HR**. The NHEJ process relies on the direct ligation of the two broken ends. HR copies the information that is missing at the break by using the homologous chromosome or the sister chromatid as a template. NHEJ occurs throughout the cell cycle but is more frequent during the Gap1 (G1) phase that precedes the start of DNA replication; HR is more prevalent after DNA replication, i.e. during the synthesis (S)/Gap 2 (G2) phases when sister chromatids are available as templates for the repair process.

Detection of the double-strand break

If the DSB is recognized by the Ku70–Ku80 heterodimer, DNA repair will proceed by NHEJ. If the break is recognized by the MRN [Mre11 (meiotic recombination 11 homolog)–Rad50 (double-strand break repair protein)–Nbs1 (Nijmegen breakage syndrome 1 or nibrin)] complex that binds to the two DNA broken ends as a heterotetramers, DNA will be repaired by the HR pathway. A long-standing question was how the MRN complex manages to beat the Ku complex to the DSB when the latter is present at a relatively high concentration throughout the cell cycle. An answer was provided by *in vivo* and single-molecule experiments that demonstrated MRN's ability to remove Ku from DSBs by endonucleolytic cuts in the nucleotide chain (Chanut et al., 2016; Myler et al., 2017).

Non-homologous end-joining repair pathway

NHEJ is the major mechanism used by cells to repair DSBs and attempt to avoid the damage that these breaks could cause. It involves three steps: detection of the break, processing of the broken ends and their ligation (Fig. 15.3). Detection of the break is the responsibility of a heterodimer consisting of Ku70 and Ku80 proteins (Mimori and Hardin, 1986). Once on the DNA, Ku recruits DNA-PK by interacting with its catalytic subunit DNA-PKcs (Lees-Miller et al., 1990). This complex protects and aligns the broken ends and may serve as a platform for the DNA repair proteins. The next step consists of preparing the DNA ends for eventual ligation which, depending on the damage that was inflicted by the break, will involve different enzymes. Among these are the DNA cross-link repair 1C (DCLRE1C) protein that has 5′ to 3′ exonuclease activity and cleaves single-stranded overhangs (Ma et al., 2002), DNA polymerases μ and λ that fill in the gaps induced by the break (Fan and Wu, 2004; Lee et al., 2004; Mahajan et al., 2002), polynucleotide kinase (PNK) that has both 3′-DNA phosphatase and 5′-DNA kinase activities (Chappell et al., 2002) and others. Ligation of the processed broken ends is carried out by DNA ligase IV. This enzyme is targeted to the broken ends by the XRCC4 (X-ray repair cross complementing 4) protein (Koch et al., 2004) that, in turn, interacts with Ku (McElhinny et al., 2000).

Recently, new interacting factors have been implicated in the NHEJ process. Their major function is to regulate the compaction of the chromatin fiber in order to make it more accessible to the NHEJ machinery. PARP-1 associates with DNA near the DSB and recruits the chromatin remodeler CHD2 (chromodomain helicase DNA-binding protein 2) that, in turn, remodels the chromatin to facilitate the recruitment of Ku and XRCC4 (Luijsterburg et al., 2016).

An alternative non-homologous end-joining (alt-EJ, also referred to as aNHEJ or A-NHEJ) pathway, which operates in the absence of Ku80 and XRCC4, can occur to repair DSBs (Kabotyanski et al., 1998; Liang and Jasin, 1996). As in the case of the HR pathway (see The homologous recombination pathway, p. 177), alt-EJ

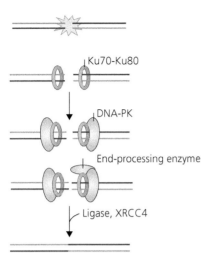

Fig. 15.3 Diagram illustrating the major steps in NHEJ repair. See the text for a discussion of the various factors involved.

(Modified from Lazzerini-Denchi and Sfeir, 2016.)

relies on the presence of 3′ single-stranded DNA over-hangs generated by the resection of the broken DNA ends. If, during the early stages of resection, the short segments of 3′ single-stranded DNA generated on both sides of the break contain a region of homology, DSBs become available to alt-EJ. Following pairing of these segments, the remaining non-complementary 3′ ends are removed by exonuclease activity, and the gaps are filled by DNA polymerase Polθ (Wang and Xu, 2017).

The homologous recombination pathway

HR requires that one of the strands of the broken DNA molecule be resected by endonucleases, in order to allow the strand invasion of the sister chromatid. In the S/G2 phases of the cell cycle, the presence of a DSB is detected by the MRN complex (Fig. 15.4). This complex recruits ATM, a protein kinase that phosphorylates checkpoint proteins such as p53 and the histone variant H2A.X (Lee and Paull, 2004).

Phosphorylated H2A.X, referred to as γH2A.X, spreads for some distance on both sides of the DSB and is thought to initiate the process of recruiting the proteins that will repair the break (Morrison and Shen, 2005). Another possible function for γH2A.X is the recruitment of cohesin, a complex that is responsible for maintaining sister chromatids together. The accumulation of cohesin at the site of DSBs, first noted in yeast (Sjogren and Nasmyth, 2001) and in human cells (Kim et al., 2002), could ensure access to the unbroken

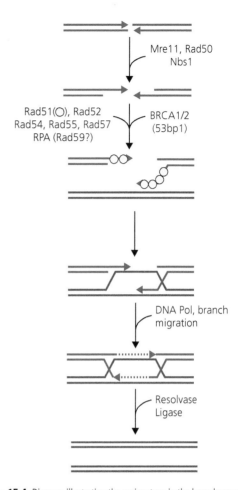

Fig. 15.4 Diagram illustrating the major steps in the homologous recombination (HR) repair pathway. See the text for a discussion of the various factors involved.

(Modified from Peterson and Cote, 2004.)

sister chromatid for HR repair. This possible role for γH2A.X may explain why this modification extends so far from the site of the break. The MRN endonucleolytic complex processes the broken ends to generate the long 3′ single-stranded DNA tails that can invade the homologous DNA strand (Dolganov et al., 1996). Resection is promoted by *BRCA1* (breast cancer 1) gene product (Zhong et al., 1999) and, as discussed in Epigenetic modifications in double-strand break repair, p. 178, is inhibited by 53BP1 (p53-binding protein 1). The single-stranded tails are bound transiently by RPA (replication protein A), which is rapidly replaced by a complex containing the strand exchange protein Rad51, a process that is stimulated by BRCA2 (Carreira et al., 2009). Rad51 monomers form a helical filament that invades the double-stranded DNA of a sister chromatid in search of a homologous region (Sung, 1994). Rad55 and Rad57 are non-functional homologs of Rad51 that, nevertheless, stabilize the Rad51-generated filament (Janke et al., 2016). The free 3′ ends are extended in a manner similar to the synthesis of the leading strand during DNA replication. DNA synthesis, which requires the dislodgement of the Rad51 filaments by Rad54 (Li and Heyer, 2009), is carried out principally by Polδ, although other polymerases (Polη and Polζ) may be involved in some instances (Li et al., 2013). Invasion of the DNA of a sister chromatid generates a configuration of four connected helices referred to as a Holliday junction (Holliday, 1964). Processing of the junction to reconstitute separate DNA molecules is carried out by a **resolvase**. These enzymes hydrolyze the sugar–phosphate backbone of the two crossed strands that flank the region of strand exchange (Fekairi et al., 2009; Ip et al., 2008).

DSBs that occur in transcribing genes are preferentially repaired by HR (Aymard et al., 2014). The presence of nascent RNA at the site of the break can give rise to R-loops (see Chapter 14) that would interfere with the repair process. In human cells, an RNA:DNA helicase is attracted to DSB-induced R-loops and significantly decreases their frequency (Cohen et al., 2018).

Epigenetic modifications in double-strand break repair

In addition to the phosphorylation of H2A.X, an important subsequent epigenetic modification of the chromatin at the site of DSBs is the ubiquitination of H2A and H2A.X (Huen et al., 2007; Kolas et al., 2007; Mattiroli et al., 2012; Wang and Elledge, 2007). This modification and others that include dimethylated H4K20 (a common mark of nucleosomes throughout the human genome) and methylated H4K79 attract DNA repair factors and complexes, including 53BP1 (Botuyan et al., 2006; Huyen et al., 2004; Mailand et al., 2007; Mattiroli et al., 2012; Sanders et al., 2004). The MOF histone acetyl transferase that is responsible for the acetylation of H4K16 (Neal et al., 2000; Smith et al., 2005) was shown to interact directly with ATM (Gupta et al., 2005) and to be required for the recruitment of MDC1, BRCA1 and 53BP1 (Li et al., 2010). The histone modification H3K36me3 is also found at the site of a DSB (Pfister et al., 2014) where it appears to promote the acetylation of histone H4 at K16 (Li and Wang, 2017).

As mentioned previously, resection cannot occur if the broken ends are bound by 53BP1 molecules, and repair is directed to the NHEJ pathway (Bunting et al., 2010). 53BP1 is recruited to DSBs by its simultaneous recognition of H3K20me2 and H2Ak15ub on the same nucleosomes (Fradet-Turcotte et al., 2013). Binding of 53BP1 is inhibited by several epigenetic modifications. Acetylation of H4 at K16 in a nucleosome interferes with the binding of 53BP1 to the neighboring H4K20me2, and the HR pathway is favored (Hsiao and Mizzen, 2013; Tang et al., 2013). Another acetylation mark that enables repair by HR is the acetylation of H2AK15. TIP60 [Tat (tyrosine amino transferase)-interacting protein of 60 kDa], an enzyme that acetylates several lysines on histone H4 in different genomic contexts, targets H4K20me at the site of DSBs and acetylates the neighboring H2AK15, thereby blocking the ubiquitination of this residue. Binding of 53BP1 is abrogated and, here again, repair is channeled to the HR pathway (Jacquet et al., 2016).

Sumoylation also plays a critical role in DSB repair (see The three-dimensional arrangement of double-strand breaks, p. 179), although its targets are a group of non-histone proteins such as BRCA1, a ubiquitin protein ligase (that participates also in transcription and recombination) and 53BP1 that enhances the p53-mediated activation of cell cycle arrest and cell death genes in response to DNA damage (Galanty et al., 2009; Morris et al., 2009).

The conformation of chromatin at the site of a double-strand break

The level of condensation of the chromatin is very dynamic and helps to orchestrate the various steps of the repair mechanism. A complex containing a histone methyl transferase, HP1 and KAP1 (KRAB-associated protein 1) binds briefly to the site of the break where it trimethylates H3K9 and induces a transient compacted state that spreads for some distance from the break (Ayrapetov et al., 2014); the nucleosome remodeling and deacetylation (NuRD) complex is also recruited to the site of the break (Chou et al., 2010), presumably by KAP1. Recently, recruitment of NuRD and of a co-repressor was shown to depend on the demethylation of H3K4me3, if this modification is present at the site of the break (Gong et al., 2017). The heterochromatinization of DSB sites may inhibit local transcription and may keep the broken ends from moving away from each other (Ayrapetov et al., 2014). But heterochromatin is refractory to the repair process and, therefore, must be decondensed. This remodeling is achieved by the action of TIP60 histone acetyl transferase that is a subunit of the human NuA4 complex and is activated by the presence of H3K9me3 in nucleosomes (Auger et al., 2008; Sun et al., 2009). The TIP60 complex targets are lysines 5, 8 and 12 on histone H4 and lysine 5 on the H2A histone; this complex also mediates the exchange of the H2A/H2A.X histones for the H2A.Z variant through its ATP-dependent remodeling activity, and it acetylates histone H4 and H2A.Z to create an open chromatin structure that facilitates the recruitment and function of the DSBR factors. In the case of HR, the INO80 remodeling complex localizes to the sites of DSBs where it is required for resection of the broken ends (Gospodinov et al., 2011). Since the presence of H2A.Z at these sites appears to be inhibitory to resection (Xu et al., 2012), INO80 may accomplish its effect by removing H2A.Z and replacing it with H2A, as it does in yeast (Papamichos-Chronakis et al., 2011).

A similar series of reactions contribute to the decondensation of constitutive heterochromatin to allow repair of DSBs. Heterochromatin is characterized by the presence of the NuRD complex and of histone methyl transferases that are recruited by KAP1; these nucleosome and histone-modifying activities result in an enrichment of H3K9me3 and of heterochromatin protein 1 (HP1), as well as histone lysine hypoacetylation. At the site of the DSB, KAP1 is phosphorylated by ATM, leading to the loss of the NuRD complex and relaxation of the heterochromatic condensed state (Goodarzi et al., 2011).

Given the importance of maintaining tissue-specific patterns of histone modifications through mitotic divisions, it is not surprising that mechanisms have evolved to restore the epigenetic landscape following the repair of DNA damage. Following a DSB, parental histones that are evicted from the damaged region are mostly conserved and associate with newly synthesized histones to reconstitute nucleosomes. This process is dependent on the presence of DDB2 (DNA damage-binding protein 2) and is thought to re-establish the chromatin status that preceded the break (Adam et al., 2016).

The three-dimensional arrangement of double-strand breaks

Constitutive pericentromeric heterochromatin consists of regions of highly repeated DNA sequences and clusters of transposable elements. In flies and human genomes, it constitutes approximately 30% of the genome (Ho et al., 2014). Because of its repetitive nature, exchanges in heterochromatin are prone to be ectopic and non-allelic, leading to translocations, duplications and deletions. Repair requires special regulation, especially in the case of HR that is the primary pathway used within these regions. A major aspect of this regulation is the relocation, before strand invasion, of the broken DNA segment away from its association with other heterochromatic segments. In *Drosophila*, while the initial steps of the DSBR pathway occur at the nuclear location where the DSBs were induced, later steps beginning with Rad51 recruitment occur after the broken ends have been relocated to the periphery of heterochromatin domains in the perinuclear region (Chiolo et al., 2011). Sumoylation of ubiquitin ligases prevents the binding of Rad51 and promotes relocation to the nuclear periphery; the relocation is stabilized by the interaction of the DSB with nuclear pore proteins (Ryu et al., 2015).

By using microscopic techniques that allow resolution at the nanometer level, in combination with genome sequencing, the architectural structure in the region of DSBs was determined and found to be altered (Natale et al., 2017). This reorganization is clearly intended to increase the efficiency of DSBR. The regions of the DNA surrounding the breakpoint that are decorated with γH2A.X form clusters flanked by CCCTC-binding factor (CTCF) and included in a chromatin loop. CTCF interacts with Rad51 and promotes Rad51 repair foci formation (Lang et al., 2017).

Chapter summary

In addition to the mistakes that can occur during DNA replication, the genome is under constant attack by a variety of external and internal agents such as UV or ionizing radiation, chemical carcinogens and oxidative radicals. A number of DNA repair pathways have evolved in order to maintain the integrity of the genetic material.

Mismatch repair (MMR) operates when an improper nucleotide is used or when an insertion or deletion occurs during DNA replication. The proteins of this repair system recognize the replication error and specifically cleave the new DNA strand to allow an exonuclease to remove the region of the error; the appropriate DNA polymerase then fills the gap, and a ligase seals the strand break. MMR is also the primary repair system for replication slippage errors. **Nucleotide excision repair** (NER) repairs damage that distorts the DNA helix such as the presence of pyrimidine dimers, induced by UV light, or of DNA adducts. Repair is achieved by removing a stretch of several nucleotides that includes the damaged area and resynthesizing the removed DNA segment. **Base excision repair** (BER) is a repair system devoted to the removal of damaged or altered DNA bases that are not bulky, i.e. that do not result in a conformational change in the chromatin. The base or bases are excised; the DNA is nicked at that site, and DNA polymerases add one or more nucleotides, as needed. **Single-strand break repair** (SSBR) uses the same enzymatic steps as BER. Double-strand break repair (DSBR) can involve either **non-homologous end-joining** (NHEJ) or **homologous recombination** (HR). NHEJ is the major mechanism used by cells to repair DSBs. Broken DNA ends are processed to facilitate ligation and joined directly. An **alternative non-homologous end-joining** (alt-EJ) relies on the presence of a region of homology in the 3′ single-stranded DNA overhangs generated by the early resection of the broken DNA ends. HR requires that one of the strands of the broken DNA molecule be resected by endonucleases in order to generate a single-stranded end that will participate in the strand invasion of the sister chromatid. The remainder of the process is analogous to meiotic recombination.

The chromatin state at the site of the DSB must be modified to allow access to the repair machinery. This modification involves remodeling complexes, as well as histone-modifying enzymes. This is particularly true in the case of breaks that occur in constitutive heterochromatin.

References

Adam, S., Dabin, J., Chevallier, O., Leroy, O., Baldeyron, C., Corpet, A., Lomonte, P., Renaud, O., Almouzni, G. & Polo, S. E. 2016. Real-time tracking of parental histones reveals their contribution to chromatin integrity following DNA damage. *Mol Cell*, 64, 65–78.

Auger, A., Galarneau, L., Altaf, M., Nourani, A., Doyon, Y., Utley, R. T., Cronier, D., Allard, S. & Cote, J. 2008. Eaf1 is the platform for NuA4 molecular assembly that evolutionarily links chromatin acetylation to ATP-dependent exchange of histone H2A variants. *Mol Cell Biol*, 28, 2257–70.

Aymard, F., Bugler, B., Schmidt, C. K., Guillou, E., Caron, P., Briois, S., Iacovoni, J. S., Daburon, V., Miller, K. M., Jackson, S. P. & Legube, G. 2014. Transcriptionally active chromatin recruits homologous recombination at DNA double-strand breaks. *Nat Struct Mol Biol*, 21, 366–74.

Ayrapetov, M. K., Gursoy-Yuzugullu, O., Xu, C., Xu, Y. & Price, B. D. 2014. DNA double-strand breaks promote methylation of histone H3 on lysine 9 and transient formation of repressive chromatin. *Proc Natl Acad Sci U S A*, 111, 9169–74.

Bergink, S., Salomons, F. A., Hoogstraten, D., Groothuis, T. A., De Waard, H., Wu, J., Yuan, L., Citterio, E., Houtsmuller, A. B., Neefjes, J., Hoeijmakers, J. H., Vermeulen, W. & Dantuma, N. P. 2006. DNA damage triggers nucleotide excision repair-dependent monoubiquitylation of histone H2A. *Genes Dev*, 20, 1343–52.

Botuyan, M. V., Lee, J., Ward, I. M., Kim, J. E., Thompson, J. R., Chen, J. & Mer, G. 2006. Structural basis for the methylation state-specific recognition of histone H4-K20 by 53BP1 and Crb2 in DNA repair. *Cell*, 127, 1361–73.

Bunting, S. F., Callen, E., Wong, N., Chen, H. T., Polato, F., Gunn, A., Bothmer, A., Feldhahn, N., Fernandez-Capetillo, O., Cao, L., Xu, X., Deng, C. X., Finkel, T., Nussenzweig, M., Stark, J. M. & Nussenzweig, A. 2010. 53BP1 inhibits homologous recombination in Brca1-deficient cells by blocking resection of DNA breaks. *Cell*, 141, 243–54.

Carreira, A., Hilario, J., Amitani, I., Baskin, R. J., Shivji, M. K., Venkitaraman, A. R. & Kowalczykowski, S. C. 2009. The BRC repeats of BRCA2 modulate the DNA-binding selectivity of RAD51. *Cell*, 136, 1032–43.

Chanut, P., Britton, S., Coates, J., Jackson, S. P. & Calsou, P. 2016. Coordinated nuclease activities counteract Ku at single-ended DNA double-strand breaks. *Nat Commun*, 7, 12889.

Chappell, C., Hanakahi, L. A., Karimi-Busheri, F., Weinfred, M. & West, S. C. 2002. Involvement of human polynucleotide kinase in double-strand break repair by non-homologous end joining. *EMBO J*, 21, 2827–32.

Chiolo, I., Minoda, A., Colmenares, S. U., Polyzos, A., Costes, S. V. & Karpen, G. H. 2011. Double-strand breaks in heterochromatin move outside of a dynamic HP1a domain to complete recombinational repair. *Cell*, 144, 732–44.

Chitale, S. & Richly, H. 2018. DICER- and MMSET-catalyzed H4K20me2 recruits the nucleotide excision repair factor XPA to DNA damage sites. *J Cell Biol*, 217, 527–40.

Chou, D. M., Adamson, B., Dephoure, N. E., Tan, X., Nottke, A. C., Hurov, K. E., Gygi, S. P., Colaiacovo, M. P. & Elledge, S. J. 2010. A chromatin localization screen reveals poly (ADP ribose)-regulated recruitment of the repressive polycomb and NuRD complexes to sites of DNA damage. *Proc Natl Acad Sci U S A*, 107, 18475–80.

Cohen, S., Puget, N., Lin, Y. L., Clouaire, T., Aguirrebengoa, M., Rocher, V., Pasero, P., Canitrot, Y. & Legube, G. 2018. Senataxin resolves RNA:DNA hybrids forming at DNA double-strand breaks to prevent translocations. *Nat Commun*, 9, 533.

Dolganov, G. M., Maser, R. S., Novikov, A., Tosto, L., Chong, S., Bressan, D. A. & Petrini, J. H. 1996. Human Rad50 is physically associated with human Mre11: identification of a conserved multiprotein complex implicated in recombinational DNA repair. *Mol Cell Biol*, 16, 4832–41.

Fan, W. & Wu, X. 2004. DNA polymerase lambda can elongate on DNA substrates mimicking non-homologous end joining and interact with XRCC4-ligase IV complex. *Biochem Biophys Res Commun*, 323, 1328–33.

Fekairi, S., Scaglione, S., Chahwan, C., Taylor, E. R., Tissier, A., Coulon, S., Dong, M. Q., Ruse, C., Yates, J. R., 3RD, Russell, P., Fuchs, R. P., Mcgowan, C. H. & Gaillard, P. H. 2009. Human SLX4 is a Holliday junction resolvase subunit that binds multiple DNA repair/recombination endonucleases. *Cell,* 138, 78–89.

Fishel, R. 2015. Mismatch repair. *J Biol Chem,* 290, 26395–403.

Fradet-Turcotte, A., Canny, M. D., Escribano-Diaz, C., Orthwein, A., Leung, C. C., Huang, H., Landry, M. C., Kitevski-Leblanc, J., Noordermeer, S. M., Sicheri, F. & Durocher, D. 2013. 53BP1 is a reader of the DNA-damage-induced H2A Lys 15 ubiquitin mark. *Nature,* 499, 50–4.

Galanty, Y., Belotserkovskaya, R., Coates, J., Polo, S., Miller, K. M. & Jackson, S. P. 2009. Mammalian SUMO E3-ligases PIAS1 and PIAS4 promote responses to DNA double-strand breaks. *Nature,* 462, 935–9.

Gong, F., Clouaire, T., Aguirrebengoa, M., Legube, G. & Miller, K. M. 2017. Histone demethylase KDM5A regulates the ZMYND8-NuRD chromatin remodeler to promote DNA repair. *J Cell Biol,* 216, 1959–74.

Goodarzi, A. A., Kurka, T. & Jeggo, P. A. 2011. KAP-1 phosphorylation regulates CHD3 nucleosome remodeling during the DNA double-strand break response. *Nat Struct Mol Biol,* 18, 831–9.

Gorman, J., Chowdhury, A., Surtees, J. A., Shimada, J., Reichman, D. R., Alani, E. & Greene, E. C. 2007. Dynamic basis for one-dimensional DNA scanning by the mismatch repair complex Msh2–Msh6. *Mol Cell,* 28, 359–70.

Gospodinov, A., Vaissiere, T., Krastev, D. B., Legube, G., Anachkova, B. & Herceg, Z. 2011. Mammalian Ino80 mediates double-strand break repair through its role in DNA end strand resection. *Mol Cell Biol,* 31, 4735–45.

Gracheva, E., Chitale, S., Wilhelm, T., Rapp, A., Byrne, J., Stadler, J., Medina, R., Cardoso, M. C. & Richly, H. 2016. ZRF1 mediates remodeling of E3 ligases at DNA lesion sites during nucleotide excision repair. *J Cell Biol,* 213, 185–200.

Gupta, A., Sharma, G. G., Young, C. S., Agarwal, M., Smith, E. R., Paull, T. T., Lucchesi, J. C., Khanna, K. K., Ludwig, T. & Pandita, T. K. 2005. Involvement of human MOF in ATM function. *Mol Cell Biol,* 25, 5292–305.

Ho, J. W., Jung, Y. L., Liu, T., Alver, B. H., Lee, S., Ikegami, K., Sohn, K. A., Minoda, A., Tolstorukov, M. Y., Appert, A., Parker, S. C., Gu, T., Kundaje, A., Riddle, N. C., Bishop, E., Egelhofer, T. A., Hu, S. S., Alekseyenko, A. A., Rechtsteiner, A., Asker, D., Belsky, J. A., Bowman, S. K., Chen, Q. B., Chen, R. A., Day, D. S., Dong, Y., Dose, A. C., Duan, X., Epstein, C. B., Ercan, S., Feingold, E. A., Ferrari, F., Garrigues, J. M., Gehlenborg, N., Good, P. J., Haseley, P., He, D., Herrmann, M., Hoffman, M. M., Jeffers, T. E., Kharchenko, P. V., Kolasinska-Zwierz, P., Kotwaliwale, C. V., Kumar, N., Langley, S. A., Larschan, E. N., Latorre, I., Libbrecht, M. W., Lin, X., Park, R., Pazin, M. J., Pham, H. N., Plachetka, A., Qin, B., Schwartz, Y. B., Shoresh, N., Stempor, P., Vielle, A., Wang, C., Whittle, C. M., Xue, H., Kingston, R. E., Kim, J. H., Bernstein, B. E., Dernburg, A. F., Pirrotta, V., Kuroda, M. I., Noble, W. S., Tullius, T. D., Kellis, M., Macalpine, D. M., Strome, S., Elgin, S. C., Liu, X. S.,

Lieb, J. D., Ahringer, J., Karpen, G. H. & Park, P. J. 2014. Comparative analysis of metazoan chromatin organization. *Nature,* 512, 449–52.

Holliday, R. 1964. A mechanism for gene conversion in fungi. *Genet Res,* 5, 282–304.

Hsiao, K. Y. & Mizzen, C. A. 2013. Histone H4 deacetylation facilitates 53BP1 DNA damage signaling and double-strand break repir. *J Mol Cell Biol,* 5, 157–65.

Huen, M. S., Grant, R., Manke, I., Minn, K., Yu, X., Yaffe, M. B. & Chen, J. 2007. RNF8 transduces the DNA-damage signal via histone ubiquitylation and checkpoint protein assembly. *Cell,* 131, 901–14.

Huyen, Y., Zgheib, O., Ditullio, R. A.,JR., Gorgoulis, V. G., Zacharatos, P., Petty, T. J., Sheston, E. A., Mellert, H. S., Stavridi, E. S. & Halazonetis, T. D. 2004. Methylated lysine 79 of histone H3 targets 53BP1 to DNA double-strand breaks. *Nature,* 432, 406–11.

Hwang, B. J., Ford, J. M., Hanawalt, P. C. & Chu, G. 1999. Expression of the p48 xeroderma pigmentosum gene is p53-dependent and is involved in global genomic repair. *Proc Natl Acad Sci U S A,* 96, 424–8.

Ip, S. C., Rass, U., Blanco, M. G., Flynn, H. R., Skehel, J. M. & West, S. C. 2008. Identification of Holliday junction resolvases from humans and yeast. *Nature,* 456, 357–61.

Jacquet, K., Fradet-Turcotte, A., Avvakumov, N., Lambert, J. P., Roques, C., Pandita, R. K., Paquet, E., Herst, P., Gingras, A. C., Pandita, T. K., Legube, G., Doyon, Y., Durocher, D. & Cote, J. 2016. The TIP60 complex regulates bivalent chromatin recognition by 53BP1 through direct H4K20me binding and H2AK15 acetylation. *Mol Cell,* 62, 409–21.

Janke, R., Kong, J., Braberg, H., Cantin, G., Yates, J. R., 3RD, Krogan, N. J. & Heyer, W. D. 2016. Nonsense-mediated decay regulates key components of homologous recombination. *Nucleic Acids Res,* 44, 5218–30.

Kabotyanski, E. B., Gomelsky, L., Han, J. O., Stamato, T. D. & Roth, D. B. 1998. Double-strand break repair in Ku86- and XRCC4-deficient cells. *Nucleic Acids Res,* 26, 5333–42.

Kapetanaki, M. G., Guerrero-Santoro, J., Bisi, D. C., Hsieh, C. L., Rapic-Otrin, V. & Levine, A. S. 2006. The DDB1-CUL4ADDB2 ubiquitin ligase is deficient in xeroderma pigmentosum group E and targets histone H2A at UV-damaged DNA sites. *Proc Natl Acad Sci U S A,* 103, 2588–93.

Khoronenkova, S. V. & Dianov, G. L. 2015. ATM prevents DSB formation by coordinating SSB repair and cell cycle progression. *Proc Natl Acad Sci U S A,* 112, 3997–4002.

Kim, J. S., Krasieva, T. B., Lamorte, V., Taylor, A. M. & Yokomori, K. 2002. Specific recruitment of human cohesin to laser-induced DNA damage. *J Biol Chem,* 277, 45149–53.

Koch, C. A., Agyei, R., Galicia, S., Metalnikov, P., O'donnell, P., Starostine, A., Weinfeld, M. & Durocher, D. 2004. Xrcc4 physically links DNA end processing by polynucleotide kinase to DNA ligation by DNA ligase IV. *EMBO J,* 23, 3874–85.

Kolas, N. K., Chapman, J. R., Nakada, S., Ylanko, J., Chahwan, R., Sweeney, F. D., Panier, S., Mendez, M., Wildenhain, J., Thomson, T. M., Pelletier, L., Jackson, S. P. & Durocher, D. 2007. Orchestration of the DNA-damage response by the RNF8 ubiquitin ligase. *Science*, 318, 1637–40.

Kolodner, R. D. & Marsischky, G. T. 1999. Eukaryotic DNA mismatch repair. *Curr Opin Genet Dev*, 9, 89–96.

Lang, F., Li, X., Zheng, W., Li, Z., Lu, D., Chen, G., Gong, D., Yang, L., Fu, J., Shi, P. & Zhou, J. 2017. CTCF prevents genomic instability by promoting homologous recombination-directed DNA double-strand break repair. *Proc Natl Acad Sci U S A*, 114, 10912–17.

Lazzerini-Denchi, E. & Sfeir, A. 2016. Stop pulling my strings—what telomeres taught us about the DNA damage response. *Nat Rev Mol Cell Biol*, 17, 364–78.

Lee, J. H. & Paull, T. T. 2004. Direct activation of the ATM protein kinase by the Mre11/Rad50/Nbs1 complex. *Science*, 304, 93–6.

Lee, J. W., Blanco, L., Zhou, T., Garcia-Diaz, M., Bebenek, K., Kunkel, T. A., Wang, Z. & Povirk, L. F. 2004. Implication of DNA polymerase lambda in alignment-based gap filling for nonhomologous DNA end joining in human nuclear extracts. *J Biol Chem*, 279, 805–11.

Lees-Miller, S. P., Chen, Y. R. & Anderson, C. W. 1990. Human cells contain a DNA-activated protein kinase that phosphorylates simian virus 40 T antigen, mouse p53, and the human Ku autoantigen. *Mol Cell Biol*, 10, 6472–81.

Li, F., Mao, G., Tong, D., Huang, J., Gu, L., Yang, W. & Li, G. M. 2013. The histone mark H3K36me3 regulates human DNA mismatch repair through its interaction with MutSα. *Cell*, 153, 590–600.

Li, L. & Wang, Y. 2017. Cross-talk between the H3K36me3 and H4K16ac histone epigenetic marks in DNA double-strand break repair. *J Biol Chem*, 292, 11951–9.

Li, X., Corsa, C. A., Pan, P. W., Wu, L., Ferguson, D., Yu, X., Min, J. & Dou, Y. 2010. MOF and H4 K16 acetylation play important roles in DNA damage repair by modulating recruitment of DNA damage repair protein Mdc1. *Mol Cell Biol*, 30, 5335–47.

Li, X. & Heyer, W. D. 2009. RAD54 controls access to the invading 3′-OH end after RAD51-mediated DNA strand invasion in homologous recombination in *Saccharomyces cerevisiae*. *Nucleic Acids Res*, 37, 638–46.

Liang, F. & Jasin, M. 1996. Ku80-deficient cells exhibit excess degradation of extrachromosomal DNA. *J Biol Chem*, 271, 14405–11.

Luijsterburg, M. S., De Krijger, I., Wiegant, W. W., Shah, R. G., Smeenk, G., De Groot, A. J. L., Pines, A., Vertegaal, A. C. O., Jacobs, J. J. L., Shah, G. M. & Van Attikum, H. 2016. PARP1 links CHD2-mediated chromatin expansion and H3.3 deposition to DNA repair by non-homologous end-joining. *Mol Cell*, 61, 547–62.

Ma, Y., Pannicke, U., Schwarz, K. & Lieber, M. R. 2002. Hairpin opening and overhang processing by an Artemis/ DNA-dependent protein kinase complex in nonhomologous end joining and V(D)J recombination. *Cell*, 108, 781–94.

Mahajan, K. N., Nick Mcelhinny, S. A., Mitchell, B. S. & Ramsden, D. A. 2002. Association of DNA polymerase mu (pol mu) with Ku and ligase IV: role for pol mu in end-joining double-strand break repair. *Mol Cell Biol*, 22, 5194–202.

Mailand, N., Bekker-Jensen, S., Faustrup, H., Melander, F., Bartek, J., Lukas, C. & Lukas, J. 2007. RNF8 ubiquitylates histones at DNA double-strand breaks and promotes assembly of repair proteins. *Cell*, 131, 887–900.

Mattiroli, F., Vissers, J. H., Van Dijk, W. J., Ikpa, P., Citterio, E., Vermeulen, W., Marteijn, J. A. & Sixma, T. K. 2012. RNF168 ubiquitinates K13-15 on H2A/H2AX to drive DNA damage signaling. *Cell*, 150, 1182–95.

Mcelhinny, N., S. A., Snowden, C. M., Mccarville, J. & Ramsden, D. A. 2000. Ku recruits the XRCC4-ligase IV complex to DNA ends. *Mol Cell Biol*, 20, 2996–3003.

Mimori, T. & Hardin, J. A. 1986. Mechanism of interaction between Ku protein and DNA. *J Biol Chem*, 261, 10375–9.

Morris, J. R., Boutell, C., Keppler, M., Densham, R., Weekes, D., Alamshah, A., Butler, L., Galanty, Y., Pangon, L., Kiuchi, T., Ng, T. & Solomon, E. 2009. The SUMO modification pathway is involved in the BRCA1 response to genotoxic stress. *Nature*, 462, 886–90.

Morrison, A. J. & Shen, X. 2005. DNA repair in the context of chromatin. *Cell Cycle*, 4, 568–71.

Myler, L. R., Gallardo, I. F., Soniat, M. M., Deshpande, R. A., Gonzalez, X. B., Kim, Y., Paull, T. T. & Finkelstein, I. J. 2017. Single-molecule imaging reveals how Mre11-Rad50-Nbs1 initiates DNA break repair. *Mol Cell*, 67, 891–8 e4.

Natale, F., Rapp, A., Yu, W., Maiser, A., Harz, H., Scholl, A., Grulich, S., Anton, T., Horl, D., Chen, W., Durante, M., Taucher-Scholz, G., Leonhardt, H. & Cardoso, M. C. 2017. Identification of the elementary structural units of the DNA damage response. *Nat Commun*, 8, 15760.

Neal, K. C., Pannuti, A., Smith, E. R. & Lucchesi, J. C. 2000. A new human member of the MYST family of histone acetyl transferases with high sequence similarity to *Drosophila* MOF. *Biochim Biophys Acta*, 1490, 170–4.

Papamichos-Chronakis, M., Watanabe, S., Rando, O. J. & Peterson, C. L. 2011. Global regulation of H2A.Z localization by the IN80 chromatin-remodeling enzyme is essential for genome integrity. *Cell*, 144, 200–13.

Peterson, C. L. & Cote, J. 2004. Cellular machineries for chromosomal DNA repair. *Genes Dev*, 18, 602–16.

Pfister, S. X., Ahrabi, S., Zalmas, L. P., Sarkar, S., Aymard, F., Bachrati, C. Z., Helleday, T., Legube, G., La Thangue, N. B., Porter, A. C. & Humphrey, T. C. 2014. SETD2-dependent histone H3K36 trimethylation is required for homologous recombination repair and genome stability. *Cell Rep*, 7, 2006–18.

Pogozelski, W. K. & Tullius, T. D. 1998. Oxidative strand scission of nucleic acids: routes initiated by hydrogen abstraction from the sugar poiety. *Chem Rev,* 98, 1089–108.

Ryu, T., Spatola, B., Delabaere, L., Bowlin, K., Hopp, H., Kunitake, R., Karpen, G. H. & Chiolo, I. 2015. Heterochromatic breaks move to the nuclear periphery to continue recombinational repair. *Nat Cell Biol,* 17, 1401–11.

Sanders, S. L., Portoso, M., Mata, J., Bahler, J., Allshire, R. C. & Kouzarides, T. 2004. Methylation of histone H4 lysine 20 controls recruitment of Crb2 to sites of DNA damage. *Cell,* 119, 603–14.

Shivji, M. K., Eker, A. P. & Wood, R. D. 1994. DNA repair defect in xeroderma pigmentosum group C and complementing factor from HeLa cells. *J Biol Chem,* 269, 22749–57.

Sjogren, C. & Nasmyth, K. 2001. Sister chromatid cohesion is required for postreplicative double-strand break repair in *Saccharomyces cerevisiae. Curr Biol,* 11, 991–5.

Smith, E. R., Cayrou, C., Huang, R., Lane, W. S., Cote, J. & Lucchesi, J. C. 2005. A human protein complex homologous to the *Drosophila* MSL complex is responsible for the majority of histone H4 acetylation at lysine 16. *Mol Cell Biol,* 25, 9175–88.

Stadler, J. & Richly, H. 2017. Regulation of DNA repair mechanisms: how the chromatin environment regulates the DNA damage response. *Int J Mol Sci,* 18, pii: E1715.

Sugasawa, K., Ng, J. M., Masutani, C., Iwai, S., Van Der Spek, P. J., Eker, A. P., Hanaoka, F., Bootsma, D. & Hoeijmakers, J. H. 1998. Xeroderma pigmentosum group C protein complex is the initiator of global genome nucleotide excision repair. *Mol Cell,* 2, 223–32.

Sun, Y., Jiang, X., Xu, Y., Ayrapetov, M. K., Moreau, L. A., Whetstine, J. R. & Price, B. D. 2009. Histone H3 methylation links DNA damage detection to activation of the tumour suppressor Tip60. *Nat Cell Biol,* 11, 1376–82.

Sung, P. 1994. Catalysis of ATP-dependent homologous DNA pairing and strand exchange by yeast RAD51 protein. *Science,* 265, 1241–3.

Tang, J., Cho, N. W., Cui, G., Manion, E. M., Shanbhag, N. M., Botuyan, M. V., Mer, G. & Greenberg, R. A. 2013. Acetylation limits 53BP1 association with damaged chromatin to promote homologous recombination. *Nat Struct Mol Biol,* 20, 317–25.

Van Hoffen, A., Natarajan, A. T., Mayne, L. V., Van Zeeland, A. A., Mullenders, L. H. & Venema, J. 1993. Deficient repair of the transcribed strand of active genes in Cockayne's syndrome cells. *Nucleic Acids Res,* 21, 5890–5.

Wang, B. & Elledge, S. J. 2007. Ubc13/Rnf8 ubiquitin ligases control foci formation of the Rap80/Abraxas/Brca1/Brcc36 complex in response to DNA damage. *Proc Natl Acad Sci U S A,* 104, 20759–63.

Wang, H. & Xu, X. 2017. Microhomology-mediated end joining: new players join the team. *Cell Biosci,* 7, 6.

Wang, H., Zhai, L., Xu, J., Joo, H. Y., Jackson, S., Erdjument-Bromage, H., Tempst, P., Xiong, Y. & Zhang, Y. 2006. Histone H3 and H4 ubiquitylation by the CUL4-DDB-ROC1 ubiquitin ligase facilitates cellular response to DNA damage. *Mol Cell,* 22, 383–94.

Whitehouse, C. J., Taylor, R. M., Thistlethwaite, A., Zhang, H., Karimi-Busheri, F., Lasko, D. D., Weinfeld, M. & Caldecott, K. W. 2001. XRCC1 stimulates human polynucleotide kinase activity at damaged DNA termini and accelerates DNA single-strand break repair. *Cell,* 104, 107–17.

Xu, Y., Ayrapetov, M. K., Xu, C., Gursoy-Yuzugullu, O., Hu, Y. & Price, B. D. 2012. Histone H2A.Z controls a critical chromatin remodeling step required for DNA double-strand break repair. *Mol Cell,* 48, 723–33.

Zhong, Q., Chen, C. F., Li, S., Chen, Y., Wang, C. C., Xiao, J., Chen, P. L., Sharp, Z. D. & Lee, W. H. 1999. Association of BRCA1 with the hRad50-hMre11-p95 complex and the DNA damage response. *Science,* 285, 747–50.

Zou, L. & Elledge, S. J. 2003. Sensing DNA damage through ATRIP recognition of RPA-ssDNA complexes. *Science,* 300, 1542–8.

CHAPTER 16

Inheritance of chromatin modifications through the cell cycle

A fundamental aspect of development is the diversification of genetically identical cells produced by mitosis during the early development of an organism. A general model for the initiation of this process is based on the presence of morphogens (transcription factors, non-coding RNAs, etc.) differentially deposited into the developing oocyte, often in the form of gradients. Distinct sets of genes respond to different levels of morphogens and, in turn, activate new genes. Additional mechanisms, such as extrinsic diffusible signals, the production of reciprocal inhibition signals between cellular domains or differential timing of gene function, are thought to refine the initial response (Bier and De Robertis, 2015). Yet, during the growth of an embryo and in many tissues of the adult, groups of cells that are able to differentiate into the particular cell types of the tissue where they are located are maintained by self-renewal. Clearly, as they divide, these different embryonic or adult **stem cells** must faithfully transmit to their daughter cells the factors that are required to achieve their specific fate. Additional examples of the transmission of established transcription programs, representing either active or silenced gene function, are the clonal inheritance of X inactivation in mammals or the persistence of developmental pathways in *Drosophila* and mammals initiated in early development by the activity of *HOX* genes (see Chapter 7). This type of cellular memory is the subject of intense investigation, and a substantial amount of progress has been achieved in elucidating some of the mechanisms that it involves.

Transcription factors and co-factors

Transcription generally ceases as a cell enters mitosis and, as expected, RNA polymerase II (RNAPII) falls off active promoters, but not all transcription factors are displaced. General transcription factors were found to remain bound to the promoters of previously active genes in HeLa cells, suggesting that these factors earmark genes for activation in daughter cells and help to reproduce the parental cell's gene expression (Christova and Oelgeschla, 2002). The erythroid-specific transcription factor GATA1 is lost at a number of its binding sites during mitosis but is retained on key genes that regulate hematopoiesis and allows the more rapid reactivation of these genes in daughter cells (Kadauke et al., 2012). During mitosis, another lineage-regulating transcription factor forkhead box protein A1 (FoxA1) is retained by approximately 15% of the genes that it binds during interphase; surprisingly, FoxA1 associates non-specifically in the vicinity of a number of genes that are not its regular targets during interphase. Following mitosis, FoxA1 target genes are reactivated either by the direct presence of the transcription factor that was retained at their promoter or, presumably, by a factor still present in their vicinity (Caravaca et al., 2013). As discussed in Chapter 17, the number of transcription factors that remain associated with mitotic chromosomes may be much greater than originally suspected.

DNA methylation

The methylation of cytosine residues in CpG dinucleotides is common to the genomes of higher eukaryotes (plants and vertebrates) where it is usually correlated with gene repression (see Chapter 4). It is the epitome of an epigenetic modification that can be transmitted from one cell generation to the next. As a cell replicates its DNA, a maintenance DNA methyl transferase

Epigenetics, Nuclear Organization and Gene Function: with implications of epigenetic regulation and genetic architecture for human development and health. John C. Lucchesi, Oxford University Press (2019). © John C. Lucchesi 2019.
DOI: 10.1093/oso/9780198831204.001.0001

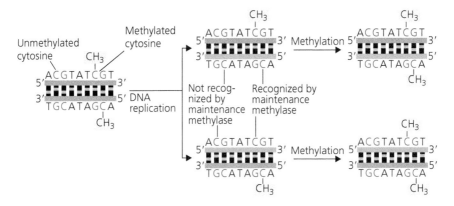

Fig. 16.1 Diagram illustrating the function of a maintenance DNA methyl transferase DNMT1.

(From *Molecular Biology of the Cell*, 4th Edition)

(DNMT1 in mammals) methylates the cytosine of a CpG dinucleotide that has been added to the new strand if the cytosine of the complementary GpC dinucleotide on the old strand is methylated (Fig. 16.1). DNMT1 is recruited to the replication fork by UHRF1 (ubiquitin-like with PHD and ring finger domains 1) where it interacts with PCNA (proliferating cell nuclear antigen), the protein that enhances the processivity of DNA polymerase (Sharif et al., 2007). UHRF1 also recruits topoisomerase II to unwind the hemi-methylated DNA in order to allow the methylation of the newly replicated strand (Lu et al., 2015). While cytosine methylation in the context of CpG dinucleotides represents an established mechanism of cellular inheritance, the role of other chromatin modifications is only now beginning to be understood.

Although the vast majority of the unmethylated CpGs on the newly synthesized strand, which are complementary to a methylated CpG on the template strand, are methylated by DNMT1 following replication, a subset of sites remains hemimethylated through more than 12 cell divisions (Xu and Corces, 2018). Hemimethylated sites occurring within gene bodies seem to lead to increased levels of transcription. Many sites occur in genomic regions that are occupied by CTCF-cohesin complexes.

Histone modifications

Given the degree of correlation between the covalent modifications of histones and the transcriptional status of chromatin regions, it is logical to expect that the panoply of modifications associated with gene activation or repression present during interphase in a particular cell type will be faithfully reproduced in daughter cells. The problem, of course, is that, during DNA replication, nucleosomes dissociate and are reassembled behind the replication fork. Approximately half of the histones that contribute to the formation of the new nucleosomes are parental and could carry the covalent modifications present prior to replication; the other half are newly synthesized histones that exhibit only the modifications necessary for their deposition. A possible solution would be for the daughter cells to use the marks present on those nucleosomes reconstituted with histones that are inherited from the parent cell as inducers of similar marks on the newly synthesized histones. This appears to be the case in *Caenorhabditis* (Gaydos et al., 2014). In this organism, sperm chromosomes are heavily decorated with the repressive mark H3K27me3 by the worm equivalent of the Polycomb repressive complex 2 (PRC2) complex (see Chapter 7). Following the fertilization of eggs from mothers (actually hermaphrodites) lacking PRC2 activity, the paternal chromosome set maintains this mark through early embryonic divisions while the maternally contributed chromosomes remain unmarked. In the presence of the maternally contributed PRC2 complex, sperm chromosomes that lack H3K27me3 do not acquire this mark. These observations lead to the conclusion that, in *Caenorhabditis*, pre-existing marks are transmitted through mitotic divisions and that the PRC2 complex maintains pre-existing marks and does not establish them. Further supporting this conclusion is the fact that many histone methyl transferases contain domains that recognize the same mark that they themselves generate, potentially leading to the replication of the old nucleosome methylation on the newly reassembled nucleosomes (Collins et al., 2008).

Similar evidence has been obtained in *Drosophila*. As is common during the development of many organisms, some *HOX* genes are kept in the repressed state, in particular regions, by the action of PRC2. If the PRE sequence associated with a particular *HOX* gene is deleted, the H3K27me3 mark responsible for repression is maintained through many cell divisions; this mark is eventually diluted, resulting in the activation of the *HOX* gene. These results indicate that transmission of H3K27me3 is responsible for the memory of the repressed state and that although PRC2 associates with this mark to further methylate chromosomes, the presence of a PRE is necessary for rendering the memory permanent (Coleman and Struhl, 2017).

In mammalian cells, using variations of a method for the isolation of protein on nascent DNA (iPOND) (Sirbu et al., 2011), PRC2 complex subunits could be found to remain on nascent DNA following passage of the replication fork (Leung et al., 2013). A different technique for the isolation of nascent chromatin (NCC), coupled with isotope labeling to distinguish newly synthesized histones, revealed that following replication, the chromatin consists of nucleosomes made up of old and new histones (Alabert et al., 2015). These nucleosomes exhibited some parental histone modifications that were diluted twofold (reflecting the deposition of new and old histones). Some of these modifications were very rapidly restored to the parental level, while others (H3K9me3 and H3K27me3) took much longer. The level of the histone variants H2A.X and H3.3 were similar in pre- and post-replicative chromatin (suggesting that they are efficiently recycled); in contrast, H2A.Z levels were very low and required an extended period during the cell cycle to be re-established (Alabert et al., 2015).

Bromodomain-containing protein 4 (BRD4) is a mammalian co-activator that recognizes the acetylated tails of histones 3 and 4. In a mouse embryonic cell line, BRD4 associates with genes that are active during the late telophase of mitosis and the early G1 cell cycle stage, suggesting that it ensures the execution of early post-mitotic gene expression in daughter cells (Dey et al., 2009). These observations imply, of course, that certain histones retain their acetylated marks through mitosis.

In summary, the fundamental aspects of cellular differentiation during animal development depend on the transmission of the state of activation or repression of specific regions of the genome and of the concomitant epigenetic modifications. This epigenetic memory must and does overcome the chromatin reorganization that occurs during cell division.

Re-establishment of the chromatin landscape in daughter cells

Nucleosome positioning following DNA replication

As previously mentioned (see Chapter 5), promoters and regulatory regions such as enhancers and CCCTC-binding factor (CTCF) sites are generally depleted of nucleosomes (Fu et al., 2008; Heintzman et al., 2007; Ozsolak et al., 2007) and are flanked by specifically positioned and phased nucleosomes. The presence of nucleosomes at promoters is considered to be repressive by preventing transcription factors from accessing their binding sites, although rare exceptions can occur where the presence of a nucleosome in a promoter may, in fact, stimulate transcription (Stunkel et al., 1997). On promoters, the level of nucleosome depletion is correlated to the rate of transcription (Valouev et al., 2011). Clearly, these aspects of chromatin organization must be transmitted from parent to daughter cells.

The changes in the nucleosomal landscape that follow DNA replication have been determined in *Drosophila* cells, by labeling nascent DNA with an adduct that allowed the isolation of newly replicated fragments bound by histone and non-histone proteins (Ramachandran and Henikoff, 2016). As expected, the progress of the replication complex removes all of the proteins that are associated with the parental DNA. Following replication, the reassembled nucleosomes are distributed evenly throughout the genome. Nucleosome loss at those promoters that were active in the parent cell is caused by the binding of cell-specific transcription factors and of remodeling complexes that facilitate transcription. A similar picture was obtained for enhancer sequences.

Transmission of genomic loop domains from parent to daughter cells

As previously discussed, the three-dimensional topological organization of the genome is instrumental in effecting the different patterns of gene expression that are responsible for cellular maintenance and differentiation, and for organismal development (see Chapter 10). One might, therefore, expect that the parameters that determine conserved, as well as tissue-specific, chromatin interactions to be faithfully transmitted from a cell to its daughter cells. The architecture of the genome during interphase is abolished, as the chromosomes condense to undergo mitosis; metaphase chromosomes exhibit homogeneous folding, consisting of a longitudinally compressed array of consecutive chromatin loops (Naumova et al., 2013). Re-establishing the interphase organization faces a number of challenges. In mammals, the binding sites that mark topologically associating domain (TAD) boundaries are only a subset of the architectural protein CTCF binding sites, which number over 10,000 in the genome and which are mainly localized within TADs. Data regarding whether CTCF remain bound through mitosis are conflicting and even if this were, in fact, the case, the few binding sites that appear to be occupied have not been identified. These considerations highlight the importance and urgency of fully understanding the factors responsible for the heritability of the nuclear organization that underpins the transcriptional states of different cells.

Epigenetic specification of centromeres

Irrespective of the maintenance or differentiation status of cells, a specialized region that must be faithfully reproduced following DNA replication and for mitosis to occur is the **centromere**. Centromeres are regions of chromosomes where a multiprotein structure—the **kinetochore**—assembles and associates with spindle microtubules to ensure the distribution of chromosomes to daughter nuclei during cell division. The position of the centromere is fixed for each chromosome of the genome and remains the same in the chromosomes of daughter cells. With some exceptions, such as the yeast *Saccharomyces* where the centromere of each chromosome consists of a unique short DNA sequence (Engel et al., 2014), the centromeres of most eukaryotes are found within large segments of repetitive DNA—α-satellite sequences in humans—and do not depend on particular DNA sequences (Melters et al., 2013).

Satellite DNA refers to highly repetitive regions in eukaryotic chromosomes that have a different nucleotide composition, and therefore a different sedimentation coefficient, than the rest of the nuclear DNA. It is the presence of a histone H3 variant—centromere protein A (CENP-A)—in a stretch of chromatin where CENP-A nucleosomes are interspersed with canonical H1- and H3-bearing nucleosomes that determines where the centromere proteins will assemble and the kinetochore will form (Fig. 16.2). This consideration strongly suggests that, although the centromeric DNA is characterized by a high content of adenine and thymine residues (i.e. AT-rich), centromeres are established epigenetically. CENP-A was first discovered as an autoantigen—a molecule that is normally present in an organism but that, in some individuals, is associated with an autoimmune disease (Earnshaw and Rothfield, 1985). CENP-A proteins have a conserved canonical histone fold at their C-terminal end (see Chapter 2) but have amino-terminal tails that vary in length and sequence in different organisms.

In contrast to the assembly of canonical nucleosomes that occurs concomitantly with DNA replication, the addition of CENP-A nucleosomes in higher eukaryotes is DNA replication-independent and takes place, for example, during telophase and the G1 portion of the cell cycle in human cells (Jansen et al., 2007). HJURP (Holiday junction recognition protein) is the dedicated CENP-A chaperone (Dunleavy et al., 2009; Foltz et al., 2009). The signals that determine the incorporation of CENP-A-bearing nucleosomes in a particular repetitive DNA region of a chromosome are still unknown.

An additional number of proteins and non-coding RNAs originating from centric and pericentromeric DNA repeats are found to associate with centromeres and appear to play a role in centromere function. Some

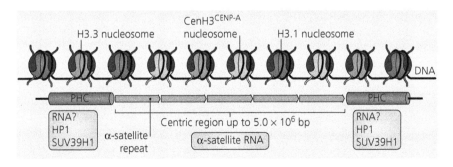

Fig. 16.2 Diagram representing the structure of a human centromere. CenH3^{CENP-A} is the authors' preferred designation of CENP-A; PHC indicates the pericentric heterochromatin.

(From Muller and Almouzni, 2017.)

of these proteins, such as CENP-N and CENP-C, interact directly with CENP-A nucleosomes (Carroll et al., 2010). In *Drosophila*, the interaction between CENP-A and CENP-C involves a long non-coding RNA transcribed from Satellite III, a particular set of DNA repeated sequences in the pericentromeric region of the X chromosome. Satellite III RNA is found on the centromeres of all of the chromosomes of the genome; it interacts with CENP-C and appears to be required for the loading or stabilization of CENP-A and CENP-C to centromeric chromatin (Rosic et al., 2014). In mice, the transcripts of tandem repeats that are present in the centromeric region associate with CENP-A and increase the activity of the kinase Aurora B which interacts with CENP-A and regulates the attachment of chromosomes to the spindle microtubules (Ferri et al., 2009). In humans, transcripts of the type of alpha satellite DNA present in the central portion of centromeres are instrumental in the assembly of CENP-C and of the Aurora B complex at the centromere (Ideue et al., 2014; Quenet and Dalal, 2014; Wong et al., 2007).

Other histone fold-containing proteins do not depend on an association with CENP-A to be targeted to centromeric DNA. Four of these proteins (CENP-T, CENP-W, CENP-S and CENP-X) are found to form stable heterotetramers (Nishino et al., 2012) that associate with a 100-base-pair DNA linker separating assembled dinucleosomes (Takeuchi et al., 2014). Whether these tetramers are actually found at centromeres has not yet been determined.

In mammals, the nucleosomes present in the pericentric heterochromatin (PHC) are marked by the repressive histone marks H3K9me2/3 and H4K20me3. The centric region chromatin exhibits some of the marks (H3K4me, H3K36me) associated with active transcription. In this region, CENP-A can be modified as well. In human CENP-A, the methylation of glycine 1 and the phosphorylation of serines 7, 16 and 18 are required for kinetochore function (Bailey et al., 2013; Goutte-Gattat et al., 2013). In addition, lysine 124 is ubiquitinated to allow CENP-A recognition by its dedicated chaperone (Bade et al., 2014; Niikura et al., 2015).

Chapter summary

While cellular differentiation involves altering the transcriptional activity of specific sets of genes, during the development and maintenance of a particular cell type or tissue, the characteristics that define particular cells must be faithfully transmitted to daughter cells. This involves reproducing, following mitosis, the particular transcriptional landscape of the parent cell.

Numerous factors and chromatin modifications have been implicated in this process.

Although transcription generally ceases as a cell enters mitosis, not all transcription factors are displaced. Among the epigenetic modifications that are implicated in the inheritance of chromatin characteristics through the cell cycle, DNA methylation has been most widely studied because of the existence of maintenance DNA methyl transferases that methylate CpG dinucleotides on the newly replicated strand if the corresponding GpC on the parent strand is methylated. With respect to histone modifications, since the nucleosomes that are deposited on the newly synthesized DNA strands are made up of old and new histones, some marks present on the old histones are maintained. These marks are necessarily diluted but are restored to parental levels, with different time frames, depending on the marks involved.

In addition to the factors just mentioned, the transcriptional profile of a cell depends on the proper distribution of nucleosomes and on the topological organization of the genome into topologically associating domains (TADs). While these aspects of chromatin organization must be transmitted from parent to daughter cells, the means of their transmission are still elusive.

Following DNA replication and for mitosis to proceed, centromeres must be specified on the daughter chromatids. In most eukaryotes, centromeres are identified by the presence of nucleosomes bearing the histone H3 variant CENP-A. In mammals, the nucleosomes present in the pericentric heterochromatin (PHC) are marked by the repressive histone marks H3K9me2/3 and H4K20me3. The centric region chromatin exhibits some of the marks (H3K4me, H3K36me) associated with active transcription. In this region, CENP-A can be modified as well. An additional number of proteins and non-coding RNAs originating from centric and pericentromeric DNA repeats are found to associate with centromeres and appear to play a role in centromere function. The signals that determine the incorporation of CENP-A-bearing nucleosomes in a particular repetitive DNA region of a chromosome are still unknown.

References

Alabert, C., Barth, T. K., Reveron-Gomez, N., Sidoli, S., Schmidt, A., Jensen, O. N., Imhof, A. & Groth, A. 2015. Two distinct modes for propagation of histone PTMs across the cell cycle. *Genes Dev*, 29, 585–90.

Bade, D., Pauleau, A. L., Wendler, A. & Erhardt, S. 2014. The E3 ligase CUL3/RDX controls centromere maintenance by ubiquitylating and stabilizing CENP-A in a CAL1-dependent manner. *Dev Cell*, 28, 508–19.

Bailey, A. O., Panchenko, T., Sathyan, K. M., Petkowski, J. J., Pai, P. J., Bai, D. L., Russell, D. H., Macara, I. G., Shabanowitz, J., Hunt, D. F., Black, B. E. & Foltz, D. R. 2013. Posttranslational modification of CENP-A influences the conformation of centromeric chromatin. *Proc Natl Acad Sci U S A*, 110, 11827–32.

Bier, E. & De Robertis, E. M. 2015. EMBRYO DEVELOPMENT. BMP gradients: A paradigm for morphogen-mediated developmental patterning. *Science*, 348, aaa5838.

Caravaca, J. M., Donahue, G., Becker, J. S., He, X., Vinson, C. & Zaret, K. S. 2013. Bookmarking by specific and nonspecific binding of FoxA1 pioneer factor to mitotic chromosomes. *Genes Dev*, 27, 251–60.

Carroll, C. W., Milks, K. J. & Straight, A. F. 2010. Dual recognition of CENP-A nucleosomes is required for centromere assembly. *J Cell Biol*, 189, 1143–55.

Christova, R. & Oelgeschlager, T. 2002. Association of human TFIID-promoter complexes with silenced mitotic chromatin in vivo. *Nat Cell Biol*, 4, 79–82.

Coleman, R. T. & Struhl, G. 2017. Causal role for inheritance of H3K27me3 in maintaining the OFF state of a *Drosophila* HOX gene. *Science*, 356, pii: eaai8236.

Collins, R. E., Northrop, J. P., Horton, J. R., Lee, D. Y., Zhang, X., Stallcup, M. R. & Cheng, X. 2008. The ankyrin repeats of G9a and GLP histone methyltransferases are mono- and dimethyllysine binding modules. *Nat Struct Mol Biol*, 15, 245–50.

Dey, A., Nishiyama, A., Karpova, T., Mcnally, J. & Ozato, K. 2009. Brd4 marks select genes on mitotic chromatin and directs postmitotic transcription. *Mol Biol Cell*, 20, 4899–909.

Dunleavy, E. M., Roche, D., Tagami, H., Lacoste, N., Ray-Gallet, D., Nakamura, Y., Daigo, Y., Nakatani, Y. & Almouzni-Pettinotti, G. 2009. HJURP is a cell-cycle-dependent maintenance and deposition factor of CENP-A at centromeres. *Cell*, 137, 485–97.

Earnshaw, W. C. & Rothfield, N. 1985. Identification of a family of human centromere proteins using autoimmune sera from patients with scleroderma. *Chromosoma*, 91, 313–21.

Engel, S. R., Dietrich, F. S., Fisk, D. G., Binkley, G., Balakrishnan, R., Costanzo, M. C., Dwight, S. S., Hitz, B. C., Karra, K., Nash, R. S., Weng, S., Wong, E. D., Lloyd, P., Skrzypek, M. S., Miyasato, S. R., Simison, M. & Cherry, J. M. 2014. The reference genome sequence of *Saccharomyces cerevisiae*: then and now. *G3 (Bethesda)*, 4, 389–98.

Ferri, F., Bouzinba-Segard, H., Velasco, G., Hube, F. & Francastel, C. 2009. Non-coding murine centromeric transcripts associate with and potentiate Aurora B kinase. *Nucleic Acids Res*, 37, 5071–80.

Foltz, D. R., Jansen, L. E., Bailey, A. O., Yates, J. R., 3RD, Bassett, E. A., Wood, S., Black, B. E. & Cleveland, D. W. 2009. Centromere-specific assembly of CENP-a nucleosomes is mediated by HJURP. *Cell*, 137, 472–84.

Fu, Y., Sinha, M., Peterson, C. L. & Weng, Z. 2008. The insulator binding protein CTCF positions 20 nucleosomes around its binding sites across the human genome. *PLoS Genet*, 4, e1000138.

Gaydos, L. J., Wang, W. & Strome, S. 2014. Gene repression. H3K27me and PRC2 transmit a memory of repression across generations and during development. *Science*, 345, 1515–18.

Goutte-Gattat, D., Shuaib, M., Ouararhni, K., Gautier, T., Skoufias, D. A., Hamiche, A. & Dimitrov, S. 2013. Phosphorylation of the CENP-A amino-terminus in mitotic centromeric chromatin is required for kinetochore function. *Proc Natl Acad Sci U S A*, 110, 8579–84.

Heintzman, N. D., Stuart, R. K., Hon, G., Fu, Y., Ching, C. W., Hawkins, R. D., Barrera, L. O., Van Calcar, S., Qu, C., Ching, K. A., Wang, W., Weng, Z., Green, R. D., Crawford, G. E. & Ren, B. 2007. Distinct and predictive chromatin signatures of transcriptional promoters and enhancers in the human genome. *Nat Genet*, 39, 311–18.

Ideue, T., Cho, Y., Nishimura, K. & Tani, T. 2014. Involvement of satellite I noncoding RNA in regulation of chromosome segregation. *Genes Cells*, 19, 528–38.

Jansen, L. E., Black, B. E., Foltz, D. R. & Cleveland, D. W. 2007. Propagation of centromeric chromatin requires exit from mitosis. *J Cell Biol*, 176, 795–805.

Kadauke, S., Udugama, M. I., Pawlicki, J. M., Achtman, J. C., Jain, D. P., Cheng, Y., Hardison, R. C. & Blobel, G. A. 2012. Tissue-specific mitotic bookmarking by hematopoietic transcription factor GATA1. *Cell*, 150, 725–37.

Leung, K. H., Abou El Hassan, M. & Bremner, R. 2013. A rapid and efficient method to purify proteins at replication forks under native conditions. *Biotechniques*, 55, 204–6.

Lu, L. Y., Kuang, H., Korakavi, G. & Yu, X. 2015. Topoisomerase II regulates the maintenance of DNA methylation. *J Biol Chem*, 290, 851–60.

Melters, D. P., Bradnam, K. R., Young, H. A., Telis, N., May, M. R., Ruby, J. G., Sebra, R., Peluso, P., Eid, J., Rank, D., Garcia, J. F., Derisi, J. L., Smith, T., Tobias, C., Ross-Ibarra, J., Korf, I. & Chan, S. W. 2013. Comparative analysis of tandem repeats from hundreds of species reveals unique insights into centromere evolution. *Genome Biol*, 14, R10.

Muller, S. & Almouzni, G. 2017. Chromatin dynamics during the cell cycle at centromeres. *Nat Rev Genet*, 18, 192–208.

Naumova, N., Imakaev, M., Fudenberg, G., Zhan, Y., Lajoie, B. R., Mirny, L. A. & Dekker, J. 2013. Organization of the mitotic chromosome. *Science*, 342, 948–53.

Niikura, Y., Kitagawa, R., Ogi, H., Abdulle, R., Pagala, V. & Kitagawa, K. 2015. CENP-A K124 ubiquitylation is required for CENP-A deposition at the centromere. *Dev Cell*, 32, 589–603.

Nishino, T., Takeuchi, K., Gascoigne, K. E., Suzuki, A., Hori, T., Oyama, T., Morikawa, K., Cheeseman, I. M. &

Fukagawa, T. 2012. CENP-T-W-S-X forms a unique centromeric chromatin structure with a histone-like fold. *Cell*, 148, 487–501.

Ozsolak, F., Song, J. S., Liu, X. S. & Fisher, D. E. 2007. High-throughput mapping of the chromatin structure of human promoters. *Nat Biotechnol*, 25, 244–8.

Quenet, D. & Dalal, Y. 2014. A long non-coding RNA is required for targeting centromeric protein A to the human centromere. *Elife*, 3, e03254.

Ramachandran, S. & Henikoff, S. 2016. Transcriptional regulators compete with nucleosomes post-replication. *Cell*, 165, 580–92.

Rosic, S., Kohler, F. & Erhardt, S. 2014. Repetitive centromeric satellite RNA is essential for kinetochore formation and cell division. *J Cell Biol*, 207, 335–49.

Sharif, J., Muto, M., Takebayashi, S., Suetake, I., Iwamatsu, A., Endo, T. A., Shinga, J., Mizutani-Koseki, Y., Toyoda, T., Okamura, K., Tajima, S., Mitsuya, K., Okano, M. & Koseki, H. 2007. The SRA protein Np95 mediates epigenetic inheritance by recruiting Dnmt1 to methylated DNA. *Nature*, 450, 908–12.

Sirbu, B. M., Couch, F. B., Feigerle, J. T., Bhaskara, S., Hiebert, S. W. & Cortez, D. 2011. Analysis of protein dynamics at active, stalled, and collapsed replication forks. *Genes Dev*, 25, 1320–7.

Stunkel, W., Kober, I. & Seifart, K. H. 1997. A nucleosome positioned in the distal promoter region activates transcription of the human U6 gene. *Mol Cell Biol*, 17, 4397–405.

Takeuchi, K., Nishino, T., Mayanagi, K., Horikoshi, N., Osakabe, A., Tachiwana, H., Hori, T., Kurumizaka, H. & Fukagawa, T. 2014. The centromeric nucleosome-like CENP-T-W-S-X complex induces positive supercoils into DNA. *Nucleic Acids Res*, 42, 1644–55.

Valouev, A., Johnson, S. M., Boyd, S. D., Smith, C. L., Fire, A. Z. & Sidow, A. 2011. Determinants of nucleosome organization in primary human cells. *Nature*, 474, 516–20.

Wong, L. H., Brettingham-Moore, K. H., Chan, L., Quach, J. M., Anderson, M. A., Northrop, E. L., Hannan, R., Saffery, R., Shaw, M. L., Williams, E. & Choo, K. H. 2007. Centromere RNA is a key component for the assembly of nucleoproteins at the nucleolus and centromere. *Genome Res*, 17, 1146–60.

Xu, C. & Corces, V. G. 2018. Nascent DNA methylome mapping reveals inheritance of hemimethylation at CTCF/cohesin sites. *Science*, 359, 1166-70.

CHAPTER 17

Stem cells

In most multicellular organisms, when the fertilized egg enters the series of mitotic divisions that initiate the development of the embryonic form and later of the adult, each of the first few cells (blastomeres) that are produced is able to give rise to a complete individual. The fertilized egg and these early cells are said to be **totipotent**. As the number of embryonic cells increases, their ability to differentiate becomes restricted to subsets of particular cell types and the cells are now considered to be **pluripotent** cells. In many adult tissues, groups of undifferentiated cells are present for the purpose of replacing the various cell types specific to the tissue. These adult stem cells are characterized as **multipotent**. A major current area of research in the stem cell field is the identification of the extrinsic environmental factors that influence stem cell renewal and differentiation and the translation of this knowledge into the use of stem cells for clinical benefit. As expected, given the fundamental role of epigenetic regulation in cellular differentiation, chromatin modifications contribute significantly to the maintenance of cellular potential, as well as the passage from one stem cell stage to another.

Characteristics of stem cells

Totipotent cells

Sperm and oocytes are highly specialized, terminally differentiated cells. In order to give rise to a zygote and early blastomeres that are totipotent, these gametes must undergo fundamental changes in chromatin composition and architecture. These changes are directed by maternal products present in the mature oocyte. In *Caenorhabditis* (Fig. 17.1), the fertilized egg is the only truly totipotent cell—the daughter cells of the very first mitotic division are unequal in size and are already destined for different developmental fates (Sulston et al., 1983).

Mammalian totipotent cells have not yet been successfully established in cell cultures; for this reason, the molecular characteristics that define totipotency have been only partially described. Before the fusion of the egg and sperm-derived pronuclei to produce the zygote, the DNA-associated protamines of the male pronucleus are replaced by nucleosomes that contain primarily the H3.3 histone variant (van der Heijden et al., 2005). Once formed, these nucleosomes exhibit various histone modifications that are very different from those present in the female pronuclei (Arney et al., 2002; Santos et al., 2005). One feature of the first few embryonic cells is a burst of transcriptional activity (Bouniol et al., 1995) that includes the activation of specific types of transposable elements (Peaston et al., 2004) and the use of endogenous retrovirus promoters to activate numerous genes (Macfarlan et al., 2012). In addition, a global demethylation of the maternal and paternal genomes begins following fertilization and proceeds to the blastocyst stage (Rougier et al., 1998).

Another distinctive feature of early totipotent cells is the absence of chromocenters, regions of the nucleus where the blocks of constitutive heterochromatin that flank the centromeres aggregate. In the mouse, centromeric regions are arranged at the periphery of nucleolar precursor bodies until the two-cell stage when they begin to associate with one another and form chromocenters (Probst et al., 2007). The significance of these chromatin rearrangements in the establishment of totipotency remains to be clarified.

Pluripotent cells

These cells are characterized by the ability to self-renew, while being poised to respond to differentiation

Epigenetics, Nuclear Organization and Gene Function: with implications of epigenetic regulation and genetic architecture for human development and health. John C. Lucchesi, Oxford University Press (2019). © John C. Lucchesi 2019.
DOI: 10.1093/oso/9780198831204.001.0001

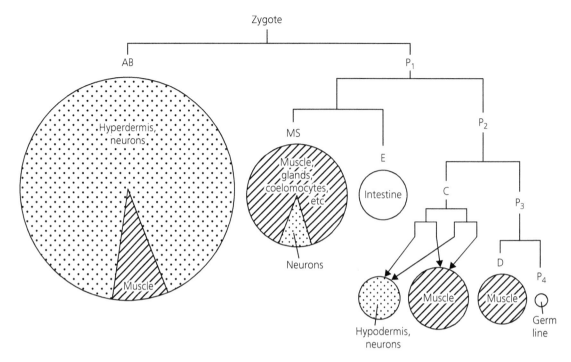

Fig. 17.1 The *Caenorhabditis* fertilized egg divides unequally into large and smaller daughter cells that differ with respect to the differentiation potential of their descendants. Areas of circles and sectors are proportional to the number of cells that are found in the different lineages.

(From Sulston et al., 1983.)

signals. In mammals, they can be defined experimentally by their ability, following injection into mice, to form cellular masses (**teratomas**) with differentiated tissues originating from all three embryonic germ layers.

As the embryo grows, the change of its cells from totipotency to pluripotency involves the activation of the zygotic genome and the breakdown of maternal transcripts. These events constitute the maternal-to-zygotic transition. In *Drosophila*, activation of the zygotic genome is initiated by the maternally deposited transcription factor Zinc-finger-early-*Drosophila*-activator (Zelda, ZLD). Discovered by its association with a sequence motif present in the promoter region of many early-acting genes (Liang et al., 2008), ZLD acts as a pioneer transcription factor that facilitates promoter access by other transcription activators (Schulz et al., 2015; Sun et al., 2015). Activation of the zygotic genome is a gradual process, leading to a large burst of transcription at the time of blastoderm formation. As mentioned previously, linker histones are responsible for the compacted structure of inactive heterochromatin regions and the silencing of repetitive sequences and transposable elements. In *Drosophila*, a single linker histone variant

(H1) is found in somatic cells from the cellular blastoderm on into the adult. During the early stages of development, when nuclear divisions are so rapid that no transcription is allowed to occur, *Drosophila* embryos possess an embryonic variant (dBigH1) that prevents zygotic genome activation. Following the cellularization process that leads to blastoderm formation, this variant is gradually replaced by histone H1 (Pérez-Montero et al., 2013). In mice, the maternal-to-zygotic transcription occurs at the two-cell stage and begins with the activation of genes involved in the basic processes of transcription, RNA processing and the synthesis of ribosomes (Zeng et al., 2004). A second wave of activation precedes the formation of the blastula (Hamatani et al., 2004). In a manner similar to *Drosophila*, Stella, a maternal protein in the egg, is necessary for the activation of a number of genes in the early mouse embryo (Huang et al., 2017). In addition, Stella activates a group of transposable elements (discussed earlier) that are required for early embryonic development, presumably by affecting the transcription of neighboring genes (Huang et al., 2017).

In mammals, pluripotent cells can assume different states characterized by differences in epigenetic modifications and gene expression. Two major types of pluripotent cells are recognized in mice and humans. Cells derived from the inner cell mass of mouse blastocysts (an embryonic stage prior to implantation into the wall of the uterus) are **naïve** pluripotent stem cells referred to as ESCs; cells derived from the epiblast [a tissue of the late blastocyst, formed soon after implantation (Fig. 17.2)] are **primed** pluripotent stem cells referred to as epiblast stem cells (EpiSCs) (Evans and Kaufman, 1981; Martin, 1981). ESCs can be operationally distinguished from EpiSCs by their ability to form germline-transmitting chimeras. Chimeras are formed by injecting ESCs of a particular genotype into the cavity of blastocysts produced by females of a different genotype. Following surgical implantation into surrogate mothers, the injected blastocysts give rise to adult mice with tissues of both genotypes that include the germline. EpiSCs injected into blastocysts lead to the formation of somatic chimeras but do not participate in germline formation (Rossello et al., 2016).

In humans, ESCs derived from blastocysts and placed in culture usually exhibit the characteristics of mouse EpiSCs (Tesar et al., 2007; Thomson et al., 1998).

ESCs are characterized by an open chromatin configuration, an elevated level of transcription and a general disorganization of inactive chromatin domains (de Wit et al., 2013; Efroni et al., 2008). The pluripotency of ESCs is the responsibility of three factors—Oct4, Sox2 and Nanog—that maintain an open chromatin state (Chambers et al., 2003; Mitsui et al., 2003; Nichols et al., 1998; Niwa et al., 2000). When induced in differentiated fibroblasts that are to be reprogrammed into pluripotent cells (see Chapter 18), Oct4, Sox2 and Nanog bind to condensed and inactive regions of the genome (as defined by a lack of sensitivity to DNase digestion) and, therefore, qualify as pioneer factors (Soufi et al., 2012). In ESCs, these three core pluripotency factors bind to the promoters of a group of genes that they activate and that are responsible for pluripotency; they also repress a set of genes poised for expression during cellular differentiation or that encode transcription factors important for differentiation into extra-embryonic, endodermal, mesodermal and ectodermal lineages (Boyer et al., 2005). The levels of Nanog and Oct4, and therefore the maintenance of pluripotency, are regulated by long non-coding RNAs (lncRNAs) (Guttman et al., 2011). Nanog and Oct4, in turn, regulate the transcription of additional lncRNAs (Sheik Mohamed et al., 2010). Binding of pluripotency factors influences the genome configuration in ESCs; for example, distantly located Nanog binding sites are brought together in intra- and inter-chromosomal high-density binding sites (de Wit et al., 2013). In fact, all three pluripotency master transcription factors occupy large enhancer domains that are rich in Mediator complex (see Chapter 3) and interact with genes that are active in maintaining the character of the ESCs. These domains, referred to as **super-enhancers**, induce higher-than-normal transcriptional activity to the genes with which they interact (Whyte et al., 2013).

Human EpiSCs can be converted to the naïve-state characteristic of ESCs by over-expressing pluripotency transcription factors and by inhibiting particular signal transduction and energy metabolism pathways (Hanna et al., 2010), or by enhancing the level of histone acetylation and, again, inhibiting signal transduction and energy metabolism pathways (Ware et al., 2014). In most naïve human ESCs derived from female blastocysts, both X chromosomes are active.

Although widely used as experimental models, it is necessary to note that cultured ESCs exhibit some characteristics, such as aberrant methylation and expression pattern of imprinted genes, which distinguish them from natural stem cells (Dean et al., 1998).

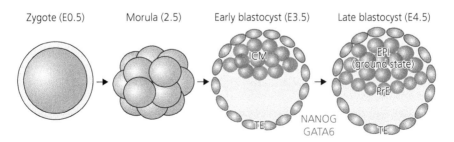

Fig. 17.2 Early stages of mammalian embryonic development highlighting the tissue sources of naïve and primed pluripotent stem cells. In the late blastocyst, "ground state" refers to the origin of all embryonic lineages.

(From Hackett and Surani, 2014.)

Multipotent adult stem cells

The identification of stem cells in an increasing number of adult tissues suggests that most tissues and organs are derived from, and renewed by, groups of specific **multipotent** or **progenitor** cells. Among the most extensively characterized adult stem cells are those responsible for the differentiation of all of the different types of blood cells [hematopoietic stem cells (HSCs) present in the bone marrow of vertebrates], those responsible for the differentiation of all the cell types found in the intestinal crypts (invaginations that occur at the base of the finger-like protruding villi), the cells present in the hair follicles of the epidermis that contribute to all of the hair cellular lineages and the germ cells responsible for the production of gametes. The demonstration of adult stem cells in these and other tissues required the identification of specific molecular markers and the development of suitable stem cell assays. Original attempts to identify hair follicle stem cells relied on the use of an antiserum raised against cytokeratin 15 (K15), one of the keratins that constitute intermediate filaments responsible for maintaining the shape of epithelial cells. The antiserum recognized a specific group of cells in hair follicles that exhibited some of the characteristics of stem cells (Lyle et al., 1998). Subsequently, follicular stem cells were isolated by cell sorting, on the basis of their unique expression of K15 (Morris et al., 2004). Another characteristic of these stem cells—the presence in their cell membrane of a specific receptor protein (Lgr6)—once again allowed their isolation by cell sorting (Snippert et al., 2010). A closely related receptor (Lgr5) has been used to identify stem cells from the small intestine (Barker et al., 2007). Lgr5 is also present in hair follicle cells and in highly restricted areas of other tissues, suggesting an expected commonality of molecular characteristics among different types of adult stem cells. HSCs were initially isolated by a laborious elimination process that utilized a large number of antisera against cell surface proteins characteristic of differentiated cells (Spangrude et al., 1988). Over the years, similar protocols have been followed to isolate these cells and study some of their characteristics until recently, when the expression of a particular homeobox gene was found to be limited to hematopoietic cells in the bone marrow of mice and could be used for their precise identification. The demonstration that the isolated cells were, in fact, stem cells made use of an *in vivo* assay—cells injected into mice that had been irradiated to destroy their hematopoietic bone marrow system were found to allow the mice to survive by restoring blood cell synthesis (Chen et al., 2016; Till and McCulloch, 1961).

The stem cell niche concept

The proper differentiation of adult stem cells into the cells characteristic of the tissue where they reside depends on their particular location. In this micro-environment, referred to as the **niche**, adult stem cells are maintained in their undifferentiated, proliferative state by signals from the somatic cells with which they associate and, when appropriate, are subjected to signals that induce them to differentiate (Schofield, 1978). A variety of factors and signals constitute the niche and determine whether stem cells should renew themselves or differentiate. Usually, stem cells adhere to an extra-cellular matrix that is synthesized by somatic niche cells and serves as an anchoring substrate. Stem cells are in direct contact with neighboring cells and are exposed to secreted factors. They are also exposed to physical factors such as oxygen tension and to physical forces generated by cell–cell contacts. The release of stem cells from their niche leading to their differentiation involves the action of secreted proteases, among which metalloproteases play a prominent role (Heissig et al., 2002).

The existence of stem cell niches was first investigated in the *Drosophila* male and female gonads. The highly complex molecular pathways and their interactions that regulate the self-renewal and the differentiation of germline stem cells have been worked out in great detail (Losick et al., 2011; Slaidina and Lehman, 2014). The female ovary is made up of a number of egg-producing chambers called ovarioles (Fig. 17.3). A few germline stem cells, located at the tip of each ovariole, are attached to cap cells that hold them into place by means of special protein bridges (adherens junctions) (Song et al., 2002). This association ensures that, following mitosis, one daughter cell remains attached to the cap cell layer, while the other no longer has this contact. Germ cells are also in contact with escort cells that aid them to differentiate. The cells forming the terminal filament of the ovariole secrete cytokines that activate the production of ligands (DPP and GBB) in the cap cells. These ligands associate with their respective receptors in the stem cells that are in contact with the cap cell layer and prevent the synthesis of differentiation factors (such as Brn2, Ascl1 and Myt1l). In addition, specific translational repressors (NOS and PUM) present in the stem cells interfere with the translation of other differentiation-promoting genes. Following the division of a stem cell, the concentration

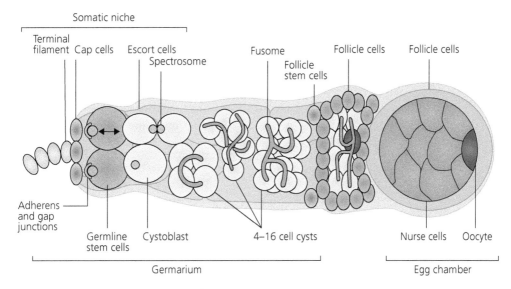

Fig. 17.3 Diagram illustrating the cellular composition of the distal tip of a *Drosophila* ovariole. The somatic niche contains two or three germline stem cells and is made up of terminal filament cells, cap cells and escort cells. Attachment of the stem cells to the cap cells ensures that when they divide, one daughter cell remains attached while the other daughter cell is no longer associated with the cap cells and is earmarked for differentiation.

(From Slaidina and Lehman, 2014.)

of the DPP and GBB ligands is significantly less in the daughter cell that is no longer associated with the cap cells because of its location and because the escort cells express receptors that serve as a sink for the diffusible ligands (Luo et al., 2015). This daughter cell, referred to as the cystoblast in *Drosophila* oogenesis, is generally known as a progenitor cell—a cell that can divide a few times but produces only differentiated cells. The cystoblast leaves the niche and undergoes four mitotic divisions, generating a cyst of 16 cells connected by ring canals. One cell of the cyst is the developing oocyte, while the other 15 cells are nurse cells that contribute cytoplasmic components to the growing oocyte.

The *Drosophila* midgut, analogous to the mammalian small intestine, is lined by two types of cells: digestive cells (enterocytes) and hormone-producing cells (enteroendocrine cells). These cells are derived from a common population of stem cells that, in contrast to most other stem cells, are not associated with niche-specific somatic cells. (Micchelli and Perrimon, 2006; Ohlstein and Spradling, 2006). A subsequent study revealed that the midgut is, in fact, subdivided into several subregions with distinct transcriptional profiles and physiological properties and that the stem cells found in each region differ as well in gene expression (Marianes and Spradling, 2013). The stem cells in each sub-region of

the midgut can, therefore, be considered to occupy different niches.

The niche concept is applicable to the proliferation and differentiation of mammalian stem cells. As development proceeds, HSCs occur in a succession of sites that can be considered to represent niches that provide the signals necessary for self-renewal and migration to the next sites. HSCs are first formed in the early embryo by mesodermal cells, some of which migrate to the yolk sac and differentiate into embryonic blood cells. Other HSCs colonize the fetal liver that provides the major source of blood cells during fetal development. The postnatal HSC niches are in the bone marrow and spleen. Adult HSCs are heterogeneous, and different types can be distinguished on the basis of their proliferation and differentiation characteristics (Guenechea et al., 2001), leading to the conclusion that HSCs must occupy different microenvironments in their niche (Kiel et al., 2005).

Maintenance of pluripotency during stem cell proliferation

Mitosis is a highly disruptive process during which genomic transcription is stopped and many epigenetic modifications of the chromatin are eliminated. As previously mentioned (see Chapter 16), transcription factors have been reported to dissociate from their genomic locations, although some, such as FoxA1 and GATA1, appear to be retained through the mitotic process. In ESCs, the pluripotency factors Sox2 and Oct4 remain bound to mitotic chromatin (Deluz et al., 2016). These experimental results support the long-held belief that only a few selected transcription factors are retained through mitosis. A *caveat* has been raised regarding this contention, suggesting that it is based on a procedural artifact (the fixation step in the experimental techniques that have been used). Using live cell imaging, many more transcription factors are found on mitotic chromosomes (Teves et al., 2016).

Some histone marks have also been found to persist on mitotic chromosomes and through cell division. Enhancers of stem cell-related genes that are marked by the presence of Oct4, Sox2 and Klf4 are marked by the presence of H3K27ac; this histone isoform is present also on the promoters of housekeeping genes (Liu et al., 2017).

Epigenetic marks of stem cell self-renewal and differentiation

As stated in Pluripotent cells, p. 191, the transcription factors Oct4, Nanog and Sox2 are responsible for the maintenance of the pluripotent state. In ESCs, these factors associate with clusters of enhancers to form super-enhancers that activate the major genes responsible for pluripotency and self-renewal. In differentiating cells, super-enhancers form at the sites of binding of the major transcription factors responsible for the specific characteristics of those cells (Whyte et al., 2013).

Binding of the three transcription factors should open the chromatin for access to other transcription factors. As expected, in mouse ESCs, the chromatin-remodeling enzyme Chd1 (that has been implicated in transcriptional activation) is highly elevated in ESCs (Gaspar-Maia et al., 2009), and Oct4 recruits the chromatin remodeler BRG1 (King and Klose, 2017). The differentiation of ESCs requires the silencing of pluripotency genes by standard epigenetic modifications (Feldman et al., 2006).

In general, the epigenome of ESCs does not contain any specific novel chromatin marks, although an exception has been reported regarding the nature of DNA methylation. In somatic cells, while DNA methylation commonly occurs on the cytosine of CpG doublets, a very low level of non-CpG methylation is present on CpH doublets (where H can be adenine, thymine or cytosine). This type of DNA methylation, mediated by the DNA methyl transferases DNMT3a and DNMT3b,

is enriched in ESCs and lost during differentiation (Gowher and Jeltsch., 2001; Laurent et al., 2010; Lister et al., 2009).

During hematopoietic or hair follicle stem cell differentiation, genes that are associated with the development and function of a particular terminally differentiated cell type exhibit significantly less DNA methylation than genes responsible for the development of other cell types (Bock et al., 2012).

Acetylated H4K16 occurs in association with gene promoters throughout the genome. The complexes responsible for targeting the histone acetyl transferase MOF to these sites are the MSL (male-specific lethal) complex (Smith et al., 2005) and the NSL (non-specific lethal) complex (Cai et al., 2010; Mendjan et al., 2006; Raja et al., 2010). Not surprisingly, MOF plays a significant role in the epigenetic regulation of mammalian ESCs. While other histone acetyl transferases are involved in the activation of lineage-specific genes, MOF regulates the expression of the key pluripotency-mediating transcription factors Oct4, Nanog and Sox2 (Li et al., 2012). In mouse ESCs and in neural progenitor cells derived from the ESCs, both MSL and NSL complexes target the same general housekeeping genes, but they target different enhancers; although both complexes seem to bind to the key pluripotency genes, only the NSL complex appears to be required for their transcription (Chelmicki et al., 2014).

Histone methylation complexes are involved in ESC self-renewal and differentiation. Polycomb group (PcG) protein complexes are responsible for the H3K27me3 repressive histone mark, while trithorax group (TrxG) protein complexes mediate the H3K4me3 activation mark (see Chapter 7). In mammals, the latter are designated as MLL/SET1 complexes that include a WDR5 (WD repeat domain 5) protein subunit. WDR5 is thought to position the N-terminal tail of histone H3 for methylation by the histone methyl transferase of the complexes. In ESCs, WDR5 interacts directly with the transcription factor Oct4 at H3K4me3-modified nucleosomes and collaborates with Nanog and Sox2 to activate a number of genes involved in maintaining pluripotency, among which are the three pluripotency genes themselves (Ang et al., 2011).

The repressive PRC complexes are involved in multiple and apparently redundant functions in the maintenance of the pluripotent state. Elimination of either the PRC1 or PRC2 complexes induces differentiation of ESCs; the elimination of both complexes results in the loss of the ability to differentiate (Leeb et al., 2010). Recent evidence indicates that, at the onset of differentiation, the re-establishment of H3K27me3 that normally follows DNA replication during ESC maintenance is delayed (Petruk et al., 2017). This window of decondensed chromatin throughout the genome is exploited by lineage-specific transcription factors that bind their cognate DNA sites.

Of equal importance in the epigenetic regulation of the self-renewal and pluripotency of ESCs is the activity of histone deacetylases and histone demethylases. The NuRD (nucleosome remodeling and deacetylase) complex is necessary for the transition from persistent self-renewal to lineage-specific differentiation (Kaji et al., 2006). Although it does not regulate the core factors Nanog, Oct4 and Sox2, NuRD binds to and represses several genes in the pluripotency network (Reynolds et al., 2012a). NuRD's association with its target genes recruits the Polycomb repressive complex PRC2 that is responsible for the H3K27me3 repressive histone isoform (see Chapter 7). A distinct feature of ESCs is the prevalence of bivalent promoters exhibiting the simultaneous presence of the activation mark H3K4me3 and the repressive mark H3K27me3 on key developmental genes (Azura et al., 2006; Bernstein et al., 2006). Abrogation of NuRD results in an increase in H3K27ac, a decrease in H3K27me3 at bivalent promoters and the induction of differentiation genes (Reynolds et al., 2012b). NuRD also appears to interact functionally with the histone demethylase LSD1 that removes the methyl groups in methylated H3K4 and H3K9. In ESCs, LSD1 demethylates the signature H3K4me1 of many bivalent poised genes enhancers, as well as those of the three core pluripotency genes, ensuring their inactivation (Whyte et al., 2012).

The question, of course, is how the epigenetic regulators mentioned above become targeted to specific genomic loci to mediate self-renewal regulatory functions or to initiate differentiation pathways. As mentioned in previous sections, transcription factors bind to promoters and enhancers and target, or interfere with the targeting of, remodeling and modifying complexes to these regulatory regions. In general, and especially in the case of differentiating stem cells, pioneer factors are specific for particular cell lineages and actually establish lineage commitment.

The role of non-coding RNAs

A large number of long non-coding RNAs (see Chapter 6) are present in mouse ESCs and a subset of these RNAs were found to be differentially expressed during the differentiation of **embryoid bodies**—*in vitro* aggregates of ESCs that exhibit regional differentiation into distinct cell types (Dinger et al., 2008). A similar correlation between the differential expression of lncRNAs and the differentiation into pluripotent cells and neurons was seen with human ESCs (Ng et al., 2012). The first comprehensive functional study consisted of inducing loss of function of over 200 long intergenic non-coding RNAs (lincRNAs) in mouse ESCs (Guttman et al., 2011). The results indicated that lincRNAs affect, both positively and negatively, a substantial portion of the genes expressed in ESCs. Loss of function of several lincRNAs resulted in a very significant decrease in the levels of pluripotency gene products, including Nanog, Oct4 and Sox2, indicating that these RNAs play a role in the maintenance of pluripotency. Several other lincRNAs were shown to repress the genes that act in the differentiation of ESCs into specific cell lineages. Interestingly, the promoters of most lncRNA coding units are occupied by the three core and other pluripotency factors that contribute to their transcription. lncRNAs are thought to guide various chromatin-modifying complexes to specific DNA sites using several possible strategies (Hung and Chang, 2010). A mechanistic role for some lncRNAs in maintaining the active chromatin state and pluripotency of ESCs, as well as directing them to differentiate, could be to associate with the WRD5 subunits of MLL complexes and guide them to their sites of gene activation (Yang et al., 2014).

Many micro RNAs (miRNAs) in mouse ESCs remain constant or increase upon the formation of embryoid bodies, while others are repressed during differentiation (Houbaviy et al., 2003). Many miRNAs are differentially expressed during the differentiation of ESCs into primed EpiSCs (Gu et al., 2016). Specific miRNAs appear in ESCs differentiating into mesoderm-derived cardiomyocytes. The induced expression of these miRNAs in undifferentiated ESCs results in enhanced mesoderm gene expression and suppression of differentiation into ectodermal or endodermal lineages (Ivey et al., 2008). Some miRNAs appear to be involved in inducing differentiation by targeting the transcripts of the core pluripotency transcription factors (Tay et al., 2008; Xu et al., 2009).

lincRNAs and miRNAs can interact directly in regulating the shift from pluripotency to differentiation. One of the many ways by which lincRNAs function to regulate gene activity is to compete with transcripts for the miRNAs that would target and degrade them or prevent their translation, thereby reducing the repressive function of the miRNAs. Such lincRNAs, sometimes referred to as competing endogenous RNAs (ceRNAs), were discovered during the differentiation of mouse myoblasts into muscle tissue (Cesana et al., 2011). A particular lincRNA, named regulator of reprogramming (lincRNA-RoR), was found to enhance the frequency of changing fibroblasts into pluripotent cells (Loewer et al., 2010). lincRNA-RoR is present in undifferentiated ESCs where its transcription is under the control of the three core pluripotency factors; induction of high levels of this RNA results in a drop in the level of the three factors, suggesting a regulatory feedback loop. In undifferentiated ESCs, lincRNA-RoR hybridizes with the miRNAs that would suppress the transcription of the core pluripotency factors; following the onset of differentiation, the level of these miRNAs is greatly increased, exceeding the competition potential of the lincRNA-RoR and leading to the reduction of the core pluripotency factors (Wang et al., 2013).

X chromosome inactivation in female stem cells

Both X chromosomes are active in mouse naïve ESCs derived from female blastocysts. Transcription of the *Xist* gene responsible for silencing (see Chapter 9) is repressed directly by Nanog, Oct4 and Sox2 which bind within the gene's first intron (Navarro et al., 2008). As added insurance, three pluripotency-associated factors (Rex1, c-Myc and Klf4) bind the promoter regulatory region of *Tsix* and activate the transcription of the Tsix antisense RNA (Navarro et al., 2010). As a consequence of the differentiation of ESCs into primed EpiSCs, one of the two X chromosomes is inactivated by the spreading of Xist RNA. PcG repressive complexes are recruited to methylate H3K27 (Rougeulle et al., 2004; Silva et al., 2003) and mono-ubiquitinate H2A (de Napoles et al., 2004). Starting early during the process of differentiation, H3K4me3 is lost from the inactive X chromosome, followed by the loss of H3K4me2 (O'Neill et al., 2008). As differentiation is proceeding, the inactive X acquires the histone variant macroH2A (Mermoud et al., 1999). Additional factors recruited during inactivation function to mediate the hypermethylation of CpG islands by *de novo* DNA methyl transferases (Gendrel et al., 2012).

Chromatin architecture of stem cells

The genome of stem cells exhibits the same features of topological organization that are found in somatic cells. Each chromosome occupies a particular territory—genomic regions that are predominantly active tend to interact within the interior of the nucleus, and regions that are mostly inactive are usually associated with the nuclear lamina at the inner surface of the nuclear membrane. Chromosomes are organized as topologically associated domains (TADs).

As expected, at the onset and throughout differentiation, the topological organization of the ESC genome changes, reflecting the changes in transcriptional activity that underlie the progression of pluripotent cells to multipotent progenitor cells and then to differentiated cells. During the transition of mouse ESCs to neural precursor cells and to terminally differentiated cells (astrocytes), although the same broad genomic regions are associated with the nuclear lamina in the different cell types, a number of single genes or of small gene clusters change their lamina interaction (Peric-Hupkes et al., 2010). A comparison of naïve human ESCs and primed ESCs (generated by treatment with fibroblast growth factor) reveals that TAD boundaries remain constant (Ji et al., 2016). Similar studies with naïve and primed mouse ESCs or with ESC-derived lineages showed that there are significant differences in the association of enhancers and promoters within the TADs, reflecting the significant changes in transcriptional activity that occur at the onset of ESC differentiation (Dixon et al., 2015; Joshi et al., 2015). Another aspect of the spatial genome organization of ESCs involves the Polycomb repressive complexes. In *Drosophila* embryonic cells, inactive chromosomal domains tend to cluster and remain within the chromosomal territories, while domains that contain predominantly active genes (as defined by high levels of H3K4me3 and open, DNase-sensitive chromatin) tend to form extraterritorial contacts with other active domains. Among these are long-range contacts between the two major *Hox* gene clusters that are approximately 10 megabases apart. Both clusters are regulated by Polycomb complexes and undergo a number of interactions with Polycomb group target genes (Sexton et al., 2012). Mouse ESCs displayed a similar association—*Hox* loci that are strongly occupied by Polycomb and enriched in H3K27me3 co-localize with distant genomic regions that are also occupied by Polycomb. In an analogous manner, genomic regions occupied by the core pluripotency factors Oct4, Sox2 and Nanog tend to cluster. Elimination of one of the key components of a Polycomb complex disrupts the long-range interactions mediated by this complex without altering the interactions effected by other factors or the organization of the TADs. Finally, a comparison of ESCs with mouse embryonic fibroblasts clearly demonstrates that long-range chromatin interactions change during differentiation (Denholtz et al., 2013). A more detailed study of the role of Polycomb in the maintenance of ESC pluripotency differentiated between promoter–promoter and enhancer–promoter contacts that involve the Polycomb repressive complex PRC1. Promoters bound by PRC1 tend to engage in long-range contacts, as eminently demonstrated by the promoters across the four *Hox* gene clusters, many of which display bivalent chromatin marks. In addition, most of these promoters are in contact with a poised enhancer (identified by H3K4me1 and H3K27me3) usually occupied by PRC1 (Schoenfelder et al., 2015).

Chapter summary

In most multicellular organisms, development is initiated by the fusion between a sperm and an egg to generate a zygote. The zygote and each of the cells that are produced by the very early mitotic divisions are able to give rise to a whole organism and, therefore, are said to be **totipotent**. During the ensuing development, the totipotency of the cells that are produced by subsequent divisions declines; these cells are nevertheless able to differentiate into different subgroups of tissues and are referred to as **pluripotent**. Such cells can be extracted from early embryonic stages and grown in culture as **embryonic stem cells** (ESCs). ESCs are characterized by an open chromatin configuration and an elevated level of transcription. Their pluripotent nature is the result of the action of the pioneer transcription factors Oct4, Sox2 and Nanog, which bind to the promoters of a group of genes that they activate and that are responsible for pluripotency; these three factors also repress genes that are activated for cellular differentiation and development.

Multipotent or **progenitor** stem cells are present in adult organisms where they are able to differentiate into the various cells present in specific tissues and, thereby, give rise to, and renew, most tissues. The differentiation of these cells depends on the microenvironment of their location, referred to as the **niche**. Within the niche, adult stem cells are subjected to signals that either direct them to proliferate or to differentiate. The proliferation of stem cells requires that the factors responsible for their multipotency and lack of differentiation be transmitted through mitotic divisions.

Differentiation of stem cells requires the silencing of the pluripotency genes and the activation of genes that are associated with the development and function of a particular terminally differentiated cell type. Standard shifts in DNA methylation and in repressive histone modifications or in modifications associated with activation accompany these changes in transcriptional programs. Long non-coding RNAs (lncRNAs) and micro RNAs (miRNAs) affect, both positively and negatively,

a substantial portion of ESC genes, including Nanog, Oct4 and Sox2.

In ESCs, both X chromosomes are active because Nanog, Oct4 and Sox2 directly repress the transcription of the *Xist* gene; other pluripotency-associated factors enhance the transcription of the *Tsix* gene. As differentiation initiates and the function of pluripotency genes is abrogated, random X inactivation is established.

The genome of stem cells exhibits the same features of topological organization that are found in somatic cells. At the onset and throughout differentiation, the topological organization of the ESC genome changes, reflecting the changes in transcriptional activity that underlie the progression of pluripotent cells to multipotent progenitor cells and then to differentiated cells.

References

Ang, Y. S., Tsai, S. Y., Lee, D. F., Monk, J., Su, J., Ratnakumar, K., Ding, J., Ge, Y., Darr, H., Chang, B., Wang, J., Rendl, M., Bernstein, E., Schaniel, C. & Lemischka, I. R. 2011. Wdr5 mediates self-renewal and reprogramming via the embryonic stem cell core transcriptional network. *Cell*, 145, 183–97.

Arney, K. L., Bao, S., Bannister, A. J., Kouzarides, T. & Surani, M. A. 2002. Histone methylation defines epigenetic asymmetry in the mouse zygote. *Int J Dev Biol*, 46, 317–20.

Azuara, V., Perry, P., Sauer, S., Spivakov, M., Jorgensen, H. F., John, R. M., Gouti, M., Casanova, M., Warnes, G., Merkenschlager, M. & Fisher, A. G. 2006. Chromatin signatures of pluripotent cell lines. *Nat Cell Biol*, 8, 532–8.

Barker, N., Van Es, J. H., Kuipers, J., Kujala, P., Van Den Born, M., Cozijnsen, M., Haegebarth, A., Korving, J., Begthel, H., Peters, P. J. & Clevers, H. 2007. Identification of stem cells in small intestine and colon by marker gene Lgr5. *Nature*, 449, 1003–7.

Bernstein, B. E., Mikkelsen, T. S., Xie, X., Kamal, M., Huebert, D. J., Cuff, J., Fry, B., Meissner, A., Wernig, M., Plath, K., Jaenisch, R., Wagschal, A., Feil, R., Schreiber, S. L. & Lander, E. S. 2006. A bivalent chromatin structure marks key developmental genes in embryonic stem cells. *Cell*, 125, 315–26.

Bock, C., Beerman, I., Lien, W. H., Smith, Z. D., Gu, H., Boyle, P., Gnirke, A., Fuchs, E., Rossi, D. J. & Meissner, A. 2012. DNA methylation dynamics during in vivo differentiation of blood and skin stem cells. *Mol Cell*, 47, 633–47.

Bouniol, C., Nguyen, E. & Debey, P. 1995. Endogenous transcription occurs at the 1-cell stage in the mouse embryo. *Exp Cell Res*, 218, 57–62.

Boyer, L. A., Lee, T. I., Cole, M. F., Johnstone, S. E., Levine, S. S., Zucker, J. P., Guenther, M. G., Kumar, R. M., Murray, H. L., Jenner, R. G., Gifford, D. K., Melton, D. A., Jaenisch, R. & Young, R. A. 2005. Core transcriptional regulatory circuitry in human embryonic stem cells. *Cell*, 122, 947–56.

Cai, Y., Jin, J., Swanson, S. K., Cole, M. D., Choi, S. H., Florens, L., Washburn, M. P., Conaway, J. W. & Conaway, R. C. 2010. Subunit composition and substrate specificity of a MOF-containing histone acetyltransferase distinct from the male-specific lethal (MSL) complex. *J Biol Chem*, 285, 4268–72.

Cesana, M., Cacchiarelli, D., Legnini, I., Santini, T., Sthandier, O., Chinappi, M., Tramontano, A. & Bozzoni, I. 2011. A long noncoding RNA controls muscle differentiation by functioning as a competing endogenous RNA. *Cell*, 147, 358–69.

Chambers, I., Colby, D., Robertson, M., Nichols, J., Lee, S., Tweedie, S. & Smith, A. 2003. Functional expression cloning of Nanog, a pluripotency sustaining factor in embryonic stem cells. *Cell*, 113, 643–55.

Chelmicki, T., Dundar, F., Turley, M. J., Khanam, T., Aktas, T., Ramirez, F., Gendrel, A. V., Wright, P. R., Videm, P., Backofen, R., Heard, E., Manke, T. & Akhtar, A. 2014. MOF-associated complexes ensure stem cell identity and Xist repression. *Elife*, 3, e02024.

Chen, J. Y., Miyanishi, M., Wang, S. K., Yamazaki, S., Sinha, R., Kao, K. S., Seita, J., Sahoo, D., Nakauchi, H. & Weissman, I. L. 2016. Hoxb5 marks long-term haematopoietic stem cells and reveals a homogenous perivascular niche. *Nature*, 530, 223–7.

Dean, W., Bowden, L., Aitchison, A., Klose, J., Moore, T., Meneses, J. J., Reik, W. & Feil, R. 1998. Altered imprinted gene methylation and expression in completely ES cell-derived mouse fetuses: association wiyh aberrant phenotypes. *Development*, 125, 2273–82.

Deluz, C., Friman, E. T., Strebinger, D., Benke, A., Raccaud, M., Callegari, A., Leleu, M., Manley, S. & Suter, D. M. 2016. A role for mitotic bookmarking of SOX2 in pluripotency and differentiation. *Genes Dev*, 30, 2538–50.

Denholtz, M., Bonora, G., Chronis, C., Splinter, E., De Laat, W., Ernst, J., Pellegrini, M. & Plath, K. 2013. Long-range chromatin contacts in embryonic stem cells reveal a role for pluripotency factors and polycomb proteins in genome organization. *Cell Stem Cell*, 13, 602–16.

De Napoles, M., Mermoud, J. E., Wakao, R., Tang, Y. A., Endoh, M., Appanah, R., Nesterova, T. B., Silva, J., Otte, A. P., Vidal, M., Koseki, H. & Brockdorff, N. 2004. Polycomb group proteins Ring1A/B link ubiquitylation of histone H2A to heritable gene silencing and X inactivation. *Dev Cell*, 7, 663–76.

De Wit, E., Bouwman, B. A., Zhu, Y., Klous, P., Splinter, E., Verstegen, M. J., Krijger, P. H., Festuccia, N., Nora, E. P., Welling, M., Heard, E., Geijsen, N., Poot, R. A., Chambers, I. & De Laat, W. 2013. The pluripotent genome in three dimensions is shaped around pluripotency factors. *Nature*, 501, 227–31.

Dinger, M. E., Amaral, P. P., Mercer, T. R., Pang, K. C., Bruce, S. J., Gardiner, B. B., Askarian-Amiri, M. E., Ru, K., Solda, G., Simons, C., Sunkin, S. M., Crowe, M. L., Grimmond, S. M., Perkins, A. C. & Mattick, J. S. 2008. Long noncoding RNAs in mouse embryonic stem cell pluripotency and differentiation. *Genome Res*, 18, 1433–45.

Dixon, J. R., Jung, I., Selvaraj, S., Shen, Y., Antosiewicz-Bourget, J. E., Lee, A. Y., Ye, Z., Kim, A., Rajagopal, N., Xie, W., Diao, Y., Liang, J., Zhao, H., Lobanenkov, V. V., Ecker, J. R., Thomson, J. A. & Ren, B. 2015. Chromatin

Efroni, S., Duttagupta, R., Cheng, J., Dehghani, H., Hoeppner, D. J., Dash, C., Bazett-Jones, D. P., Le Grice, S., Mckay, R. D., Buetow, K. H., Gingeras, T. R., Misteli, T. & Meshorer, E. 2008. Global transcription in pluripotent embryonic stem cells. *Cell Stem Cell*, 2, 437–47.

Evans, M. J. & Kaufman, M. H. 1981. Establishment in culture of pluripotential cells from mouse embryos. *Nature*, 292, 154–6.

Feldman, N., Gerson, A., Fang, J., Li, E., Zhang, Y., Shinkai, Y., Cedar, H. & Bergman, Y. 2006. G9a-mediated irreversible epigenetic inactivation of Oct-3/4 during early embryogenesis. *Nat Cell Biol*, 8, 188–94.

Gaspar-Maia, A., Alajem, A., Polesso, F., Sridharan, R., Mason, M. J., Heidersbach, A., Ramalho-Santos, J., Mcmanus, M. T., Plath, K., Meshorer, E. & Ramalho-Santos, M. 2009. Chd1 regulates open chromatin and pluripotency of embryonic stem cells. *Nature*, 460, 863–8.

Gendrel, A. V., Apedaile, A., Coker, H., Termanis, A., Zvetkova, I., Godwin, J., Tang, Y. A., Huntley, D., Montana, G., Taylor, S., Giannoulatou, E., Heard, E., Stancheva, I. & Brockdorff, N. 2012. Smchd1-dependent and -independent pathways determine developmental dynamics of CpG island methylation on the inactive X chromosome. *Dev Cell*, 23, 265–79.

Gowher, H. & Jeltsch, A. 2001. Enzymatic properties of recombinant Dnmt3a DNA methyltransferase from mouse: the enzyme modifies DNA in a non-processive manner and also methylates non-CpG [correction of non-CpA] sites. *J Mol Biol*, 309, 1201–8.

Gu, K. L., Zhang, Q., Yan, Y., Li, T. T., Duan, F. F., Hao, J., Wang, X. W., Shi, M., Wu, D. R., Guo, W. T. & Wang, Y. 2016. Pluripotency-associated miR-290/302 family of microRNAs promote the dismantling of naive pluripotency. *Cell Res*, 26, 350–66.

Guenechea, G., Gan, O. I., Dorrell, C. & Dick, J. E. 2001. Distinct classes of human stem cells that differ in proliferative and self-renewal potential. *Nat Immunol*, 2, 75–82.

Guttman, M., Donaghey, J., Carey, B. W., Garber, M., Grenier, J. K., Munson, G., Young, G., Lucas, A. B., Ach, R., Bruhn, L., Yang, X., Amit, I., Meissner, A., Regev, A., Rinn, J. L., Root, D. E. & Lander, E. S. 2011. lincRNAs act in the circuitry controlling pluripotency and differentiation. *Nature*, 477, 295–300.

Hackett, J. A. & Surani, M. A. 2014. Regulatory principles of pluripotency: from the ground state up. *Cell Stem Cell*, 15, 416–30.

Hamatani, T., Carter, M. G., Sharov, A. A. & Ko, M. S. 2004. Dynamics of global gene expression changes during mouse preimplantation development. *Dev Cell*, 6, 117–31.

Hanna, J., Cheng, A. W., Saha, K., Kim, J., Lengner, C. J., Soldner, F., Cassady, J. P., Muffat, J., Carey, B. W. & Jaenisch, R. 2010. Human embryonic stem cells with biological and epigenetic characteristics similar to those of mouse ESCs. *Proc Natl Acad Sci U S A*, 107, 9222–7.

Heissig, B., Hattori, K., Dias, S., Friedrich, M., Ferris, B., Hackett, N. R., Crystal, R. G., Besmer, P., Lyden, D., Moore, M. A., Werb, Z. & Rafii, S. 2002. Recruitment of stem and progenitor cells from the bone marrow niche requires MMP-9 mediated release of kit-ligand. *Cell*, 109, 625–37.

Houbaviy, H. B., Murray, M. F. & Sharp, P. A. 2003. Embryonic stem cell-specific microRNAs. *Dev Cell*, 5, 351–8.

Huang, Y., Kim, J. K., Do, D. V., Lee, C., Penfold, C. A., Zylicz, J. J., Marioni, J. C., Hackett, J. A. & Surani, M. A. 2017. Stella modulates transcriptional and endogenous retrovirus programs during maternal-to-zygotic transition. *Elife*, 6, pii: e22345.

Hung, T. & Chang, H. Y. 2010. Long noncoding RNA in genome regulation: prospects and mechanisms. *RNA Biol*, 7, 582–5.

Ivey, K. N., Muth, A., Arnold, J., King, F. W., Yeh, R. F., Fish, J. E., Hsiao, E. C., Schwartz, R. J., Conklin, B. R., Bernstein, H. S. & Srivastava, D. 2008. MicroRNA regulation of cell lineages in mouse and human embryonic stem cells. *Cell Stem Cell*, 2, 219–29.

Ji, X., Dadon, D. B., Powell, B. E., Fan, Z. P., Borges-Rivera, D., Shachar, S., Weintraub, A. S., Hnisz, D., Pegoraro, G., Lee, T. I., Misteli, T., Jaenisch, R. & Young, R. A. 2016. 3D Chromosome regulatory landscape of human pluripotent cells. *Cell Stem Cell*, 18, 262–75.

Joshi, O., Wang, S. Y., Kuznetsova, T., Atlasi, Y., Peng, T., Fabre, P. J., Habibi, E., Shaik, J., Saeed, S., Handoko, L., Richmond, T., Spivakov, M., Burgess, D. & Stunnenberg, H. G. 2015. Dynamic reorganization of extremely long-range promoter–promoter interactions between two states of pluripotency. *Cell Stem Cell*, 17, 748–57.

Kaji, K., Caballero, I. M., Macleod, R., Nichols, J., Wilson, V. A. & Hendrich, B. 2006. The NuRD component Mbd3 is required for pluripotency of embryonic stem cells. *Nat Cell Biol*, 8, 285–92.

Kiel, M. J., Yilmaz, O. H., Iwashita, T., Yilmaz, O. H., Terhorst, C. & Morrison, S. J. 2005. SLAM family receptors distinguish hematopoietic stem and progenitor cells and reveal endothelial niches for stem cells. *Cell*, 121, 1109–21.

King, H. W. & Klose, R. J. 2017. The pioneer factor OCT4 requires the chromatin remodeller BRG1 to support gene regulatory element function in mouse embryonic stem cells. *Elife*, 6, pii: e22631.

Laurent, L., Wong, E., Li, G., Huynh, T., Tsirigos, A., Ong, C. T., Low, H. M., Kin Sung, K. W., Rigoutsos, I., Loring, J. & Wei, C. L. 2010. Dynamic changes in the human methylome during differentiation. *Genome Res*, 20, 320–31.

Leeb, M., Pasini, D., Novatchkova, M., Jaritz, M., Helin, K. & Wutz, A. 2010. Polycomb complexes act redundantly to repress genomic repeats and genes. *Genes Dev*, 24, 265–76.

Li, X., Li, L., Pandey, R., Byun, J. S., Gardner, K., Qin, Z. & Dou, Y. 2012. The histone acetyltransferase MOF is a key regulator of the embryonic stem cell core transcriptional network. *Cell Stem Cell*, 11, 163–78.

Liang, H. L., Nien, C. Y., Liu, H. Y., Metzstein, M. M., Kirov, N. & Rushlow, C. 2008. The zinc-finger protein Zelda is a key activator of the early zygotic genome in *Drosophila*. *Nature*, 456, 400–3.

Lister, R., Pelizzola, M., Dowen, R. H., Hawkins, R. D., Hon, G., Tonti-Filippini, J., Nery, J. R., Lee, L., Ye, Z., Ngo, Q. M., Edsall, L., Antosiewicz-Bourget, J., Stewart, R., Ruotti, V., Millar, A. H., Thomson, J. A., Ren, B. & Ecker, J. R. 2009. Human DNA methylomes at base resolution show widespread epigenomic differences. *Nature*, 462, 315–22.

Liu, Y., Pelham-Webb, B., Di Giammartino, D. C., Li, J., Kim, D., Kita, K., Saiz, N., Garg, V., Doane, A., Giannakakou, P., Hadjantonakis, A. K., Elemento, O. & Apostolou, E. 2017. Widespread mitotic bookmarking by histone marks and transcription factors in pluripotent stem cells. *Cell Rep*, 19, 1283–93.

Loewer, S., Cabili, M. N., Guttman, M., Loh, Y. H., Thomas, K., Park, I. H., Garber, M., Curran, M., Onder, T., Agarwal, S., Manos, P. D., Datta, S., Lander, E. S., Schlaeger, T. M., Daley, G. Q. & Rinn, J. L. 2010. Large intergenic non-coding RNA-RoR modulates reprogramming of human induced pluripotent stem cells. *Nat Genet*, 42, 1113–17.

Losick, V. P., Morris, L. X., Fox, D. T. & Spradling, A. 2011. *Drosophila* stem cell niches: a decade of discovery suggests a unified view of stem cell regulation. *Dev Cell*, 21, 159–71.

Luo, L., Wang, H., Fan, C., Liu, S. & Cai, Y. 2015. Wnt ligands regulate Tkv expression to constrain Dpp activity in the *Drosophila* ovarian stem cell niche. *J Cell Biol*, 209, 595–608.

Lyle, S., Christofidou-Solomidou, M., Liu, Y., Elder, D. E., Albelda, S. & Cotsarelis, G. 1998. The C8/144B monoclonal antibody recognizes cytokeratin 15 and defines the location of human hair follicle stem cells. *J Cell Sci*, 111 (Pt 21), 3179–88.

Macfarlan, T. S., Gifford, W. D., Driscoll, S., Lettieri, K., Rowe, H. M., Bonanomi, D., Firth, A., Singer, O., Trono, D. & Pfaff, S. L. 2012. Embryonic stem cell potency fluctuates with endogenous retrovirus activity. *Nature*, 487, 57–63.

Marianes, A. & Spradling, A. C. 2013. Physiological and stem cell compartmentalization within the *Drosophila* midgut. *Elife*, 2, e00886.

Martin, G. R. 1981. Isolation of a pluripotent cell line from early mouse embryos cultured in medium conditioned by teratocarcinoma stem cells. *Proc Natl Acad Sci U S A*, 78, 7634–8.

Mendjan, S., Taipale, M., Kind, J., Holz, H., Gebhardt, P., Schelder, M., Vermeulen, M., Buscaino, A., Duncan, K., Mueller, J., Wilm, M., Stunnenberg, H. G., Saumweber, H. & Akhtar, A. 2006. Nuclear pore components are involved in the transcriptional regulation of dosage compensation in *Drosophila*. *Mol Cell*, 21, 811–23.

Mermoud, J. E., Costanzi, C., Pehrson, J. R. & Brockdorff, N. 1999. Histone macroH2A1.2 relocates to the inactive X chromosome after initiation and propagation of X-inactivation. *J Cell Biol*, 147, 1399–408.

Micchelli, C. A. & Perrimon, N. 2006. Evidence that stem cells reside in the adult *Drosophila* midgut epithelium. *Nature*, 439, 475–9.

Mitsui, K., Tokuzawa, Y., Itoh, H., Segawa, K., Murakami, M., Takahashi, K., Maruyama, M., Maeda, M. & Yamanaka, S. 2003. The homeoprotein Nanog is required for maintenance of pluripotency in mouse epiblast and ES cells. *Cell*, 113, 631–42.

Morris, R. J., Liu, Y., Marles, L., Yang, Z., Trempus, C., Li, S., Lin, J. S., Sawicki, J. A. & Cotsarelis, G. 2004. Capturing and profiling adult hair follicle stem cells. *Nat Biotechnol*, 22, 411–17.

Navarro, P., Chambers, I., Karwacki-Neisius, V., Chureau, C., Morey, C., Rougeulle, C. & Avner, P. 2008. Molecular coupling of Xist regulation and pluripotency. *Science*, 321, 1693–5.

Navarro, P., Oldfield, A., Legoupi, J., Festuccia, N., Dubois, A., Attia, M., Schoorlemmer, J., Rougeulle, C., Chambers, I. & Avner, P. 2010. Molecular coupling of Tsix regulation and pluripotency. *Nature*, 468, 457–60.

Ng, S. Y., Johnson, R. & Stanton, L. W. 2012. Human long non-coding RNAs promote pluripotency and neuronal differentiation by association with chromatin modifiers and transcription factors. *EMBO J*, 31, 522–33.

Nichols, J., Zevnik, B., Anastassiadis, K., Niwa, H., Klewe-Nebenius, D., Chambers, I., Scholer, H. & Smith, A. 1998. Formation of pluripotent stem cells in the mammalian embryo depends on the POU transcription factor Oct4. *Cell*, 95, 379–91.

Niwa, H., Miyazaki, J. & Smith, A. G. 2000. Quantitative expression of Oct-3/4 defines differentiation, dedifferentiation or self-renewal of ES cells. *Nat Genet*, 24, 372–6.

O'neill, L. P., Spotswood, H. T., Fernando, M. & Turner, B. M. 2008. Differential loss of histone H3 isoforms mono-, di- and tri-methylated at lysine 4 during X-inactivation in female embryonic stem cells. *Biol Chem*, 389, 365–70.

Ohlstein, B. & Spradling, A. 2006. The adult *Drosophila* posterior midgut is maintained by pluripotent stem cells. *Nature*, 439, 470–4.

Peaston, A. E., Evsikov, A. V., Graber, J. H., De Vries, W. N., Holbrook, A. E., Solter, D. & Knowles, B. B. 2004. Retrotransposons regulate host genes in mouse oocytes and preimplantation embryos. *Dev Cell*, 7, 597–606.

Perez-Montero, S., Carbonell, A., Moran, T., Vaquero, A. & Azorin, F. 2013. The embryonic linker histone H1 variant of *Drosophila*, dBigH1, regulates zygotic genome activation. *Dev Cell*, 26, 578–90.

Peric-Hupkes, D., Meuleman, W., Pagie, L., Bruggeman, S. W., Solovei, I., Brugman, W., Graf, S., Flicek, P., Kerkhoven, R. M., Van Lohuizen, M., Reinders, M., Wessels, L. & Van Steensel, B. 2010. Molecular maps of the reorganization of genome-nuclear lamina interactions during differentiation. *Mol Cell*, 38, 603–13.

Petruk, S., Cai, J., Sussman, R., Sun, G., Kovermann, S. K., Mariani, S. A., Calabretta, B., Mcmahon, S. B., Brock, H. W., Iacovitti, L. & Mazo, A. 2017. Delayed accumulation of H3K27me3 on nascent DNA is essential for recruitment of transcription factors at early stages of stem cell differentiation. *Mol Cell*, 66, 247–57 e5.

Probst, A. V., Santos, F., Reik, W., Almouzni, G. & Dean, W. 2007. Structural differences in centromeric heterochromatin are spatially reconciled on fertilisation in the mouse zygote. *Chromosoma*, 116, 403–15.

Raja, S. J., Charapitsa, I., Conrad, T., Vaquerizas, J. M., Gebhardt, P., Holz, H., Kadlec, J., Fraterman, S., Luscombe, N. M. & Akhtar, A. 2010. The nonspecific lethal complex is a transcriptional regulator in *Drosophila*. *Mol Cell*, 38, 827–41.

Reynolds, N., Latos, P., Hynes-Allen, A., Loos, R., Leaford, D., O'shaughnessy, A., Mosaku, O., Signolet, J., Brennecke, P., Kalkan, T., Costello, I., Humphreys, P., Mansfield, W., Nakagawa, K., Strouboulis, J., Behrens, A., Bertone, P. & Hendrich, B. 2012. NuRD suppresses pluripotency gene expression to promote transcriptional heterogeneity and lineage commitment. *Cell Stem Cell*, 10, 583–94.

Reynolds, N., Salmon-Divon, M., Dvinge, H., Hynes-Allen, A., Balasooriya, G., Leaford, D., Behrens, A., Bertone, P. & Hendrich, B. 2012. NuRD-mediated deacetylation of H3K27 facilitates recruitment of Polycomb repressive complex 2 to direct gene repression. *EMBO J*, 31, 593–605.

Rossello, R. A., Pfenning, A., Howard, J. T. & Hochgeschwender, U. 2016. Characterization and genetic manipulation of primed stem cells into a functional naive state with ESRRB. *World J Stem Cells*, 8, 355–66.

Rougeulle, C., Chaumeil, J., Sarma, K., Allis, C. D., Reinberg, D., Avner, P. & Heard, E. 2004. Differential histone H3 Lys-9 and Lys-27 methylation profiles on the X chromosome. *Mol Cell Biol*, 24, 5475–84.

Rougier, N., Bourc'his, D., Gomes, D. M., Niveleau, A., Plachot, M., Paldi, A. & Viegas-Pequignot, E. 1998. Chromosome methylation patterns during mammalian preimplantation development. *Genes Dev*, 12, 2108–13.

Santos, F., Peters, A. H., Otte, A. P., Reik, W. & Dean, W. 2005. Dynamic chromatin modifications characterise the first cell cycle in mouse embryos. *Dev Biol,* 280, 225–36.

Schoenfelder, S., Sugar, R., Dimond, A., Javierre, B. M., Armstrong, H., Mifsud, B., Dimitrova, E., Matheson, L., Tavares-Cadete, F., Furlan-Magaril, M., Segonds-Pichon, A., Jurkowski, W., Wingett, S. W., Tabbada, K., Andrews, S., Herman, B., Leproust, E., Osborne, C. S., Koseki, H., Fraser, P., Luscombe, N. M. & Elderkin, S. 2015. Polycomb repressive complex PRC1 spatially constrains the mouse embryonic stem cell genome. *Nat Genet,* 47, 1179–86.

Schofield, R. 1978. The relationship between the spleen colony-forming cell and the haemopoietic stem cell. *Blood Cells,* 4, 7–25.

Schulz, K. N., Bondra, E. R., Moshe, A., Villalta, J. E., Lieb, J. D., Kaplan, T., Mckay, D. J. & Harrison, M. M. 2015. Zelda is differentially required for chromatin accessibility, transcription factor binding, and gene expression in the early *Drosophila* embryo. *Genome Res,* 25, 1715–26.

Sexton, T., Yaffe, E., Kenigsberg, E., Bantignies, F., Leblanc, B., Hoichman, M., Parrinello, H., Tanay, A. & Cavalli, G. 2012. Three-dimensional folding and functional organization principles of the *Drosophila* genome. *Cell,* 148, 458–72.

Sheik Mohamed, J., Gaughwin, P. M., Lim, B., Robson, P. & Lipovich, L. 2010. Conserved long noncoding RNAs transcriptionally regulated by Oct4 and Nanog modulate pluripotency in mouse embryonic stem cells. *RNA,* 16, 324–37.

Silva, J., Mak, W., Zvetkova, I., Appanah, R., Nesterova, T. B., Webster, Z., Peters, A. H., Jenuwein, T., Otte, A. P. & Brockdorff, N. 2003. Establishment of histone h3 methylation on the inactive X chromosome requires transient recruitment of Eed–Enx1 polycomb group complexes. *Dev Cell,* 4, 481–95.

Slaidina, M. & Lehmann, R. 2014. Translational control in germline stem cell development. *J Cell Biol,* 207, 13–21.

Smith, E. R., Cayrou, C., Huang, R., Lane, W. S., Cote, J. & Lucchesi, J. C. 2005. A human protein complex homologous to the *Drosophila* MSL complex is responsible for the majority of histone H4 acetylation at lysine 16. *Mol Cell Biol,* 25, 9175–88.

Snippert, H. J., Haegebarth, A., Kasper, M., Jaks, V., Van Es, J. H., Barker, N., Van De Wetering, M., Van Den Born, M., Begthel, H., Vries, R. G., Stange, D. E., Toftgard, R. & Clevers, H. 2010. Lgr6 marks stem cells in the hair follicle that generate all cell lineages of the skin. *Science,* 327, 1385–9.

Song, X., Zhu, C. H., Doan, C. & Xie, T. 2002. Germline stem cells anchored by adherens junctions in the *Drosophila* ovary niches. *Science,* 296, 1855–7.

Soufi, A., Donahue, G. & Zaret, K. S. 2012. Facilitators and impediments of the pluripotency reprogramming factors' initial emngagement with the genome. *Cell,* 151, 994–1004.

Spangrude, G. J., Heimfeld, S. & Weissman, I. L. 1988. Purification and characterization of mouse hematopoietic stem cells. *Science,* 241, 58–62.

Sulston, J. E., Schierenberg, E., White, J. G. & Thomson, J. N. 1983. The embryonic cell lineage of the nematode *Caenorhabditis elegans. Dev Biol,* 100, 64–119.

Sun, Y., Nien, C. Y., Chen, K., Liu, H. Y., Johnston, J., Zeitlinger, J. & Rushlow, C. 2015. Zelda overcomes the high intrinsic nucleosome barrier at enhancers during *Drosophila* zygotic genome activation. *Genome Res,* 25, 1703–14.

Tay, Y., Zhang, J., Thomson, A. M., Lim, B. & Rigoutsos, I. 2008. MicroRNAs to Nanog, Oct4 and Sox2 coding regions modulate embryonic stem cell differentiation. *Nature,* 455, 1124–8.

Tesar, P. J., Chenoweth, J. G., Brook, F. A., Davies, T. J., Evans, E. P., Mack, D. L., Gardner, R. L. & Mckay, R. D. 2007. New cell lines from mouse epiblast share defining features with human embryonic stem cells. *Nature,* 448, 196–9.

Teves, S. S., An, L., Hansen, A. S., Xie, L., Darzacq, X. & Tjian, R. 2016. A dynamic mode of mitotic bookmarking by transcription factors. *Elife,* 5, pii: e22280.

Thomson, J. A., Itskovitz-Eldor, J., Shapiro, S. S., Waknitz, M. A., Swiergiel, J. J., Marshall, V. S. & Jones, J. M. 1998. Embryonic stem cell lines derived from human blastocysts. *Science,* 282, 1145–7.

Till, J. E. & Mc, C. E. 1961. A direct measurement of the radiation sensitivity of normal mouse bone marrow cells. *Radiat Res,* 14, 213–22.

Van Der Heijden, G. W., Dieker, J. W., Derijck, A. A., Muller, S., Berden, J. H., Braat, D. D., Van Der Vlag, J. & De Boer, P. 2005. Asymmetry in histone H3 variants and lysine methylation between paternal and maternal chromatin of the early mouse zygote. *Mech Dev,* 122, 1008–22.

Wang, Y., Xu, Z., Jiang, J., Xu, C., Kang, J., Xiao, L., Wu, M., Xiong, J., Guo, X. & Liu, H. 2013. Endogenous miRNA sponge lincRNA-RoR regulates Oct4, Nanog, and Sox2 in human embryonic stem cell self-renewal. *Dev Cell,* 25, 69–80.

Ware, C. B., Nelson, A. M., Mecham, B., Hesson, J., Zhou, W., Jonlin, E. C., Jimenez-Caliani, A. J., Deng, X., Cavanaugh, C., Cook, S., Tesar, P. J., Okada, J., Margaretha, L., Sperber, H., Choi, M., Blau, C. A., Treuting, P. M., Hawkins, R. D., Cirulli, V. & Ruohola-Baker, H. 2014. Derivation of naive human embryonic stem cells. *Proc Natl Acad Sci U S A,* 111, 4484–9.

Whyte, W. A., Bilodeau, S., Orlando, D. A., Hoke, H. A., Frampton, G. M., Foster, C. T., Cowley, S. M. & Young, R. A. 2012. Enhancer decommissioning by LSD1 during embryonic stem cell differentiation. *Nature,* 482, 221–5.

Whyte, W. A., Orlando, D. A., Hnisz, D., Abraham, B. J., Lin, C. Y., Kagey, M. H., Rahl, P. B., Lee, T. I. & Young, R.

A. 2013. Master transcription factors and mediator establish super-enhancers at key cell identity genes. *Cell,* 153, 307–19.

Xu, N., Papagiannakopoulos, T., Pan, G., Thomson, J. A. & Kosik, K. S. 2009. MicroRNA-145 regulates OCT4, SOX2, and KLF4 and represses pluripotency in human embryonic stem cells. *Cell,* 137, 647–58.

Yang, Y. W., Flynn, R. A., Chen, Y., Qu, K., Wan, B., Wang, K. C., Lei, M. & Chang, H. Y. 2014. Essential role of lncRNA binding for WDR5 maintenance of active chromatin and embryonic stem cell pluripotency. *Elife,* 3, e02046.

Zeng, F., Baldwin, D. A. & Schultz, R. M. 2004. Transcript profiling during preimplantation mouse development. *Dev Biol,* 272, 483–96.

CHAPTER 18

Nuclear reprogramming and induced pluripotency

The ability to reprogram a nucleus, in other words to change the transcriptional landscape of an entire genome, was first demonstrated by John Gurdon who, using a technique developed by Robert Briggs and Thomas King (Briggs and King, 1952; King and Briggs, 1955), transplanted nuclei from the gut of swimming tadpoles into enucleated frog eggs. Approximately one-third of these eggs underwent some level of development, with a few reaching the normal-feeding tadpole stage (Gurdon, 1960). Gurdon's experiments established that factors in the egg cytoplasm, which normally ensure the pluripotency of embryonic cells during early development, are able to reverse most of the genomic modifications responsible for cellular differentiation.

These pioneering studies have led to the modern practice of somatic cell nuclear transfer (commonly referred to as **cloning**) used to generate animals that are genetically identical. Successful cloning requires that the cytoplasm of the recipient oocyte erases the epigenetic modifications that are present in the donor nucleus, that the DNA methylation pattern of imprinted genes is re-established correctly and, in mammals, if the cloned embryo is genetically female, that normal X inactivation occurs.

One of the early questions that was raised following the successful cloning of the first mammal (Campbell et al., 1996; Wilmut et al., 1997) was whether an organism derived from a female somatic cell nucleus had the same inactive X chromosome in all of its tissues as the donor nucleus from which it arose, or whether X inactivation was random. This latter possibility would require that the inactive X be reactivated early during the development of the cloned organism. Evidence in favor of this scenario was obtained by determining that both X chromosomes are active during the cleavage stages of cloned mouse embryos, eventually leading to random X inactivation (Eggan et al., 2000). In contrast, the same inactive X chromosome present in the donor somatic cell nucleus was inactivated in extra-embryonic tissues, suggesting that an epigenetic mark identifying that chromosome is maintained in these tissues. All of the other imprinted genes in the genome of the donor nucleus are maintained, so that the uniparental gene expression of the nuclear donor is conserved in the cloned individual (Inoue et al., 2002).

A corollary of Gurdon's demonstration that somatic nuclei can be reprogrammed is the successful induction of somatic cells to return to the pluripotent state. The primary goal of this research has been to develop techniques for cell therapy, such as tissue regeneration using cells derived from the same individual, or the modeling of diseases by deriving cell lines from patient somatic cells that recapitulate the characteristics of the disease.

Methods for reprogramming somatic cells

The first successful attempt to force somatic cells to re-enter the pluripotent state was carried out by Shinya Yamanaka. Early experiments had shown that the introduction of single transcription factors could transform one type of differentiated somatic cells into another type (Davis et al., 1987; Kulessa et al., 1995). Numerous studies had provided evidence that, in addition to the core transcription factors Oct4, Sox2 and Nanog, a number of factors, including several that are often

Epigenetics, Nuclear Organization and Gene Function: with implications of epigenetic regulation and genetic architecture for human development and health. John C. Lucchesi, Oxford University Press (2019). © John C. Lucchesi 2019.
DOI: 10.1093/oso/9780198831204.001.0001

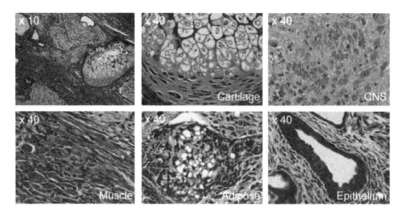

Fig. 18.1 Various tissues are present in teratomas formed by the subcutaneous injection of iPSCs into mice, demonstrating the pluripotent nature of these cells.

(From Takahashi and Yamanaka, 2006.)

up-regulated in tumors, participate in the maintenance of the pluripotent state in ESCs. Reasoning that some of these factors might induce pluripotency in somatic cells, Yamanaka tested the effect of expressing them individually and in combinations in embryonic and adult mouse fibroblast cultured cells. Four factors—Oct4, Sox2, Klf4 and c-Myc—were sufficient to induce pluripotency in the fibroblasts; surprisingly, Nanog was not required (Takahashi and Yamanaka, 2006). That the cells had been reprogrammed was demonstrated by monitoring the expression of characteristic pluripotency genes and, following injection of the cells into mice, by the formation of cellular masses (**teratomas**) with differentiated tissues originating from all three embryonic germ layers (Fig. 18.1). The reprogrammed cells were termed **induced pluripotent stem cells** (iPSCs) to distinguish their origin from that of ESCs. The same four transcription factors were found to induce pluripotency in human fibroblasts (Takahashi et al, 2007). In these cells, the pluripotent state could also be induced by a set of factors that included Nanog and Lin28, in addition to Oct4 and Sox2 (Yu et al., 2007). Although these minimal sets of transcription factors are sufficient to force somatic cells to re-enter the pluripotent state, their efficiency can be greatly enhanced by the expression of additional pluripotency-associated genes, as well as cell cycle enhancers. There is also some experimental evidence that manipulating the acetylation or methylation state of histones or altering the level of histone variants facilitates the reprogramming of somatic cells to iPSCs (for example, Ding et al., 2014; Huangfu et al., 2008; Liang et al., 2010; Shinagawa et al., 2014). Given the broad effects on gene action of

these modifications, their specific contribution to the reprogramming process has been difficult to establish.

Recently, the successful induction of iPSCs from mouse embryonic fibroblasts was obtained by the specific epigenetic remodeling (p300-mediated acetylation) of the *Sox2* gene promoter or of the *Oct4* gene promoter and its enhancer (Liu et al., 2018).

A completely different method of reprogramming somatic cells consists of treating them with micro RNAs (miRNAs). These RNAs are highly expressed in ESCs (Houbaviy et al., 2003; Suh et al., 2004) where they are involved in regulating pluripotency genes (Judson et al., 2009). Induction of specific miRNAs, which are particularly abundant in ESCs but disappear following differentiation, resulted in the transformation of cancer cells into ESC-like pluripotent cells (Lin et al., 2008) or of mouse and human somatic cells into iPSCs (Anokye-Danso et al., 2011; Miyoshi et al., 2011). Vectors that express miRNAs can be added to vectors expressing the reprogramming transcription factors in order to increase the number of iPSC colonies induced (Howden et al., 2015). The levels and function of miRNAs are modulated by specific long intergenic non-coding RNAs (lincRNAs) whose expression is elevated in iPSCs.

Molecular aspects of reprogramming

A variety of cell types can be reprogrammed, including fibroblasts, keratinocytes, liver cells and intestinal cells (Fig. 18.2). The reprogramming process is protracted, taking several days, and is usually successful in only a relatively small percentage of the treated somatic cells

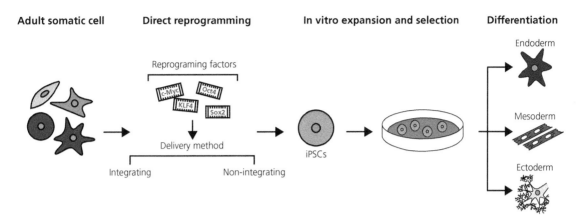

Fig. 18.2 Different types of somatic cells, such as fibroblasts, keratinocytes, liver and intestinal cells, can be induced to re-enter the pluripotent state by treatment with the pioneer transcription factors Oct4, Sox2, Klf4 and c-Myc. These factors can be introduced into cells by integrating viral vectors, such as lentiviruses, or by non-integrating vectors such as adeno- or Sendai viruses. The use of integrating viruses can lead to concerns that the process may lead to insertional mutations in the treated cells.

(Modified from Menon et al., 2016.)

that must be selected and expanded. During the onset of the process, Oct4, Sox2 and Klf4 act as pioneer transcription factors and bind together to the enhancer elements of many genes that are involved in reprogramming and that are present in closed, condensed chromatin; c-Myc is recruited by the pioneer factors and facilitates reprogramming by repressing the differentiated gene expression of the somatic cells (Sridharan et al., 2009) and by its role as a general enhancer of proliferation and growth-related gene promoters (Nie et al., 2012). Binding of the three factors to silent chromatin regions increases as a function of time (Soufi et al., 2012).

The **initial reprogramming phase** is marked by the loss of somatic cell-specific markers, a change in cell shape and an increase in the rate of cell division caused by the binding of c-Myc to cell cycle gene promoters. With the exception of the specific genes involved in these early events, there are no large-scale epigenetic chromatin modifications at the onset of dedifferentiation. Exposure to the four pioneer transcription factors for at least 10–12 days is necessary for the cells to proceed through the reprogramming process (Brambrink et al., 2008; Stadtfeld et al., 2008); early withdrawal of these factors causes the cells to stop the reprogramming process or to regress to their somatic state.

The Oct4, Sox2 and Klf4 transgene products induce the expression of genes mainly involved in DNA replication and the process of cell division; among the down-regulated genes are cell surface genes involved in cell adhesion and intercellular contacts (Polo et al.,

2012). Following a period of time, referred to as the **maturation phase**, a small proportion of the treated cells enter a **stabilization phase**. During the maturation phase, a specific set of endogenous genes that include core pluripotency factors, such as Oct4, Sox2 and Nanog, are activated to establish the stabilization phase and maintain pluripotency (Golipour et al., 2012; Okita et al., 2007; Polo et al., 2012). As expected, the promoters of pluripotency-associated genes become demethylated and those of genes that are normally methylated in pluripotent cells, but demethylated in fibroblasts, become methylated (Lee et al., 2014; Polo et al., 2012). Also, as expected, the interaction of the core reprogramming factors with traditional histone-modifying complexes results in histone marks that occur throughout the genome during the process of reprogramming (Ang et al., 2012; Mansour et al., 2012).

In mice, the reprogramming of female somatic cells to iPSCs is usually accompanied by a reactivation of the inactive X chromosome (Maherali et al., 2007). This process involves the loss of Xist RNA and of DNA methylation, late during reprogramming when the pluripotency genes have been activated (Pasque et al., 2014). In humans, the inactive X chromosome is not reactivated, and iPSCs retain the same Xi that was present in the somatic cells from which they were derived (Tchieu et al., 2010).

In iPSCs, long-range inter-TAD chromatin contacts occur among the genes that encode the inducing transcription factors Oct4, Sox2, c-Myc and Klf4 and the many genes responsible for pluripotency that they activate (Apostolou et al., 2013; Denholtz et al., 2013; de Wit et al., 2013). These contacts are different from those found in the differentiated cells that gave rise to the iPSCs and similar to those present in pluripotent embryonic stem cells (ESCs).

As mentioned above, reprogramming is usually successful in a small percentage of treated cells. Many of the other cells exhibit some characteristics that suggest that different types of pluripotent cells may exist. In addition, it appears that treatment of somatic cells with the traditional cocktail of reprogramming transcription factors may lead to different end points, depending on the culture conditions (Han et al., 2011). Using a resource established as a collaboration by a group of laboratories and consisting of a set of cell samples obtained at different times during the course of reprogramming mouse embryonic fibroblasts, two distinct types of pluripotent cells can be established. Constant high levels of the traditional reprogramming factors lead to a type of iPSCs, called F-class to reflect the fuzzy appearance of the colonies that they form; lower levels of reprogramming factors lead to distinct types of compact colony-forming cells. F-class cells require the constant presence of the reprogramming factors and reproduce much faster than ESCs; they can be converted to cells that are similar to ESCs by inhibiting the activity of histone deacetylases. In comparison to ESCs and the somatic cells from which they were derived, F-class cells undergo a global increase in gene activation (Tonge et al., 2014).

Epigenetic changes during reprogramming

Histone modifications

ESCs and iPSCs have less condensed and more transcriptionally active chromatin than differentiated cells that exhibit highly condensed heterochromatin domains. Therefore, it is not surprising that the genomes of pluripotent cells are enriched in histone marks associated with active chromatin, such as H3K4me3, H3K36me2, H3K9ac and H3K27ac, and have lower levels of H3K27me3 and H3K9me3 than differentiated cells. The change in levels of H3K9me3 and its associated heterochromatin protein HP1 occurs early during the reprogramming process (Mattout et al., 2011).

A significant number of genes in ESCs reside in regions marked with methylated H3K27, within which there are smaller regions of methylated H3K4. These bivalent regions overlap the transcription start sites of many genes that encode transcription factors and developmental genes (Bernstein et al., 2006). The occurrence of bivalent promoters increases gradually after the first wave of transcriptional activity during the reprogramming process, reflecting the switch of differentiation-specific active genes to an inactive, but poised, status (Polo et al., 2012).

DNA methylation

The levels of DNA methyl transferases and demethylases are increased during the stabilization phase, reflecting the changes in the pattern of DNA methylation that occur mostly late during reprogramming. As expected, the CpG islands surrounding the endogenous pluripotency genes that are activated at this time become demethylated; the reverse is true of genes that are normally demethylated in somatic cells (Polo et al., 2012).

Therapeutic applications of iPSCs

The application of iPSCs in the clinical setting offers new and exciting potential approaches for the prevention, diagnosis and treatment of disease. One direction that has been taken in the use of iPSCs is the establishment of master cell banks that can provide the basic material for studies on differentiation and for preclinical studies (Baghbaderani et al., 2015). The second direction is the development of **autologous** cell therapy techniques whereby a patient's own cells are reprogrammed.

A number of human diseases have been reproduced in laboratory animal models, an approach that has yielded valuable information on the etiology, symptoms and potential therapies. Nevertheless, the extrapolation of such information to humans is subjected to the fact that fundamental biological differences exist between distant species. The use of disease models in the form of iPSCs derived from patients circumvents the interspecies problem and potentially allows the study of disease progression during the process of tissue differentiation. The first example of this approach was the demonstration that iPSCs derived from a patient with spinal muscular atrophy differentiated into neurons typically defective (Ebert et al., 2009). Similarly, iPSCs derived from a patient with Rett syndrome that is caused by mutations in the gene encoding a methylated CpG dinucleotide binding protein MeCP2 (see Chapter 8) differentiate into abnormal neurons; early drug treatment during differentiation alleviated the neuronal defects, suggesting that successful therapies may be applied before disease onset (Marchetto et al., 2010).

Another potential use of iPSCs is in regenerative medicine. One aspect of this area of medicine is the delivery of specific cells to diseased regions of the body where they will restore tissue or organ function. A major advantage of using iPSCs generated from a patient's own somatic cells is that it would circumvent the potential immunological incompatibility that could result from the use of **heterologous** cells, i.e. cells derived from an unrelated individual. A major problem is that the iPSCs derived from a patient may carry the genetic mutation that is the cause of disease. This problem can be addressed with one of the available techniques for genome editing such as the CRISPR/Cas9 system. This system consists of targeting a double-stranded break at a precise location in the genome; the break can be repaired by the non-homologous end-joining pathway (NHEJ) that often results in insertions or deletions, thereby disrupting the site; it can also be repaired by homologous recombination (HR) that, if a homologous template is provided, can generate precise mutations or restore the normal DNA sequence. Examples of the numerous instances of the successful use of this technique include the correction of the mutation in the adult β-globin gene in iPSCs from patients with sickle cell disease (Huang et al., 2015) or the correction of the dystrophin gene in iPSCs from patients with Duchenne muscular dystrophy (Li et al., 2015). Although there are no particular reasons to exclude, a priori, any pathological condition from the use of iPSC-mediated regenerative treatment, areas that currently offer the most promise are spinal cord injury (Priest et al., 2015), retinal blindness (Schwartz et al., 2016), heart failure (Burridge et al., 2014), diabetes (Pagliuca et al., 2014; Rezania et al., 2014) and Parkinson's disease (Kriks et al., 2011).

A number of *caveats* should be considered when planning or implementing the use of iPSCs for clinical purposes. Long-term culturing often results in a variety of genomic aberrations such as deletions and duplications, often involving tumor suppressors or oncogenic genes (Laurent et al., 2011). In human iPSCs that retain the inactive X chromosome of the reprogrammed somatic cells, many of the genes on the Xi become reactivated over time (Mekhoubad et al., 2012). Similarly, iPSCs undergo changes in the DNA methylation status of imprinted loci, with concomitant loss of allele-specific gene expression (Nazor et al., 2012).

iPSCs generated from somatic cells of old individuals exhibit a loss of aging characteristics such as telomere shortening, specific gene expression profile and chromatin epigenetic marks; furthermore, when induced to differentiate, iPSCs maintain their rejuvenated phenotype (Lapasset et al., 2011). This presents a significant problem in attempting to establish models for diseases with an age-related onset such as Alzheimer's disease. The solution is provided by the ability to induce somatic cells (such as fibroblasts) to re-differentiate directly into a number of cell types, including neurons (Pang et al., 2011; Vierbuchen et al., 2010). Fibroblasts from old individuals give rise to neurons that exhibit all of the aging signatures of the parent cells (Mertens et al., 2015).

Chapter summary

The first demonstration of nuclear reprogramming was provided by John Gurdon who obtained swimming tadpoles following the transplantation of tadpole gut nuclei into enucleated frog eggs. This pioneering work eventually culminated in nuclear transfers in mammals, leading to the successful cloning of different species.

A corollary of this work was the induction of somatic cells to return to the pluripotent state. In the initial experiments, four core transcription factors known to maintain the pluripotent state in ESCs—Oct4, Sox2, Klf4 and c-Myc—were used to induce pluripotency in adult-derived fibroblasts. These cells are referred to as **induced pluripotent stem cells** (iPSCs). Since those early experiments, reprogramming has also been achieved by the induction of specific miRNAs that are abundant in ESCs but disappear after differentiation. Currently, favored methodologies involve the use of pluripotency transcription factors whose effect is enhanced by miRNAs.

Reprogramming consists of an **initial phase** that is marked by the loss of somatic cell-specific markers. Following a **maturation phase**, during which a specific set of endogenous genes that include core pluripotency factors are activated, a small number of cells enter a **stabilization phase**. iPSCs, like embryonic stem cells, have less condensed and more transcriptionally active chromatin than differentiated cells. The number of genes with bivalent promoter marks increases during reprogramming, reflecting the switch of differentiation-specific active genes to an inactive, but poised, status. The levels of DNA methyl transferases and demethylases are increased during the stabilization phase, underlying the changes in the pattern of DNA methylation that occur mostly late during reprogramming.

The potential therapeutic applications of iPSCs are varied and exciting. For example, a patient's own cells could be reprogrammed, thereby avoiding the problem of rejection following the use of these cells to restore tissue or organ function. iPSCs derived from individuals at risk of developing late-onset neurological diseases could be differentiated in culture to predict the future occurrence of the disease. Numerous *caveats* should be considered when planning the use of iPSCs for clinical purposes such as the fact that long-term culturing often results in genomic mutations that may, by chance, involve tumor suppressors or oncogenes.

References

Ang, Y. S., Tsai, S. Y., Lee, D. F., Monk, J., Su, J., Ratnakumar, K., Ding, J., Ge, Y., Darr, H., Chang, B., Wang, J., Rendl, M., Bernstein, E., Schaniel, C. & Lemischka, I. R. 2011. Wdr5 mediates self-renewal and reprogramming via the embryonic stem cell core transcriptional network. *Cell,* 145, 183–97.

Anokye-Danso, F., Trivedi, C. M., Juhr, D., Gupta, M., Cui, Z., Tian, Y., Zhang, Y., Yang, W., Gruber, P. J., Epstein, J. A. & Morrisey, E. E. 2011. Highly efficient miRNA-mediated reprogramming of mouse and human somatic cells to pluripotency. *Cell Stem Cell,* 8, 376–88.

Apostolou, E. & Hochedlinger, K. 2013. Chromatin dynamics during cellular reprogramming. *Nature,* 502, 462–71.

Baghbaderani, B. A., Tian, X., Neo, B. H., Burkall, A., Dimezzo, T., Sierra, G., Zeng, X., Warren, K., Kovarcik, D. P., Fellner, T. & Rao, M. S. 2015. cGMP-manufactured human induced pluripotent stem cells are available for pre-clinical and clinical applications. *Stem Cell Reports,* 5, 647–59.

Bernstein, B. E., Mikkelsen, T. S., Xie, X., Kamal, M., Huebert, D. J., Cuff, J., Fry, B., Meissner, A., Wernig, M., Plath, K., Jaenisch, R., Wagschal, A., Feil, R., Schreiber, S. L. & Lander, E. S. 2006. A bivalent chromatin structure marks key developmental genes in embryonic stem cells. *Cell,* 125, 315–26.

Brambrink, T., Foreman, R., Welstead, G. G., Lengner, C. J., Wernig, M., Suh, H. & Jaenisch, R. 2008. Sequential expression of pluripotency markers during direct reprogramming of mouse somatic cells. *Cell Stem Cell,* 2, 151–9.

Briggs, R. & King, T. J. 1952. Transplantation of living nuclei from blastula cells into enucleated frogs' eggs. *Proc Natl Acad Sci U S A,* 38, 455–63.

Burridge, P. W., Matsa, E., Shukla, P., Lin, Z. C., Churko, J. M., Ebert, A. D., Lan, F., Diecke, S., Huber, B., Mordwinkin, N. M., Plews, J. R., Abilez, O. J., Cui, B., Gold, J. D. & Wu, J. C. 2014. Chemically defined generation of human cardiomyocytes. *Nat Methods,* 11, 855–60.

Campbell, K. H., Mcwhir, J., Ritchie, W. A. & Wilmut, I. 1996. Sheep cloned by nuclear transfer from a cultured cell line. *Nature,* 380, 64–6.

Davis, R. L., Weintraub, H. & Lassar, A. B. 1987. Expression of a single transfected cDNA converts fibroblasts to myoblasts. *Cell,* 51, 987–1000.

Denholtz, M., Bonora, G., Chronis, C., Splinter, E., De Laat, W., Ernst, J., Pellegrini, M. & Plath, K. 2013. Long-range chromatin contacts in embryonic stem cells reveal a role for pluripotency fsctors and polycomb proteins in genome organization. *Cell Stem Cell,* 13, 602–16.

De Wit, E., Bouwman, B. A., Zhu, Y., Klous, P., Splinter, E., Verstegen, M. J., Krijger, P. H., Festuccia, N., Nora, E. P., Welling, M., Heard, E., Geijsen, N., Poot, R. A., Cambers, I. & De Laat, W. 2013. The pluripotent genome in three dimensions is shaped around pluripotency fctors. *Nature,* 501, 227–31.

Ding, X., Wang, X., Sontag, S., Qin, J., Wanek, P., Lin, Q. & Zenke, M. 2014. The polycomb protein Ezh2 impacts on induced pluripotent stem cell generation. *Stem Cells Dev,* 23, 931–40.

Ebert, A. D., Yu, J., Rose, F. F.,JR., Mattis, V. B., Lorson, C. L., Thomson, J. A. & Svendsen, C. N. 2009. Induced pluripotent stem cells from a spinal muscular atrophy patient. *Nature,* 457, 277–80.

Eggan, K., Akutsu, H., Hochedlinger, K., Rideout, W., 3RD, Yanagimachi, R. & Jaenisch, R. 2000. X-Chromosome inactivation in cloned mouse embryos. *Science,* 290, 1578–81.

Golipour, A., David, L., Liu, Y., Jayakumaran, G., Hirsch, C. L., Trcka, D. & Wrana, J. L. 2012. A late transition in somatic cell reprogramming requires regulators distinct from the pluripotency network. *Cell Stem Cell,* 11, 769–82.

Gurdon, J. B. 1960. The developmental capacity of nuclei taken from differentiating endoderm cells of *Xenopus laevis. J Embryol Exp Morphol,* 8, 505–26.

Han, D. W., Greber, B., Wu, G., Tapia, N., Arauzo-Bravo, M. J., Ko, K., Bernemann, C., Stehling, M. & Scholer, H. R. 2011. Direct reprogramming of fibroblasts into epiblast stem cells. *Nat Cell Biol,* 13, 66–71.

Houbaviy, H. B., Murray, M. F. & Sharp, P. A. 2003. Embryonic stem cell-specific microRNAs. *Dev Cell,* 5, 351–8.

Howden, S. E., Maufort, J. P., Duffin, B. M., Elefanty, A. G., Stanley, E. G. & Thomson, J. A. 2015. Simultaneous reprogramming and gene correction of patient fibroblasts. *Stem Cell Reports,* 5, 1109–18.

Huang, X., Wang, Y., Yan, W., Smith, C., Ye, Z., Wang, J., Gao, Y., Mendelsohn, L. & Cheng, L. 2015. Production of gene-corrected adult beta globin protein in human erythrocytes differentiated from patient iPSCs after genome editing of the sickle point mutation. *Stem Cells,* 33, 1470–9.

Huangfu, D., Maehr, R., Guo, W., Eijkelenboom, A., Snitow, M., Chen, A. E. & Melton, D. A. 2008. Induction of pluripotent stem cells by defined factors is greatly improved by small-molecule compounds. *Nat Biotechnol,* 26, 795–7.

Inoue, K., Kohda, T., Lee, J., Ogonuki, N., Mochida, K., Noguchi, Y., Tanemura, K., Kaneko-Ishino, T., Ishino, F. & Ogura, A. 2002. Faithful expression of imprinted genes in cloned mice. *Science,* 295, 297.

Judson, R. L., Babiarz, J. E., Venere, M. & Blelloch, R. 2009. Embryonic stem cell-specific microRNAs promote induced pluripotency. *Nat Biotechnol,* 27, 459–61.

King, T. J. & Briggs, R. 1955. Changes in the nuclei of differentiating gastrula cells, as demonstrated by nuclear transplantation. *Proc Natl Acad Sci U S A,* 41, 321–5.

Kriks, S., Shim, J. W., Piao, J., Ganat, Y. M., Wakeman, D. R., Xie, Z., Carrillo-Reid, L., Auyeung, G., Antonacci, C., Buch, A., Yang, L., Beal, M. F., Surmeier, D. J., Kordower, J. H., Tabar, V. & Studer, L. 2011. Dopamine neurons derived from human ES cells efficiently engraft in animal models of Parkinson's disease. *Nature,* 480, 547–51.

Kulessa, H., Frampton, J. & Graf, T. 1995. GATA-1 reprograms avian myelomonocytic cell lines into eosinophils, thromboblasts, and erythroblasts. *Genes Dev,* 9, 1250–62.

Lapasset, L., Milhavet, O., Prieur, A., Besnard, E., Babled, A., Ait-Hamou, N., Leschik, J., Pellestor, F., Ramirez, J. M., De Vos, J., Lehmann, S. & Lemaitre, J. M. 2011. Rejuvenating senescent and centenarian human cells by reprogramming through the pluripotent state. *Genes Dev,* 25, 2248–53.

Laurent, L. C., Ulitsky, I., Slavin, I., Tran, H., Schork, A., Morey, R., Lynch, C., Harness, J. V., Lee, S., Barrero, M. J., Ku, S., Martynova, M., Semechkin, R., Galat, V., Gottesfeld, J., Izpisua Belmonte, J. C., Murry, C., Keirstead, H. S., Park, H. S., Schmidt, U., Laslett, A. L., Muller, F. J., Nievergelt, C. M., Shamir, R. & Loring, J. F. 2011. Dynamic changes in the copy number of pluripotency and cell proliferation genes in human ESCs and iPSCs during reprogramming and time in culture. *Cell Stem Cell,* 8, 106–18.

Lee, D. S., Shin, J. Y., Tonge, P. D., Puri, M. C., Lee, S., Park, H., Lee, W. C., Hussein, S. M., Bleazard, T., Yun, J. Y., Kim, J., Li, M., Cloonan, N., Wood, D., Clancy, J. L., Mosbergen, R., Yi, J. H., Yang, K. S., Kim, H., Rhee, H., Wells, C. A., Preiss, T., Grimmond, S. M., Rogers, I. M., Nagy, A. & Seo, J. S. 2014. An epigenomic roadmap to induced pluripotency reveals DNA methylation as a reprogramming modulator. *Nat Commun,* 5, 5619.

Li, H. L., Fujimoto, N., Sasakawa, N., Shirai, S., Ohkame, T., Sakuma, T., Tanaka, M., Amano, N., Watanabe, A., Sakurai, H., Yamamoto, T., Yamanaka, S. & Hotta, A. 2015. Precise correction of the dystrophin gene in duchenne muscular dystrophy patient induced pluripotent stem cells by TALEN and CRISPR-Cas9. *Stem Cell Reports,* 4, 143–54.

Liang, G., Taranova, O., Xia, K. & Zhang, Y. 2010. Butyrate promotes induced pluripotent stem cell generation. *J Biol Chem,* 285, 25516–21.

Lin, S. L., Chang, D. C., Chang-Lin, S., Lin, C. H., Wu, D. T., Chen, D. T. & Ying, S. Y. 2008. Mir-302 reprograms human skin cancer cells into a pluripotent ES-cell-like state. *RNA,* 14, 2115–24.

Liu, P., Chen, M., Liu, Y., Qi, L. S. & Ding, S. 2018. CRISPR-based chromatin remodeling of the endogenous Oct4 or

Sox2 locus enables reprogramming to pluripotency. *Cell Stem Cell,* 22, 252–61 e4.

Maherali, N., Sridharan, R., Xie, W., Utikal, J., Eminli, S., Arnold, K., Stadtfeld, M., Yachechko, R., Tchieu, J., Jaenisch, R., Plath, K. & Hochedlinger, K. 2007. Directly reprogrammed fibroblasts show global epigenetic remodeling and widespread tissue contribution. *Cell Stem Cell,* 1, 55–70.

Mansour, A. A., Gafni, O., Weinberger, L., Zviran, A., Ayyash, M., Rais, Y., Krupalnik, V., Zerbib, M., Amann-Zalcenstein, D., Maza, I., Geula, S., Viukov, S., Holtzman, L., Pribluda, A., Canaani, E., Horn-Saban, S., Amit, I., Novershtern, N. & Hanna, J. H. 2012. The H3K27 demethylase Utx regulates somatic and germ cell epigenetic reprogramming. *Nature,* 488, 409–13.

Marchetto, M. C., Carromeu, C., Acab, A., Yu, D., Yeow, G. W., Mu, Y., Chen, G., Gage, F. H. & Muotri, A. R. 2010. A model for neural development and treatment of Rett syndrome using human induced pluripotent stem cells. *Cell,* 143, 527–39.

Mattout, A., Biran, A. & Meshorer, E. 2011. Global epigenetic changes during somatic cell reprogramming to iPS cells. *J Mol Cell Biol,* 3, 341–50.

Mekhoubad, S., Bock, C., De Boer, A. S., Kiskinis, E., Meissner, A. & Eggan, K. 2012. Erosion of dosage compensation impacts human iPSC disease modeling. *Cell Stem Cell,* 10, 595–609.

Menon, S., Shailendra, S., Renda, A., Longaker, M. & Quarto, N. 2016. An overview of direct somatic reprogramming: the ins and outs of iPSCs. *Int J Mol Sci,* 17, 141.

Mertens, J., Paquola, A. C. M., Ku, M., Hatch, E., Bohnke, L., Ladjevardi, S., Mcgrath, S., Campbell, B., Lee, H., Herdy, J. R., Goncalves, J. T., Toda, T., Kim, Y., Winkler, J., Yao, J., Hetzer, M. W. & Gage, F. H. 2015. Directly reprogrammed human neurons retain aging-associated transcriptomic signatures and reveal age-related nucleocytoplasmic defects. *Cell Stem Cell,* 17, 705–18.

Miyoshi, N., Ishii, H., Nagano, H., Haraguchi, N., Dewi, D. L., Kano, Y., Nishikawa, S., Tanemura, M., Mimori, K., Tanaka, F., Saito, T., Nishimura, J., Takemasa, I., Mizushima, T., Ikeda, M., Yamamoto, H., Sekimoto, M., Doki, Y. & Mori, M. 2011. Reprogramming of mouse and human cells to pluripotency using mature microRNAs. *Cell Stem Cell,* 8, 633–8.

Nazor, K. L., Altun, G., Lynch, C., Tran, H., Harness, J. V., Slavin, I., Garitaonandia, I., Muller, F. J., Wang, Y. C., Boscolo, F. S., Fakunle, E., Dumevska, B., Lee, S., Park, H. S., Olee, T., D'lima, D. D., Semechkin, R., Parast, M. M., Galat, V., Laslett, A. L., Schmidt, U., Keirstead, H. S., Loring, J. F. & Laurent, L. C. 2012. Recurrent variations in DNA methylation in human pluripotent stem cells and their differentiated derivatives. *Cell Stem Cell,* 10, 620–34.

Nie, Z., Hu, G., Wei, G., Cui, K., Yamane, A., Resch, W., Wang, R., Green, D. R., Tessarollo, L., Casellas, R., Zhao, K.

& Levens, D. 2012. c-Myc is a universal amplifier of expressed genes in lymphocytes and embryonic stem cells. *Cell*, 151, 68–79.

Okita, K., Ichisaka, T. & Yamanaka, S. 2007. Generation of germline-competent induced pluripotent stem cells. *Nature*, 448, 313–17.

Pagliuca, F. W., Millman, J. R., Gurtler, M., Segel, M., Van Dervort, A., Ryu, J. H., Peterson, Q. P., Greiner, D. & Melton, D. A. 2014. Generation of functional human pancreatic beta cells *in vitro*. *Cell*, 159, 428–39.

Pang, Z. P., Yang, N., Vierbuchen, T., Ostermeier, A., Fuentes, D. R., Yang, T. Q., Citri, A., Sebastiano, V., Marro, S., Sudhof, T. C. & Wernig, M. 2011. Induction of human neuronal cells by defined transcription factors. *Nature*, 476, 220–3.

Pasque, V., Tchieu, J., Karnik, R., Uyeda, M., Sadhu Dimashkie, A., Case, D., Papp, B., Bonora, G., Patel, S., Ho, R., Schmidt, R., Mckee, R., Sado, T., Tada, T., Meissner, A. & Plath, K. 2014. X chromosome reactivation dynamics reveal stages of reprogramming to pluripotency. *Cell*, 159, 1681–97.

Polo, J. M., Anderssen, E., Walsh, R. M., Schwarz, B. A., Nefzger, C. M., Lim, S. M., Borkent, M., Apostolou, E., Alaei, S., Cloutier, J., Bar-Nur, O., Cheloufi, S., Stadtfeld, M., Figueroa, M. E., Robinton, D., Natesan, S., Melnick, A., Zhu, J., Ramaswamy, S. & Hochedlinger, K. 2012. A molecular roadmap of reprogramming somatic cells into iPS cells. *Cell*, 151, 1617–32.

Priest, C. A., Manley, N. C., Denham, J., Wirth, E. D., 3RD & Lebkowski, J. S. 2015. Preclinical safety of human embryonic stem cell-derived oligodendrocyte progenitors supporting clinical trials in spinal cord injury. *Regen Med*, 10, 939–58.

Rezania, A., Bruin, J. E., Arora, P., Rubin, A., Batushansky, I., Asadi, A., O'dwyer, S., Quiskamp, N., Mojibian, M., Albrecht, T., Yang, Y. H., Johnson, J. D. & Kieffer, T. J. 2014. Reversal of diabetes with insulin-producing cells derived *in vitro* from human pluripotent stem cells. *Nat Biotechnol*, 32, 1121–33.

Schwartz, S. D., Tan, G., Hosseini, H. & Nagiel, A. 2016. Subretinal transplantation of embryonic stem cell-derived retinal pigment epithelium for the treatment of macular degeneration: an assessment at 4 years. *Invest Ophthalmol Vis Sci*, 57, ORSFc1-9.

Shinagawa, T., Takagi, T., Tsukamoto, D., Tomaru, C., Huynh, L. M., Sivaraman, P., Kumarevel, T., Inoue, K., Nakato, R., Katou, Y., Sado, T., Takahashi, S., Ogura, A., Shirahige, K. & Ishii, S. 2014. Histone variants enriched in

oocytes enhance reprogramming to induced pluripotent stem cells. *Cell Stem Cell*, 14, 217–27.

Soufi, A., Donahue, G. & Zaret, K. S. 2012. Facilitators and impediments of the pluripotency reprogramming factors' initial engagement with the genome. *Cell*, 151, 994–1004.

Sridharan, R., Tchieu, J., Mason, M. J., Yachechko, R., Kuoy, E., Horvath, S., Zhou, Q. & Plath, K. 2009. Role of the murine reprogramming factors in the induction of pluripotency. *Cell*, 136, 364–77.

Stadtfeld, M., Maherali, N., Breault, D. T. & Hochedlinger, K. 2008. Defining molecular cornerstones during fibroblast to iPS cell reprogramming in mouse. *Cell Stem Cell*, 2, 230–40.

Suh, M. R., Lee, Y., Kim, J. Y., Kim, S. K., Moon, S. H., Lee, J. Y., Cha, K. Y., Chung, H. M., Yoon, H. S., Moon, S. Y., Kim, V. N. & Kim, K. S. 2004. Human embryonic stem cells express a unique set of microRNAs. *Dev Biol*, 270, 488–98.

Takahashi, K., Tanabe, K., Ohnuki, M., Narita, M., Ichisaka, T., Tomoda, K. & Yamanaka, S. 2007. Induction of pluripotent stem cells from adult human fibroblasts by defined factors. *Cell*, 131, 861–72.

Takahashi, K. & Yamanaka, S. 2006. Induction of pluripotent stem cells from mouse embryonic and adult fibroblast cultures by defined factors. *Cell*, 126, 663–76.

Tchieu, J., Kuoy, E., Chin, M. H., Trinh, H., Patterson, M., Sherman, S. P., Aimiuwu, O., Lindgren, A., Hakimian, S., Zack, J. A., Clark, A. T., Pyle, A. D., Lowry, W. E. & Plath, K. 2010. Female human iPSCs retain an inactive X chromosome. *Cell Stem Cell*, 7, 329–42.

Tonge, P. D., Corso, A. J., Monetti, C., Hussein, S. M., Puri, M. C., Michael, I. P., Li, M., Lee, D. S., Mar, J. C., Cloonan, N., Wood, D. L., Gauthier, M. E., Korn, O., Clancy, J. L., Preiss, T., Grimmond, S. M., Shin, J. Y., Seo, J. S., Wells, C. A., Rogers, I. M. & Nagy, A. 2014. Divergent reprogramming routes lead to alternative stem-cell states. *Nature*, 516, 192–7.

Vierbuchen, T., Ostermeier, A., Pang, Z. P., Kokubu, Y., Sudhof, T. C. & Wernig, M. 2010. Direct conversion of fibroblasts to functional neurons by defined factors. *Nature*, 463, 1035–41.

Wilmut, I., Schnieke, A. E., Mcwhir, J., Kind, A. J. & Campbell, K. H. 1997. Viable offspring derived from fetal and adult mammalian cells. *Nature*, 385, 810–13.

Yu, J., Vodyanik, M. A., Smuga-Otto, K., Antosiewicz-Bourget, J., Frane, J. L., Tian, S., Nie, J., Jonsdottir, G. A., Ruotti, V., Stewart, R., Slukvin, Ii & Thomson, J. A. 2007. Induced pluripotent stem cell lines derived from human somatic cells. *Science*, 318, 1917–20.

CHAPTER 19

Transgenerational inheritance of epigenetic traits

It may not be an overstatement to say that most, if not all, salient environmental stimuli in early life will have an impact on the future behavior and health of an individual. This contention forms the basis of the developmental origins of health and disease (DOHaD) hypothesis that can trace its inception to the initial observations of David Barker and his colleagues (Barker et al., 1989). Differential gene activation and the associated epigenetic chromatin modifications that occur during embryonic development and in adult life arise through the influence of a wide range of factors: gradients of maternal morphogens, intercellular signals and extra-embryonic environmental factors. Exposure to environmental influences during embryogenesis or during early postnatal life can cause epigenetic modifications that alter gene expression during critical periods of development. Very often, these epigenetic changes do not manifest themselves as abnormal physiologies until adulthood when they can promote adult-onset diseases, ranging from tumors and reproductive problems to metabolic defects.

In certain instances, these effects are transmitted to descendants who have not been exposed to the same environmental factors as their parents. The persistence of an epigenetic trait, induced by exposure to some environmental factors, through subsequent generations that have not been exposed to these factors is referred to as transgenerational epigenetic inheritance.

Adult effects of embryonic exposure to particular environmental factors

The nutritional status during pregnancy or during early postnatal development can affect adult physiology.

The induced epigenetic changes that regulate gene expression during these critical developmental periods may initiate a series of effects on cellular differentiation that do not manifest as abnormal physiologies and disease until adulthood. Some of the more common end products are obesity (Ost et al., 2014; Ravelli et al., 1999), diabetes (Dörner and Plagemann, 1994; Wei et al., 2014) and cardiovascular disease, including hypertension (Benediktsson et al., 1993; Bertram et al., 2008; Langley and Jackson, 1994). In addition, early-life exposure to malnutrition has a demonstrable effect on adult cognition (Lucas et al., 1998) and on the susceptibility to develop psychopathologies (St Clair et al., 2005).

At the present time, whether any of these environmentally induced metabolic diseases or neurological dysfunctions are transmitted to subsequent generations is experimentally difficult to ascertain in humans. Such a demonstration requires data on the nutritional status during pregnancy of the great-grandmothers or of the grandfathers of adult individuals, as well as nutritional and clinical information on their grandparents and parents (Box 19.1). Still, some studies have relied on historical records of specific populations undergoing periods of famine or of food abundance to suggest transgenerational inheritance—the inheritance of phenotypes by subsequent generations, even after the initiating event is no longer present. In an example of such a study, over-nutrition of paternal grandfathers led to an increase in diabetes-related mortality in their grandsons, while malnutrition of the father, and perhaps also of the paternal grandmothers, diminished their sons' or grandsons' risk of cardiovascular disease (Kaati et al., 2002).

Epigenetics, Nuclear Organization and Gene Function: with implications of epigenetic regulation and genetic architecture for human development and health. John C. Lucchesi, Oxford University Press (2019). © John C. Lucchesi 2019. DOI: 10.1093/oso/9780198831204.001.0001

Box 19.1 The parameters of transgenerational epigenetic inheritance

Transgenerational epigenetic inheritance is the persistence of an epigenetic trait, induced by exposure to some environmental factors, through subsequent generations that have not been exposed to these factors. To be considered a case of transgenerational inheritance, the epigenetic phenotype, induced in a gravid female parent (F0) must manifest itself in the F3 progeny and subsequent generations (Fig. 19B.1). The reason is that the F1 progeny was exposed to the environmental conditions while *in utero*, and the F2 progeny is produced by the F1's developing gametes that were also exposed to the conditions. To qualify for transgenerational inheritance, transmission from an F0 exposed male requires that the epigenetic trait manifests itself in the F2 progeny, since it is only the F1 that were exposed to the environmental conditions in the form of gametes in the F0 male.

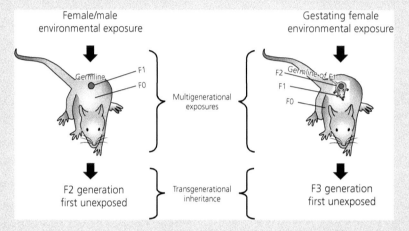

Fig. 19B.1 Inheritance of an epigenetic phenotype induced by exposing a male or a female (left) or a gravid female (right) to an environmental factor.

(From Skinner, 2014.)

Epigenetic transgenerational transmission

The transgenerational inheritance of epigenetic modifications represents a fascinating, albeit disputed, aspect of modern genetics. An important parameter of transgenerational inheritance is that, in order to be transmitted, the epigenetic changes incurred as a result of environmental challenges, in either the soma or the germline, must be transmitted from the soma to the germline in subsequent generations. In interpreting data and observations that test the occurrence of this type of inheritance, some of the caveats are that the transmission of certain epigenetic traits is not as robust as that of purely genetic traits and that it is difficult to completely eliminate single nucleotide polymorphisms or the movement of transposable elements as the cause of the new phenotype (Heard and Martienssen, 2014).

Operational concepts sometimes used in discussing this subject are the occurrence of "epimutations" that consist of environmentally induced changes in the normal epigenetic modifications present in some functioning area of the genome, and "epialleles" representing the particular set of modifications that result from epimutations.

Physiological stress and chronic metabolic diseases

In contrast to the relative paucity of clear cases of transgenerational transmission of the effects of physiological stress in humans, a growing body of experimental data, obtained by using different model organisms, indicates the existence of, and suggests the possible universality of, this phenomenon. Some examples that illustrate the effects of diets with too much fat or sugar on the metabolism of later generations are discussed in the following paragraphs.

The F3 descendants of malnourished female rats, although raised on a standard diet, presented with an abnormal glucose metabolism (Benyshek et al., 2006). In an analogous study, *Drosophila* virgin females fed on a high-sucrose diet displayed a marked increase in the percentage of body fat that persisted in the F2 generation (Buescher et al., 2013). In *Caenorhabditis*, starving individuals during the first larval stage (L1) led to increased longevity in the F3 generation (Rechavi et al., 2014). This effect was mediated by the induction of small RNA molecules in the starved L1 individuals. These small RNAs, some of which target genes involved in nutrition, were transmitted to progeny for at least three generations.

In an effort to ensure exclusive inheritance through the gametes, sperm and oocytes isolated from mice fed on a high-fat or a control diet were used in different combinations for *in vitro* fertilization. The resulting F1 females and males from fertilizations involving gametes from parents fed on a high-fat diet developed obesity and diabetes-like phenotypes, with some sex-specific differences (Huypens et al., 2016).

One of the earliest cases of epigenetic transmission of an obesity phenotype is associated with a coat color mutation. In mice, the *Agouti* gene is expressed for a short period of time during the early stages of hair growth and causes a switch from black or dark brown pigmentation to yellow. The result is dark hair with a yellow tip that gives the mouse its typical coat color. The *Agouti* gene product is a signaling protein whose constitutive expression causes a number of phenotypes, including obesity, tumor susceptibility and premature infertility (Duhl et al., 1994). Some mutations in the *Agouti* gene are caused by the 5′ insertion of an endogenous retrotransposon (IAP) and are characterized by a mottled phenotype, as the IAP element is transcribed in some clones of cells of the developing mice but is inactive in other clones. Different members of a litter exhibit a range of phenotypes, depending on the overall expression of the *Agouti* gene, which, in turn, depends on the epigenetic status of the retrotransposons. Significantly, females characterized by a particular level of *Agouti* expression produce progeny with a similar phenotype, suggesting the inheritance of a particular level of epigenetic modifications affecting the transcription of the retrotransposons (Morgan et al., 1999).

Another physiological stress that can result in a transgenerational phenotype is heat shock. Organisms that are exposed to a sudden temperature increase exhibit specific genomic alterations referred to as the heat-shock response. First discovered in *Drosophila* (Ritossa, 1962), this response includes a general abrogation of transcription and translation, accompanied by the induction of a specific group of proteins, many of which are molecular chaperones or components of the proteolytic pathways. One of the proteins induced by heat shock in *Caenorhabditis* is present for up to 14 generations that have been raised at a normal temperature. The presence of the protein is correlated with a decrease in the silencing H3K9me3 modified histone in both the eggs and sperm, although the heat-shock protein is not expressed in either germline (Klosin et al., 2017).

Endocrine disruptors

Instances of exposure to environmental toxicants have bolstered the case for the existence of the transgenerational inheritance of some epigenetic traits. As discussed in the following paragraphs, these chemicals mediate their phenotypes by affecting one or more of the traditional epigenetic regulatory pathways: DNA or histone covalent modifications, non-coding RNAs or the enzymatic pathways responsible for their biogenesis. Although the epigenetic alterations of chromatin correlated with exposure to these endocrine disruptors can often be precisely documented, the biochemical connections between exposure to a chemical and epigenetic effect are often poorly understood.

In an early study, pregnant female mice exposed to benzo(a)pyrene—an oncogenic chemical found in coal tar—gave rise to five successive generations of offspring with a high frequency of tumors (Turusov et al., 1990). More recent studies, considered seminal in validating the occurrence of transgenerational inheritance, involved exposure to vinclozolin, a fungicide, and methoxychlor, a pesticide that replaced DDT on the market. These compounds, with estrogenic and anti-androgenic effects, respectively, cause a severe degeneration of spermatogenic cells in the gonads of exposed male rats, leading to infertility (Fig. 19.1). F0 gestating mothers were exposed, and their male progeny and subsequent descendants, assayed until the F4 generation, exhibited various aspects of abnormal spermatogenesis (Anway et al., 2005), as well as an increased frequency of cancer, kidney disease and immune cell defects (Anway et al., 2006). The effect of vinclozolin on spermatogenesis in rats was later disputed (Schneider et al., 2013), but its occurrence was confirmed in mice (Brieno-Enriquez et al., 2015).

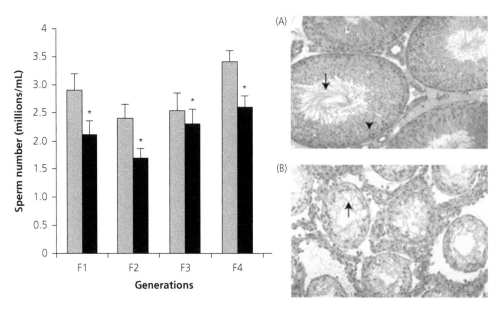

Fig. 19.1 Left: transgenerational inheritance of vinclozolin treatment of F0 gestating female rats, measured as a function of sperm numbers in the epididymis. Right: testis histology of an F3-generation control male (A) or an F3-generation male descendant from an F0 vinclozolin-treated female (B). (Modified from Anway et al., 2005.)

Additional examples of the transgenerational inheritance of endocrine disruptors' effects are exposure to bisphenol-A (BPA), which can act as an estrogen substitute (Wolstenholme et al., 2013), or exposure to phthalates, which are anti-androgenic (Doyle et al., 2013); these compounds are used in the manufacture of certain plastic products; they can leach out and contaminate the food with which they may be in contact. Exposure to phthalates in F0 mice led to F3 and F4 males with testes exhibiting a significant decrease in sperm count and stem cell colonization, as well as obvious germ cell disorganization reminiscent of the studies described earlier.

Neurodevelopmental, behavioral and psychiatric effects

A case of a behavioral pattern in rats has been used in the classic literature as an example of the transmission of an epigenetic maternal effect (Francis et al., 1999). Lines of laboratory rats can be selected on the basis of the extent of the nurturing and caring that females display towards their progeny. Daughters of females that exhibit a high level of this behavior mimic their mothers; daughters of females that have a much less developed caring behavior produce daughters with a similar characteristic. High level of nurturing and caring result in epigenetic changes in the transcriptional regulation of genes in neuronal pathways that reduce stress levels; this neurological state is maintained through development and influences adult behavior. Some of the gene products involved are the glucocorticoid receptor (Liu et al., 1997), the gamma amino butyric acid receptor (Caldji et al., 2003) and the estrogen receptors (Champagne et al., 2006). The nurturing and caring behavior of females could be simply due to genetic differences between the selected lines; yet, cross-fostering of newborn rats—allowing the biological offspring of females from one line to be reared by females of the other line—leads to an adult behavior characteristic of the foster mothers. These results indicate that the nurturing and caring behavior is an epigenetic trait.

Male mice subjected to a series of randomly ordered stress-inducing conditions sired offspring that exhibited abnormalities in the hypothalamic–pituitary–adrenal axis ("axis" refers to a set of interactions between these three endocrine glands) and a significantly reduced stress response (Rodgers et al., 2013).

A possible bridge between the rat experiments just discussed and human behavior is provided by the observation that suicide victims who had been subjected to childhood abuse or severe neglect exhibited high levels of methylation in the glucocorticoid gene promoter, decreasing the expression of the gene in comparison to controls (McGowan et al., 2009).

Worms exposed to an attractive odorant during the first larval stage migrate more rapidly towards that odor as adults, suggesting that an olfactory imprint can be established. If such imprinting is reinforced by repeating the larval exposure in four successive generations, it becomes transmitted for over 40 generations, in the absence of the stimulus (Remy, 2010). Another striking example of the transgenerational inheritance of a similar neurological input was provided by the demonstration that mice could transmit an olfactory fear-conditioned memory to their descendants. Males repeatedly exposed to a particular odor coupled with a mild electric shock developed a fear towards the odor. F1 and F2 descendants of these males exhibited an abnormal sensitivity when exposed to the odor alone (Dias and Ressler, 2014).

A large number of studies with animal models have begun to document that the administration of drugs such as alcohol, opioids, marijuana, cocaine and nicotine can result in epigenetic modifications that may be transmitted to unexposed descendants (reviewed in Yohn et al., 2015). An example of these studies consists of determining the effect of nicotine exposure of male mice on their offspring. Mice whose fathers were administered nicotine in their drinking water exhibit enhanced resistance to nicotine toxicity, as well as other toxic substances such as cocaine (Vallaster et al., 2017). These resistant offspring have increased levels of transcription of a number of genes encoding detoxifying enzymes in their liver cells.

Mechanisms of transgenerational inheritance

Specific environment-induced chromatin changes that affect the biogenesis of proteins or of non-coding RNAs can be regarded as epigenetic mutations or epialleles (Rakyan et al., 2002). The epigenetic modifications that give rise to such alleles involve DNA or histone covalent modifications, non-coding RNAs or the enzymatic pathways responsible for their biogenesis (see the chapters in Part II).

DNA methylation

Genomic imprinting is one of the first identified examples of the involvement of DNA methylation in the transgenerational transmission of epigenetic regulatory signals. As previously discussed (see Chapter 8), some genes in the genomes of multicellular organisms are expressed through the allele inherited from the male parent; others are expressed through the maternal allele. This uniparental expression is constant over generations and involves DNA methylation of the imprinting control region associated with the transmitted silenced allele.

Many examples of transgenerational inheritance elicited by the types of exposure to environmental influences, discussed in Epigenetic transgenerational transmission, p. 214, have been correlated with changes in the pattern of DNA methylation. Mice exposed to early stress in the form of chronic maternal separation produced sons presenting with behaviors correlated with changes in the methylation of candidate genes involved with depression or emotional balance. These behavioral and epigenetic traits were maintained, in part, in subsequent generations (Franklin et al., 2010). In another mouse experiment, sperm were obtained from males that had been subjected to malnutrition *in utero* but that had been fed normally during postnatal development and into adulthood; specific loci in these males were found to exhibit a significant decrease in DNA methylation. When these F1 males were bred to well-fed females, they produced F2 males with normal methylation of those loci in somatic tissues, although their level of expression was affected (Radford et al., 2014). In a study with pigs, F0 boars were fed a diet supplemented with high amounts of micronutrients involved in the metabolic pathway leading to DNA methylation; F1 and F2 boars were fed a standard diet. A number of genes involved in the development of physiological and anatomical traits were expressed differently in F2 individuals descending from treated F0 boars and exhibited significant differences in DNA methylation (Braunschweig et al., 2012).

Histone modifications

In worms, flies and mammals, facultative transcriptional repression involves the methylation of histone H3 at lysine 27 (H3K27me) by the Polycomb repressive complex PRC2. In *Caenorhabditis*, this mark is present at many sites in the genome but is particularly concentrated on the X chromosomes that become inactivated during germ cell development (Kelly et al., 2002). In the absence of an active PRC2 complex in males, sperm chromosomes lack H3K27me during early embryonic development; surprisingly, this histone H3 isoform is not added to the paternal chromosome complement following fertilization, even though maternal PRC2 is present in the eggs (Gaydos et al., 2014). This result shows that the maternal PRC2 can only preserve the H3K27me pattern that already exists on the paternal chromosome complement. In the absence of an active

PRC2 complex in the fertilized egg, the H3K27me that is normally present in the sperm is maintained during the first embryonic cell divisions, then disappears. Therefore, the presence of PRC2 is necessary for the re-establishment of H3K27me-mediated inactivation of the X chromosome and of the other repressed chromosomal sites in the larval germline. These observations indicate that the memory of repression is transmitted during development and across generations.

Caenorhabditis mutants that lack the ability to demethylate H3K4me2, a mark associated with genes that are active or poised for activity, display a progressive transgenerational increase in sterility (Katz et al., 2009). The occurrence of this fertility defect can be accelerated by eliminating a histone H3K9 methyl transferase and can be reverted by loss of function of a H3K9me3 demethylase (Greer et al., 2014; Kerr et al., 2014), suggesting that the transgenerational effect on fertility depends on a balance between activating and repressive histone methylation marks. The methylation of H3K4 by trithorax-group histone methyl transferases also serves as a regulator of life span in *Caenorhabditis* (Greer et al., 2010). Altering the level of H3K4me3 by modulating the methyl transferases in one generation leads to changes in longevity in subsequent generations (Greer et al., 2011).

The transgenerational transmission of histone covalent modifications is not as well documented in mammals. This is somewhat surprising, given that the basis for this type of inheritance through the male germline was established by the observation that a small proportion of nucleosomes are retained when the haploid genome of spermatids is packaged into compacted chromatin (Wykes and Krawetz, 2003). These nucleosomes are enriched in covalently modified histones that were initially mapped to early embryonic developmental genes and regulators, regions generally characterized by a high CpG content and a low level of DNA methylation (Brykczynska et al., 2010; Hammoud et al., 2009). Using a more sensitive mapping technique, it now appears that, although a small subset of nucleosomes are present on CpG-rich promoters, the majority of the nucleosomes that have been retained in sperm are located in gene-poor regions (Carone et al., 2014). One set of experimental results suggests that the epigenetic modifications characteristic of constitutive heterochromatin in somatic cells are present in human sperm and transmitted to the early embryo (van de Werken et al., 2014). In experimentally decondensed sperm, H3K9me3 could be seen to localize with centromeres, indicating that it was marking pericentromeric constitutive heterochromatin. In eggs fertilized *in vitro* by two sperms,

therefore that contained three pronuclei, dense chromatin bodies, similar to constitutive heterochromatin seen in somatic cells, were present in the male pronuclei. These bodies contained high levels of H3K9me3, H4K20me3 and HP1 heterochromatin proteins and hybridized with several repeat sequences probes.

Epigenetic inheritance must overcome germline reprogramming

There are two significant hurdles that must be overcome for the transmission of epigenetic patterns from one generation to the next. The first is the extensive remodeling of the genetic material during the formation of the highly differentiated sperm and egg, and, following fertilization, of the zygote that will produce the next generation. The mature sperm chromatin is highly condensed, with most of its histones replaced by protamines—small arginine-rich nuclear proteins (Gatewood et al., 1990). The chromatin of oocytes becomes more condensed and quiescent as well during oogenesis.

The transmission of methylated DNA signals in transgenerational inheritance faces the same problem that was discussed with regard to the transmission of imprinted alleles (see Chapter 8): the epigenetic reprogramming that normally occurs in the germline (Fig. 19.2). In mammals, the genome is demethylated soon after fertilization and is remethylated following implantation. Some genomic regions, such as centromeric repeated DNA sequences or imprinted genes, escape the wave of somatic demethylation (Magaraki et al., 2017). As they enter the genital ridges (gonadal precursors) early during embryonic development, primordial germ cells become demethylated, including all imprinted loci, transposable elements (with some exceptions) and, in females, the inactive X chromosome (Chuva de Sousa Lopes et al., 2008); germ cells are remethylated later in development in males or after birth in females (Monk et al., 1987). Demethylation is achieved by the repression of *de novo* and maintenance DNA methyl transferases and the enhanced activities of demethylases (Hackett et al., 2013; Irie et al., 2015). An important function of demethylation in germ cells is that it allows for the establishment of imprinted marks according to the sex of the individual; for example, a developing embryo that receives a particular imprinted locus from its father and an active homologous locus from its mother should transmit only imprinted copies of the locus to the next generation if it is a male, and only active copies if it is a female. Some of the signals that protect imprinted genes from

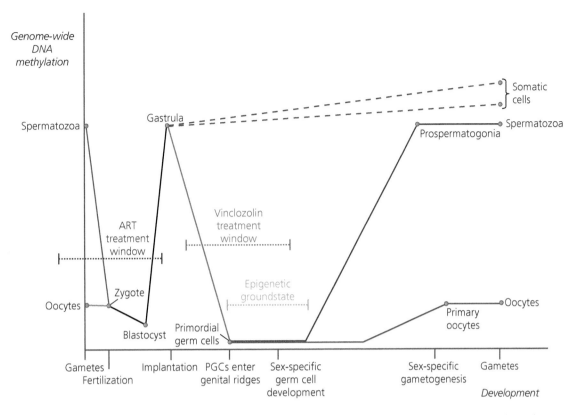

Fig. 19.2 Diagram of epigenetic reprogramming of the mammalian germline in terms of the level of CpG methylation (*y*-axis): male germline levels (blue lines), female germline levels (pink lines), methylation in all embryonic cells prior to germline specification (black lines) and methylation in the germline cells during specification (green line). A general loss of DNA methylation occurs during the development of the zygote into the blastocyst, but imprinted alleles remain methylated. *De novo* methylation of the genome occurs by the gastrula stage. As primordial germ cells are formed in the gonad primordia, the methylation of their genomes, including imprinted genes, is erased. During the early stages of gametogenesis, *de novo* methylation occurs, leading to the levels of methylation found in mature gametes and including the methylation of imprinted genes according to the developing embryo's sex.

(From McCarrey, 2014.)

demethylation during the genome-wide reprogramming that occurs in the early embryo have been generally elusive. PGC7 (also known as Stella; see Chapter 17) is a protein that associates with maternal chromatin by binding H3K9me2 and was thought to prevent the demethylation of the concomitant 5-methyl cytosines; this suggestion was based on the accumulation of 5-hydroxymethyl cytosine in the absence of Stella (Nakamura et al., 2007; Nakamura et al., 2012). As discussed in Chapter 17, a likely role for Stella is in gene activation. A more straightforward candidate is the transcription repressor TRIM28 that binds to the methylated allele of imprinted loci and is necessary for maintaining their methylation during and after early embryonic reprogramming (Alexander et al., 2015).

Another potential function of methylation erasure in the germline may be to eliminate the epigenetic modifications that were induced during early development by environmental stimuli and thereby to ensure that development and function follow closely the genetic blueprint (Heard and Martienssen, 2014). This suggestion implies that, in order to be transmitted, epigenetic modifications must escape the reprogramming process.

In contrast to the necessary differences in imprinted loci, many differences exist in the extent of methylation of numerous other loci between mature sperm and oocytes. These differences must be reconciled so that, in the early developing embryo, the two parental genomes become concordant with respect to non-imprinted gene expression. Recent results obtained with zebrafish suggest the existence of a mechanism that may achieve this goal in vertebrates. In the zebrafish, the sperm is significantly hypermethylated, in comparison to the oocyte; the paternal methylation pattern is maintained during early embryogenesis, and the maternal pattern eventually becomes similar to that of the sperm (Jiang et al., 2013; Potock et al., 2013). Maintenance of the sperm methylation pattern is achieved by the presence of so-called **placeholder** nucleosomes in all regions of the sperm and early embryos where the DNA is not methylated (Murphy et al., 2018). Placeholder nucleosomes contain the histone variant H2A.Z, as well as H3K4me1, and prevent DNA methylation, thereby ensuring that parental gene expression is recapitulated during development.

Small non-coding RNAs

Although, as discussed previously, many studies on transgenerational epigenetic transmission in mammals have focused on DNA methylation, small non-coding RNAs are clearly involved at some level in the transmission of epialleles. In worms and flies, where DNA methylation is either non-existent or minimal and restricted to a small developmental window, transgenerational epigenetic inheritance relies extensively on small RNA-mediated interference. In *Drosophila*, tandem arrays of transposons can silence similar arrays elsewhere in the genome by generating piwi-interacting RNAs (piRNAs) (de Vanssay et al., 2012). piRNAs are transcribed from specific genomic sequence clusters that exhibit high levels of histone H3 methylated at lysine 9 (H3K9me3). Transgenerational inheritance occurs by the transmission of piRNAs through the egg; these maternal piRNAs induce their own amplification by transcription from homologous genomic sequences, leading to the silencing of the same loci that were silenced in the previous generation (Brennecke et al., 2008; Le Thomas et al., 2014).

In *Caenorhabditis*, small interfering RNAs (siRNAs), derived from injected double-stranded RNAs, can induce multigenerational silencing of endogenous genes through the heritable deposition of H3K9me (Burton et al., 2011). Multigenerational epigenetic silencing of transgenes can also be generated by piRNAs that normally target endogenous and transposon transcripts. DNA sequences injected into the syncytial gonad of hermaphrodites are rarely integrated into the genome and usually form large contiguous arrays that can be inherited as extra-chromosomal structures (Mello et al., 1991). These arrays become silenced and are present in a repressed state in the germline of many

successive generations (Kelly et al., 1997). A sequence homology between transgenes and one of the numerous different piRNAs produced by the genome elicits the synthesis of secondary endogenous small interfering RNAs (endo-siRNAs) complementary to the transgene. The synthesis of these endo-siRNAs can persist for many generations, even after the removal of the piRNA trigger (Ashe et al., 2012; Bagijn et al., 2012). As mentioned previously, the transgenerational inheritance of a starvation-induced phenotype in *Caenorhabditis* is based on the induction of small RNAs that were transmitted to the F3 generation and on the targeting of these RNAs to genes with roles in nutrition (Rechavi et al., 2014).

Experiments with mice led to somewhat similar results. Stressed mothers produced F1 male mice that exhibited modified responses to various behavioral tests and significant differences in the micro RNA (miRNA) populations found in their sperm and in certain regions of the brain (Gapp et al., 2014). These differences persisted in the brains, but not the sperm, of F2 males. All tissues were normal in F3 males, although they still displayed modified behaviors, suggesting that the changes in sperm miRNAs may be transformed into other epigenetic modifications. In experiments on the transmission of a stress response induced in mice described earlier (Rodgers et al., 2013), a number of miRNAs were found to be present at significantly high levels in the treated males. Nine of these miRNAs injected into single-cell zygotes implanted into surrogate mothers led to adults that recapitulated the results of the original experiment, i.e. exhibited abnormalities in the hypothalamic–pituitary–adrenal axis and a significantly reduced stress response; furthermore, the miRNAs in the developing embryos led to the degradation of cognate maternal messenger RNAs (Rodgers et al., 2015).

A different type of small non-coding RNAs—transfer RNA-derived small RNAs (tsRNAs)—has been implicated in sperm-mediated regulation in early embryos. Following experiments demonstrating that male mice fed a low-protein diet produced F1s with enhanced expression of genes involved in lipid and cholesterol synthesis (Carone et al., 2010), the small RNAs present in the sperm of the diet-treated males were analyzed. Most of these RNAs were around 32 nucleotides long and were derived from the 5′ end of tRNAs (Peng et al., 2012). Males fed a high-fat diet for a long period of time became obese and diabetic. Sperm heads from these males injected into normal females' oocytes gave rise to offspring presenting with glucose intolerance and insulin resistance. Similar results were obtained by injecting a sperm RNA fraction of 30–40 nucleotides in length; chemical analysis of this fraction revealed features characteristic of tRNAs (Chen et al., 2016). These results provide a strong correlation between the distribution of sperm tsRNAs determined by diet in the F0 generation and the phenotype of the F1 progeny.

Somewhat surprisingly, recent evidence suggests that much of the small non-coding RNA endowment of mature sperm is contributed by somatic cells of the epididymis. As the sperm continue to mature after their formation in the testis, they fuse with small extracellular vesicles that contain and deliver short non-coding RNAs (sncRNAs), among which over 80% were tRNA fragments (Sharma et al., 2016). In embryonic stem cells in culture, as well as in early embryos, a fragment derived from one of the glycine tRNAs was shown to down-regulate a number of genes that are normally under the control of an endogenous retroviral element.

Chapter summary

Different environmental stimuli during gestation or in early life often lead to epigenetic modifications, with phenotypic consequences that do not become apparent until adult life. Aberrant nutrition during pregnancy or in early postnatal development can result in obesity, diabetes, cardiovascular disease, defective cognition or psychopathologies.

In some cases, the epigenetic modifications responsible for these effects can be transmitted to descendants who have not been exposed to the same environmental factors as their parents. Although the transgenerational inheritance of epigenetic traits is difficult to establish in humans, a growing body of experimental data on a variety of phenotypes, obtained by using different model organisms, indicates the existence of, and suggests the possible universality of, this phenomenon. Examples range from the transgenerational inheritance of conditions caused by physiological stress to the consequences of exposure to endocrine disruptors. Other examples consist of neurodevelopmental behavioral and psychiatric effects such as the inheritance of nurturing and caring behavior towards newborn and the transmission of epigenetic changes caused by addictive substances.

Many examples of transgenerational inheritance have been correlated with changes in the pattern of DNA or histone methylation. The transmission of methylated DNA signals in transgenerational inheritance faces the problem of the epigenetic reprogramming that normally occurs in the germline. Although the solutions used by organisms to circumvent this problem are still being determined, a few of the signals that protect imprinted regions for demethylation have been identified.

Small non-coding RNAs such as piRNAs and fragments of tRNAs are clearly involved at some level in transgenerational epigenetic inheritance. This is particularly the case in organisms such as flies and worms that lack DNA methylation.

References

Alexander, K. A., Wang, X., Shibata, M., Clark, A. G. & Garcia-Garcia, M. J. 2015. TRIM28 controls genomic imprinting through distinct mechanisms during and after early genome-wide reprogramming. *Cell Rep, 13*, 1194–205.

Anway, M. D., Cupp, A. S., Uzumcu, M. & Skinner, M. K. 2005. Epigenetic transgenerational actions of endocrine disruptors and male fertility. *Science, 308*, 1466–9.

Anway, M. D., Leathers, C. & Skinner, M. K. 2006. Endocrine disruptor vinclozolin induced epigenetic transgenerational adult-onset disease. *Endocrinology, 147*, 5515–23.

Ashe, A., Sapetschnig, A., Weick, E. M., Mitchell, J., Bagijn, M. P., Cording, A. C., Doebley, A. L., Goldstein, L. D., Lehrbach, N. J., Le Pen, J., Pintacuda, G., Sakaguchi, A., Sarkies, P., Ahmed, S. & Miska, E. A. 2012. piRNAs can trigger a multigenerational epigenetic memory in the germline of *C. elegans. Cell, 150*, 88–99.

Bagijn, M. P., Goldstein, L. D., Sapetschnig, A., Weick, E. M., Bouasker, S., Lehrbach, N. J., Simard, M. J. & Miska, E. A. 2012. Function, targets, and evolution of *Caenorhabditis elegans* piRNAs. *Science, 337*, 574–8.

Barker, D. J., Osmond, C., Golding, J., Kuh, D. & Wadsworth, M. E. 1989. Growth *in utero*, blood pressure in childhood and adult life, and mortality from cardiovascular disease. *BMJ, 298*, 564–7.

Benediktsson, R., Lindsay, R. S., Noble, J., Seckl, J. R. & Edwards, C. R. 1993. Glucocorticoid exposure in utero: new model for adult hypertension. *Lancet, 341*, 339–41.

Benyshek, D. C., Johnston, C. S. & Martin, J. F. 2006. Glucose metabolism is altered in the adequately-nourished grand-offspring (F3 generation) of rats malnourished during gestation and perinatal life. *Diabetologia, 49*, 1117–19.

Bertram, C., Khan, O., Ohri, S., Phillips, D. I., Matthews, S. G. & Hanson, M. A. 2008. Transgenerational effects of prenatal nutrient restriction on cardiovascular and hypothalamic-pituitary-adrenal function. *J Physiol, 586*, 2217–29.

Braunschweig, M., Jagannathan, V., Gutzwiller, A. & Bee, G. 2012. Investigations on transgenerational epigenetic response down the male line in F2 pigs. *PLoS One, 7*, e30583.

Brennecke, J., Malone, C. D., Aravin, A. A., Sachidanandam, R., Stark, A. & Hannon, G. J. 2008. An epigenetic role for maternally inherited piRNAs in transposon silencing. *Science, 322*, 1387–92.

Brieno-Enriquez, M. A., Garcia-Lopez, J., Cardenas, D. B., Guibert, S., Cleroux, E., Ded, L., Hourcade Jde, D., Peknicova, J., Weber, M. & Del Mazo, J. 2015. Exposure to endocrine disruptor induces transgenerational epigenetic deregulation of microRNAs in primordial germ cells. *PLoS One*, 10, e0124296.

Brykczynska, U., Hisano, M., Erkek, S., Ramos, L., Oakeley, E. J., Roloff, T. C., Beisel, C., Schubeler, D., Stadler, M. B. & Peters, A. H. 2010. Repressive and active histone methylation mark distinct promoters in human and mouse spermatozoa. *Nat Struct Mol Biol*, 17, 679–87.

Buescher, J. L., Musselman, L. P., Wilson, C. A., Lang, T., Keleher, M., Baranski, T. J. & Duncan, J. G. 2013. Evidence for transgenerational metabolic programming in *Drosophila*. *Dis Model Mech*, 6, 1123–32.

Burton, N. O., Burkhart, K. B. & Kennedy, S. 2011. Nuclear RNAi maintains heritable gene silencing in *Caenorhabditis elegans*. *Proc Natl Acad Sci U S A*, 108, 19683–8.

Caldji, C., Diorio, J. & Meaney, M. J. 2003. Variations in maternal care alter GABA(A) receptor subunit expression in brain regions associated with fear. *Neuropsychopharmacology*, 28, 1950–9.

Carone, B. R., Fauquier, L., Habib, N., Shea, J. M., Hart, C. E., Li, R., Bock, C., Li, C., Gu, H., Zamore, P. D., Meissner, A., Weng, Z., Hofmann, H. A., Friedman, N. & Rando, O. J. 2010. Paternally induced transgenerational environmental reprogramming of metabolic gene expression in mammals. *Cell*, 143, 1084–96.

Carone, B. R., Hung, J. H., Hainer, S. J., Chou, M. T., Carone, D. M., Weng, Z., Fazzio, T. G. & Rando, O. J. 2014. High-resolution mapping of chromatin packaging in mouse embryonic stem cells and sperm. *Dev Cell*, 30, 11–22.

Champagne, F. A., Weaver, I. C., Diorio, J., Dymov, S., Szyf, M. & Meaney, M. J. 2006. Maternal care associated with methylation of the estrogen receptor-alpha1b promoter and estrogen receptor-alpha expression in the medial pre-optic area of female offspring. *Endocrinology*, 147, 2909–15.

Chen, Q., Yan, M., Cao, Z., Li, X., Zhang, Y., Shi, J., Feng, G. H., Peng, H., Zhang, X., Zhang, Y., Qian, J., Duan, E., Zhai, Q. & Zhou, Q. 2016. Sperm tsRNAs contribute to intergenerational inheritance of an acquired metabolic disorder. *Science*, 351, 397–400.

Chuva De Sousa Lopes, S. M., Hayashi, K., Shovlin, T. C., Mifsud, W., Surani, M. A. & Mclaren, A. 2008. X chromosome activity in mouse XX primordial germ cells. *PLoS Genet*, 4, e30.

De Vanssay, A., Bouge, A. L., Boivin, A., Hermant, C., Teysset, L., Delmarre, V., Antoniewski, C. & Ronsseray, S. 2012. Paramutation in *Drosophila* linked to emergence of a piRNA-producing locus. *Nature*, 490, 112–15.

Dias, B. G. & Ressler, K. J. 2014. Parental olfactory experience influences behavior and neural structure in subsequent generations. *Nat Neurosci*, 17, 89–96.

Dorner, G. & Plagemann, A. 1994. Perinatal hyperinsulinism as possible predisposing factor for diabetes mellitus,

obesity and enhanced cardiovascular risk in later life. *Horm Metab Res*, 26, 213–21.

Doyle, T. J., Bowman, J. L., Windell, V. L., Mclean, D. J. & Kim, K. H. 2013. Transgenerational effects of di-(2-ethyl-hexyl) phthalate on testicular germ cell associations and spermatogonial stem cells in mice. *Biol Reprod*, 88, 112.

Duhl, D. M., Vrieling, H., Miller, K. A., Wolff, G. L. & Barsh, G. S. 1994. Neomorphic agouti mutations in obese yellow mice. *Nat Genet*, 8, 59–65.

Francis, D., Diorio, J., Liu, D. & Meaney, M. J. 1999. Nongenomic transmission across generations of maternal behavior and stress responses in the rat. *Science*, 286, 1155–8.

Franklin, T. B., Russig, H., Weiss, I. C., Graff, J., Linder, N., Michalon, A., Vizi, S. & Mansuy, I. M. 2010. Epigenetic transmission of the impact of early stress across generations. *Biol Psychiatry*, 68, 408–15.

Gapp, K., Jawaid, A., Sarkies, P., Bohacek, J., Pelczar, P., Prados, J., Farinelli, L., Miska, E. & Mansuy, I. M. 2014. Implication of sperm RNAs in transgenerational inheritance of the effects of early trauma in mice. *Nat Neurosci*, 17, 667–9.

Gatewood, J. M., Cook, G. R., Balhorn, R., Schmid, C. W. & Bradbury, E. M. 1990. Isolation of four core histones from human sperm chromatin representing a minor subset of somatic histones. *J Biol Chem*, 265, 20662–6.

Gaydos, L. J., Wang, W. & Strome, S. 2014. Gene repression. H3K27me and PRC2 transmit a memory of repression across generations and during development. *Science*, 345, 1515–18.

Greer, E. L., Beese-Sims, S. E., Brookes, E., Spadafora, R., Zhu, Y., Rothbart, S. B., Aristizabal-Corrales, D., Chen, S., Badeaux, A. I., Jin, Q., Wang, W., Strahl, B. D., Colaiacovo, M. P. & Shi, Y. 2014. A histone methylation network regulates transgenerational epigenetic memory in *C. elegans*. *Cell Rep*, 7, 113–26.

Greer, E. L., Maures, T. J., Hauswirth, A. G., Green, E. M., Leeman, D. S., Maro, G. S., Han, S., Banko, M. R., Gozani, O. & Brunet, A. 2010. Members of the H3K4 trimethylation complex regulate lifespan in a germline-dependent manner in *C. elegans*. *Nature*, 466, 383–7.

Greer, E. L., Maures, T. J., Ucar, D., Hauswirth, A. G., Mancini, E., Lim, J. P., Benayon, B. A., Shi, Y. & Brunet, A. 2011. Transgenerational epigenetic inheritance of longevity in *Caenorhabditis elegans*. *Nature*, 479, 365–71.

Hackett, J. A., Sengupta, R., Zylicz, J. J., Murakami, K., Lee, C., Down, T. A. & Surani, M. A. 2013. Germline DNA demethylation dynamics and imprint erasure through 5-hydroxymethylcytosine. *Science*, 339, 448–52.

Hammoud, S. S., Nix, D. A., Zhang, H., Purwar, J., Carrell, D. T. & Cairns, B. R. 2009. Distinctive chromatin in human sperm packages genes for embryo development. *Nature*, 460, 473–8.

Heard, E. & Martienssen, R. A. 2014. Transgenerational epigenetic inheritance: myths and mechanisms. *Cell*, 157, 95–109.

Huypens, P., Sass, S., Wu, M., Dyckhoff, D., Tschop, M., Theis, F., Marschall, S., Hrabe De Angelis, M. & Beckers, J. 2016. Epigenetic germline inheritance of diet-induced obesity and insulin resistance. *Nat Genet*, 48, 497–9.

Irie, N., Weinberger, L., Tang, W. W., Kobayashi, T., Viukov, S., Manor, Y. S., Dietmann, S., Hanna, J. H. & Surani, M. A. 2015. SOX17 is a critical specifier of human primordial germ cell fate. *Cell*, 160, 253–68.

Jiang, L., Zhang, J., Wang, J. J., Wang, L., Zhang, L., Li, G., Yang, X., Ma, X., Sun, X., Cai, J., Zhang, J., Huang, X., Yu, M., Wang, X., Liu, F., Wu, C. I., He, C., Zhang, B., Ci, W. & Liu, J. 2013. Sperm, but not oocyte, DNA methylome is inherited by zebrafish early embryos. *Cell*, 153, 773–84.

Kaati, G., Bygren, L. O. & Edvinsson, S. 2002. Cardiovascular and diabetes mortality determined by nutrition during parents' and grandparents' slow growth period. *Eur J Hum Genet*, 10, 682–8.

Katz, D. J., Edwards, T. M., Reinke, V. & Kelly, W. G. 2009. A *C. elegans* LSD1 demethylase contributes to germline immortality by reprogramming epigenetic memory. *Cell*, 137, 308–20.

Kelly, W. G., Schaner, C. E., Dernburg, A. F., Lee, M. H., Kim, S. K., Villeneuve, A. M. & Reinke, V. 2002. X-chromosome silencing in the germline of *C. elegans*. *Development*, 129, 479–92.

Kelly, W. G., Xu, S., Montgomery, M. K. & Fire, A. 1997. Distinct requirements for somatic and germline expression of a generally expressed *Caernorhabditis elegans* gene. *Genetics*, 146, 227–38.

Kerr, S. C., Ruppersburg, C. C., Francis, J. W. & Katz, D. J. 2014. SPR-5 and MET-2 function cooperatively to reestablish an epigenetic ground state during passage through the germ line. *Proc Natl Acad Sci U S A*, 111, 9509–14.

Klosin, A., Casas, E., Hidalgo-Carcedo, C., Vavouri, T. & Lehner, B. 2017. Transgenerational transmission of environmental information in *C. elegans*. *Science*, 356, 320–3.

Langley, S. C. & Jackson, A. A. 1994. Increased systolic blood pressure in adult rats induced by fetal exposure to maternal low protein diets. *Clin Sci (Lond)*, 86, 217–22; discussion 121.

Le Thomas, A., Marinov, G. K. & Aravin, A. A. 2014. A transgenerational process defines piRNA biogenesis in *Drosophila virilis*. *Cell Rep*, 8, 1617–23.

Liu, D., Diorio, J., Tannenbaum, B., Caldji, C., Francis, D., Freedman, A., Sharma, S., Pearson, D., Plotsky, P. M. & Meaney, M. J. 1997. Maternal care, hippocampal glucocorticoid receptors, and hypothalamic-pituitary-adrenal responses to stress. *Science*, 277, 1659–62.

Lucas, A., Morley, R. & Cole, T. J. 1998. Randomised trial of early diet in preterm babies and later intelligence quotient. *BMJ*, 317, 1481–7.

Magaraki, A., Van Der Heijden, G., Sleddens-Linkels, E., Magarakis, L., Van cappellen, W. A., Peters, A. H. F. M., Gribnau, J., Baarends, W. M. & Elope, M. 2017. Silencing markers are retained on pericentromeric heterochromatin during murine primordial germ cell development. *Epigenetics Chromatin*, 10. doi: 1186/s13072-017-0119-3.

Mccarrey, J. R. 2014. Distinctions between transgenerational and non-transgenerational epimutations. *Mol Cell Endocrinol*, 398, 13–23.

Mcgowan, P. O., Sasaki, A., D'alessio, A. C., Dymov, S., Labonte, B., Szyf, M., Turecki, G. & Meaney, M. J. 2009. Epigenetic regulation of the glucocorticoid receptor in human brain associates with childhood abuse. *Nat Neurosci*, 12, 342–8.

Mello, C. C., Kramer, J. M., Stinchcomb, D. & Ambros, V. 1991. Efficient gene transfer in *C. elegans*: extrachromosomal maintenance and integration of transforming sequences. *EMBO J*, 10, 3959–70.

Monk, M., Boubelik, M. & Lehnert, S. 1987. Temporal and regional changes in DNA methylation in the embryonic, extraembryonic and germ cell lineages during mouse embryo development. *Development*, 99, 371–82.

Morgan, H. D., Sutherland, H. G., Martin, D. I. & Whitelaw, E. 1999. Epigenetic inheritance at the agouti locus in the mouse. *Nat Genet*, 23, 314–18.

Murphy, P. J., Wu, S. F., James, C. R., Wike, C. L. & Cairns, B. R. 2018. Placeholder Nucleosomes underlie germline-to-embryo DNA methylation reprogramming. *Cell*, 172, 993–1006 e13.

Nakamura, T., Arai, Y., Umehara, H., Masuhara, M., Kimura, T., Taniguchi, H., Sekimoto, T., Ikawa, M., Yoneda, Y., Okabe, M., Tanaka, S., Shiota, K. & Nakano, T. 2007. PGC7/Stella protects against DNA demethylation in early embryogenesis. *Nat Cell Biol*, 9, 64–71.

Nakamura, T., Liu, Y. J., Nakashima, H., Umehara, H., Inoue, K., Matoba, S., Tachibana, M., Ogura, A., Shinkai, Y. & Nakano, T. 2012. PGC7 binds histone H3K9me2 to protect against conversion of 5mC to 5hmC in early embryos. *Nature*, 486, 415–19.

Ost, A., Lempradl, A., Casas, E., Weigert, M., Tiko, T., Deniz, M., Pantano, L., Boenisch, U., Itskov, P. M., Stoeckius, M., Ruf, M., Rajewsky, N., Reuter, G., Iovino, N., Ribeiro, C., Alenius, M., Heyne, S., Vavouri, T. & Pospisilik, J.a. 2014. Paternal diet defines offspring chromatin state and intergenerational obesity. *Cell*, 159, 1352–6.

Peng, H., Shi, J., Zhang, Y., Zhang, H., Liao, S., Li, W., Lei, L., Han, C., Ning, L., Cao, Y., Zhou, Q., Chen, Q. & Duan, E. 2012. A novel class of tRNA-derived small RNAs extremely enriched in mature mouse sperm. *Cell Res*, 22, 1609–12.

Potok, M. E., Nix, D. A., Parnell, T. J. & Cairns, B. R. 2013. Reprogramming the maternal zebrafish genome after fertilization to match the paternal methylation pattern. *Cell*, 153, 759–72.

Radford, E. J., Ito, M., Shi, H., Corish, J. A., Yamazawa, K., Isganaitis, E., Seisenberger, S., Hore, T. A., Reik, W., Erkek, S., Peters, A., Patti, M. E. & Ferguson-Smith, A. C. 2014. *In utero* effects. *In utero* undernourishment perturbs the adult sperm methylome and intergenerational metabolism. *Science*, 345, 1255903.

Rakyan, V. K., Blewitt, M. E., Druker, R., Preis, J. I. & Whitelaw, E. 2002. Metastable epialleles in mammals. *Trends Genet*, 18, 348–51.

Ravelli, A. C., Van Der Meulen, J. H., Osmond, C., Barker, D. J. & Bleker, O. P. 1999. Obesity at the age of 50 y in men and women exposed to famine prenatally. *Am J Clin Nutr*, 70, 811–16.

Rechavi, O., Houri-Ze'evi, L., Anava, S., Goh, W. S. S., Kerk, S. Y., Hannon, G. J. & Hobert, O. 2014. Starvation-induced transgenerational inheritance of small RNAs in *C. elegans*. *Cell*, 158, 277–87.

Remy, J. J. 2010. Stable inheritance of an acquired behavior in *Caenorhabditis elegans*. *Curr Biol*, 20, R877–8.

Ritossa, F. 1962. A new puffing pattern induced by temperature shock and DNP in *Drosophila*. *Experientia*, 18, 571–3.

Rodgers, A. B., Morgan, C. P., Bronson, S. L., Revello, S. & Bale, T. L. 2013. Paternal stress exposure alters sperm microRNA content and reprograms offspring HPA stress axis regulation. *J Neurosci*, 33, 9003–12.

Rodgers, A. B., Morgan, C. P., Leu, N. A. & Bale, T. L. 2015. Transgenerational epigenetic programming via sperm microRNA recapitulates effects of paternal stress. *Proc Natl Acad Sci U S A*, 112, 13699–704.

Schneider, S., Marxfeld, H., Groters, S., Buesen, R. & Van Ravenzwaay, B. 2013. Vinclozolin—no transgenerational inheritance of anti-androgenic effects after maternal exposure during organogenesis via the intraperitoneal route. *Reprod Toxicol*, 37, 6–14.

Sharma, U., Conine, C. C., Shea, J. M., Boskovic, A., Derr, A. G., Bing, X. Y., Belleannee, C., Kucukural, A., Serra, R. W., Sun, F., Song, L., Carone, B. R., Ricci, E. P., Li, X. Z., Fauquier, L., Moore, M. J., Sullivan, R., Mello, C. C., Garber, M. & Rando, O. J. 2016. Biogenesis and function of tRNA fragments during sperm maturation and fertilization in mammals. *Science*, 351, 391–6.

Skinner, M. K. 2014. Environmental stress and epigenetic transgenerational inheritance. *BMC Med*, 12, 153.

St Clair, D., Xu, M., Wang, P., Yu, Y., Fang, Y., Zhang, F., Zheng, X., Gu, N., Feng, G., Sham, P. & He, L. 2005. Rates of adult schizophrenia following prenatal exposure to the Chinese famine of 1959–1961. *JAMA*, 294, 557–62.

Tamaru, H. & Selker, E. U. 2001. A histone H3 methyltransferase controls DNA methylation in *Neurospora crassa*. *Nature*, 414, 277–83.

Turusov, V. S., Nikonova, T. V. & Parfenov Yu, D. 1990. Increased multiplicity of lung adenomas in five generations of mice treated with benz(a)pyrene when pregnant. *Cancer Lett*, 55, 227–31.

Vallaster, M. P., Kukreja, S., Bing, X. Y., Ngolab, J., Zhao-Shea, R., Gardner, P. D., Tapper, A. R. & Rando, O. J. 2017. Paternal nicotine exposure alters hepatic xenobiotic metabolism in offspring. *Elife*, 6, pii: e24771.

Van De Werken, C., Van Der Heijden, G. W., Eleveld, C., Teeuwssen, M., Albert, M., Baarends, W. M., Laven, J. S., Peters, A. H. & Baart, E. B. 2014. Paternal heterochromatin formation in human embryos is H3K9/HP1 directed and primed by sperm-derived histone modifications. *Nat Commun*, 5, 5868.

Wei, Y., Yang, C. R., Wei, Y. P., Zhao, Z. A., Hou, Y., Schatten, H. & Sun, Q. Y. 2014. Paternally induced transgenerational inheritance of susceptibility to diabetes in mammals. *Proc Natl Acad Sci U S A*, 111, 1873–8.

Wolstenholme, J. T., Goldsby, J. A. & Rissman, E. F. 2013. Transgenerational effects of prenatal bisphenol A on social recognition. *Horm Behav*, 64, 833–9.

Wykes, S. M. & Krawetz, S. A. 2003. The structural organization of sperm chromatin. *J Biol Chem*, 278, 29471–7.

Yohn, N. L., Bartolomei, M. S. & Blendy, J. A. 2015. Multigenerational and transgenerational inheritance of drug exposure: the effects of alcohol, opiates, cocaine, marijuana, and nicotine. *Prog Biophys Mol Biol*, 118, 21–33.

PART V

Epigenetics, Human Health and Development

The different aspects of genetic regulation that we have discussed—the retrieval of genomic information, its manifestation and regulation via epigenetic modifications and its dependence on specific subnuclear domains—are responsible for normal development and for normal cellular differentiation. Individual cells and whole multicellular organisms can tolerate a relatively broad level of variation in the molecular steps of developmental pathways that result from allele variations in the genetic blueprint and from environmental changes. Nevertheless, it is clear that dysfunctions at any level of genetic regulation have the potential to result in an increased susceptibility to disease or actually give rise to overt pathologies. A number of specific examples were provided in previous chapters in the context of particular epigenetic modifications (Chapter 4, Box 4.1; Chapter 8, Boxes 8.1 and 8.2; Chapter 9, Box 9.1).

In Part V, the development and dysfunction of some major physiological systems are discussed.

As stated previously, two important considerations are to be kept in mind while evaluating the association between epigenetic modifications and disease. The first is whether experimental protocols have been used that ensure the comparison of appropriately selected tissue samples (that consider, for example, the possible cell composition differences between patient and control tissues) by methods that are reasonably free of bias (such as ambient conditions and operator influence) and by the use of statistical methods that normalize the data and detect and account for batch effects. The second consideration is the frequent difficulty in determining if the modifications cause or contribute to the symptoms or are a consequence of the disease process.

CHAPTER 20

Aging, cellular senescence and cancer: the role of genomic instability, cellular homeostasis and telomeres

Aging is a progressive functional decline that impacts multicellular organisms and that begins soon after the completion of the developmental phase of the life cycle. Because it affects most anatomical and physiological systems, as well as the integrity of the genome, aging is a highly complex phenomenon. It is, therefore, operationally useful to parse the aging process and its usually successive phenotypes into a series of hallmarks that include genomic instability, mitochondrial dysfunction, telomere attrition, epigenetic alterations, loss of proteostasis, cellular senescence and stem cell exhaustion (Lopez-Otin et al., 2013). Aging has been studied in various experimental organisms, most prevalently in yeast, *Caenorhabditis, Drosophila* and mice (given the latter's relatively long life span, mouse models that reproduce premature aging syndromes are often used). Although these studies have provided valuable insights, the *caveat* that they may not always be directly applicable to human aging must be considered.

Clearly, a number of the aging hallmarks are also causative factors of oncogenesis, highlighting the intimate connection between aging and cancer, a connection that can be summarized by the observation that aging is the most important risk factor for cancer. Cancer cells differ from normal cells in that: (1) they evade the signals that regulate the rate of cell division or that prevent cell division in some differentiated tissues, (2) they evade the signals that program normal cells for death and (3) they acquire the ability to metastasize, i.e. to leave the tissue where they first form and invade other tissues (Hanahan and Weinberg, 2000; Hanahan and Weinberg, 2011).

In this chapter and in Chapter 21, various parameters of the aging process and their similarity and impact on cancer development are discussed.

Genomic instability

Genomic instability results from the accumulation of DNA damage over time. This damage consists of errors that occur during DNA replication or as a consequence of exposure to endogenous or environmental insults. Reactive oxygen species that are the byproduct of oxidative phosphorylation carried out by mitochondria, exposure to ultraviolet (UV) and ionizing radiation, exposure to certain chemicals such as alkylating agents and aromatic amines that generate DNA adducts will interfere with replication and may lead to mutations. Most of these events are repaired by a variety of mechanisms (see Chapter 15). Nevertheless, some errors escape repair and their number increases over time (Faggioli et al., 2012; Forsberg et al., 2012; Moskalev et al., 2013). Mutations, chromosomal rearrangements or instances of partial aneuploidy can lead to deregulated gene expression that may cause or contribute to the onset of many of the other aging hallmarks, including cancer.

The first breakthrough in identifying the causative factors of cancer was the identification of **oncogenes**

Epigenetics, Nuclear Organization and Gene Function: with implications of epigenetic regulation and genetic architecture for human development and health. John C. Lucchesi, Oxford University Press (2019). © John C. Lucchesi 2019. DOI: 10.1093/oso/9780198831204.001.0001

and their counterparts **tumor suppressor genes**. Onco-genes are genes that, in their wild type form (referred to as **proto-oncogenes**), are responsible for the occurrence of normal cell division and differentiation (Stehelin et al., 1976). Mutations, or chromosomal rearrangements such as translocations, that lead to their constitutive expression or to abnormally high levels of expression transform proto-oncogenes into oncogenes and result in an increase in the rate of cell division or prevent timely cell death. As expected, these mutations are dominant. Tumor suppressor genes regulate cell division and thereby limit cell growth and proliferation. The normal cell cycle includes a series of checkpoints that determine if the cell has successfully completed a particular stage and allow it to proceed to the next one (Fig. 20.1). Loss-of-function mutations of tumor suppressor genes are usually recessive; in homozygous or hemizygous (a single allele is present in the genome) conditions, these mutations result in failure of one or more of the cell division checkpoints to operate.

In addition to their effect on checkpoints, oncogenes and tumor suppressor genes can impact every step of the cell division process. For example, over-expression of oncogenes, such as *cyclin E1*, *Ras* (rat sarcoma) and *Myc* (avian myelocytomatosis viral oncogene homolog), can affect DNA replication by interfering with the normal pattern of origin of replication firing (Dominguez-Sola et al., 2007; Jones et al., 2013). Loss of the tumor suppressor p53 protein, which can occur by mutation of the *TP53* gene or by deacetylation of its protein product leading to ubiquitination and degradation, results in failure to arrest the cell cycle or to induce apoptosis in cells that have incurred DNA damage or replication errors (Baker et al., 1989). In addition to these effects on the cell cycle, oncogenes induce the silencing of genes that characterize differentiated cells, and activate stem cell transcriptional programs (Poli et al., 2018). Other consequences of the action of some oncogenes include the loss of cell adhesion molecules, allowing cancer cells to metastasize and invade other tissues (Cavallaro and Christofori, 2004).

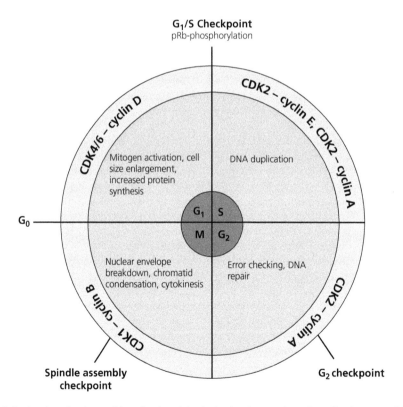

Fig. 20.1 Diagram indicating the major phases of the cell cycle and the checkpoints that stop the cycle or allow it to proceed. The different cyclins and their cognate cyclin-dependent kinases are indicated in the periphery.

(From Tan et al., 2017.)

Given the critical role that DNA repair plays in reversing the numerous mutational events that impact the genome, the majority of the genes that are responsible for the numerous factors involved in the cell's different DNA repair pathways can, in fact, be considered to be tumor suppressors—loss of function of many of these factors leads to an increase in the incidence of mutations that, of course, can occur in proto-oncogenes and other cancer-related genes.

One of the first connections between a DNA damage repair mechanism and oncogenesis was the discovery that the cells of some hereditary colorectal cancers were characterized by a high number of somatic mutations in microsatellites—regions of short tandem repeats (Thibodeau et al., 1993). This genotypic feature of cancer cells, referred to as "microsatellite instability," is caused by mutations or epigenetic alterations that inactivate the mismatch repair (MMR) pathway (Liu et al., 1995). Microsatellite instability has been found in other cancer types, including endometrial, ovarian, gall bladder, gastric and prostate cancers (Gelsomino et al., 2016; Lee et al., 2016).

Xeroderma pigmentosum (XP) is a hereditary form of cancer caused by a defect in a different DNA repair pathway. Although this disease affects a variety of organ systems, it is predominantly characterized by a hypersensitivity to the UV rays present in sunlight and a propensity to develop skin cancer. Exposure to UV rays induces the formation of pyrimidine dimers (covalent linkage between two successive thymines or cytosines) that are normally removed and replaced by the nucleotide excision repair (NER) pathway. Individuals with XP carry mutations that inactivate one of the enzymatic steps involved in NER (Cleaver, 1968; Setlow et al., 1969).

Mitochondrial DNA (mtDNA)

A significant component of the cellular DNA is extranuclear and is found in the mitochondria. mtDNA has retained the genes for the ribosomal RNAs and the transfer RNAs that are required for mitochondrial protein synthesis and for a number of polypeptides of the mitochondrial energy-generating process such as the electron transport complexes. Other genes necessary for mitochondrial function are present in the nuclear chromosomal DNA. Evidence that alterations in mtDNA may be responsible for age-related diseases was first obtained in studying the etiology of maternally inherited pathological conditions such as some particular forms of myopathy (Holt et al., 1989), of optic neuropathy (Singh et al., 1989) or of epilepsy (Shoffner et al., 1990).

Direct evidence that mtDNA mutations may undergo clonal expansion and result in a disease phenotype was provided by a study of the side effects of retroviral drugs. Treatment with these drugs causes an increase in somatic mutations that are apparently not due to higher levels of genomic mutagenesis; rather they are due to increased mtDNA turnover, leading to the clonal expansion of existing mtDNA mutations (Payne et al., 2011). As in the case of numerous nuclear genes, mtDNA genes depend on histone acetylation for their transcription. MOF, the acetyl transferase responsible for the acetylation of histone H4 at lysine 16, plays a key role in dosage compensation in flies as a member of the MSL (male-specific lethal) complex (see Chapter 9); as a member of a different complex, the NSL (non-specific lethal) complex, MOF targets a large number of genes located on all chromosomes, from flies to mammals (Smith et al., 2005; Taipale et al., 2005). In addition to these nuclear chromatin modifications, MOF binds mtDNA and is required for the transcription of the mitochondrial genome (Chatterjee et al., 2016). The targeted loss of MOF in mouse cardiomyocytes leads to mitochondrial dysfunction and cardiomyopathy.

The role that mitochondrial mutations may play in cancer development is somewhat paradoxical. Presumably because of their increased metabolic requirements, cancers appear to exert selective pressure for normal function of their mitochondrial component (Ju et al., 2014; Stewart et al., 2015). In fact, the cells of some tumors have higher-than-normal numbers of mitochondria (Hasumi et al., 2016).

Changes in cellular homeostasis

Cellular homeostasis is the maintenance of biochemical reactions and processes within an acceptable range; organisms exhibit homeostasis during their early development and, with respect to some aspects of physiology, during adult life. Given that all metabolic pathways depend on the action of proteins, a major component of cellular homeostasis is **proteostasis**, the proper concentration folding and positioning of proteins within the different cellular and nuclear compartments. During the aging process, alterations in proteostasis result in the disruption of a variety of metabolic pathways that connect with environmental factors. An example is the age-related variation in the mTOR (mechanistic target of rapamycin) signaling pathway. Because it can sense and integrate many different environmental and hormonal clues, this signaling pathway plays a major role in the regulation of metabolism. Many of the pathway's functions have been elucidated by the discovery that it is inhibited by rapamycin, an antibiotic that has been extensively used experimentally as an immunosuppressant in mammalian systems. In yeast, rapamycin, bound to its receptor, targets the products of two genes *TOR1* and *TOR2* (target of rapamycin 1 and 2) and stops dividing cells in the G1 phase (Heitman et al., 1991). These observations led to the isolation of a related protein mTOR (mammalian TOR) from human cells (Sabers et al., 1995). Subsequent studies have demonstrated that mTOR (now renamed **mechanistic** TOR) is a serine/threonine protein kinase that exists in two complexes (mTORC1 and mTORC2). These complexes regulate many different pathways involved in cell growth and metabolism, including the insulin/insulin-like growth factor 1 (IGF-1) pathway. Among other functions, one of the major roles of mTORC1 is to regulate mRNA translation and protein synthesis. The rate of translation depends on the presence of factors that bind the mRNA cap (see Chapter 3) and their concentration relative to an inhibitor (4E-BP) that prevents their binding (Pause et al., 1994). mTOR inactivates this inhibitor by phosphorylating it and maintains the appropriate protein levels, ensuring cellular proteostasis. Recently, another function of the mTORC1 complex was shown to be the phosphorylation and activation of the p300 acetyl transferase, implicating this complex in the regulation of transcription (Wan et al., 2017). An important function of the mTORC2 complex is the regulation of lipid biogenesis by increasing the stability of SREBP1 (sterol regulatory element-binding protein 1), a transcription factor that activates the transcription of several genes involved in lipid synthesis (Li et al., 2016).

Loss-of-function mutations of mTOR lengthen the life span of *C. elegans, Drosophila* and mice (Kapahi et al., 2004; Vellai et al., 2003; Wu et al., 2013). Consistent with these results, the cells of patients with premature aging diseases exhibit an increase in mTOR signaling (whether it increases during normal aging has not been fully determined). Rapamycin treatment extends the life span in all of the mentioned model organisms (for example, Bjedov et al., 2010) and in mice delays the onset of age-related diseases (Flynn et al., 2013).

The aberrant activation of the mTOR pathway is characteristic of different human cancers. Its oncogenic effects include an increase in ribosome biogenesis that is crucial for enhanced cell proliferation.

While defects in proteostasis would be expected during oncogenesis, the relationship that exists between molecular chaperones and cancer was somewhat surprising. Chaperones are proteins that bind to newly synthesized polypeptides to ensure that they are properly folded and to avoid inappropriate interactions with other proteins that could lead to the formation of deleterious aggregates. The most evolutionarily conserved chaperones are the heat-shock proteins that are induced following exposure to increases in ambient temperature, as well as the result of some other environmental stresses such as anoxia, exposure to heavy metals and some drugs (Ritossa, 1964; Sorger, 1991). Induction is achieved by the activation of transcription factors termed heat-shock factors (HSFs) that bind to cognate elements within the promoters of heat-shock protein genes. The first evidence for a correlation between molecular chaperones and cancer was provided by the observation that loss-of-function mutations leading to the absence of HSF1, the major regulator of the heat-shock response, protected mice from the oncogenic activity of the *Ras* gene (Dai et al., 2007). In fact, over-expression of HSF1 is found in a variety of cancers, for example in liver, breast and uterus cancers (Engerud et al., 2014; Fang et al., 2012; Santagata et al., 2011). A possible explanation for the oncogenic effect of HSF1 could be that the chaperones that it induces protect proteins that are important for malignant transformation by preventing their misfolding and degradation. In addition, HSF1 has been shown to have a regulatory role in several aspects of cellular metabolism, including glycolysis (Zhao et al., 2009) and lipid metabolism (Jin et al., 2011), two pathways that are modulated differently in normal and cancer cells.

Telomeres

Telomeres were first discovered in *Tetrahymena* (Blackburn and Gall, 1978). DNA replication leaves two single-stranded regions at the 3′ end of the newly synthesized DNA strands (see Chapter 14). The ends of the chromosomes of most organisms contain a highly repeated sequence rich in guanine (G) nucleotides that constitutes the **telomere**. In vertebrates, including humans, the sequence is 5′-TTAGGG-3′ on the end of one strand of the DNA molecule of a chromatid (and, as expected, 3′-AATCCC-5′ on the other strand), repeated several thousand times. Following each cycle of replication, the length of the repeated region is shorter on the 5′ ends of the newly replicated DNA than on the 3′ ends, on average, by a number of nucleotides that corresponds to the length of the RNA primer used to generate the terminal Okazaki fragment. If these gaps were left unfilled, the chromosomes would get shorter at every round of DNA replication. The problem is solved by the enzyme **telomerase** (Greider and Blackburn, 1985), a reverse transcriptase that synthesizes DNA using RNA as a template (Fig. 20.2).

Following an overview of the function of telomerase in establishing the telomere length, its contrasting role in the aging process and in oncogenesis will be discussed.

Telomere biogenesis

Telomerase associates with several accessory proteins and with an RNA molecule that can pair with the last repeats at the 3′ end of the template DNA strand, and serves as a template to add additional repeats (Greider and Blackburn, 1989) (Fig. 20.2). The RNA in question [TR (telomerase RNA) or TERC (telomerase RNA component)] is constitutively expressed in cells (Feng et al., 1995). Following a certain lengthening of the 3′ end, a DNA polymerase (the Polα/primase, most likely followed by Polδ) synthesizes the complementary sequence on the lagging strand. The small gap left by the eventual removal of the necessary RNA primer at the 5′ end of this newly synthesized DNA segment is inconsequential in light of the lengthening performed by the telomerase.

In the absence of telomerase, telomere elongation can occur by an alternative lengthening of telomeres (ALT)

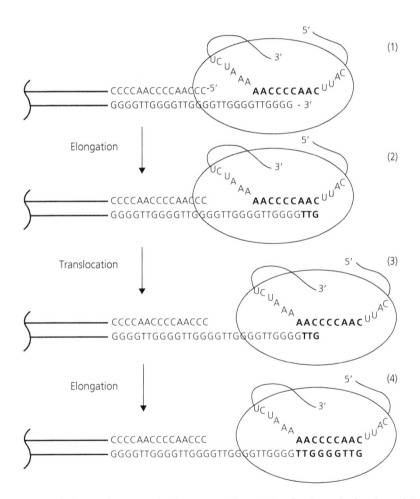

Fig. 20.2 Telomere elongation. The large oval represents the telomerase complex consisting of an RNA molecule and a catalytic subunit. (1) The RNA has a segment that is complementary to the repeated sequences at the end of the template strand (TTGGGG, in this example). (2) The catalytic subunit has a reverse transcriptase activity that it uses to elongate the template strand. (3) The telomerase then translocates and (4) repeats the process of strand elongation. Following the release of the telomerase, the elongated strand is used as a template by a DNA polymerase to fill in the gap in the newly replicated strand (not shown).

(From Greider and Blackburn, 1989.)

pathway that involves homologous recombination (HR). First discovered in yeast (Lundblad and Blackburn, 1993), this pathway is active in different types of cancers (Henson et al., 2005). The ALT pathway appears to be activated by mutations in ATRX (alpha thalassemia/mental retardation syndrome X-linked) and its binding partner DAXX (death domain-associated protein) [Heaphy et al., 2011; Lovejoy et al., 2012]. Inactivation of this complex, which is responsible for the deposition of the histone variant H3.3 at telomeres, has been correlated with the occurrence of pediatric glioblastoma and other cancers (Heaphy et al., 2011; Jiao et al., 2011).

Telomere protection

An important molecular event that occurs immediately following completion of replication involves protecting the end of the DNA molecule and preventing its misidentification as one end of a double-stranded break. This is the responsibility of a six-protein subunit complex called **shelterin**. Shelterin associates specifically with telomere DNA through the affinity of two of its subunits for the TTAGGG repeated sequence (Broccoli et al., 1997; Chong et al., 1995); it prevents the triggering of the DNA repair pathways by inhibiting the activation of the ATM and ATR kinases (Denchi and de Lange, 2007) and by inducing the telomeric DNA to form a loop (T-loop) that hides the single-stranded 3′ overhang (Griffith et al., 1999). The T-loop is formed when the 3′ overhang invades the double-stranded repeat-containing region that precedes it (Fig. 20.3). Whether telomerase is able to access the 3′ end once it is in a T-loop, or whether it must wait until the T-loop is dissolved to allow the next round of DNA replication, is not yet determined. In addition, two of the shelterin subunits increase the activity and the processivity of the telomerase (Wang et al., 2007).

Control of telomerase levels and telomere length

An obvious reason for controlling telomerase levels is provided by the fact that telomerase activity is up-regulated in most cancers (Counter et al., 1994; Kim et al., 1994). A complex (CST), consisting of three subunits and conserved from yeast to humans, associates with the single-stranded 3′ overhang of the telomeric DNA (Gao et al., 2007; Miyake et al., 2009). In both yeast and humans, this complex controls the access of telomerase to the chromosome ends and, thereby, regulates the extent of telomere DNA lengthening (Lin and Zakian., 1996). It also functions to recruit the Polα/primase for the synthesis of the missing 5′ segment (Grossi et al., 2004; Qi and Zakian, 2000). Telomerase can also be regulated by a telomeric repeat-containing RNA (TERRA) that is transcribed from subtelomeric DNA sequences and extends into the telomere repeats (Schoeftner and Blasco, 2008). *In vitro*, telomerase has a higher affinity for TERRA than for the single-stranded 3′ DNA overhang, suggesting that the RNA may have an inhibitory effect on telomerase and limit the extent of 5′ strand elongation. A third means of telomerase regulation is suggested by the observation that, in mice, subtelomeric regions exhibit high levels of CpG methylation and that loss of DNA methyl transferases leads to greatly elongated telomeres (Gonzalo et al., 2006).

The role of telomeres in aging and cancer

Telomerase is present in a number of embryonic tissues, in the germline and in adult stem cells (Wright et al., 1996). It is absent, or is present at very low levels and decreases over time, in most adult human tissues, resulting in a sustained, progressive shortening of all telomeres (Counter et al., 1994; Kim et al., 1994). Cells obtained from living tissues and grown *in vitro* eventually lose

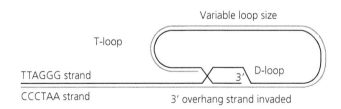

Fig. 20.3 Diagram of a T-loop. The 3′ single-stranded overhang invades an upstream double-stranded region and pairs with complementary sequences in that region.

(From Doksani and de Lange, 2014.)

their ability to divide and subsequently die. This phenomenon, termed **replicative senescence** or **cellular senescence**, was thought to reflect the existence of some factor required for cell survival that would diminish in concentration at each cell division (Hayflick and Moorhead, 1961). Although different inducers of senescence, such as DNA damage and reactive oxygen species, have been identified, the major factors that count the number of cell divisions allowed to particular somatic cell types and that limit the life span of cells in culture are telomerase activity and telomere length (Bodnar et al., 1998). The cumulative effects of somatic cell senescence are responsible for the aging process of the whole organism. Experimental evidence supporting the correlation between telomere length and aging was provided by the observation that aged mice deficient in telomerase activity can be rejuvenated by reactivating telomerase (Jaskelioff et al., 2011). As telomeres shorten throughout the life span, the risk of common age-related pathologies, such as cardiovascular disease, diabetes, neurodegenerative changes (Alzheimer's disease, Parkinson's disease, dementia) and cancer, increases. In contrast to these effects that are deleterious to the organism, cellular senescence is also an intrinsic tumor suppressor mechanism (see below).

Cellular senescence typically leads to different aspects of chromatin reorganization (Herranz and Gil, 2018). For example, senescent cells exhibit increased number and larger PML bodies; in fact, forced PML expression in cultured fibroblasts promotes premature senescence. Another chromatin change involves the formation of senescence-associated heterochromatin foci (Chapter 21).

Telomerase activity is present only in cells that tend to proliferate during the life span such as somatic stem cells, immune system cells and germ cells. As previously mentioned, somatic tissues, in general, lack telomerase activity. It is, therefore, imperative for cancer cells of somatic origin to activate telomerase in order to avoid cellular senescence and to gain "immortality." In these cells, among several transcription factors that have been implicated in the regulation of telomerase levels, an important regulator is Myc. The transformation of *Myc* from a proto-oncogene to an oncogene, resulting in an increased level of Myc protein, leads to the presence of telomerase (Klapper et al., 2003; Schneider-Stock et al., 2003; Wu et al., 1999). Another transcription factor responsible for an increase in telomerase levels is GABP (GA-binding protein). Mutations that occur most frequently in two specific sites in the promoter of the reverse transcriptase gene create a binding site for GABP (Bell et al., 2015). These mutations have been detected in numerous cancers (Horn et al., 2013; Huang et al., 2013; Killela et al., 2013).

Cells can avoid senescence and the cessation of cell growth, without requiring telomerase reactivation, through loss-of-function mutations of the regulatory proteins involved in the G1 to S phase checkpoint (Fig. 20.1). One of the major inducers of cell cycle arrest is CDKN2A (usually referred to as p16), and mutations that inactivate it allow the cell cycle to proceed (Rayess et al., 2012). These cells can divide repeatedly, but without telomerase activity, they eventually reach a "crisis" that leads to death, unless they are able to reactivate that enzyme and become immortalized. Most cancers harbor mutations in the *p53* gene (Hollstein et al.,1999), in the p16 pathway (Puig et al., 2005) or in both.

Chapter summary

Aging is a progressive functional decline that impacts multicellular organisms and that begins soon after the completion of the developmental phase of the life cycle. Some of the common hallmarks of aging are genomic instability, mitochondrial dysfunction, telomere attrition, epigenetic alterations, loss of proteostasis, cellular senescence and stem cell exhaustion. A number of these aging hallmarks are also causative factors of oncogenesis, highlighting the intimate connection between aging and cancer. Cancer cells undergo unregulated division and are able to leave their tissue of origin and invade other tissues, i.e. to **metastasize**.

Genomic instability results from the accumulation of DNA damage over time. This damage consists of errors that occur during DNA replication or, as a consequence of exposure, to endogenous or environmental insults. Although most of these events are repaired, some of them escape and accumulate. The genome contains a number of genes that are responsible for the occurrence of normal cell division and differentiation (**oncogenes**), and genes that regulate cell division and thereby limit cell growth and proliferation (**tumor suppressor genes**). Mutations that cause the over-expression of oncogenes or that inactivate tumor suppressors result in uncontrolled cell proliferation and cancer. The genes that encode factors responsible for the different DNA repair pathways can be considered to have tumor suppressor functions.

Mitochondria contain a significant component of the cellular DNA. Some mutations in mitochondrial DNA can undergo clonal expansion and have been correlated to age-related diseases, including cancers.

A major contribution to cellular homeostasis—the maintenance of molecular processes within an acceptable range—is proteostasis whereby proteins are maintained in the proper concentration, folding and localization. During the aging process, alterations in proteostasis result in the disruption of a variety of metabolic pathways that connect with environmental factors. An example is the increase in concentration during aging and in different cancers of a particular factor (mTOR), a general regulator of mRNA translation and protein synthesis.

Telomeres are terminal regions of chromosomes that protect the DNA from attack by exonucleases, prevent end-to-end fusions and prevent the shortening of the DNA molecules at each replication cycle. The enzyme responsible for these effects is **telomerase**, a reverse transcriptase that synthesizes DNA using RNA as a template. Telomerase is present in a number of embryonic tissues, in the germline, in immune system cells and in adult stem cells. It is absent, or is present at very low levels and decreases over time, in most adult human tissues, resulting in a sustained, progressive shortening of all telomeres. As telomeres shorten throughout the life span, cells eventually lose their ability to divide (replicative or cellular senescence) and the risk of common age-related pathologies increases. Cancer cells of somatic origin must activate telomerase in order to avoid cellular senescence and to gain "immortality."

References

Baker, S. J., Fearon, E. R., Nigro, J. M., Hamilton, S. R., Preisinger, A. C., Jessup, J. M., Vantuinen, P., Ledbetter, D. H., Barker, D. F., Nakamura, Y., White, R. & Vogelstein, B. 1989. Chromosome 17 deletions and *p53* gene mutations in colorectal carcinomas. *Science*, 244, 217–21.

Bell, R. J., Rube, H. T., Kreig, A., Mancini, A., Fouse, S. D., Nagarajan, R. P., Choi, S., Hong, C., He, D., Pekmezci, M., Wiencke, J. K., Wrensch, M. R., Chang, S. M., Walsh, K. M., Myong, S., Song, J. S. & Costello, J. F. 2015. Cancer. The transcription factor GABP selectively binds and activates the mutant TERT promoter in cancer. *Science*, 348, 1036–9.

Bjedov, I., Toivonen, J. M., Kerr, F., Slack, C., Jacobson, J., Foley, A. & Partridge, L. 2010. Mechanisms of life span extension by rapamycin in the fruit fly *Drosophila melanogaster*. *Cell Metab*, 11, 35–46.

Blackburn, E. H. & Gall, J. G. 1978. A tandemly repeated sequence at the termini of the extrachromosomal ribosomal RNA genes in *Tetrahymena*. *J Mol Biol*, 120, 33–53.

Bodnar, A. G., Ouellette, M., Frolkis, M., Holt, S. E., Chiu, C. P., Morin, G. B., Harley, C. B., Shay, J. W., Lichtsteiner, S. & Wright, W. E. 1998. Extension of life-span by introduction of telomerase into normal human cells. *Science*, 279, 349–52.

Broccoli, D., Smogorzewska, A., Chong, L. & De Lange, T. 1997. Human telomeres contain two distinct Myb-related proteins, TRF1 and TRF2. *Nat Genet*, 17, 231–5.

Cavallaro, U. & Christofori, G. 2004. Cell adhesion and signalling by cadherins and Ig-CAMs in cancer. *Nat Rev Cancer*, 4, 118–32.

Chatterjee, A., Seyfferth, J., Lucci, J., Gilsbach, R., Preissl, S., Bottinger, L., Martensson, C. U., Panhale, A., Stehle, T., Kretz, O., Sahyoun, A. H., Avilov, S., Eimer, S., Hein, L., Pfanner, N., Becker, T. & Akhtar, A. 2016. MOF acetyl transferase regulates transcription and respiration in mitochondria. *Cell*, 167, 722–38 e23.

Chong, L., Van Steensel, B., Broccoli, D., Erdjument-Bromage, H., Hanish, J., Tempst, P. & De Lange, T. 1995. A human telomeric protein. *Science*, 270, 1663–7.

Cleaver, J. E. 1968. Defective repair replication of DNA in xeroderma pigmentosum. *Nature*, 218, 652–6.

Counter, C. M., Hirte, H. W., Bacchetti, S. & Harley, C. B. 1994. Telomerase activity in human ovarian carcinoma. *Proc Natl Acad Sci U S A*, 91, 2900–4.

Dai, C., Whitesell, L., Rogers, A. B. & Lindquist, S. 2007. Heat shock factor 1 is a powerful multifaceted modifier of carcinogenesis. *Cell*, 130, 1005–18.

Denchi, E. L. & De Lange, T. 2007. Protection of telomeres through independent control of ATM and ATR by TRF2 and POT1. *Nature*, 448, 1068–71.

Doksani, Y. & De Lange, T. 2014. The role of double-strand break repair pathways at functional and dysfunctional telomeres. *Cold Spring Harb Perspect Biol*, 6, a016576.

Dominguez-Sola, D., Ying, C. Y., Grandori, C., Ruggiero, L., Chen, B., Li, M., Galloway, D. A., Gu, W., Gautier, J. & Dalla-Favera, R. 2007. Non-transcriptional control of DNA replication by c-Myc. *Nature*, 448, 445–51.

Engerud, H., Tangen, I. L., Berg, A., Kusonmano, K., Halle, M. K., Oyan, A. M., Kalland, K. H., Stefansson, I., Trovik, J., Salvesen, H. B. & Krakstad, C. 2014. High level of HSF1 associates with aggressive endometrial carcinoma and suggests potential for HSP90 inhibitors. *Br J Cancer*, 111, 78–84.

Faggioli, F., Wang, T., Vijg, J. & Montagna, C. 2012. Chromosome-specific accumulation of aneuploidy in the aging mouse brain. *Hum Mol Genet*, 21, 5246–53.

Fang, F., Chang, R. & Yang, L. 2012. Heat shock factor 1 promotes invasion and metastasis of hepatocellular carcinoma *in vitro* and *in vivo*. *Cancer*, 118, 1782–94.

Feng, J., Funk, W. D., Wang, S. S., Weinrich, S. L., Avilion, A. A., Chiu, C. P., Adams, R. R., Chang, E., Allsopp, R. C., Yu, J. & Et al. 1995. The RNA component of human telomerase. *Science*, 269, 1236–41.

Flynn, J. M., O'leary, M. N., Zambataro, C. A., Academia, E. C., Presley, M. P., Garrett, B. J., Zykovich, A., Mooney, S. D., Strong, R., Rosen, C. J., Kapahi, P., Nelson, M. D., Kennedy, B. K. & Melov, S. 2013. Late-life rapamycin treatment reverses age-related heart dysfunction. *Aging Cell*, 12, 851–62.

Forsberg, L. A., Rasi, C., Razzaghian, H. R., Pakalapati, G., Waite, L., Thilbeault, K. S., Ronowicz, A., Wineinger, N. E., Tiwari, H. K., Boomsma, D., Westerman, M. P., Harris, J. R., Lyle, R., Essand, M., Eriksson, F., Assimes, T. L., Iribarren, C., Strachan, E., O'hanlon, T. P., Rider, L. G., Miller, F. W., Giedraitis, V., Lannfelt, L., Ingelsson, M., Piotrowski, A., Pedersen, N. L., Absher, D. & Dumanski, J. P. 2012. Age-related somatic structural changes in the nuclear genome of human blood cells. *Am J Hum Genet*, 90, 217–28.

Gao, H., Cervantes, R. B., Mandell, E. K., Otero, J. H. & Lundblad, V. 2007. RPA-like proteins mediate yeast telomere function. *Nat Struct Mol Biol*, 14, 208–14.

Gelsomino, F., Barbolini, M., Spallanzani, A., Pugliese, G. & Cascinu, S. 2016. The evolving role of microsatellite instability in colorectal cancer: a review. *Cancer Treat Rev*, 51, 19–26.

Gonzalo, S., Jaco, I., Fraga, M. F., Chen, T., Li, E., Esteller, M. & Blasco, M. A. 2006. DNA methyltransferases control telomere length and telomere recombination in mammalian cells. *Nat Cell Biol*, 8, 416–24.

Greider, C. W. & Blackburn, E. H. 1985. Identification of a specific telomere terminal transferase activity in *Tetrahymena* extracts. *Cell*, 43, 405–13.

Greider, C. W. & Blackburn, E. H. 1989. A telomeric sequence in the RNA of *Tetrahymena* telomerase required for telomere repeat synthesis. *Nature*, 337, 331–7.

Griffith, J. D., Comeau, L., Rosenfield, S., Stansel, R. M., Bianchi, A., Moss, H. & De Lange, T. 1999. Mammalian telomeres end in a large duplex loop. *Cell*, 97, 503–14.

Grossi, S., Puglisi, A., Dmitriev, P. V., Lopes, M. & Shore, D. 2004. Pol12, the B subunit of DNA polymerase alpha, functions in both telomere capping and length regulation. *Genes Dev*, 18, 992–1006.

Hanahan, D. & Weinberg, R. A. 2000. The hallmarks of cancer. *Cell*, 100, 57–70.

Hanahan, D. & Weinberg, R. A. 2011. Hallmarks of cancer: the next generation. *Cell*, 144, 646–74.

Hasumi, H., Baba, M., Hasumi, Y., Furuya, M. & Yao, M. 2016. Birt-Hogg-Dube syndrome: clinical and molecular aspects of recently identified kidney cancer syndrome. *Int J Urol*, 23, 204–10.

Hayflick, L. & Moorhead, P. S. 1961. The serial cultivation of human diploid cell strains. *Exp Cell Res*, 25, 585–621.

Heaphy, C. M., De Wilde, R. F., Jiao, Y., Klein, A. P., Edil, B. H., Shi, C., Bettegowda, C., Rodriguez, F. J., Eberhart, C. G., Hebbar, S., Offerhaus, G. J., Mclendon, R., Rasheed, B. A., He, Y., Yan, H., Bigner, D. D., Oba-Shinjo, S. M., Marie, S. K., Riggins, G. J., Kinzler, K. W., Vogelstein, B., Hruban, R. H., Maitra, A., Papadopoulos, N. & Meeker, A. K. 2011. Altered telomeres in tumors with ATRX and DAXX mutations. *Science*, 333, 425.

Heitman, J., Movva, N. R. & Hall, M. N. 1991. Targets for cell cycle arrest by the immunosuppressant rapamycin in yeast. *Science*, 253, 905–9.

Henson, J. D., Hannay, J. A., Mccarthy, S. W., Royds, J. A., Yeager, T. R., Robinson, R. A., Wharton, S. B., Jellinek, D. A., Arbuckle, S. M., Yoo, J., Robinson, B. G., Learoyd, D. L., Stalley, P. D., Bonar, S. F., Yu, D., Pollock, R. E. & Reddel, R. R. 2005. A robust assay for alternative lengthening of telomeres in tumors shows the significance of alternative lengthening of telomeres in sarcomas and astrocytomas. *Clin Cancer Res*, 11, 217–25.

Herranz, N. & Gil, J. 2018. Mechanisms and function of cellular senescence. *J Clin Invest*, 128, 1238-46.

Hollstein, M., Hergenhahn, M., Yang, Q., Bartsch, H., Wang, Z. Q. & Hainaut, P. 1999. New approaches to understanding *p53* gene tumor mutation spectra. *Mutat Res*, 431, 199–209.

Holt, I. J., Harding, A. E., Cooper, J. M., Schapira, A. H., Toscano, A., Clark, J. B. & Morgan-Hughes, J. A. 1989. Mitochondrial myopathies: clinical and biochemical features of 30 patients with major deletions of muscle mitochondrial DNA. *Ann Neurol*, 26, 699–708.

Horn, S., Figl, A., Rachakonda, P. S., Fischer, C., Sucker, A., Gast, A., Kadel, S., Moll, I., Nagore, E., Hemminki, K., Schadendorf, D. & Kumar, R. 2013. TERT promoter mutations in familial and sporadic melanoma. *Science*, 339, 959–61.

Huang, F. W., Hodis, E., Xu, M. J., Kryukov, G. V., Chin, L. & Garraway, L. A. 2013. Highly recurrent TERT promoter mutations in human melanoma. *Science*, 339, 957–9.

Jaskelioff, M., Muller, F. L., Paik, J. H., Thomas, E., Jiang, S., Adams, A. C., Sahin, E., Kost-Alimova, M., Protopopov, A., Cadinanos, J., Horner, J. W., Maratos-Flier, E. & Depinho, R. A. 2011. Telomerase reactivation reverses tissue degeneration in aged telomerase-deficient mice. *Nature*, 469, 102–6.

Jiao, Y., Shi, C., Edil, B. H., De Wilde, R. F., Klimstra, D. S., Maitra, A., Schulick, R. D., Tang, L. H., Wolfgang, C. L., Choti, M. A., Velculescu, V. E., Diaz, L. A.,JR., Vogelstein, B., Kinzler, K. W., Hruban, R. H. & Papadopoulos, N. 2011. DAXX/ATRX, MEN1, and mTOR pathway genes are frequently altered in pancreatic neuroendocrine tumors. *Science*, 331, 1199–203.

Jin, X., Moskophidis, D. & Mivechi, N. F. 2011. Heat shock transcription factor 1 is a key determinant of HCC development by regulating hepatic steatosis and metabolic syndrome. *Cell Metab*, 14, 91–103.

Jones, R. M., Mortusewicz, O., Afzal, I., Lorvellc, M., Garcia, P., Helleday, T. & Petermann, E. 2013. Increased replication initiation and conflicts with transcription underlie Cyclin E-induced replication stress. *Oncogene*, 32, 3744–53.

Ju, Y. S., Alexandrov, L. B., Gerstung, M., Martincorena, I., Nik-Zainal, S., Ramakrishna, M., Davies, H. R., Papaemmanuil, E., Gundem, G., Shlien, A., Bolli, N., Behjati, S., Tarpey, P. S., Nangalia, J., Massie, C. E., Butler, A. P., Teague, J. W., Vassiliou, G. S., Green, A. R., Du, M. Q., Unnikrishnan, A., Pimanda, J. E., Teh, B. T., Munshi, N., Greaves, M., Vyas, P., El-Naggar, A. K., Santarius, T., Collins, V. P., Grundy, R., Taylor, J. A., Hayes, D. N., Malkin, D., Group, I. B. C., Group, I. C. M. D., Group, I. P. C., Foster, C. S., Warren, A. Y., Whitaker, H. C., Brewer, D.,

Eeles, R., Cooper, C., Neal, D., Visakorpi, T., Isaacs, W. B., Bova, G. S., Flanagan, A. M., Futreal, P. A., Lynch, A. G., Chinnery, P. F., Mcdermott, U., Stratton, M. R. & Campbell, P. J. 2014. Origins and functional consequences of somatic mitochondrial DNA mutations in human cancer. *Elife*, 3. doi: 10.7554/eLife.02935.

Kapahi, P., Zid, B. M., Harper, T., Koslover, D., Sapin, V. & Benzer, S. 2004. Regulation of lifespan in *Drosophila* by modulation of genes in the TOR signaling pathway. *Curr Biol*, 14, 885–90.

Killela, P. J., Reitman, Z. J., Jiao, Y., Bettegowda, C., Agrawal, N., Diaz, L. A., JR., Friedman, A. H., Friedman, H., Gallia, G. L., Giovanella, B. C., Grollman, A. P., He, T. C., He, Y., Hruban, R. H., Jallo, G. I., Mandahl, N., Meeker, A. K., Mertens, F., Netto, G. J., Rasheed, B. A., Riggins, G. J., Rosenquist, T. A., Schiffman, M., Shih Ie, M., Theodorescu, D., Torbenson, M. S., Velculescu, V. E., Wang, T. L., Wentzensen, N., Wood, L. D., Zhang, M., Mclendon, R. E., Bigner, D. D., Kinzler, K. W., Vogelstein, B., Papadopoulos, N. & Yan, H. 2013. TERT promoter mutations occur frequently in gliomas and a subset of tumors derived from cells with low rates of self-renewal. *Proc Natl Acad Sci U S A*, 110, 6021–6.

Kim, N. W., Piatyszek, M. A., Prowse, K. R., Harley, C. B., West, M. D., Ho, P. L., Coviello, G. M., Wright, W. E., Weinrich, S. L. & Shay, J. W. 1994. Specific association of human telomerase activity with immortal cells and cancer. *Science*, 266, 2011–15.

Klapper, W., Krams, M., Qian, W., Janssen, D. & Parwaresch, R. 2003. Telomerase activity in B-cell non-Hodgkin lymphomas is regulated by hTERT transcription and correlated with telomere-binding protein expression but uncoupled from proliferation. *Br J Cancer*, 89, 713–19.

Lee, H. S., Park, K. U., Kim, D. W., Lhn, M. H., Kim, W. H., Seo, A, N., CHANG, H. E., NAM, S. K., LEE, S. Y., OH, H. K. & KANG, S. B. 2016. Elevated microsatellite alterations at selected tetranucleotide repeats (EMAST) and microsatellite instabilty in patients with colorectal cancer and its clinical features. *Curr Mol Med*, 16, 829–39.

Li, S., Oh, Y. T., Yue, P., Khuri, F. R. & Sun, S. Y. 2016. Inhibition of mTOR complex 2 induces GSK3/FBXW7-dependent degradation of sterol regulatory element-binding protein 1 (SREBP1) and suppresses lipogenesis in cancer cells. *Oncogene*, 35, 642–50.

Lin, J. J. & Zakian, V. A. 1996. The Saccharomyces CDC13 is a singlre-strand TG1-3 telomeric DNA-binding protein in vitro that affects telomere behavior *in vivo*. *Proc Natl Acad Sci U S A*, 93, 13760–5.

Liu, B., Nicolaides, N. C., Markowitz, S., Willson, J. K., Parsons, R. E., Jen, J., Papadopolous, N., Peltomaki, P., De La Chapelle, A., Hamilton, S. R. & Et Al. 1995. Mismatch repair gene defects in sporadic colorectal cancers with microsatellite instability. *Nat Genet*, 9, 48–55.

Lopez-Otin, C., Blasco, M. A., Partridge, L., Serrano, M. & Kroemer, G. 2013. The hallmarks of aging. *Cell*, 153, 1194–217.

Lovejoy, C. A., Li, W., Reisenweber, S., Thongthip, S., Bruno, J., De Lange, T., De, S., Petrini, J. H., Sung, P. A., Jasin, M., Rosenbluh, J., Zwang, Y., Weir, B. A., Hatton, C., Ivanova, E., Macconaill, L., Hanna, M., Hahn, W. C., Lue, N. F., Reddel, R. R., Jiao, Y., Kinzler, K., Vogelstein, B., Papadopoulos, N., Meeker, A. K. & Consortium, A. L. T. S. C. 2012. Loss of ATRX, genome instability, and an altered DNA damage response are hallmarks of the alternative lengthening of telomeres pathway. *PLoS Genet*, 8, e1002772.

Lundblad, V. & Blackburn, E. H. 1993. An alternative pathway for yeast telomere maintenance rescues est1- senescence. *Cell*, 73, 347–60.

Miyake, Y., Nakamura, M., Nabetani, A., Shimamura, S., Tamura, M., Yonehara, S., Saito, M. & Ishikawa, F. 2009. RPA-like mammalian Ctc1-Stn1-Ten1 complex binds to single-stranded DNA and protects telomeres independently of the Pot1 pathway. *Mol Cell*, 36, 193–206.

Moskalev, A. A., Shaposhnikov, M. V., Plyusnina, E. N., Zhavoronkov, A., Budovsky, A., Yanai, H. & Fraifeld, V. E. 2013. The role of DNA damage and repair in aging through the prism of Koch-like criteria. *Ageing Res Rev*, 12, 661–84.

Pause, A., Belsham, G. J., Gingras, A. C., Donze, O., Lin, T. A., Lawrence, J. C., Jr. & Sonenberg, N. 1994. Insulin-dependent stimulation of protein synthesis by phosphorylation of a regulator of 5′-cap function. *Nature*, 371, 762–7.

Payne, B. A., Wilson, I. J., Hateley, C. A., Horvath, R., Santibanez-Koref, M., Samuels, D. C., Price, D. A. & Chinnery, P. F. 2011. Mitochondrial aging is accelerated by anti-retroviral therapy through the clonal expansion of mtDNA mutations. *Nat Genet*, 43, 806–10.

Poli, V., Fagnocchi, L., Fasciani, A., Cherubini, A., Mazzoleni, S., Ferillo, S., Miluzio, A., Gaudioso, G., Vaira, V., Turdo, A., Giaggianesi, M., Chinnici, A., Lipari, E., Bicciato, S., Bosari, S., Todaro, M. & Zippo, A. 2018. MYC-driven epigenetic reprogramming favors the onset of tumorigenesis by inducing a stem-like state. *Nat Commun*, 9, 1024-39.

Puig, S., Malvehy, J., Badenas, C., Ruiz, A., Jimenez, D., Cuellar, F., Azon, A., Gonzalez, U., Castel, T., Campoy, A., Herrero, J., Marti, R., Brunet-Vidal, J. & Mila, M. 2005. Role of the CDKN2A locus in patients with multiple primary melanomas. *J Clin Oncol*, 23, 3043–51.

Qi, H. & Zakian, V. A. 2000. The *Saccharomyces* telomere-binding protein Cdc13p interacts with both the catalytic subunit of DNA polymerase alpha and the telomerase-associated est1 protein. *Genes Dev*, 14, 1777–88.

Rayess, H., Wang, M. B. & Srivatsan, E. S. 2012. Cellular senescence and tumor suppressor gene *p16*. *Int J Cancer*, 130, 1715–25.

Ritossa, F. M. 1964. Experimental activation of specific loci in polytene chromosomes of *Drosophila*. *Exp Cell Res*, 35, 601–7.

Sabers, C. J., Martin, M. M., Brunn, G. J., Williams, J. M., Dumont, F. J., Wiederrecht, G. & Abraham, R. T. 1995. Isolation of a protein target of the FKBP12-rapamycin complex in mammalian cells. *J Biol Chem*, 270, 815–22.

Santagata, S., Hu, R., Lin, N. U., Mendillo, M. L., Collins, L. C., Hankinson, S. E., Schnitt, S. J., Whitesell, L., Tamimi, R. M., Lindquist, S. & Ince, T. A. 2011. High levels of nuclear heat-shock factor 1 (HSF1) are associated with poor prognosis in breast cancer. *Proc Natl Acad Sci U S A,* 108, 18378–83.

Schneider-Stock, R., Boltze, C., Jager, V., Epplen, J., Landt, O., Peters, B., Rys, J. & Roessner, A. 2003. Elevated telomerase activity, c-MYC-, and hTERT mRNA expression: association with tumour progression in malignant lipomatous tumours. *J Pathol,* 199, 517–25.

Schoeftner, S. & Blasco, M. A. 2008. Developmentally regulated transcription of mammalian telomeres by DNA-dependent RNA polymerase II. *Nat Cell Biol,* 10, 228–36.

Setlow, R. B., Regan, J. D., German, J. & Carrier, W. L. 1969. Evidence that xeroderma pigmentosum cells do not perform the first step in the repair of ultraviolet damage to their DNA. *Proc Natl Acad Sci U S A,* 64, 1035–41.

Shoffner, J. M., Lott, M. T., Lezza, A. M., Seibel, P., Ballinger, S. W. & Wallace, D. C. 1990. Myoclonic epilepsy and ragged-red fiber disease (MERRF) is associated with a mitochondrial DNA tRNA(Lys) mutation. *Cell,* 61, 931–7.

Singh, G., Lott, M. T. & Wallace, D. C. 1989. A mitochondrial DNA mutation as a cause of Leber's hereditary optic neuropathy. *N Engl J Med,* 320, 1300–5.

Smith, E. R., Cayrou, C., Huang, R., Lane, W. S., Cote, J. & Lucchesi, J. C. 2005. A human protein complex homologous to the *Drosophila* MSL complex is responsible for the majority of histone H4 acetylation at lysine 16. *Mol Cell Biol,* 25, 9175–88.

Sorger, P. K. 1991. Heat shock factor and the heat shock response. *Cell,* 65, 363–6.

Stehelin, D., Varmus, H. E., Bishop, J. M. & Vogt, P. K. 1976. DNA related to the transforming gene(s) of avian sarcoma viruses is present in normal avian DNA. *Nature,* 260, 170–3.

Stewart, J. B., Alaei-Mahabadi, B., Sabarinathan, R., Samuelsson, T., Gorodkin, J., Gustafsson, C. M. & Larsson, E. 2015. Simultaneous DNA and RNA mapping of somatic mitochondrial mutations across diverse human cancers. *PLoS Genet,* 11, e1005333.

Taipale, M., Rea, S., Richter, K., Vilar, A., Lichter, P., Imhof, A. & Akhtar, A. 2005. hMOF histone acetyltransferase is required for histone H4 lysine 16 acetylation in mammalian cells. *Mol Cell Biol,* 25, 6798–810.

Tan, E. P., Duncan, F. E. & Slawson, C. 2017. The sweet side of the cell cycle. *Biochem Soc Trans,* 45, 313–22.

Thibodeau, S. N., Bren, G. & Schaid, D. 1993. Microsatellite instability in cancer of the proximal colon. *Science,* 260, 816–19.

Vellai, T., Takacs-Vellai, K., Zhang, Y., Kovacs, A. L., Orosz, L. & Muller, F. 2003. Genetics: influence of TOR kinase on lifespan in *C. elegans. Nature,* 426, 620.

Wan, W., You, Z., Xu, Y., Zhou, L., Guan, Z., Peng, C., Wong, C. C. L., Su, H., Zhou, T., Xia, H. & Liu, W. 2017. mTORC1 phosphorylates acetyltransferase p300 to regulate autophagy and lipogenesis. *Mol Cell,* 68, 323–35 e6.

Wang, F., Podell, E. R., Zaug, A. J., Yang, Y., Baciu, P., Cech, T. R. & Lei, M. 2007. The POT1–TPP1 telomere complex is a telomerase processivity factor. *Nature,* 445, 506–10.

Wright, W. E., Piatyszek, M. A., Rainey, W. E., Byrd, W. & Shay, J. W. 1996. Telomerase activity in human germline and embryonic tissues and cells. *Dev Genet,* 18, 173–9.

Wu, J. J., Liu, J., Chen, E. B., Wang, J. J., Cao, L., Narayan, N., Fergusson, M. M., Rovira, Ii, Allen, M., Springer, D. A., Lago, C. U., Zhang, S., Dubois, W., Ward, T., Decabo, R., Gavrilova, O., Mock, B. & Finkel, T. 2013. Increased mammalian lifespan and a segmental and tissue-specific slowing of aging after genetic reduction of mTOR expression. *Cell Rep,* 4, 913–20.

Wu, K. J., Grandori, C., Amacker, M., Simon-Vermot, N., Polack, A., Lingner, J. & Dalla-Favera, R. 1999. Direct activation of TERT transcription by c-MYC. *Nat Genet,* 21, 220–4.

Zhao, Y. H., Zhou, M., Liu, H., Ding, Y., Khong, H. T., Yu, D., Fodstad, O. & Tan, M. 2009. Upregulation of lactate dehydrogenase A by ErbB2 through heat shock factor 1 promotes breast cancer cell glycolysis and growth. *Oncogene,* 28, 3689–701.

CHAPTER 21

Aging, cellular senescence and cancer: epigenetic alterations and nuclear remodeling

In recent years, a major emphasis has been placed on identifying the many epigenetic changes that are correlated with the aging process and with oncogenesis, and in deciphering whether these changes are a consequence of senescence and cancer development or whether they play an active role in the onset of the processes. The evidence to date suggests that both instances occur and that the genetic and epigenetic aspects of aging and of neoplastic cellular transformation are highly interdependent.

Changes in DNA methylation

Early observations established that an increase in methylation of particular genes occurs as a function of age (Ahuja et al., 1998; Baylin et al., 1998; Vertino et al., 1994). A landmark study of the differences that develop over time between identical twins established the fact that the genomic pattern of DNA methylation within an individual changes with age (Fraga et al., 2005). These changes involve repeated sequences, as well as single-copy genes. Whole genome analysis using the widely used single-nucleotide bisulfite sequencing technique confirmed the occurrence of widespread DNA demethylation and specific site hypermethylation in replicative senescent human cells (Cruickshanks et al., 2013). (Bisulfite sequencing involves treatment of DNA with bisulfite that converts non-methylated cytosines to uracil and leaves the methylated cytosines; sequencing of treated and untreated samples reveals the position of methylated cytosines.)

Using a set of specific CpG genomic sites, it was possible to develop a predictor with which the age of many cell types and heterogeneous tissues can be estimated with remarkable accuracy and is highly correlated with chronological age (Horvath, 2013). In this study, approximately half of the CpG methylation clock sites became methylated and half lost their methylation with age. As expected, embryonic stem (ES) cells have a DNA methylation age close to zero. Further providing proof of principle, the age of induced pluripotent stem (iPS) cells, derived from adult cells, is the same as that of ES cells. A similar study, involving a smaller number of predictive methylation markers, revealed that the DNA methylation pattern ages more rapidly in men than in women and that age-related changes in this pattern are reflected in accelerated changes of gene expression (Hannum et al., 2013). More recently, the methylation of a group of CpG sites was used to predict with accuracy the gestational age, i.e. the time spent *in utero*, and the developmental maturity of newly born infants (Knight et al., 2016).

A predictable consequence of the age-related demethylation of some regions of the genome is the reactivation of previously silenced transposable elements (see Chapter 2, Box 2.1). Retrotransposons that are no longer inactivated can be transcribed, and the resulting RNAs reverse-transcribed to produce copies that can insert throughout the genome, constituting a form of genetic instability. An increase in the level of retrotransposition as a function of age has been documented from yeast to mammals (Dennis et al., 2012; Li et al., 2013; Maxwell et al., 2011; Wang et al., 2011).

Epigenetics, Nuclear Organization and Gene Function: with implications of epigenetic regulation and genetic architecture for human development and health. John C. Lucchesi, Oxford University Press (2019). © John C. Lucchesi 2019.
DOI: 10.1093/oso/9780198831204.001.0001

DNA methylation is the first epigenetic modification reported to be correlated to cancer. Relative to normal cells, colon cancer cells tended to exhibit hypomethylation of the genome (Feinberg and Vogelstein, 1983). The general conclusions, derived from a number of subsequent studies, were that most cancers exhibited hyper- and hypomethylation of CpG islands associated with specific gene promoters, and hypomethylation of lamina-associated domains (LADs) and of large genomic blocks that included repetitive elements (Baylin et al., 1998; Berman et al., 2011; Hansen et al., 2011; Timp et al., 2014; Vertino et al., 1994). Such genome-wide hypomethylation has been associated with the aberrant activation of repetitive elements and endogenous retroviruses, with the ectopic activation of non-lineage-appropriate enhancers and with cryptic activation of intragenic antisense transcription. In addition, normally unmethylated CpG islands associated with specific gene promoters can become hypermethylated, leading to the inactivation of tumor suppressor genes. Although some differences in DNA methylation between normal tissues and the specific cancers that arose from each tissue appear to be tissue-specific, there are significant similarities across cancer types (Chen et al., 2016; Perez et al., 2018).

In some cases, changes in DNA methylation related to oncogenesis can be ascribed directly to the occurrence of mutations in DNA methyl transferases or demethylases. Loss-of-function mutations in *DNMT3* or *TET2* lead to hematologic cancers (Genovese et al., 2014; Jaiswal et al., 2014; Xie et al., 2014).

DNA methylation profiling has been used to establish prognostic biomarkers for many tumors. For example, in neuroblastoma, a childhood cancer, a number of genomic regions that are either hyper- or hypomethylated have been used to subdivide different subtypes of the disease in terms of survival and biological variables (Decock et al., 2016; Henrich et al., 2016).

An unexpected correlation appears to exist between particular histone marks in normal cells and those methylated DNA sites that are specific to the cancers that originate from these cells. In particular, the presence of H3K27me3 in gene promoter regions is a highly predictive mark for genes that will exhibit hypomethylation when these cells undergo transformation to cancer cells (Chen et al., 2016).

Changes in histone modifications

The many post-translational histone modifications that are implicated in the maintenance and modulation of transcription during cellular and organismal differentiation, and in the processes of chromatin replication and DNA repair, exhibit significant changes in the aging process. Most of these changes are the result of alterations in the level or activity of the enzymes responsible for the deposition or removal of the modifications. These conclusions, derived principally from studies with model organisms, contribute to the contention that a fundamental cause of aging is the accumulation of somatic genomic mutations (Garcia et al., 2010; Vijg et al., 2002).

Many of the earlier observations on the age-related changes in epigenetic modifications were made by noting the effects of abrogating or over-expressing particular histone-modifying enzymes on the longevity of an organism or the replicative performance of a cell line. The changes are often characteristic of a particular species and, within that species, they can be tissue-specific (Larson et al., 2012; Wood et al., 2010). In *Caenorhabditis elegans*, for example, senescence is correlated with a global increase in H3K4me3 (the mark of active genes) and a decrease in H3K27me3 (the silencing mark laid down by Polycomb repressive complexes) and in H3K36me3 (a modification associated with transcription elongation that occurs along gene bodies and prevents the use of cryptic transcription initiation sites) (Greer et al., 2010; McColl et al., 2008; Pu et al., 2015; Sen et al., 2015). In *Drosophila*, a similar result is obtained with respect to H3K4me3, while a decrease in H3K27me3 enhances longevity (Siebold et al., 2010). In a specific tissue—the adult *Drosophila* brain—decreases in both H3K4me3 and H3K36me3 and an increase in H3K9me3 (a mark of repressed chromatin) were observed as a function of age (Wood et al., 2010). In mammals, the patterns of age-related histone modifications are similarly diverse. In the brain tissue of a mouse strain with accelerated senescence, the levels of H4K20me1 (a mark associated with actively transcribing genes) and H3K36me3 are decreased and that of H3K27me3 is increased (Wang et al., 2010). Cells collected from an individual with a premature aging syndrome show a decrease in H3K9me3 and H3K27me3 and an increase in H4K20me3 (Shumaker et al., 2006). In spite of such variation among different organisms and their tissues, the general impression emerges that histone methylation marks involved in repression diminish as a function of age and those associated with active gene function are increased. Histone acetylation, a chromatin modification usually associated with open chromatin and gene activation, appears to follow the same general pattern.

Yeast provides a very useful example of the age-associated changes in histone acetylation. Cells that are undergoing replicative senescence exhibit an increase in H4K16ac, and a decrease in H3K56ac located in the globular domain of histone H3 (Dang et al., 2009; Feser et al., 2010). While the induction of a modest increase in H3K56ac (present at enhancers and transcription start sites) retards the aging process, a large increase in the level of this modification has the opposite effect, suggesting that a narrow range is optimal for replicative longevity (Lamming and Sinclair, 2008). Similarly, senescing human fibroblasts exhibit an increase in H4K16ac and a decrease in H3K56ac (O'Sullivan et al., 2010).

The **sirtuins** (SIRTs) are a set of enzymes of particular significance to the aging process. First discovered in yeast as gene products that are required for silencing mating-type loci and telomeres (Rine and Herskowitz, 1987), these enzymes were implicated in the aging process (Guarente, 1999). In *Caenorhabditis* and *Drosophila*, SIRT over-expression leads to significant increases in longevity (Banerjee et al., 2012; Rizki et al., 2011). SIRTs are highly conserved protein deacetylases represented in mammals, including humans, by seven members (SIRT1–7). SIRTs differ from other deacetylases in that they transfer the acetyl group from acetylated lysines to NAD+ (the oxidized form of nicotinamide adenine dinucleotide). In humans, SIRT1–3 are active deacetylases. SIRT1, to date the most extensively studied member of the group, promotes silencing and the formation of heterochromatin by deacetylating histones H1K26ac, H3K9ac and H4K16ac (Imai et al., 2000; Landry et al., 2000; Smith et al., 2000). In addition, SIRT1 deacetylates and suppresses the activity of transcription factors such as FOXO (forkhead transcription factor O). Normally, the FOXO factors activate the transcription of target genes that trigger DNA repair or cause cell cycle arrest or cell death. SIRT1 also deacetylates and inactivates the tumor suppressor factor p53, thereby reducing cell death in cases of DNA damage or oxidative stress.

A particular aspect of the function of SIRTs is their interaction with caloric restriction, a physiological state known to affect longevity (Lee et al., 1999; McCay et al., 1935; Weindruch et al., 1986). Caloric restriction reduces the level of insulin and of insulin-like growth factor 1 (IGF-1) and extends longevity in worms, flies and mice (Bluher et al., 2003); it increases the concentration of cellular NAD+, which, in turn, induces the synthesis of SIRT1 (Rodgers et al., 2004).

Because of its influence on chromatin structure and its association with gene activity, histone acetylation plays a prominent role in the onset and maintenance of the oncogenic process. Widespread loss of histone H4

lysine 16 acetylation (H4K16ac) and of lysine 20 tri-methylation (H4K20me3) is a characteristic of lymphoma and adenocarcinoma cells (Fraga et al., 2005). Consistent with these observations, the levels of different histone deacetylases were usually found to be elevated in cancers, including prostate (Patra et al., 2001), colon (Della Ragione et al., 2001) and lung (Osada et al., 2001) cancers. In contrast, in gliomas (aggressive brain tumors), the level of H4K16ac is elevated. Surprisingly, the presence of the SIRT that specifically deacetylates this histone isoform is also increased. In this case, the SIRT deacetylates MOF, the acetyl transferase responsible for the acetylation of H4 at lysine 16, thereby increasing its affinity for this histone (Saidi et al., 2018).

Histone H3 methylated at lysine 4 (H3K4me3) is associated with active promoters and is catalyzed by the methyl transferase MLL (mixed-lineage leukemia)/KMT2, a homolog of the *Drosophila Trx* gene (see Chapter 7) and of yeast *Set1*. Many cancer types are found to carry mutations in genes that encode chromatin-modifying proteins. A particular example are mutations in the genes that encode MLL enzymes (Lawrence et al., 2014). The link between MLL function and cancer was first established by the observation that the *MLL* gene was involved in translocations found in acute myeloid leukemias (Ziemin-van der Poel et al., 1991). In these translocations, the portion of the MLL protein that lacks the methyl transferase domain is fused to a partner protein, resulting in a chimeric protein with no methyl transferase activity (Mohan et al., 2010). The chimeric proteins often recruit the DOT1L/KMT4 methyl transferase that methylates H3K79, a modification correlated with gene transcription (Schubeler et al., 2004). Other examples of mutations in chromatin-modifying genes are mutations in the EZH2 methyl transferase subunit of the PRC2 complex that is responsible for the H3K27me3 repressive mark. Some mutations in the EZH2 active site that increase its enzymatic activity are often found in some lymphomas (Morin et al., 2010; Sneeringer et al., 2010). Loss-of-function mutations in EZH2 or in other core subunits of the PRC2 complex are associated with other cancers such as certain leukemias (see, for example, Ntziachristos et al., 2012). Other histone-modifying enzymes are implicated in tumorigenesis by the observation that lower levels of H3K18ac and of H3K4me2 in cancer cells predict a worse prognosis for prostate cancer survival (Seligson et al., 2005).

In some cancers, the histone modifications that normally characterize enhancers—H3K4me1[-] and H3K27ac—are altered. Mutations in the particular MLL methyl transferases responsible for H3K4me1 are found in a variety of cancers, including, for example, breast cancer (Sjoblom et al., 2006), childhood medulloblastoma (Parsons et al., 2011) and bladder cancer (Gui et al., 2011). The transcriptional profile of some cancers is significantly similar to the profile of ES cells. These ES genes are activated by the *Myc* gene product and include the transcription factors responsible for the maintenance of pluripotency (Stergachis et al., 2013; Wong et al., 2008). In a more recent study, changes in the methylation status of enhancers during the differentiation of ES cells to hematopoietic cells and in acute leukemia cells revealed a significant similarity between groups of enhancers in the ES and cancer cells (Aran et al., 2016).

Changes in histone variants

A change in the levels of two histone variants is associated with senescence. Whereas the canonical histone H3 is incorporated when nucleosomes are re-established following DNA replication, H3.3 is deposited in chromatin in a replication-independent manner and is found at active enhancers and promoters and along gene bodies, as well as in telomeric and pericentric heterochromatin (see Chapter 4). Senescent cells accumulate abnormal amounts of H3.3 and of a truncated form of this variant, perhaps due to the slow-down and eventual cessation of DNA replication (Duarte et al., 2014). The second histone variant macroH2A is normally present in silenced regions of the genome and in heterochromatin (see Chapter 4). Histone macroH2A is prevalent in the senescence-associated heterochromatic foci (SAHFs; see Changes in nuclear architecture) and accumulates in senescent cells of a number of mammalian species (Kreiling et al., 2011; Zhang et al., 2005).

A number of non-canonical histones are implicated in oncogenesis and cancer. Mutations that occur in the H3.3 gene are very strongly correlated with particular cancers. For example, a mutation that changes the lysine at position 27 to a methionine (H3.3K27M) is found in pediatric glioblastomas (Schwartzentruber et al., 2012; Wu et al., 2012), and a substitution of lysine 36 for methionine (H3.3K36M) is present in most chondroblastomas (Behjati et al., 2013). These H3.3 mutations exert their action in heterozygous condition, i.e. in the presence of normal H3.3 alleles; evidence that they are a causative factor of the cancers is provided by the observation that when they are expressed in mesenchymal progenitor cells or ES cells, they induce neoplastic transformations (Funato et al., 2014; Lu et al., 2016).

Another histone variant macroH2A, which occurs in regions of facultative repressed chromatin, appears to

be associated with the occurrence of melanomas. This variant normally is involved in the suppression of the *CDK8* oncogene; absence of macroH2A isoforms leads to an increase in the level of CDK8 and to the progression of malignant melanomas (Kapoor et al., 2010). CDK8 has cancer tissue-type specific expression and is a selective regulator of gene expression levels (Rzymski et al., 2015). For this reason, macroH2A has been implicated in numerous tumor types.

In addition to its role in the repair of double-strand DNA breaks, the H2A variant H2A.Z is over-expressed in a variety of different cancers (see, for example, Hua et al., 2008; Kim et al., 2013; Vardabasso et al., 2015). H2A.Z is enriched at gene promoters and other regulatory regions and is usually associated with the modulation of transcription.

Chromatin remodeling

There are several distinct families of remodelers (Henikoff, 2016). Some of these families function in DNA repair and recombination (the RAD complexes); others facilitate activation (for example, SNF2) or the regular spacing of nucleosomes (ISWI), or help during the elongation process of transcription (CHD). The changes in the chromatin organization and in the transcriptional programs exhibited by senescent cells obviously involve changes in nucleosomal positioning. Surprisingly, the expected concomitant changes in the distribution and function of ATP-dependent remodeling complexes, with a few exceptions, have been less obvious. The nucleosome remodeling and deacetylase (NuRD) complex is down-regulated in cells from normally aging individuals and in cells from prematurely aging patients (Pegoraro et al., 2009). Down-regulation is achieved by the reduced transcription of some of the complex subunits. A similar involvement of NuRD in the regulation of longevity was found in *Caenorhabditis* and *Drosophila* (De Vaux et al., 2013).

Mutations in several subunits of the SWI/SNF complexes (also called BAF complexes in humans; see Chapter 7) are found in different cancers with significantly high frequencies, and certain subunits are mutated more frequently in particular cancer types (Kadoch et al., 2013). SWI/SNF subunit mutations occur at a higher frequency in cancers than do mutations in the subunits of other remodeling complexes, such as TIP60, INO80, NuRD, ISWI, or of histone-modifying complexes such as histone deacetylases and methyl transferases (Kadoch et al., 2013). Rhabdoid tumors are predominantly associated with the loss of the SNF5 (also known as SMARCB1) subunit of the SWI/SNF complex. The loss of this subunit causes a decrease in the number of complexes present on enhancers that are involved in differentiation, while retaining those complexes that are responsible for current cell identity (Wang et al., 2017).

CHD4, the ATP-dependent subunit of NuRD complexes, is required for the growth of leukemic cells in childhood acute myeloid leukemia (Heshmati et al., 2018). The specificity of this requirement is demonstrated by the observation that CHD4 is not necessary for the growth of normal hematopoietic cells.

Changes in nuclear architecture

Changes in the size and shape of nuclei

The shape of nuclei is determined by their DNA content, the general organization of the chromatin, the composition of the nuclear membrane and the underlying nuclear lamina. As previously discussed, the nuclear lamina is instrumental in conferring mechanical stability to the nucleus and plays important roles in DNA replication and in the organization of chromatin and gene expression during interphase. In mammals, four different types of lamins are found in the lamina, and their relative expression levels are characteristic of different cell types (see Chapter 11). A number of diseases, termed **laminopathies**, have been associated with mutations in the *LMNA* gene that encodes the two splice variant isoforms lamins A and C. These diseases include certain muscular dystrophies, certain cardiomyopathies and, of particular relevance here, premature aging syndromes (De Sandre-Giovannoli et al., 2003; Eriksson et al., 2003).

The role of lamins on the nuclear organization of chromatin can be studied by tracking the movement of tagged regions in living cells. These observations can be converted into diffusion parameters that range from normal or random diffusion, comparable to Brownian movement, to abnormal or constrained diffusion. Using this approach, telomeres were found to exhibit abnormal diffusion in normal mouse cells and to switch to normal diffusion if the cells were deficient in lamin A or expressing the *LMNA* gene mutations causing various diseases (Bronshtein et al., 2015). Whether there is any overlap in the physiological consequences of changes in telomere localization caused by these lamin A defects or by uncapping of the telomere ends has not been determined.

Changes in the expression of lamins have been seen in different cancers. For example, the expression of

lamins A and C are decreased in colon cancer (Belt et al., 2011), gastric carcinoma (Wu et al., 2009) and ovarian cancer (Gong et al., 2015); in contrast, these lamins are over-expressed in prostate cancer (Kong et al., 2012) and colorectal cancer (Willis et al., 2008). A number of early observations revealed that in cells from laminopathy patients, some of the heterochromatic portions of the genome that are normally associated with the lamina are displaced, leading to a modification of the pattern of gene expression (Goldman et al., 2004; Sabatelli et al., 2001; Scaffidi and Misteli, 2006).

Changes in the arrangement of the chromosomes

In the nuclei of eukaryotes, the location of chromosomes is not random, with each chromosome occupying a particular position or chromosome territory. In general, gene-rich chromosomes are more centrally located while gene-poor chromosomes are more peripheral (see Chapter 10). Changes in the relative positions of particular chromosomes were detected in a number of tumor cells (Cremer et al., 2003; Wiech et al., 2005). In a different study, the pair-wise association of six chromosomes was calculated. While the probabilities of these associations were characteristic of different normal tissues, significant changes existed between a malignant breast cell line and the normal breast cell line control (Marella et al., 2009).

Changes in the A (active) and B (inactive) compartments

Another aspect of the intra-nuclear reorganization that occurs as a function of age is a change in the distribution of the heterochromatic regions of the genome. Human fibroblasts induced to undergo senescence exhibit a number of punctate elements, termed senescence-associated heterochromatic foci (SAHFs), which are enriched in heterochromatin proteins such as HP1 and H3K9me3 (Narita et al., 2003). These foci are not the result of a re-distribution of the canonical heterochromatic regions of the genome, such as those associated with centromeric or telomeric repetitive sequences, and can be considered to be facultative heterochromatin (see Chapter 2). Although SAHFs occur in many senescent human cell types, they are not found in the cells of patients with some premature aging disease or in mouse cells. In contrast, all senescent cells examined undergo a significant level of decondensation of their pericentromeric regions (Swanson et al., 2013). These senescence-associated distention of satellites

(SADS) regions, made up predominantly of repeated DNA sequences, retain the H3K9me3 and H3K27me3 marks of heterochromatin, indicating that this unfolding is different from that involved in gene activation.

The canonical, universal nucleosomal organization of eukaryotic heterochromatin is altered in aging cells by a progressive reduction in histone synthesis. First reported in yeast (Dang et al., 2009), a reduction in the level of histones occurs in *C. elegans* (Ni et al., 2012) and in human cultured cells (Ivanov et al., 2013; O'Sullivan et al., 2010). An intriguing causal correlation was found between telomere shortening and nucleosome depletion during the replicative senescence of human fibroblasts in culture (O'Sullivan et al., 2010), suggesting that reduction in histone levels may be as common a cause of aging as telomere shortening.

Not surprisingly, the nuclear position of individual genes is also altered in cancer cells. One of the first detailed studies of this occurrence made use of normal mammary epithelial cells that could mimic breast cancer cells following the over-expression of an epidermal growth factor receptor (ERBB2). Several cancer-associated genes—*AKT1* (alpha serine/threonine protein kinase), *BCL2* (apoptosis regulator) and *VEGFA* (vascular endothelial growth factor A)—moved to a more peripheral position during tumorigenesis, while the *ErbB2* gene became more internal. Interestingly, the transcriptional activity of these genes was not correlated to their new position (Meaburn and Misteli, 2008). Two of four genes identified in the cell culture experiment (*AKT1* and *ERBB2*) were repositioned in breast cancer tissue specimens, and some genes, such as *TGFB3* (transforming growth factor beta 3), that did not reposition in cell culture were repositioned in breast cancer tissues (Meaburn et al., 2009).

Changes in nucleolar size

Evidence obtained with *Caenorhabditis, Drosophila*, mice and humans supports the conclusion that nucleolar size is inversely correlated with longevity. Long-lived mutant strains of the three model organisms exhibited smaller nucleoli and a concomitant reduction in the synthesis of ribosomal RNA, ribosomal proteins and some of the factors and enzymes (such as fibrillarin) involved in ribosome biogenesis (Tiku et al., 2016). In a complementary observation, individuals affected by the Hutchinson–Gilford progeria premature aging syndrome exhibit elevated global protein synthesis, enhanced ribosome biogenesis and increased nucleolar size (Buchwalter and Hetzer, 2017).

Changes in the topological organization of the genome

Some of the observations just discussed may have resulted from a change in the topologically associating domain (TAD) organization of the genome, altering the normal relationship of enhancers and target genes. Specific examples of this occurrence in cancer are the demonstration that defects in two elements that are critical in establishing chromatin architecture—CTCF (CCCTC-binding factor) and cohesin—are functionally correlated with carcinogenesis. Mutations in the *IDH* (isocitrate dehydrogenase) gene often occur in gliomas (Cairns et al., 2013; Yan et al., 2009). These mutations cause the formation of an inhibitor that interferes with DNA demethylases, leading to an over-methylation of the genome (Noushmehr et al., 2010) and potentially interfering with CTCF binding and genome topology. In fact, the experimental elimination of a CTCF domain boundary resulted in the aberrant activation of a particular glioma oncogene by a constitutive enhancer (Flavahan et al., 2016). Mutations in different cohesin components in human hematopoietic stem and progenitor cells lead to an increase in the occupancy of a number of transcription factor binding sites in the genome and to the expression of myeloid leukemia-specific genes (Mazumdar et al., 2015). Consistent with these observations is the presence of point mutations at CTCF/cohesin sites in a number of human colorectal cancers (Katainen et al., 2015). In another study, the activation of proto-oncogenes responsible for the onset of myeloid leukemia was correlated with deletions that spanned a neighboring loop boundary CTCF site (Hnisz et al., 2016).

The role of non-coding RNAs

First reported in *Caenorhabditis* (Lee et al., 1993), the level and pattern of short non-coding RNAs and a decrease in some of the processing enzymes, such as Dicer, are altered in aging worms, flies and mice (Boehm and Slack, 2005; Ibanez-Ventoso et al., 2006; Liu et al., 2012; Mori et al., 2012). The levels of specific micro RNAs (miRNAs) were found to be increased in the brain of aging primates and to be correlated with age-related neurodegeneration (Persengiev et al., 2011). Long non-coding RNAs (lncRNAs) are also correlated with the aging process. A particular lncRNA increases in mice brains as a function of age and induces a decline in learning ability (Meier et al., 2010). Additional lncRNAs are present in different levels as cells enter into replicative senescence (Abdelmohsen et al., 2013; Kour and Rath, 2015).

Given the relatively recent, extensive documentation of the role of non-coding RNAs in most of the mechanisms and functions of the genetic material (see Chapter 6), it is expected that these RNAs will be implicated in the maintenance of genomic stability, thereby mitigating the effects of aging and oncogenesis. The hypothesis that non-coding RNAs may participate in the repair of double-stranded breaks was suggested by the correlated observations that: (1) cellular senescence, a state in which cells that normally proliferate have stopped dividing, yet remain alive and metabolically active, can be caused by induced DNA damage (Di Leonardo et al., 1994) or the activation of oncogenes (Bartkova et al., 2006; Di Micco et al., 2006); (2) the nuclei of oncogene-induced senescent cells exhibit a large number of dispersed foci that are heterochromatic in nature (Narita et al., 2003); (3) non-coding RNAs are instrumental in heterochromatin formation (Martienssen and Moazed, 2015). The involvement of DNA repair-specific miRNAs was indicated by the dependence of DNA repair foci (containing 53BP1 and phosphorylated ATM) on the presence of the RNases Dicer and Drosha that are responsible for miRNA synthesis (see Chapter 6). These RNAs, referred to as diRNAs (double-strand break-induced small RNAs), are produced from sequences that flank the break (Francia et al., 2012; Wei et al., 2012). How the transcription of these diRNAs is induced and whether their role is to target particular proteins to the site of the break, to bridge the break and serve as a template for repair or to perform some other function related to DNA repair and stability is not determined at this time.

In contrast, a large number of observations have associated miRNAs with cancer. A small chromosomal deletion that is often present in patients with a particular type of leukemia was shown to contain two miRNAs (Calin et al., 2002). Specific miRNAs are over- or under-expressed in tumors, suggesting that they can act as oncogenic or tumor suppressor factors (Volinia et al., 2006). A single nucleotide change in a miRNA increases its affinity for the BRCA1 and BRCA2 mRNAs and indicates an earlier age of onset for breast and ovarian cancers (Shen et al., 2008). The presence of particular miRNAs in metastatic lesions can be used to identify their tissue of origin (Rosenfeld et al., 2008). CTCF has been implicated in the regulation of miRNA synthesis (Saito and Saito, 2012), and in glioblastoma, a particular miRNA is down-regulated by the loss of

CTCF from its promoter, leading to DNA methylation and inactivation (Ayala-Ortega et al., 2016).

Like miRNAs, tRNA-derived non-coding RNAs are misregulated in cancers where they can act as oncogenes or tumor suppressors (Pekarsky et al., 2016).

In yeast, there is evidence that long antisense transcripts complementary to the region of the break can serve as templates for double-stranded break repair, once again helping to maintain genomic stability (Keskin et al., 2014; Keskin et al., 2016). Whether the same process operates in flies or mammals is not known, but a large number of observations have implicated the misregulation of lncRNAs in a variety of cancers (Iyer et al., 2015; Yan et al., 2015). For example, the expression of lncRNAs is increased in pancreatic cancer (Previdi et al., 2017); some lncRNAs are over-expressed, while others are down-regulated in head and neck cancers (Li et al., 2017) and in skin melanoma (Leucci et al., 2016). MALAT1 (metastasis-associated lung adenocarcinoma transcript 1), first identified as a marker for lung adenocarcinoma (Ji et al, 2003), performs a number of transcription regulator functions, including the regulation of the alternative splicing of oncogenes and of tumor suppressor genes (Malakar et al, 2017; Tano et al., 2010; Tripathi et al., 2010).

Cellular (replicative) senescence as a native defense against cancer

Senescent cells exhibit increased levels of active p53 (a protein that responds to cellular stress by inducing cell cycle arrest, DNA repair and apoptosis), p21 (a protein that is induced by p53 and is responsible for the p53-mediated cell cycle arrest) and p16 (a protein that prevents the degradation of p53 but also induces cell cycle arrest independently of p53). Among the first indications that senescence is a mechanism of tumor suppression was the observation that the *RAS* and *RAF* oncogenes induce a cell cycle arrest that is practically identical to the arrest that occurs in regular cellular senescence (Serrano et al., 1997; Zhu et al., 1998). However, in addition to this beneficial role played by senescence as an intrinsic defense against the proliferation of potentially cancerous cells, this cellular state can also contribute to the transformation of neighboring cells that are predisposed towards tumorigenesis, i.e. premalignant cells into malignant cells (Krtolica et al., 2001). This is accomplished through the SASP (senescence-associated secretory phenotype) that is triggered by activation of the DNA damage response, upstream and independently of the function of p53 (Rodier et al., 2009). This phenotype consists of a variety of secreted factors associated with the extracellular matrix such as the insulin-like growth factor-binding protein 3 (IGFBP-3) that suppresses insulin-like growth factor (IGF) signaling, interleukin 6 (IL6) (a cytokine that functions in inflammation), PAI-1 (plasminogen activator inhibitor 1) and a number of additional growth factors and proteases (Goldstein et al., 1994; Kuilman et al., 2008; Liu and Hornsby, 2007).

The histone methyl transferase MLL1/KMT2 provides a direct connection between SASP and epigenetic chromatin modifications. MLL1 directly methylates and promotes the expression of a number of cancer-promoting kinase and transcription factor genes; its abrogation prevents the occurrence of SASP and its effect on neighboring premalignant cells (Capell et al., 2016).

Chapter summary

In recent years, a major emphasis has been placed on identifying the many epigenetic changes that are correlated with the aging process and with oncogenesis. One of the first modifications to be implicated with aging and with cancer is a change in the pattern of DNA methylation. Age-related and cancer-related changes consist of specific site hypermethylation, potentially inactivating tumor suppressor genes, and generally widespread DNA demethylation. The consequences of DNA demethylation include the activation of repetitive elements and retroviruses, the ectopic activation of non-lineage-appropriate enhancers and the cryptic activation of intragenic antisense transcription.

Age-related changes in histone modifications are generally characteristic of a particular species and, within that species, they can be tissue-specific. In spite of such variation, the general impression emerges that histone methylation marks involved in repression diminish as a function of age and those related to active gene function are increased. Histone acetylation, a chromatin modification usually associated with open chromatin and gene activation, appears to follow the same general pattern. The **sirtuins** are a set of highly conserved protein deacetylases of particular significance to the aging process. These enzymes, which deacetylate several histones and other proteins, including transcription and tumor suppressor factors, increase the life span in model organisms. Many cancer types are found to carry mutations in chromatin-modifying genes, such as those encoding methyl or acetyl transferases, affecting the histone modifications of promoters and enhancers.

Two histone variants—H3.3 and macroH2A—normally present in cells are increased in concentration as a function of age. Mutations that occur in the H3.3 gene are very strongly correlated with particular cancers. The histone variant H2A.Z is over-expressed in a variety of different cancers.

The aging process and oncogenesis present with a number of changes in nuclear architecture. Mutations in the lamina-coding genes lead to diseases termed laminopathies, among which are premature aging syndromes. Changes in the distribution of the heterochromatic regions of the genome occur as a function of age. Mutations in the subunits of some ATP-dependent remodeling complexes are found in different cancers with significant frequency. Changes in the relative position of chromosome territories have been detected in a number of tumor cells. The size of the nucleoli and the level of their function in ribosome biogenesis are inversely correlated with longevity. Epigenetic DNA modifications affect the architectural protein binding sites at TAD borders and therefore can cause the merging of neighboring TADs.

The level and pattern of short and long non-coding RNAs are altered in aging worms, flies and mice, and a large number of observations have associated miRNAs with cancer.

The positive correlation between age-related and cancer-related epigenetic changes is striking. Nevertheless, cellular or replicative senescence, whereby the total number of divisions that a particular cell will undergo is limited mostly by the absence of telomerase, represents a mechanism of tumor suppression.

References

Abdelmohsen, K., Panda, A., Kang, M. J., Xu, J., Selimyan, R., Yoon, J. H., Martindale, J. L., De, S., Wood, W. H., 3Rd, Becker, K. G. & Gorospe, M. 2013. Senescence-associated lncRNAs: senescence-associated long noncoding RNAs. *Aging Cell,* 12, 890–900.

Ahuja, N., Li, Q., Mohan, A. L., Baylin, S. B. & Issa, J. P. 1998. Aging and DNA methylation in colorectal mucosa and cancer. *Cancer Res,* 58, 5489–94.

Aran, D., Abu-Remaileh, M., Levy, R., Meron, N., Toperoff, G., Edrei, Y., Bergman, Y. & Hellman, A. 2016. Embryonic stem cell (ES)-specific enhancers specify the expression potential of ES genes in cancer. *PLoS Genet,* 12, e1005840.

Ayala-Ortega, E., Arzate-Mejia, R., Perez-Molina, R., Gonzalez-Buendia, E., Meier, K., Guerrero, G. & Recillas-Targa, F. 2016. Epigenetic silencing of miR-181c by DNA methylation in glioblastoma cell lines. *BMC Cancer,* 16, 226.

Banerjee, K. K., Ayyub, C., Ali, S. Z., Mandot, V., Prasad, N. G. & Kolthur-Seetharam, U. 2012. dSir2 in the adult fat body, but not in muscles, regulates life span in a diet-dependent manner. *Cell Rep*, 2, 1485–91.

Bartkova, J., Rezaei, N., Liontos, M., Karakaidos, P., Kletsas, D., Issaeva, N., Vassiliou, L. V., Kolettas, E., Niforou, K., Zoumpourlis, V. C., Takaoka, M., Nakagawa, H., Tort, F., Fugger, K., Johansson, F., Sehested, M., ANDAndersen, C. L., Dyrskjot, L., Orntoft, T., Lukas, J., Kittas, C., Helleday, T., Halazonetis, T. D., Bartek, J. & Gorgoulis, V. G. 2006. Oncogene-induced senescence is part of the tumorigenesis barrier imposed by DNA damage checkpoints. *Nature*, 444, 633–7.

Baylin, S. B., Herman, J. G., Graff, J. R., Vertino, P. M. & Issa, J. P. 1998. Alterations in DNA methylation: a fundamental aspect of neoplasia. *Adv Cancer Res*, 72, 141–96.

Behjati, S., Tarpey, P. S., Presneau, N., Scheipl, S., Pillay, N., Van Loo, P., Wedge, D. C., Cooke, S. L., Gundem, G., Davies, H., Nik-Zainal, S., Martin, S., Mclaren, S., Goodie, V., Robinson, B., Butler, A., Teague, J. W., Halai, D., Khatri, B., Myklebost, O., Baumhoer, D., Jundt, G., Hamoudi, R., Tirabosco, R., Amary, M. F., Futreal, P. A., Stratton, M. R., Campbell, P. J. & Flanagan, A. M. 2013. Distinct H3F3A and H3F3B driver mutations define chondroblastoma and giant cell tumor of bone. *Nat Genet*, 45, 1479–82.

Belt, E. J., Fijneman, R. J., Van Den Berg, E. G., Bril, H., Delis-Van Diemen, P. M., Tijssen, M., Van Essen, H. F., De Lange-De Klerk, E. S., Belien, J. A., Stockmann, H. B., Meijer, S. & Meijer, G. A. 2011. Loss of lamin A/C expression in stage II and III colon cancer is associated with disease recurrence. *Eur J Cancer*, 47, 1837–45.

Berman, B. P., Weisenberger, D. J., Aman, J. F., Hinoue, T., Ramjan, Z., Liu, Y., Noushmehr, H., Lange, C. P., Van Dijk, C. M., Tollenaar, R. A., Van Den Berg, D. & Laird, P. W. 2011. Regions of focal DNA hypermethylation and long-range hypomethylation in colorectal cancer coincide with nuclear lamina-associated domains. *Nat Genet*, 44, 40–6.

Bluher, M., Kahn, B. B. & Kahn, C. R. 2003. Extended longevity in mice lacking the insulin receptor in adipose tissue. *Science*, 299, 572–4.

Boehm, M. & Slack, F. 2005. A developmental timing micro-RNA and its target regulate life span in *C. elegans*. *Science*, 310, 1954-7.

Bronshtein, I., Kepten, E., Kanter, I., Berezin, S., Lindner, M., Redwood, A. B., Mai, S., Gonzalo, S., Foisner, R., Shav-Tal, Y. & Garini, Y. 2015. Loss of lamin A function increases chromatin dynamics in the nuclear interior. *Nat Commun*, 6, 8044.

Buchwalter, A. & Hetzer, M. W. 2017. Nucleolar expansion and elevated protein translation in premature aging. *Nat Commun*, 8, 328.

Cairns, R. A. & Mak, T. W. 2013. Oncogenic isocitrate dehydrogenase mutations: mechanisms, models, and clinical opportunities. *Cancer Discov*, 3, 730–41.

Calin, G. A., Dumitru, C. D., Shimizu, M., Bichi, R., Zupo, S., Noch, E., Aldler, H., Rattan, S., Keating, M., Rai, K., Rassenti, L., Kipps, T., Negrini, M., Bullrich, F. & Croce, C. M. 2002. Frequent deletions and down-regulation of micro-RNA genes miR15 and miR16 at 13q14 in chronic lymphocytic leukemia. *Proc Natl Acad Sci U S A*, 99, 15524–9.

Capell, B. C., Drake, A. M., Zhu, J., Shah, P. P., Dou, Z., Dorsey, J., Simola, D. F., Donahue, G., Sammons, M., Rai, T. S., Natale, C., Ridky, T. W., Adams, P. D. & Berger, S. L. 2016. MLL1 is essential for the senescence-associated secretory phenotype. *Genes Dev*, 30, 321–36.

Chen, Y., Breeze, C. E., Zhen, S., Beck, S. & Teschendorff, A. E. 2016. Tissue-independent and tissue-specific patterns of DNA methylation alteration in cancer. *Epigenetics Chromatin*, 9, 10.

Cremer, M., Kupper, K., Wagler, B., Wizelman, L., Von Hase, J., Weiland, Y., Kreja, L., Diebold, J., Speicher, M. R. & Cremer, T. 2003. Inheritance of gene density-related higher order chromatin arrangements in normal and tumor cell nuclei. *J Cell Biol*, 162, 809–20.

Cruickshanks, H. A., Mcbryan, T., Nelson, D. M., Vanderkraats, N. D., Shah, P. P., Van Tuyn, J., Singh Rai, T., Brock, C., Donahue, G., Dunican, D. S., Drotar, M. E., Meehan, R. R., Edwards, J. R., Berger, S. L. & Adams, P. D. 2013. Senescent cells harbour features of the cancer epigenome. *Nat Cell Biol*, 15, 1495–506.

Dang, W., Steffen, K. K., Perry, R., Dorsey, J. A., Johnson, F. B., Shilatifard, A., Kaeberlein, M., Kennedy, B. K. & Berger, S. L. 2009. Histone H4 lysine 16 acetylation regulates cellular lifespan. *Nature*, 459, 802–7.

Decock, A., Ongenaert, M., Cannoodt, R., Verniers, K., De Wilde, B., Laureys, G., Van Roy, N., Berbegall, A. P., Bienertova-Vasku, J., Bown, N., Clement, N., Combaret, V., Haber, M., Hoyoux, C., Murray, J., Noguera, R., Pierron, G., Schleiermacher, G., Schulte, J. H., Stallings, R. L., Tweddle, D. A., Children's, C., Leukaemia, G., De Preter, K., Speleman, F. & Vandesompele, J. 2016. Methyl-CpG-binding domain sequencing reveals a prognostic methylation signature in neuroblastoma. *Oncotarget*, 7, 1960–72.

Della Ragione, F., Criniti, V., Della Pietra, V., Borriello, A., Oliva, A., Indaco, S., Yamamoto, T. & Zappia, V. 2001. Genes modulated by histone acetylation as new effectors of butyrate activity. *FEBS Lett*, 499, 199–204.

Dennis, S., Sheth, U., Feldman, J. L., English, K. A. & Priess, J. R. 2012. *C. elegans* germ cells show temperature and age-dependent expression of Cer1, a Gypsy/Ty3-related retrotransposon. *PLoS Pathog*, 8, e1002590031.

De Sandre-Giovannoli, A., Bernard, R., Cau, P., Navarro, C., Amiel, J., Boccaccio, I., Lyonnet, S., Stewart, C. L., Munnich, A., Le Merrer, M. & Levy, N. 2003. Lamin a truncation in Hutchinson-Gilford progeria. *Science*, 300, 2055.

De Vaux, V., Pfefferli, C., Passannante, M., Belhaj, K., Von Essen, A., Sprecher, S. G., Muller, F. & Wicky, C. 2013. The

Caenorhabditis elegans LET-418/Mi2 plays a conserved role in lifespan regulation. *Aging Cell,* 12, 1012–20.

Di Leonardo, A., Linke, S. P., Clarkin, K. & Wahl, G. M. 1994. DNA damage triggers a prolonged p53-dependent G1 arrest and long-term induction of Cip1 in normal human fibroblasts. *Genes Dev,* 8, 2540–51.

Di Micco, R., Fumagalli, M., Cicalese, A., Piccinin, S., Gasparini, P., Luise, C., Schurra, C., Garre, M., Nuciforo, P. G., Bensimon, A., Maestro, R., Pelicci, P. G. & D'adda Di Fagagna, F. 2006. Oncogene-induced senescence is a DNA damage response triggered by DNA hyper-replication. *Nature,* 444, 638–42.

Duarte, L. F., Young, A. R., Wang, Z., Wu, H. A., Panda, T., Kou, Y., Kapoor, A., Hasson, D., Mills, N. R., MA'AYAN, A., NARITA, M. & BERNSTEIN, E. 2014. Histone H3.3 and its proteolytically p[rocessed form drive a cellular senescence programme. 2014. *Nat Commun,* 5. doi: 10.1038/ncommons6210.

Eriksson, M., Brown, W. T., Gordon, L. B., Glynn, M. W., Singer, J., Scott, L., Erdos, M. R., Robbins, C. M., Moses, T. Y., Berglund, P., Dutra, A., Pak, E., Durkin, S., Csoka, A. B., Boehnke, M., Glover, T. W. & Collins, F. S. 2003. Recurrent *de novo* point mutations in lamin A cause Hutchinson-Gilford progeria syndrome. *Nature,* 423, 293–8.

Feinberg, A. P. & Vogelstein, B. 1983. Hypomethylation distinguishes genes of some human cancers from their normal counterparts. *Nature,* 301, 89–92.

Feser, J., Truong, D., Das, C., Carson, J. J., Kieft, J., Harkness, T. & Tyler, J. K. 2010. Elevated histone expression promotes life span extension. *Mol Cell,* 39, 724–35.

Flavahan, W. A., Drier, Y., Liau, B. B., Gillespie, S. M., Venteicher, A. S., Stemmer-Rachamimov, A. O., Suva, M. L. & Bernstein, B. E. 2016. Insulator dysfunction and oncogene activation in IDH mutant gliomas. *Nature,* 529, 110–14.

Fraga, M. F., Ballestar, E., Paz, M. F., Ropero, S., Setien, F., Ballestar, M. L., Heine-Suner, D., Cigudosa, J. C., Urioste, M., Benitez, J., Boix-Chornet, M., Sanchez-Aguilera, A., Ling, C., Carlsson, E., Poulsen, P., Vaag, A., Stephan, Z., Spector, T. D., Wu, Y. Z., Plass, C. & Esteller, M. 2005. Epigenetic differences arise during the lifetime of monozygotic twins. *Proc Natl Acad Sci U S A,* 102, 10604–9.

Francia, S., Michelini, F., Saxena, A., Tang, D., De Hoon, M., Anelli, V., Mione, M., Carninci, P. & D'adda Di Fagagna, F. 2012. Site-specific DICER and DROSHA RNA products control the DNA-damage response. *Nature,* 488, 231–5.

Funato, K., Major, T., Lewis, P. W., Allis, C. D. & Tabar, V. 2014. Use of human embryonic stem cells to model pediatric gliomas with H3.3K27M histone mutation. *Science,* 346, 1529–33.

Garcia, A. M., Calder, R. B., Dolle, M. E., Lundell, M., Kapahi, P. & Vijg, J. 2010. Age- and temperature-dependent somatic mutation accumulation in *Drosophila melanogaster. PLoS Genet,* 6, e1000950.

Genovese, G., Kahler, A. K., Handsaker, R. E., Lindberg, J., Rose, S. A., Bakhoum, S. F., Chambert, K., Mick, E., Neale,

B. M., Fromer, M., Purcell, S. M., Svantesson, O., Landen, M., Hoglund, M., Lehmann, S., Gabriel, S. B., Moran, J. L., Lander, E. S., Sullivan, P. F., Sklar, P., Gronberg, H., Hultman, C. M. & Mccarroll, S. A. 2014. Clonal hematopoiesis and blood-cancer risk inferred from blood DNA sequence. *N Engl J Med,* 371, 2477–87.

Goldman, R. D., Schumaker, D. K., Erdos, M. R., Eriksson, M., Goldman, A. E., Gordon, L. B., Gruenbaum, Y., Khuon, S., Mendez, M., Varga, R. & Collins, F. S. 2004. Accumulation of mutant lamina A causes progressive changes in nuclear architecture in Hutchinson-Gilford progeris syndrome. *Proc Natl Acad Sci U S A,* 101, 8963–8.

Goldstein, S., Moerman, E. J., Fujii, S. & Sobel, B. E. 1994. Overexpression of plasminogen activator inhibitor type-1 in senescent fibroblasts from normal subjects and those with Werner syndrome. *J Cell Physiol,* 161, 571–9.

Gong, G., Chen, P., Li, L., Tan, H., Zhou, J., Zhou, Y., Yang, X. & Wu, X. 2015. Loss of lamin A but not lamin C expression in epithelial ovarian cancer cells is associated with metastasis and poor prognosis. *Pathol Res Pract,* 211, 175–82.

Greer, E. L., Maures, T. J., Hauswirth, A. G., Green, E. M., Leeman, D. S., Maro, G. S., Han, S., Banko, M. R., Gozani, O. & Brunet, A. 2010. Members of the H3K4 trimethylation complex regulate lifespan in a germline-dependent manner in *C. elegans. Nature,* 466, 383–7.

Guarente, L. 1999. Diverse and dynamic functions of the Sir silencing complex. *Nat Genet,* 23, 281–5.

Gui, Y., Guo, G., Huang, Y., Hu, X., Tang, A., Gao, S., Wu, R., Chen, C., Li, X., Zhou, L., He, M., Li, Z., Sun, X., Jia, W., Chen, J., Yang, S., Zhou, F., Zhao, X., Wan, S., Ye, R., Liang, C., Liu, Z., Huang, P., Liu, C., Jiang, H., Wang, Y., Zheng, H., Sun, L., Liu, X., Jiang, Z., Feng, D., Chen, J., Wu, S., Zou, J., Zhang, Z., Yang, R., Zhao, J., Xu, C., Yin, W., Guan, Z., Ye, J., Zhang, H., Li, J., Kristiansen, K., Nickerson, M. L., Theodorescu, D., Li, Y., Zhang, X., Li, S., Wang, J., Yang, H., Wang, J. & Cai, Z. 2011. Frequent mutations of chromatin remodeling genes in transitional cell carcinoma of the bladder. *Nat Genet,* 43, 875–8.

Hannum, G., Guinney, J., Zhao, L., Zhang, L., Hughes, G., Sadda, S., Klotzle, B., Bibikova, M., Fan, J. B., Gao, Y., Deconde, R., Chen, M., Rajapakse, I., Friend, S., Ideker, T. & Zhang, K. 2013. Genome-wide methylation profiles reveal quantitative views of human aging rates. *Mol Cell,* 49, 359–67.

Hansen, K. D., Timp, W., Bravo, H. C., Sabunciyan, S., Langmead, B., Mcdonald, O. G., Wen, B., Wu, H., Liu, Y., Diep, D., Briem, E., Zhang, K., Irizarry, R. A. & Feinberg, A. P. 2011. Increased methylation variation in epigenetic domains across cancer types. *Nat Genet,* 43, 768–75.

Henikoff, S. 2016. Mechanisms of nucleosome dynamics *in vivo. Cold Spring Harb Perspect Med,* 6. doi: 10.1101/cshperspect.a026666.

Henrich, K. O., Bender, S., Saadati, M., Dreidax, D., Gartlgruber, M., Shao, C., Herrmann, C., Wiesenfarth, M.,

Parzonka, M., Wehrmann, L., Fischer, M., Duffy, D. J., Bell, E., Torkov, A., Schmezer, P., Plass, C., Hofer, T., Benner, A., Pfister, S. M. & Westermann, F. 2016. Integrative genome-scale analysis identifies epigenetic mechanisms of transcriptional deregulation in unfavorable neuroblastomas. *Cancer Res,* 76, 5523–37.

Heshmati, Y., Turkoz, G., Harisankar, A., Kharazi, S., Bostrom, J., Dolatabadi, E. K., Krstic, A., Chang, D., Mansson, R., Altun, M., Qian, H. & Walfridsson, J. 2018. The Chromatin-Remodeling Factor CHD4 Is Required For Maintenance Of Childhood Acute Myeloid Leukemia. *Haematologica,* 103, 1169–1181.

Hnisz, D., Weintraub, A. S., Day, D. S., Valton, A. L., Bak, R. O., Li, C. H., Goldmann, J., Lajoie, B. R., Fan, Z. P., Sigova, A. A., Reddy, J., Borges-Rivera, D., Lee, T. I., Jaenisch, R., Porteus, M. H., Dekker, J. & Young, R. A. 2016. Activation of proto-oncogenes by disruption of chromosome neighborhoods. *Science,* 351, 1454–8.

Horvath, S. 2013. DNA methylation age of human tissues and cell types. *Genome Biol,* 14, R115.

Hua, S., Kallen, C. B., Dhar, R., Baquero, M. T., Mason, C. E., Russell, B. A., Shah, P. K., Liu, J., Khramtsov, A., Tretiakova, M. S., Krausz, T. N., Olopade, O. I., Rimm, D. L. & White, K. P. 2008. Genomic analysis of estrogen cascade reveals histone variant H2A.Z associated with breast cancer progression. *Mol Syst Biol,* 4, 188.

Ibanez-Ventoso, C., Yang, M., Guo, S., Robins, H., Padgett, R. W. & Driscoll, M. 2006. Modulated microRNA expression during adult lifespan in *Caenorhabditis elegans. Aging Cell,* 5, 235–46.

Imai, S., Armstrong, C. M., Kaeberlein, M. & Guarente, L. 2000. Transcriptional silencing and longevity protein Sir2 is an NAD-dependent histone deacetylase. *Nature,* 403, 795–800.

Ivanov, A., Pawlikowski, J., Manoharan, I., Van Tuyn, J., Nelson, D. M., Rai, T. S., Shah, P. P., Hewitt, G., Korolchuk, V. I., Passos, J. F., WU. H., BERGER, S. L. & ADAMS, P. D. 2013. Lysosome-mediated processing of chromatin in senescence. *J Cell Biol,* 202, 129–43.

Iyer, M. K., Niknafs, Y. S., Malik, R., Singhal, U., Sahu, A., Hosono, Y., Barrette, T. R., Prensner, J. R., Evans, J. R., Zhao, S., Poliakov, A., Cao, X., Dhanasekaran, S. M., Wu, Y. M., Robinson, D. R., Beer, D. G., Feng, F. Y., Iyer, H. K. & Chinnaiyan, A. M. 2015. The landscape of long noncoding RNAs in the human transcriptome. *Nat Genet,* 47, 199–208.

Jaiswal, S., Fontanillas, P., Flannick, J., Manning, A., Grauman, P. V., Mar, B. G., Lindsley, R. C., Mermel, C. H., Burtt, N., Chavez, A., Higgins, J. M., Moltchanov, V., Kuo, F. C., Kluk, M. J., Henderson, B., Kinnunen, L., Koistinen, H. A., Ladenvall, C., Getz, G., Correa, A., Banahan, B. F., Gabriel, S., Kathiresan, S., Stringham, H. M., Mccarthy, M. I., Boehnke, M., Tuomilehto, J., Haiman, C., Groop, L., Atzmon, G., Wilson, J. G., Neuberg, D., Altshuler, D. & Ebert, B. L. 2014. Age-related clonal hematopoiesis associated with adverse outcomes. *N Engl J Med,* 371, 2488–98.

Ji, P., Diederichs, S., Wang, W., Boing, S., Metzger, R., Schneider, P. M., Tidow, N., Brandt, B., Buerger, H., Bulk, E., Thomas, M., Berdel, W. E., Serve, H. & Muller-Tidow, C. 2003. MALAT-1, a novel noncoding RNA, and thymosin beta4 predict metastasis and survival in early-stage non-small cell lung cancer. *Oncogene,* 22, 8031–41.

Kadoch, C., Hargreaves, D. C., Hodges, C., Elias, L., Ho, L., Ranish, J. & Crabtree, G. R. 2013. Proteomic and bioinformatic analysis of mammalian SWI/SNF complexes identifies extensive roles in human malignancy. *Nat Genet,* 45, 592–601.

Kapoor, A., Goldberg, M. S., Cumberland, L. K., Ratnakumar, K., Segura, M. F., Emanuel, P. O., Menendez, S., Vardabasso, C., Leroy, G., Vidal, C. I., Polsky, D., Osman, I., Garcia, B. A., Hernando, E. & Bernstein, E. 2010. The histone variant macroH2A suppresses melanoma progression through regulation of CDK8. *Nature,* 468, 1105–9.

Katainen, R., Dfave, K., Pitkanen, E., Palin, K., Kivioja, T., Valimaki, N., Gylfe, A. E., Ristolainen, H., Hanninen, U. A., Cajuso, T., Kondelin, J., Tanskanen, T., Mecklin, J. P., Jarvinen, H., Renkonen-Sinisalo, L., Lepisto, A., Kaasinen, E., Kilpivaara, O., Tuupanen, S., Enge, M., Taipale, J. & Aaltonen, L. A. 2015. CTCF/cohesin-binding sites are frequently mutated in cancer. *Nat Genet,* 47, 818–21.

Keskin, H., Meers, C. & Storici, F. 2016. Transcript RNA supports precise repair of its own DNA gene. *RNA Biol,* 13, 157–65.

Keskin, H., Shen, Y., Huang, F., Patel, M., Yang, T., Ashley, K., Mazin, A. V. & Storici, F. 2014. Transcript-RNA-templated DNA recombination and repair. *Nature,* 515, 436–9.

Kim, K., Punj, V., Choi, J., Heo, K., Kim, J. M., Laird, P. W. & An, W. 2013. Gene dysregulation by histone variant H2A.Z in bladder cancer. *Epigenetics Chromatin,* 6, 34.

Knight, A. K., Conneely, K. N. & Smith, A. K. 2017. Gestational age predicted by DNA methylation: potential clinical and researchutility. *Epigenomics.* doi: 2217/epi-2016-0157.

Kong, L., Schafer, G., Bu, H., Zhang, Y., Zhang, Y. & Klocker, H. 2012. Lamin A/C protein is overexpressed in tissue-invading prostate cancer and promotes prostate cancer cell growth, migration and invasion through the PI3K/AKT/PTEN pathway. *Carcinogenesis,* 33, 751–9.

Kour, S. & Rath, P. C. 2015. Age-dependent differential expression profile of a novel intergenic long non-coding RNA in rat brain. *J Dev Neurosci,* 47, 286–97.

Kreiling, J. A., Tamamori-Adachi, M., Sexton, A. N., Jeyapalan, J. C., Munoz-Najar, U., Peterson, A. L., Manivannan, J., Rogers, E. S., Pchelintsev, N. A., Adams, P. D. & Sedivy, J. M. 2011. Age-associated increase in heterochromatic marks in murine and primate tissues. *Aging Cell,* 10, 292–304.

Krtolica, A., Parrinello, S., Lockett, S., Desprez, P. Y. & Campisi, J. 2001. Senescent fibroblasts promote epithelial cell growth and tumorigenesis: a link between cancer and aging. *Proc Natl Acad Sci U S A,* 98, 12072–7.

Kuilman, T., Michaloglou, C., Vredeveld, L. C., Douma, S., Van Doorn, R., Desmet, C. J., Aarden, L. A., Mooi, W. J. & Peeper, D. S. 2008. Oncogene-induced senescence relayed by an interleukin-dependent inflammatory network. *Cell,* 133, 1019–31.

Lamming, D. W., Sinclair, D. A. 2008. *The Regulation of Lifespan by Sirtuins in Saccharomyces cerevisiae.* University dissertation. Cambridge, MA: Harvard University.

Landry, J., Sutton, A., Tafrov, S. T., Heller, R. C., Stebbins, J., Pillus, L. & Sternglanz, R. 2000. The silencing protein SIR2 and its homologs are NAD-dependent protein deacetylases. *Proc Natl Acad Sci U S A,* 97, 5807–11.

Larson, K., Yan, S. J., Tsurumi, A., Liu, J., Zhou, J., Gaur, K., Guo, D., Eickbush, T. H. & Li, W. X. 2012. Heterochromatin formation promotes longevity and represses ribosomal RNA synthesis. *PLoS Genet,* 8, e1002473.

Lawrence, M. S., Stojanov, P., Mermel, C. H., Robinson, J. T., Garraway, L. A., Golub, T. R., Meyerson, M., Gabriel, S. B., Lander, E. S. & Getz, G. 2014. Discovery and saturation analysis of cancer genes across 21 tumour types. *Nature,* 505, 495–501.

Lee, C. K., Klopp, R. G., Weindruch, R. & Prolla, T. A. 1999. Gene expression profile of aging and its retardation by caloric restriction. *Science,* 285, 1390–3.

Lee, R. C., Feinbaum, R. L. & Ambros, V. 1993. The *C. elegans* heterochronic gene lin-4 encodes small RNAs with antisense complementarity to lin-14. *Cell,* 75, 843–54.

Leucci, E., Coe, E. A., Marine, J. C. & Vance, K. W. 2016. The emerging role of long non-coding RNAs in cutaneous melanoma. *Pigment Cell Melanoma Res,* 29, 619–26.

Li, W., Prazak, L., Chatterjee, N., Gruninger, S., Krug, L., Theodorou, D. & Dubnau, J. 2013. Activation of transposable elements during aging and neuronal decline in *Drosophila. Nat Neurosci,* 16, 529–31.

Li, X., Cao, Y., Gong, X. & Li, H. 2017. Long noncoding RNAs in head and neck cancer. *Oncotarget,* 8, 10726–40.

Liu, D. & Hornsby, P. J. 2007. Senescent human fibroblasts increase the early growth of xenograft tumors via matrix metalloproteinase secretion. *Cancer Res,* 67, 3117–26.

Liu, N., Landreh, M., Cao, K., Abe, M., Hendriks, G. J., Kennerdell, J. R., Zhu, Y., Wang, L. S. & Bonini, N. M. 2012. The microRNA miR-34 modulates ageing and neurodegeneration in *Drosophila. Nature,* 482, 519–23.

Lu, C., Jain, S. U., Hoelper, D., Bechet, D., Molden, R. C., Ran, L., Murphy, D., Venneti, S., Hameed, M., Pawel, B. R., Wunder, J. S., Dickson, B. C., Lundgren, S. M., Jani, K. S., De Jay, N., Papillon-Cavanagh, S., Amdrulis, I. L., Sawyer, S. L., Grynspan, D., Turcotte, R. E., Nadaf, J., Fahiminiyah, S., Muir, T. W., Majewski, J., Thompson, C. B., Chi, P., Garcia, B. A., Allis, C. D., Jabado, N. & Lewis, P. W. 2016. Histone H3K36 mutations promote sarcomagenesis through altered histone methylation landscape. *Science,* 352, 844–9.

Malakar, P., Shilo, A., Mogilevsky, A., Stein, I., Pikarsky, E., Nevo, Y., Benyamini, H., Elgavish, S., Zong, X., Prasanth, K. V. & Karni, R. 2017. Long noncoding RNA MALAT1 promotes hepatocellular carcinoma development by SRSF1 upregulation and mTOR activation. *Cancer Res,* 77, 1155–67.

Marella, N. V., Bhattacharya, S., Mukherjee, L., Xu, J. & Berezney, R. 2009. Cell type specific chromosome territory organization in the interphase nucleus of normal and cancer cells. *J Cell Physiol,* 221, 130–8.

Martienssen, R. & Moazed, D. 2015. RNAi and heterochromatin assembly. *Cold Spring Harb Perspect Biol,* 7, a019323.

Maxwell, P. H., Burhans, W. C. & Curcio, M. J. 2011. Retrotransposition is associated with genome instability during chronological aging. *Proc Natl Acad Sci U S A,* 108, 20376–81.

Mazumdar, C., Shen, Y., Xavy, S., Zhao, F., Reinisch, A., Li, R., Corces, M. R., Flynn, R. A., Buenrostro, J. D., Chan, S. M., Thomas, D., Koenig, J. L., Hong, W. J., Chang, H. Y. & Majeti, R. 2015. Leukemia-associated cohesin mutants dominantly enforce stem cell programs and impair human hematopoietic progenitor differentiation. *Cell Stem Cell,* 17, 675–88.

Mccay, C. M., Crowell, M. F. & Maynard, L. A. 1935. The effect of retarded growth upon the length of life span and upon the ultimate body size. *J Nutrition,* 10, 63–79.

Mccoll, G., Killilea, D. W., Hubbard, A. E., Vantipalli, M. C., Melov, S. & Lithgow, G. J. 2008. Pharmacogenetic analysis of lithium-induced delayed aging in *Caenorhabditis elegans. J Biol Chem,* 283, 350–7.

Meaburn, K. J., Gudha, P. R., Khan, S., Lockett, S. J. & Misteli, T. 2009. Disease-specific gene repositioning in breast cancer. *J Cell Biol,* 187, 801–12.

Meaburn, K. J. & Misteli, T. 2008. Locus-specific and activity-independent gene repositioning during early tumorigenesis. *J Cell Biol,* 180, 39–50.

Meier, I., Fellini, L., Jakovcevski, M., Schachner, M. & Morellini, F. 2010. Expression of the snoRNA host gene *gas5* in the hippocampus is upregulated by age and psychogenic stress and correlates with reduced novelty-induced behavior in C57BL/6 mice. *Hippocampus,* 20, 1027–36.

Mohan, M., Lin, C., Guest, E. & Shilatifard, A. 2010. Licensed to elongate: a molecular mechanism for MLL-based leukaemogenesis. *Nat Rev Cancer,* 10, 721–8.

Mori, M. A., Raghavan, P., Thomou, T., Boucher, J., Robida-Stubbs, S., Macotela, Y., Russell, S. J., Kirkland, J. L., Blackwell, T. K. & Kahn, C. R. 2012. Role of microRNA processing in adipose tissue in stress defense and longevity. *Cell Metab,* 16, 336–47.

Morin, R. D., Johnson, N. A., Severson, T. M., Mungall, A. J., An, J., Goya, R., Paul, J. E., Boyle, M., Woolcock, B. W., Kuchenbauer, F., Yap, D., Humphries, R. K., Griffith, O. L., Shah, S., Zhu, H., Kimbara, M., Shashkin, P., Charlot, J. F., Tcherpakov, M., Corbett, R., Tam, A., Varhol, R., Smailus, D., Moksa, M., Zhao, Y., Delaney, A., Qian, H., Birol, I., Schein, J., Moore, R., Holt, R., Horsman, D. E., Connors, J.

M., Jones, S., Aparicio, S., Hirst, M., Gascoyne, R. D. & Marra, M. A. 2010. Somatic mutations altering EZH2 (Tyr641) in follicular and diffuse large B-cell lymphomas of germinal-center origin. *Nat Genet,* 42, 181–5.

Narita, M., Nunez, S., Heard, E., Narita, M., Lin, A. W., Hearn, S. A., Spector, D. L., Hannon, G. J. & Lowe, S. W. 2003. Rb-mediated heterochromatin formation and silencing of E2F target genes during cellular senescence. *Cell,* 113, 703–16.

Ni, Z., Ebata, A., Alipanahiramandi, E. & Lee, S. S. 2012. Two SET domain containing genes link epigenetic changes and aging in *Caenorhabditis elegans*. *Aging Cell,* 11, 315–25.

Noushmehr, H., Weisenberger, D. J., Diefes, K., Phillips, H. S., Pujara, K., Berman, B. P., Pan, F., Pelloski, C. E., Sulman, E. P., Bhat, K. P., Verhaak, R. G., Hoadley, K. A., Hayes, D. N., Perou, C. M., Schmidt, H. K., Ding, L., Wilson, R. K., Van Den Berg, D., Shen, H., Bengtsson, H., Neuvial, P., Cope, L. M., Buckley, J., Herman, J. G., Baylin, S. B., Laird, P. W., Aldape, K. & Cancer Genome Atlas Research, N. 2010. Identification of a CpG island methylator phenotype that defines a distinct subgroup of glioma. *Cancer Cell,* 17, 510–22.

Ntziachristos, P., Tsirigos, A., Van Vlierberghe, P., Nedjic, J., Trimarchi, T., Flaherty, M. S., Ferres-Marco, D., Da Ros, V., Tang, Z., Siegle, J., Asp, P., Hadler, M., Rigo, I., De Keersmaecker, K., Patel, J., Huynh, T., Utro, F., Poglio, S., Samon, J. B., Paietta, E., Racevskis, J., Rowe, J. M., Rabadan, R., Levine, R. L., Brown, S., Pflumio, F., Dominguez, M., Ferrando, A. & Aifantis, I. 2012. Genetic inactivation of the polycomb repressive complex 2 in T cell acute lympho-blastic leukemia. *Nat Med,* 18, 298–301.

Osada, H., Tatematsu, Y., Masuda, A., Saito, T., Sugiyama, M., Yanagisawa, K. & Takahashi, T. 2001. Heterogeneous transforming growth factor (TGF)-beta unresponsiveness and loss of TGF-beta receptor type II expression caused by histone deacetylation in lung cancer cell lines. *Cancer Res,* 61, 8331–9.

O'sullivan, R. J., Kubicek, S., Schreiber, S. L. & Karlseder, J. 2010. Reduced histone biosynthesis and chromatin changes arising from a damage signal at telomeres. *Nat Struct Mol Biol,* 17, 1218–25.

Parsons, D. W., Li, M., Zhang, X., Jones, S., Leary, R. J., Lin, J. C., Boca, S. M., Carter, H., Samayoa, J., Bettegowda, C., Gallia, G. L., Jallo, G. I., Binder, Z. A., Nikolsky, Y., Hartigan, J., Smith, D. R., Gerhard, D. S., Fults, D. W., Vandenberg, S., Berger, M. S., Marie, S. K., Shinjo, S. M., Clara, C., Phillips, P. C., Minturn, J. E., Biegel, J. A., Judkins, A. R., Resnick, A. C., Storm, P. B., Curran, T., He, Y., Rasheed, B. A., Friedman, H. S., Keir, S. T., Mclendon, R., Northcott, P. A., Taylor, M. D., Burger, P. C., Riggins, G. J., Karchin, R., Parmigiani, G., Bigner, D. D., Yan, H., Papadopoulos, N., Vogelstein, B., Kinzler, K. W. & Velculescu, V. E. 2011. The genetic landscape of the childhood cancer medulloblas-toma. *Science,* 331, 435–9.

Patra, S. K., Patra, A. & Dahiya, R. 2001. Histone deacetylase and DNA methyltransferase in human prostate cancer. *Biochem Biophys Res Commun,* 287, 705–13.

Pegoraro, G., Kubben, N., Wickert, U., Gohler, H., Hoffmann, K. & Misteli, T. 2009. Ageing-related chromatin defects through loss of the NURD complex. *Nat Cell Biol,* 11, 1261–7.

Pekarsky, Y., Balatti, V., Palamarchuk, A., Rizzotto, L., Veneziano, D., Nigita, G., Rassenti, L. Z., Pass, H. I., Kipps, T. J., Liu, C. G. & Croce, C. M. 2016. Dysregulation of a family of short noncoding RNAs, tsRNAs, in human cancer. *Proc Natl Acad Sci U S A,* 113, 5071–6.

Perez, R. F., Tejedor, J. R., Bayon, G. F., Fernandez, A. F. & Fraga, M. F. 2018. Distinct chromatin signatures of DNA hypomethylation in aging and cancer. *Aging Cell,* 17, e12744.

Persengiev, S., Kondova, I., Otting, N., Koeppen, A. H. & Bontrop, R. E. 2011. Genome-wide analysis of miRNA expression reveals a potential role for miR-144 in brain aging and spinocerebellar ataxia pathogenesis. *Neurobiol Aging,* 32, 2316 e17–27.

Previdi, M. C., Carotenuto, P., Zito, D., Pandolfo, R. & Braconi, C. 2017. Noncoding RNAs as novel biomarkers in pancreatic cancer: what do we know? *Future Oncol,* 13, 443–53.

Pu, M., Ni, Z., Wang, M., Wang, X., Wood, J. G., Helfand, S. L., Yu, H. & Lee, S. S. 2015. Trimethylation of Lys36 on H3 restricts gene expression change during aging and impacts life span. *Genes Dev,* 29, 718–31.

Rine, J. & Herskowitz, I. 1987. Four genes responsible for a position effect on expression from HML and HMR in *Saccharomyces cerevisiae*. *Genetics,* 116, 9–22.

Rizki, G., Iwata, T. N., Li, J., Riedel, C. G., Picard, C. L., Jan, M., Murphy, C. T. & Lee, S. S. 2011. The evolutionarily conserved longevity determinants HCF-1 and SIR-2.1/SIRT1 collaborate to regulate DAF-16/FOXO. *PLoS Genet,* 7, e1002235.

Rodgers, M., Zhan, X. M. & Burke, M. D. 2004. Nutrient removal in a sequencing batch biofilm reactor (SBBR) using a vertically moving biofilm system. *Environ Technol,* 25, 211–18.

Rodier, F., Coppe, J. P., Patil, C. K., Hoeijmakers, W. A., Munoz, D. P., Raza, S. R., Freund, A., Campeau, E., Davalos, A. R. & Campisi, J. 2009. Persistent DNA damage signalling trig-gers senescence-associated inflammatory cytokine secre-tion. *Nat Cell Biol,* 11, 973–9.

Rosenfeld, N., Aharonov, R., Meiri, E., Rosenwald, S., Spector, Y., Zepeniuk, M., Benjamin, H., Shabes, N., Tabak, S., Levy, A., Lebanony, D., Goren, Y., Silberschein, E., Targan, N., Ben-Ari, A., Gilad, S., Sion-Vardy, N., Tobar, A., Feinmesser, M., Kharenko, O., Nativ, O., Nass, D., Perelman, M., Yosepovich, A., Shalmon, B., Polak-Charcon, S., Fridman, E., Avniel, A., Bentwich, I., Bentwich, Z., Cohen, D., Chajut, A. & Barshack, I. 2008. MicroRNAs accurately identify cancer tissue origin. *Nat Biotechnol,* 26, 462–9.

Rzymski, T., Mikula, M., Wiklik, K. & Brzozka, K. 2015. CDK8 kinase—an emerging target in targeted cancer therapy. *Biochim Biophys Acta*, 1854, 1617–29.

Sabatelli, P., Lattanzi, G., Ognibene, A., Columbaro, M., Capanni, C., Merlini, L., Maraldi, N. M. & Squarzoni, S. 2001. Nuclear alterations in autosomal-dominant Emery-Dreifuss muscular dystrophy. *Muscle Nerve*, 24, 826–9.

Saidi, D., Cheray, M., Osman, A. M., Stratoulias, V., Lindberg, O. R., Shen, X., Blomgren, K. & Joseph, B. 2018. Glioma-induced SIRT1-dependent activation of hMOF histone H4 lysine 16 acetyltransferase in microglia promotes a tumor supporting phenotype. *Oncoimmunology*, 7, e1382790.

Saito, Y. & Saito, H. 2012. Role of CTCF in the regulation of microRNA expression. 2012. *Front Genet*, 3, doi: 10.3389/fgene.2012.00186.

Scaffidi, P. & Misteli, T. 2006. Lamin A-dependent nuclear defects in human aging. *Science*, 312, 1059–63.

Schubeler, D., Macalpine, D. M., Scalzo, D., Wirbelauer, C., Kooperberg, C., Van Leeuwen, F., Gottschling, D. E., O'neill, L. P., Turner, B. M., Delrow, J., Bell, S. P. & Groudine, M. 2004. The histone modification pattern of active genes revealed through genome-wide chromatin analysis of a higher eukaryote. *Genes Dev*, 18, 1263–71.

Schwartzentruber, J., Korshunov, A., Liu, X. Y., Jones, D. T., Pfaff, E., Jacob, K., Sturm, D., Fontebasso, A. M., Quang, D. A., Tonjes, M., Hovestadt, V., Albrecht, S., Kool, M., Nantel, A., Konermann, C., Lindroth, A., Jager, N., Rausch, T., Ryzhova, M., Korbel, J. O., Hielscher, T., Hauser, P., Garami, M., Klekner, A., Bognar, L., Ebinger, M., Schuhmann, M. U., Scheurlen, W., Pekrun, A., Fruhwald, M. C., Roggendorf, W., Kramm, C., Durken, M., Atkinson, J., Lepage, P., Montpetit, A., Zakrzewska, M., Zakrzewski, K., Liberski, P. P., Dong, Z., Siegel, P., Kulozik, A. E., Zapatka, M., Guha, A., Malkin, D., Felsberg, J., Reifenberger, G., Von Deimling, A., Ichimura, K., Collins, V. P., Witt, H., Milde, T., Witt, O., Zhang, C., Castelo-Branco, P., Lichter, P., Faury, D., Tabori, U., Plass, C., Majewski, J., Pfister, S. M. & Jabado, N. 2012. Driver mutations in histone H3.3 and chromatin remodelling genes in paediatric glioblastoma. *Nature*, 482, 226–31.

Seligson, D. B., Horvath, S., Shi, T., Yu, H., Tze, S., Grunstein, M. & Kurdistani, S. K. 2005. Global histone modification patterns predict risk of prostate cancer recurrence. *Nature*, 435, 1262–6.

Sen, P., Dang, W., Donahue, G., Dai, J., Dorsey, J., Cao, X., Liu, W., Cao, K., Perry, R., Lee, J. Y., Wasko, B. M., Carr, D. T., He, C., Robison, B., Wagner, J., Gregory, B. D., Kaeberlein, M., Kennedy, B. K., Boeke, J. D. & Berger, S. L. 2015. H3K36 methylation promotes longevity by enhancing transcriptional fidelity. *Genes Dev*, 29, 1362–76.

Serrano, M., Lin, A. W., Mccurrach, M. E., Beach, D. & Lowe, S. W. 1997. Oncogenic ras provokes premature cell senescence associated with accumulation of p53 and p16INK4a. *Cell*, 88, 593–602.

Shen, J., Ambrosone, C. B., Dicioccio, R. A., Odunsi, K., Lele, S. B. & Zhao, H. 2008. A functional polymorphism in the miR-146a gene and age of familial breast/ovarian cancer diagnosis. *Carcinogenesis*, 29, 1963–6.

Shumaker, D. K., Dechat, T., Kohlmaier, A., Adam, S. A., Bozovsky, M. R., Erdos, M. R., Eriksson, M., Goldman, A. E., Khuon, S., Collins, F. S., Jenuwein, T. & Goldman, R. D. 2006. Mutant nuclear lamin A leads to progressive alterations of epigenetic control in premature aging. *Proc Natl Acad Sci U S A*, 103, 8703–8.

Siebold, A. P., Banerjee, R., Tie, F., Kiss, D. L., Moskowitz, J. & Harte, P. J. 2010. Polycomb repressive complex 2 and trithorax modulate *Drosophila* longevity and stress resistance. *Proc Natl Acad Sci U S A*, 107, 169–74.

Sjoblom, T., Jones, S., Wood, L. D., Parsons, D. W., Lin, J., Barber, T. D., Mandelker, D., Leary, R. J., Ptak, J., Silliman, N., Szabo, S., Buckhaults, P., Farrell, C., Meeh, P., Markowitz, S. D., Willis, J., Dawson, D., Willson, J. K., Gazdar, A. F., Hartigan, J., Wu, L., Liu, C., Parmigiani, G., Park, B. H., Bachman, K. E., Papadopoulos, N., Vogelstein, B., Kinzler, K. W. & Velculescu, V. E. 2006. The consensus coding sequences of human breast and colorectal cancers. *Science*, 314, 268–74.

Smith, J. S., Brachmann, C. B., Celic, I., Kenna, M. A., Muhammad, S., Starai, V. J., Avalos, J. L., Escalante-Semerena, J. C., Grubmeyer, C., Wolberger, C. & Boeke, J. D. 2000. A phylogenetically conserved NAD+-dependent protein deacetylase activity in the Sir2 protein family. *Proc Natl Acad Sci U S A*, 97, 6658–63.

Sneeringer, C. J., Scott, M. P., Kuntz, K. W., Knutson, S. K., Pollock, R. M., Richon, V. M. & Copeland, R. A. 2010. Coordinated activities of wild-type plus mutant EZH2 drive tumor-associated hypertrimethylation of lysine 27 on histone H3 (H3K27) in human B-cell lymphomas. *Proc Natl Acad Sci U S A*, 107, 20980–5.

Stergachis, A. B., Neph, S., Reynolds, A., Humbert, R., Miller, B., Paige, S. L., Vernot, B., Cheng, J. B., Thurman, R. E., Sandstrom, R., Haugen, E., Heimfeld, S., Murry, C. E., Akey, J. M. & Stamatoyannopoulos, J. A. 2013. Developmental fate and cellular maturity encoded in human regulatory DNA landscapes. *Cell*, 154, 888–903.

Swanson, E. C., Manning, B., Zhang, H. & Lawrence, J. B. 2013. Higher-order unfolding of satellite heterochromatin is a consistent and early event in cell senescence. *J Cell Biol*, 203, 929–42.

Tano, K., Mizuno, R., Okada, T., Rakwal, R., Shibato, J., Masuo, Y., Ijiri, K. & Akimitsu, N. 2010. MALAT-1 enhances cell motility of lung adenocarcinoma cells by influencing the expression of motility-related genes. *FEBS Lett*, 584, 4575–80.

Tiku, V., Jain, C., Raz, Y., Nakamura, S., Heestand, B., Liu, W., Spath, M., Suchiman, H. E. D., Muller, R. U., Slagboom, P. E., Partridge, L. & Antebi, A. 2016. Small nucleoli are a cellular hallmark of longevity. *Nat Commun*, 8, 16083.

Timp, W., Bravo, H. C., Mcdonald, O. G., Goggins, M., Umbricht, C., Zeiger, M., Feinberg, A. P. & Irizarry, R. A. 2014. Large hypomethylated blocks as a universal defining epigenetic alteration in human solid tumors. *Genome Med*, 6, 61.

Tripathi, V., Ellis, J. D., Shen, Z., Song, D. Y., Pan, Q., Watt, A. T., Freier, S. M., Bennett, C. F., Sharma, A., Bubulya, P. A., Blencowe, B. J., Prasanth, S. G. & Prasanth, K. V. 2010. The nuclear-retained noncoding RNA MALAT1 regulates alternative splicing by modulating SR splicing factor phosphorylation. *Mol Cell*, 39, 925–38.

Vardabasso, C., Gaspar-Maia, A., Hasson, D., Punzeler, S., Valle-Garcia, D., Straub, T., Keilhauer, E. C., Strub, T., Dong, J., Panda, T., Chung, C. Y., Yao, J. L., Singh, R., Segura, M. F., Fontanals-Cirera, B., Verma, A., Mann, M., Hernando, E., Hake, S. B. & Bernstein, E. 2015. Histone variant H2A.Z.2 mediates proliferation and drug sensitivity of malignant melanoma. *Mol Cell*, 59, 75–88.

Vertino, P. M., Issa, J. P., Pereira-Smith, O. M. & Baylin, S. B. 1994. Stabilization of DNA methyltransferase levels and CpG island hypermethylation precede SV40-induced immortalization of human fibroblasts. *Cell Growth Differ*, 5, 1395–402.

Vijg, J. & Dolle, M. E. 2002. Large genome rearrangements as a primary cause of aging. *Mech Ageing Dev*, 123, 907–15.

Volinia, S., Calin, G. A., Liu, C. G., Ambs, S., Cimmino, A., Petrocca, F., Visone, R., Iorio, M., Roldo, C., Ferracin, M., Prueitt, R. L., Yanaihara, N., Lanza, G., Scarpa, A., Vecchione, A., Negrini, M., Harris, C. C. & Croce, C. M. 2006. A microRNA expression signature of human solid tumors defines cancer gene targets. *Proc Natl Acad Sci U S A*, 103, 2257–61.

Wang, C. M., Tsai, S. N., Yew, T. W., Kwan, Y. W. & Ngai, S. M. 2010. Identification of histone methylation multiplicities patterns in the brain of senescence-accelerated prone mouse 8. *Biogerontology*, 11, 87–102.

Wang, J., Geesman, G. J., Hostikka, S. L., Atallah, M., Blackwell, B., Lee, E., Cook, P. J., Pasaniuc, B., Shariat, G., Halperin, E., Dobke, M., Rosenfeld, M. G., Jordan, I. K. & Lunyak, V. V. 2011. Inhibition of activated pericentromeric SINE/Alu repeat transcription in senescent human adult stem cells reinstates self-renewal. *Cell Cycle*, 10, 3016–30.

Wang, X., Lee, R. S., Alver, B. H., Haswell, J. R., Wang, S., Mieczkowski, J., Drier, Y., Gillespie, S. M., Archer, T. C., Wu, J. N., Tzvetkov, E. P., Troisi, E. C., Pomeroy, S. L., Biegel, J. A., Tolstorukov, M. Y., Bernstein, B. E., Park, P. J. & Roberts, C. W. 2017. SMARCB1-mediated SWI/SNF complex function is essential for enhancer regulation. *Nat Genet*, 49, 289–95.

Wei, W., Ba, Z., Gao, M., Wu, Y., Ma, Y., Amiard, S., White, C. I., Rendtlew Danielsen, J. M., Yang, Y. G. & Qi, Y. 2012. A role for small RNAs inDNA double-strand break repair. *Cell*, 149, 101–12.

Weindruch, R., Walford, R. L., Fligiel, S. & Guthrie, D. 1986. The retardation of aging in mice by dietary restriction: longevity, cancer, immunity and lifetime energy intake. *J Nutr*, 116, 641–54.

Wiech, T., Timme, S., Riede, F., Stein, S., Schuricke, M., Cremer, C., Werner, M., Hausmann, M. & Walch, A. 2005. Human archival tissues provide a valuable source for the analysis of spatial genome organization. *Histochem Cell Biol*, 123, 229–38.

Willis, N. D., Cox, T. R., Rahman-Casans, S. F., Smits, K., Przyborski, S. A., Van Den Brandt, P., Van Engeland, M., Weijenberg, M., Wilson, R. G., De Bruine, A. & Hutchison, C. J. 2008. Lamin A/C is a risk biomarker in colorectal cancer. *PLoS One*, 3, e2988.

Wong, D. J., Liu, H., Ridky, T. W., Cassarino, D., Segal, E. & Chang, H. Y. 2008. Module map of stem cell genes guides creation of epithelial cancer stem cells. *Cell Stem Cell*, 2, 333–44.

Wood, J. G., Hillenmeyer, S., Lawrence, C., Chang, C., Hosier, S., Lightfoot, W., Mukherjee, E., Jiang, N., Schorl, C., Brodsky, A. S., Neretti, N. & Helfand, S. L. 2010. Chromatin remodeling in the aging genome of *Drosophila*. *Aging Cell*, 9, 971–8.

Wu, G., Broniscer, A., Mceachron, T. A., Lu, C., Paugh, B. S., Becksfort, J., Qu, C., Ding, L., Huether, R., Parker, M., Zhang, J., Gajjar, A., Dyer, M. A., Mullighan, C. G., Gilbertson, R. J., Mardis, E. R., Wilson, R. K., Downing, J. R., Ellison, D. W., Zhang, J., Baker, S. J. & St. Jude Children's Research Hospital-Washington University Pediatric Cancer Genome, P. 2012. Somatic histone H3 alterations in pediatric diffuse intrinsic pontine gliomas and non-brainstem glioblastomas. *Nat Genet*, 44, 251–3.

Wu, Z., Wu, L., Weng, D., Xu, D., Geng, J. & Zhao, F. 2009. Reduced expression of lamin A/C correlates with poor histological differentiation and prognosis in primary gastric carcinoma. *J Exp Clin Cancer Res*, 28, 8.

Xie, M., Lu, C., Wang, J., Mclellan, M. D., Johnson, K. J., Wendl, M. C., Mcmichael, J. F., Schmidt, H. K., Yellapantula, V., Miller, C. A., Ozenberger, B. A., Welch, J. S., Link, D. C., Walter, M. J., Mardis, E. R., Dipersio, J. F., Chen, F., Wilson, R. K., Ley, T. J. & Ding, L. 2014. Age-related mutations associated with clonal hematopoietic expansion and malignancies. *Nat Med*, 20, 1472–8.

Yan, H., Parsons, D. W., Jin, G., Mclendon, R., Rasheed, B. A., Yuan, W., Kos, I., Batinic-Haberle, I., Jones, S., Riggins, G. J., Friedman, H., Friedman, A., Reardon, D., Herndon, J., Kinzler, K. W., Velculescu, V. E., Vogelstein, B. & Bigner, D. D. 2009. IDH1 and IDH2 mutations in gliomas. *N Engl J Med*, 360, 765–73.

Yan, X., Hu, Z., Feng, Y., Hu, X., Yuan, J., Zhao, S. D., Zhang, Y., Yang, L., Shan, W., He, Q., Fan, L., Kandalaft, L. E., Tanyi, J. L., Li, C., Yuan, C. X., Zhang, D., Yuan, H., Hua, K., Lu, Y., Katsaros, D., Huang, Q., Montone, K., Fan, Y., Coukos, G., Boyd, J., Sood, A. K., Rebbeck, T., Mills, G. B., Dang, C. V. & Zhang, L. 2015. Comprehensive genomic characterization of long non-coding RNAs across human cancers. *Cancer Cell*, 28, 529–40.

Zhang, R., Poustovoitov, M. V., Ye, X., Santos, H. A., Chen, W., Daganzo, S. M., Erzberger, J. P., Serebriiskii, I. G., Canutescu, A. A., Dunbrack, R. L., Pehrson, J. R., Berger, J. M., Kaufman, P. D. & Adams, P. D. 2005. Formation of MacroH2A-containing senescence-associated heterochromatin foci and senescence driven by ASF1a and HIRA. *Dev Cell*, 8, 19–30.

Zhu, J., Woods, D., Mcmahon, M. & Bishop, J. M. 1998. Senescence of human fibroblasts induced by oncogenic *Raf. Genes Dev*, 12, 2997–3007.

Ziemin-Van Der Poel, S., Mccabe, N. R., Gill, H. J., Espinosa, R., 3RD, Patel, Y., Harden, A., Rubinelli, P., Smith, S. D., Lebeau, M. M., Rowley, J. D. & Et Al. 1991. Identification of a gene, MLL, that spans the breakpoint in 11q23 translocations associated with human leukemias. *Proc Natl Acad Sci U S A*, 88, 10735–9.

CHAPTER 22

Developmental systems and their dysfunction

As indicated in previous chapters, development and cellular specialization in multicellular organisms rely on signals that originate from the genome and that, in turn, generate epigenetic modifications. A substantial body of evidence exists to support the assertion that environmental factors impact pre- and postnatal development (see Chapter 19). These factors affect gene expression, sometimes by inducing somatic mutations, but often by influencing the epigenetic regulatory programs initiated by the genome. A comprehensive review of the genetic and epigenetic interactions responsible for organismal development and of the consequences of the misregulation of these interactions is beyond the scope of this book. Therefore, the discussion will be limited to examples selected from the nervous, cardiovascular and immune systems.

Neuronal regulation of learning and memory formation

Development of the nervous system

In a landmark study, the transcriptional landscape in specific regions of the human brain was determined in tissue samples from early embryos to adults (Kang et al., 2011). The results showed that a very large proportion of all protein-coding genes are expressed in the brain and that all of these active genes are differentially expressed in various regions and at different times during the life cycle. Most of these distinct expression patterns occur in prenatal brains and are progressively less marked after birth. As expected, the differential expression of genes in time and space is primarily the responsibility of enhancers that are tissue-specific in function. Enhancers not only generate the great diversity of cell types present in the brain, but also respond to external stimuli and transduce neural activity into gene activation (Kim et al., 2010; Malik et al., 2014). Again, as expected, non-coding RNAs are highly expressed and play a critical role in brain development (Kuss and Chen, 2008; Mercer et al., 2008; Ponjavic et al., 2009). It is estimated that 40% of the long non-coding RNAs present in humans are expressed specifically in the brain (Derrien et al., 2012). This is a staggering number, given that there may be upwards of 50,000 long non-coding RNAs (lncRNAs) in human tissues.

Learning and memory

Learning involves the acquisition of new information representing facts or events. This information is stored as short-term memory that, in turn, may be consolidated into long-term memory. Retrieving or modifying long-term memory can cause its decay or its reinforcement. Memory can be subdivided into two major types: **explicit memory** for objects, living things and physical settings, and **implicit memory** for subconscious physical or mental skills. Short-term storage of implicit memory involves modulating the strength of synaptic transmission (the stimulation of one neuron by another) in the appropriate neuronal pathway. Changes in the strength of a synapse are referred to as **synaptic plasticity**; an increase in strength mediated by repeated use of the synapse can result in **long-term potentiation**. Consolidation of implicit short-term memory into long-term memory involves gene expression and protein synthesis (Kandel et al., 2014). The formation of long-term memory occurs in the hippocampus region of the brain.

Epigenetics, Nuclear Organization and Gene Function: with implications of epigenetic regulation and genetic architecture for human development and health. John C. Lucchesi, Oxford University Press (2019). © John C. Lucchesi 2019.
DOI: 10.1093/oso/9780198831204.001.0001

The following sections will discuss the changes in transcriptional activation and the concomitant epigenetic changes that underlie learning and memory formation.

Molecular events underlying memory establishment and consolidation

Much of this information has been obtained in studies with model organisms, especially the sea slug *Aplysia* and *Drosophila*. The consolidation of implicit memory involves cyclic adenosine monophosphate (cAMP), protein kinase A (PKA), MAPK (mitogen-activated protein kinase) and cAMP response element binding protein (CREB), a transcription factor that binds to cAMP response elements (CREs) (Fig. 22.1). A burst of serotonin [5-hydroxytryptamine (5-HT)] increases the synthesis of cAMP in a sensory neuron; cAMP activates PKA, which, in turn, increases the release of the neurotransmitter and enhances the synapse to the motor neuron. If successive bursts of serotonin occur,

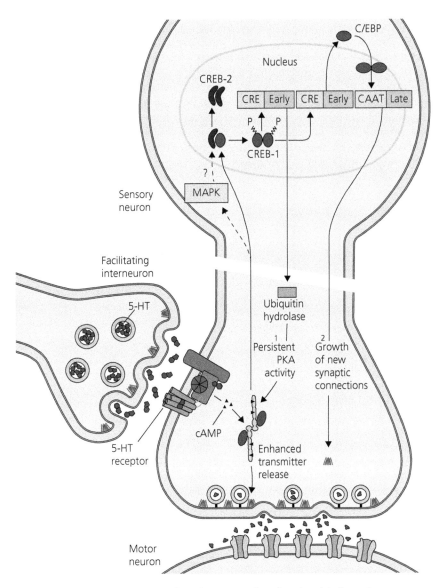

Fig. 22.1 Diagram illustrating long-term potentiation in *Aplysia*. Please see text for a discussion of the figure elements.

(From Kandel, ER, JH Schwartz and TM Jessell (2000) *Principles of Neural Science*. New York: McGraw-Hill.)

PKA recruits MAPK and both kinases enter the nucleus where PKA phosphorylates and activates CREB1 and, thereby, allows it to bind to specific DNA sequences in the regulatory region of numerous genes; MAPK phosphorylates and inactivates CRB2, an inhibitor of CREB1-activated genes (Kandel, 2012). Genes that are activated early during memory establishment encode c-Fos and EGR1 (early growth response 1) transcription factors (Guzowski, 2002). Additional transcription factors, such as tumor necrosis factor (TNF) and nuclear factor kappa B (NFκB), have also been implicated in the genomic activation that is involved in memory consolidation (Albensi and Mattson, 2000). The MEF2C (myocyte enhancer factor 2) transcription factor, which participates in a variety of developmental pathways, plays a key role in learning and memory (Barbosa et al., 2008). More recently, XBP1 (X-box binding protein 1), a transcription factor known to initiate the regulatory pathway that responds to the presence of unfolded or misfolded proteins in the cell, and whose function has been associated with neurodegenerative diseases, was shown to play a role in normal memory consolidation (Martinez et al., 2016).

Epigenetic modifications associated with learning and memory

As mentioned previously, these processes involve gene activation and protein synthesis. Therefore, the challenge is to try to distinguish those epigenetic changes that are specifically related to long-term synapsis facilitation and memory from the canonical epigenetic changes that accompany the onset and maintenance of transcription. This problem is encountered in some of the experiments that have been considered seminal in demonstrating the role of histone acetylation in the formation of long-term memory. For example, the establishment of the memory of a new taste in mice was correlated with an increase in activated mitogen-activated protein (MAP) kinase (not in its total level) and in the lysine acetylation of histones and of several non-histone proteins in a particular region of the cerebral cortex (Swank and Sweatt, 2001). Similar results were obtained with contextual fear conditioning (an experimental procedure that consists of placing an animal in a novel environment and subjecting it to an unpleasant stimulus such as a mild electric shock; when the animal is returned to the same environment, it will freeze if it remembers to associate it with the unpleasant stimulus). Using such a protocol led to an increase in MAP kinase with a concomitant increase in histone H3 acetylation; furthermore, generating an experimental increase in histone acetylation by using histone deacetylase inhibitors facilitated the formation of long-term memory (Levenson et al., 2004). As mentioned previously, MAP kinase activates a number of transcription factors, many of which target histone acetyl transferases to their sites of action. A prime example is the association of CBP (CREB-binding protein), a lysine acetyl transferase that acetylates histones and a number of other proteins. In addition to histone acetylation, the levels of methylated histones, in particular H3K4me2 or H3K4me3 (associated with gene activation) and H3K9me3 (associated with silencing), are regulated during fear conditioning and memory consolidation in mice (Gupta et al., 2010; Kerimoglu et al., 2013).

The histone variant H2A.Z plays a role in memory consolidation. Following fear conditioning in mice, memory consolidation involves the loss of H2A.Z bound to the +1 nucleosome of memory-associated genes (the first nucleosome immediately downstream of the transcription start site); consistent with this observation, experimentally induced depletion of H2A.Z in the hippocampus results in improved fear memory (Zovkic et al., 2014). Loss of H2A.Z occurs especially in the promoters of genes that are up-regulated by fear conditioning and memory formation, indicating that, although H2A.Z remains associated with steady-state genes (where its level of occupancy reflects the level of gene expression), it is a suppressor of stimulus-induced gene activity (Stefanelli et al., 2018).

DNA methylation

DNA methylation usually occurs in two forms: 5-methyl cytosine (5mC) and 5-hydroxymethyl cytosine (5hmC), both found in CpG dinucleotides (see Chapter 4). One of the first demonstrations of the involvement of methylated DNA in memory consolidation was obtained, once again, with the fear-conditioning paradigm (Miller and Sweatt, 2007). Conditioning resulted in an increase in the level of the *de novo* methyl transferases, in the methylation and repression of a specific phosphatase gene known to suppress learning and memory (*PP1*) (Koshibu et al., 2009) and in the demethylation and increased transcription of a gene that promotes memory formation (*Reelin*) (Weeber et al., 2002). Follow-up experiments correlated the persistence of memory with DNA methylation—thirty days after fear training, the fear memory was retained, unless the DNA methyl transferases were inhibited (Miller et al., 2010). In these experiments, no distinction could be made between 5mC and 5hmC. The 5hmC form is present in all tissues and cell types but is most abundant in the neurons of the central nervous system (Globisch et al., 2010). The reason for this methylated DNA distribution, i.e. the function of 5hmC, in these instances is not yet understood.

Methyl groups can be added to cytosines that are not present in CpG sites. This modification occurs in many mouse and human tissues (Varley et al., 2013; Xie et al., 2012). Non-CpG cytosines that can be methylated are referred to as CpH where H can be A, C or T (see Chapter 8). In glial cells (the support cells of the nervous system), the frequency of CpHs is similar to that in other tissues, while it is relatively high in neurons (Lister et al., 2013). Methylated CpHs occur more frequently in gene bodies than in regulatory or intergenic regions (Lister et al., 2009). Many genes that are enriched for CpH are methylated in glial cells, but not in neurons, indicating that active neuronal genes are repressed in the glia. An example is the gene that encodes the MeF2C transcription factor, which as mentioned in Molecular events underlying memory establishment and consolidation, p. 255, plays a key role in facilitating the translation of memory into learning by regulating synaptic transmission (Barbosa et al., 2008). Mutations and deletions at the MeF2C gene have been associated with severe mental retardation, stereotypic movements, epilepsy and cerebral malformation.

Chromatin remodeling

ATP-dependent nucleosome remodeling factors play an important role in neural development, and their absence or dysfunction can lead to a number of neurodevelopmental disorders and cognitive deficits. An example is provided by the BAF (BRG1- or hBRM-associated factors) complexes (see Chapter 4). A particular BAF complex, designated nBAF, contains a subunit found only in neurons (Olave et al., 2002). Mice carrying mutations responsible for higher- or lower-than-normal levels of this subunit in the hippocampus scored differently than normal mice in tests for long- and short-term memory (Vogel-Ciernia et al., 2013). Mutations in another subunit of the nBAF complex generated by conditional knockout in the hippocampus prevented the formation of particular cell types and resulted in severe effects on spatial learning and memory (Tuoc et al., 2017).

Non-coding RNAs

The first indications that micro RNAs (miRNAs) are involved in the development and function of the nervous system were based on experiments that

impaired steps in the common biosynthetic pathway of these RNAs. In order to investigate directly the neurological role of these regulatory RNAs, it was necessary to identify candidates expressed in the nervous system and to focus on specific examples. Using this approach, several miRNAs were implicated in various steps involved in learning and memory such as synapse formation and expansion. For example, over-expression or absence of miR-132 causes a deficit in synaptic plasticity and in recognition memory (Hernandez-Rapp et al., 2015; Scott et al., 2012). Inhibition of miR-124 leads to enhanced spatial learning and memory, while inhibition of miR-34 has the opposite effects (Malmevik et al., 2016). In many instances, the regulatory targets of the miRNAs have been identified. In the case of miR-132, the target is the methyl-CpG-binding protein MeCP2 and BDNF (brain-derived neurotrophic factor), involved in many aspects of nervous system development and in synaptic plasticity in the adult brain (Klein et al., 2007). The synthesis of miR-124 RNA is under the control of cAMP; its target is the mRNA of the *EGR1* gene that is essential for synaptic transmission, learning and memory (Yang et al., 2012).

A large number of long non-coding RNAs are expressed at different levels in different regions of the brain (Mercer et al., 2008). As was the case with miRNAs, the role played by lncRNAs in the nervous system can be ascertained by focusing on individual candidates. One of the first examples is Malat1 (metastasis-associated lung adenocarcinoma transcript 1), first identified as an over-expressed lncRNA in a number of cancers. Malat1 is present in several tissues and in several regions of the brain where it is associated with nuclear speckles (see Chapter 13). In the brain, it regulates the transcription of genes involved in synapse formation, function and long-term potentiation. This is presumably accomplished by recruiting and delivering primary transcript splicing factors to the site of transcription of these genes (Bernard et al., 2010).

Genome topology

Not surprisingly, the influence of the three-dimensional organization of chromatin on transcription, which has been characterized in different cell types and tissues, is implicated in the regulation of neuronal development and function (Won et al., 2016). The increase in protein synthesis that accompanies repeated neuronal stimulation and neurotransmission is underscored by an increase in the number of nucleoli, leading to an increase in the number of ribosomes and potential enhancement of protein synthesis (Jordan et al., 2007). While the boundaries of topologically associating domains (TADs) (see Chapter 10) are not significantly affected, neuronal activity-dependent genes, such as *BDNF*, are spatially relocated to transcription factories (Crepaldi et al., 2013). Recently, using mice in which CTCF can be conditionally knocked out during adulthood in specific regions of the brain, the presence of this factor was shown to be necessary for long-term but not for recent memory (Kim et al., 2018). Although CTCF plays a critical role in the formation of TADs, the topological effects of its depletion were not determined.

An unexpected modification of the topology of stimulated neurons is the occurrence of double-stranded breaks. The first report of this occurrence was based on observing an increase in γH2A.X foci in the nuclei of hippocampal neurons in mice that had been made to explore a new environment (Suberbielle et al., 2013). γH2A.X foci form at the site of double-stranded breaks and are diagnostic of incipient repair (see Chapter 15). The breaks occur in the promoter region of some of the genes that are first induced by neuronal activity such as *c-Fos* and *EGR1*. The experimental generation of double-stranded breaks in these promoter regions induces their transcription, even in the absence of neuronal stimulation. The sites of the breaks are enriched in the presence of topoisomerase IIβ (Madabhushi et al., 2015). Topoisomerase IIβ is an enzyme that creates DNA supercoils during chromosome condensation and relaxes DNA supercoils during replication and transcription; it accomplishes these functions by breaking both strands of a DNA molecule, allowing another region of the DNA to pass through the break and then resealing it (Wang, 2002). These observations suggest that the early-acting genes in non-stimulated neurons are repressed because of topological constraints; following stimulation, topoisomerase IIβ induces double-stranded breaks that release the torsional stress and, following repair, allow transcription to occur (Madabhushi et al., 2015).

Cognitive diseases

Epigenetic alterations have been correlated with the etiologies of most neuropathies, including neurodegenerative disorders such as Alzheimer's, Parkinson's and Huntington's diseases or Rubinstein–Taybi and Rett syndromes (see Chapter 4, Box 4.1 and Chapter 8, Box 8.1), as well as developmental disorders such as fragile X syndrome. In addition to anatomical and physiological malfunctions characteristic of each of these conditions, all of them have in common progressively severe cognitive deficits.

Fragile X syndrome is the most frequent form of hereditary mental retardation. It is inherited as an X-linked dominant trait. Because it is X-linked, and because of the occurrence of random X inactivation, it is therefore more common and more severe in males than females. The existence of this syndrome was probably reported for the first time in a pedigree of two generations presenting with 11 mentally retarded males and two mildly affected females (Martin and Bell, 1943); it was found to be associated with a "secondary constriction" in the X chromosome of affected males, referred to as a "fragile site" because of its very thin appearance in metaphase chromosomes (Lubs, 1969). The responsible gene *FMR1* (fragile X mental retardation 1) was cloned (Verkerk et al., 1991) and found to include a CpG island that is methylated in fragile X patients (Bell et al., 1991). A DNA segment consisting of a number of repeats of the nucleotide triplet CGG was found within the CpG island, in the 5′ untranslated region (Verkerk et al., 1991). The number of repeats of this triplet varies from 6 to 54 in normal individuals (the most common length is 30 repeats) and from 52 to 200 in individuals whose progeny is at high risk of developing the fragile X syndrome; these individuals are said to carry a premutation (Fu et al., 1991). Carriers of the premutation are at risk of developing a late-onset neurodegenerative condition and, if female, ovarian problems. Premutation alleles have a tendency to increase in length during meiosis, especially in females, giving rise to the full mutation, which consists of more than 200 CGGs.

Following synaptic activation, the *FMR1* gene mRNA is translated in dendrites, branched projections of neurons that receive stimulation by synapsing with other neurons (Weiler et al., 1997). The protein product of the gene is RNA-binding, associates with a number of mRNAs in the mammalian brain and regulates their translation The premutation causes a reduction in the fragile X mental retardation protein (FMRP) that is proportional to the level of CGG expansion; surprisingly, the premutation also causes a proportional increase in the level of FMR1 transcript (Kenneson et al., 2001). The reason is that transcripts with an excess of CGG repeats fail to be translated (Feng et al., 1995). These RNAs may lead to the late-onset symptoms by titrating out specific RNA-binding proteins or by forming a secondary structure that may sequester miRNA synthesis factors. Some recent experiments suggest that the FMR1 RNAs inhibit the translation of other mRNAs present in the same cytoplasmic location (Rovozzo et al., 2016).

Individuals carrying the full mutation fail to produce FMRP because of the loss of histones H3 and H4 acetylation in the 5′ region (Coffee et al., 1999) and methylation-based transcriptional silencing of the *FMR1* gene (Sutcliffe et al., 1992). The nature of the trigger that induces methylation of the CpG island and silencing of the gene when the CGG region exceeds a certain length is not yet understood. A possible mechanism is suggested by the observation that the FMR1 transcript binds to the gene's promoter region when the CGG repeat is of full mutation length, but not when it is within the normal or premutation range (Colak et al., 2014). The CGG portion of the mRNA binds to its complement (CCG) on the gene's DNA; this interaction, reminiscent of R-loop formation (see Chapter 14), may prevent the completion and reinitiation of transcription. The question still remains as to why a CGG repeat length needs to be greater than 200 to trigger such an event. The absence of FMRP causes a loss of translational regulation and a dysfunctional excess of proteins, leading to defects in synaptic plasticity (Zalfa et al., 2003).

Development and dysfunction of the cardiovascular system

In vertebrates, the heart develops from cells of the mesoderm, one of the three embryonic germ layers, through cascades of transcription factors that initiate and effect the concurrent differentiation of the many cell types that make up the components of this organ. Much of the fine-grained understanding of this aspect of development in mammals comes from studies in the mouse. The cardiac precursor cells, which will give rise to the entire cardiovascular system, are first identified by the presence of TBR-2 (T-box brain protein 2, also known as Eomesodermin), a transcription factor that directly activates the *Mesp1* (mesoderm posterior BHLH transcription factor 1) gene (Costello et al., 2011). Mesp1, in turn, regulates the transcription of several hundred genes, among which are key cardiac transcription factors such as GATA4 (GATA-binding protein 4) which, with the related protein GATA6, controls the onset of cardiomyocyte differentiation, NKX 2-5 (NK2 homeobox 2-5) expressed in cardiac progenitor cells and necessary for the development of the muscle walls of the heart, Hand2 (heart and neural crest derivative-expressed protein), TBX5 (T-box protein 5) that play a role in heart chambers morphogenesis, MEF2 (myocyte enhancer factor 2) that cooperates with the core cardiac transcription factors to activate contractile protein gene expression and MYOCD (myocardin), a co-activator that is necessary for the

differentiation of the smooth muscle cell lineage, to name a few (Bondue et al., 2008; Soibam et al., 2015).

Epigenetic regulation of cardiac differentiation

Histone-modifying enzymes have significant effects on cardiovascular development. In some cases, specific enzymes have been shown to interact with specific transcription factors and to act as co-activators of transcription for the factors' targeted genes. For example, during cardiomyocyte differentiation, the GATA4 factor recruits the histone acetyl transferase p300 (Dai and Markham, 2001), and TBX5 factor associates with p300 and with PCAF (P300/CBP-associated factor) that, itself, has histone acetyl transferase activity (Murakami et al., 2005). TBX5 is a transcription factor used to reprogram fibroblasts into induced cardiomyocytes. In other cases, the general involvement of modifying enzymes has been demonstrated—the simultaneous knock-down of histone deacetylases (HDAC1 and 2 or HDAC5 and 9) or of Ezh2, the histone methyl transferase of the PRC2 repressive complex, induces cardiovascular damage (Chang et al., 2004; Montgomery et al., 2007; Surface et al., 2010), but the specific steps where these enzymes exert their action are not defined.

A study of genomic patterns of gene expression and selected chromatin modifications was carried out on four stages of mouse embryonic stem cells induced to differentiate into cardiomyocytes (Wamstad et al., 2012). Distinct groups of genes, miRNAs and lncRNAs were expressed in each of the four stages (Fig. 22.2). Active genes were grouped on the basis of the histone modifications present around the transcription start site. A set of genes involved in basic cellular functions showed no variation in all four stages. Among the embryonic stem cell (ESC)-specific genes that are silenced during differentiation, there were genes that lost the H4K4me3 and H3K27ac active marks and acquired H3K27me3, while others simply lost the active marks, indicating that pluripotency genes are regulated differently. Some genes down-regulated in the mesoderm stage acquired silencing histone marks only in the cardiomyocyte stage; other genes showed typical suppressive histone marks in the mesoderm stage and maintained these marks in the cardiac precursor stage. A group of genes involved in contractile protein synthesis exhibited H3K4me1 at the mesoderm stage prior to their activation and switched to H3K4me3 in the cardiac precursor and cardiomyocyte stages. Enhancers marked by both H3K27ac and H3K4me1, or just H3K27ac, were classified as active, and those with

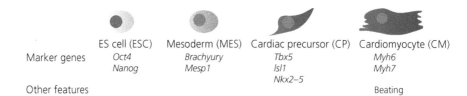

Fig. 22.2 Four stages of differentiation of ESCs into cardiomyocytes. Brachyury is a transcription factor involved in the transcriptional regulation of genes required for mesoderm formation and differentiation. Isl1 is a transcription factor that binds the promoters of the glucagon, insulin and somatostatin genes and is found *in vivo* in populations of cardiac precursors. The other genes are described in the text.

(From Wamstad et al., 2012.)

only H3K4me1 as poised. Different sets of active and poised enhancers were present during the successive stages of cardiomyocyte differentiation; the number of genes involved in general cellular functions associated with active enhancers were progressively fewer, while the number of genes responsible for specific aspects of cardiomyocyte function and heart development increased.

In addition to histone covalent modifications, ATP-dependent remodeling complexes have been implicated in specific steps during cardiovascular differentiation. A subunit of the BAF complexes Baf60c is expressed in the heart primordium of early mouse embryos where it serves as a bridge between one of the complexes and the GATA4, NKX2-5 and TBX5 transcription factors (Lickert et al., 2004). In fact, GATA4, TBX5 and Baf60c, when expressed in mouse mesodermal cells, are sufficient to induce their differentiation into contracting cardiomyocytes (Takeuchi and Bruneau, 2009).

Not surprisingly, CTCF plays an important role in heart formation. Targeted deletion of this genome architectural protein in differentiating mouse cardiomyocytes leads to a change in the level of expression of genes involved in mitochondrial function and in heart development, and results in cardiac defects and embryonic death (Gomez-Velazquez et al., 2017).

Cardiovascular diseases

A number of early studies reported cases of epigenetic misregulation correlated with specific cardiopathic phenotypes. Wolf–Hirschhorn syndrome (WHS), which presents with a number of developmental abnormalities, including heart defects, is caused by a deletion that invariably includes the *WHSC1* (Wolf–Hirschhorn syndrome candidate 1) gene. *WHSC1* encodes a histone methyl transferase responsible for the tri-methylation of H3K36 that, in embryonic heart extracts, is physically associated with the transcription factor NKX2-5 (Nimura et al., 2009). Another complex disorder

involving multiple malformations is Kabuki syndrome. Genomic sequencing of affected individuals revealed the presence of nonsense mutations in MLL2, also known as KMT2D (lysine methyl transferase 2D) (Ng et al., 2010). A study based on sequencing only the portion of the genome that encodes proteins, i.e. exons, of several hundred individuals with congenital heart disease revealed a high proportion of mutations of genes involved in the methylation or demethylation of H3K4, or the ubiquitination of H2BK120, a modification required for H3K4 methylation (Zaidi et al., 2013).

Changes in the expression of various miRNAs have been correlated to different types of heart failure. In a mouse strain developed to exhibit heart hypertrophy, a number of miRNAs were up-regulated and others down-regulated (van Rooij et al., 2006). Ectopic over-expression of some individual miRNAs from the up-regulated group was sufficient to induce hypertrophy of mouse cardiomyocytes in culture and *in vivo*. Several of these miRNAs showed similar over-expression in failing human hearts. The levels of a number of specific miRNAs circulating in the bloodstream were found to decline in patients with acute heart failure, and the extent of their decrease could be used to predict mortality (Ovchinnkova et al., 2016).

As one would expect, various epigenetic changes involving covalent modifications of DNA and histones, as well as alterations in miRNA levels, have been correlated to cardiovascular diseases such as hypertension (Friso et al., 2015) and atherosclerosis (Grimaldi et al., 2015).

Epigenetics and immune system disorders

Immune systems have evolved in order to protect organisms from the harmful effects of pathogens. The innate immune system consists of cells that are present throughout the tissues of the body and that can either ingest invading bacteria (phagocytes) or recognize and destroy virus-infected cells [natural killer (NK) cells]. Another component of the innate immune system consists of circulating proteins that can be activated by cell surface characteristics of some invading bacterial or fungal pathogens (complement system); these proteins attach themselves to the surface of the pathogens and elicit their destruction. The adaptive immune system consists of three major types of cells: macrophages and dendritic cells, T lymphocytes and B lymphocytes. Macrophages and dendritic cells acquire a foreign antigen from an invading pathogen, process it and present it to T lymphocytes, now said to be activated (Fig. 22.3).

One type of T cells (cytotoxic) recognizes the antigen on target cells and destroys them. These T cells are characterized by the presence of a transmembrane protein [cluster of differentiation 8 (CD8)] that serves as a co-receptor of the T cell receptor (TCR) (see the following paragraph). A second type of T cells (helper) associates specifically with B lymphocytes that had acquired and processed the same foreign antigen and stimulates them. Helper T cells display on their surface a different transmembrane protein (CD4) that helps them to communicate with an antigen-presenting cell. The stimulated B cells, referred to as plasma cells,

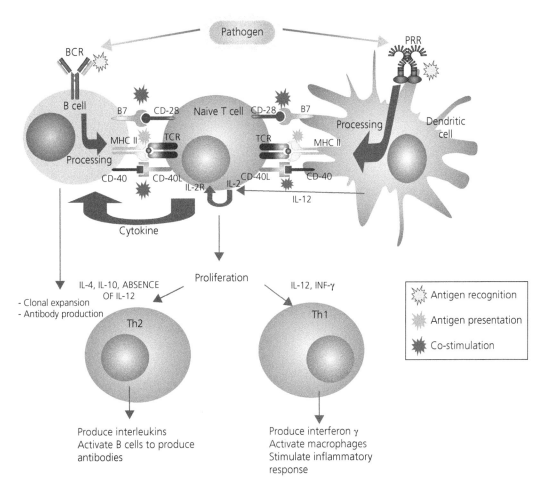

Fig. 22.3 Diagram representing the activation of B cells by activated T cells. BCRs are transmembrane receptor proteins located on B cells that can bind a particular antigen for which they are specific. PRRs are pattern recognition receptors that recognize molecular characteristics of pathogens. CD28 (cluster of differentiation 28) is one of the proteins expressed on T cells that provide co-stimulatory signals required for T cell activation; CD40 is a co-stimulatory protein found on antigen-presenting cells and is required for their activation; B7 is a membrane protein found on activated antigen-presenting cells (like dendritic cells) that pairs with CD28 on a T cell and modulates the activity of the MHC–TCR signal. Th1 and Th2 are two types of proliferating helper T cells. Interferon-γ is a cytokine with antiviral activity and is a strong activator of microphages.

(Modified from iGEM, Team Slovenia 2008.)

secrete large quantities of the antibodies (immuno-globulins) specific to the antigen. A third type of T cells (regulatory) is necessary to prevent autoimmunity (failure to recognize antigens as one's own), which they accomplish by targeting other cells of the immune system that are self-reactive, i.e. that present the organism's own antigens. Regulatory T cells are derived from naïve helper T cells and, therefore also express the CD4 cell surface marker. Cytotoxic, helper and regulatory T cells, as a group, are referred to as effector T cells.

The presentation and recognition of self or foreign antigens are mediated by the products of the two major histocompatibility complexes MHC class I (MHC I) and MHC class II (MHC II), referred to in humans as the human leukocyte antigen (HLA) system. MHC I proteins are on the surface of all cells, including, of course, macrophages and dendritic cells, and are used to present antigens, such as those generated by a viral infection, to cytotoxic T cells. MHC II proteins are on the surface of macrophages and dendritic and B cells that use them to present foreign antigens to helper T cells. The latter recognize the antigen by means of cell surface receptors (TCRs). B cells have their own receptors with which they acquire antigens that they process and present on their surface by means of MHC II proteins in order to attract, and to be stimulated by, helper T cells.

A subset of B cells and T cells are retained as memory cells to be expanded in the case of the organism's re-exposure to the same antigen. Therefore, one of the major characteristics of the immune systems is the ability to remember all past exposures to antigens. The other major characteristic is the ability to distinguish foreign antigens from those that are the organism's own. During the development of the immune system, lymphocytes that recognize self-antigens are eliminated by programmed cell death (Kappler et al., 1987; Kisielow et al., 1988). The absence of components of the immune systems can lead to immune deficiency syndromes characterized by a failure in defense against pathogens. Mistaking an indigenous or self-antigen as foreign leads to a variety of autoimmune diseases. Although a few such diseases have been ascribed to single-gene mutations, most have multiple genetic factors that determine susceptibility. In addition, the degree of concordance among identical twins varies between 15% and 75%, indicating that environmental factors play an important role in disease onset. These factors include infectious agents, tobacco smoke, alcohol consumption and exposure to ultraviolet irradiation or to environmental pollutants (Miller et al., 2012; Selmi et al., 2012).

As is true of all cases of cellular differentiation, the conversion of naïve CD8+ T cells into effector or memory T cells involves extensive changes in gene activity mediated by groups of enhancers and super-enhancers that target specific groups of genes, accompanied by the concomitant epigenetic changes. A number of naïve T cell enhancers become inactivated, and a new set of enhancers become active in effector T cells; memory T cells retain a subset of effector T cell enhancers and reactivate some enhancers from naïve T cells (He et al., 2016). Concordant with changes in their transcriptome, the different stages of CD8+ T cells exhibit dynamic differences in chromatin accessibility. Using ATAC-seq, a technique that relies on the insertion of transposons to detect regions of open chromatin (Buenrostro et al., 2013), a set of such regions is common to the different cell types, while naïve, effector and memory cells exhibit specific, differentially accessible sites (Scott-Browne et al., 2016).

Systemic lupus erythematosus (SLE)

This is an autoimmune disease that affects many organ systems. It is significantly more prevalent in females than males. Studies to determine the concordance of affected identical twins or the frequency of SLE in the families of patients reveal a definite genetic predisposition for the disease (Alarcon-Segovia et al., 2005; Deapen et al., 1992). Genome-wide association studies have identified over 50 genes with alleles that confer susceptibility to SLE (Deng and Tsao, 2014). Not surprisingly, the strongest associations were with HLA genes, especially in the class II region of the locus which, together with the class I region, contains genes for the proteins that present peptides to T cells. Strong associations were also with genes involved in type I interferon production and function (Box 22.1). It is reasonable to suggest that SLE most often is triggered in genetically susceptible individuals by particular environmental factors.

A hallmark of SLE is the high level of anti-DNA antibodies thought to originate from the presence of necrotic cells or of cells undergoing apoptosis (Ceppellini et al., 1957; Robbins et al., 1957; Miescher and Strassle, 1957). These DNA–antibody complexes activate an excessive production of interferon α (IFN-α) (Box 22.1) (Vallin et al., 1999), an occurrence that is characteristic of other autoimmune diseases. High IFN-α levels result in increased cytolysis and release of large amounts of autoantigens, leading B cells to produce autoantibodies that destroy more cells, generating a positive feedback on the process (Rönnblom, 2016). Anti-DNA antibodies

---

Box 22.1 Interferons and interleukins

Interferons (IFNs) represent the main cellular defense mechanism against infection by viruses and other pathogens. They derive their name from their ability to "interfere" with viral proliferation. Functionally, they are characterized as **cytokines**, small secreted proteins that bind to cell surface receptors and elicit the occurrence of some biochemical pathway or cellular proliferation. There are two main types of IFNs. Type I IFNs include many different proteins, among which the more important are IFN-α and IFN-β; these IFNs can be produced by many different cells, with the most effective being plasmacytoid dendritic cells found in blood and lymph nodes (Christiakov et al., 2014). The type II class includes a single member—IFN-γ—that is produced by T lymphocytes and natural killer (NK) cells (Qu et al., 2013; Talaat et al., 2015).

Following infection, a cytoplasmic receptor (IFIH1 or f MDA-5) recognizes replicating RNA viruses by the special 5′ end cap of their mRNAs or by the presence of long double-stranded RNAs. This association triggers the activation of IFN regulatory factors that stimulate the transcription of IFN-α and IFN-β (Kawai et al., 2005). DNA viruses are sensed by IFI16 (interferon inducible protein 16) (Unterholzner et al., 2010).

Interleukins (ILs) are another type of cytokines. First identified as factors that were expressed by leukocytes, it is now known that ILs are produced by many different cell types. Several ILs play specific roles in the development and function of the immune system. IL2 is produced by T cells, following their activation by antigen-presenting cells, and auto-stimulates their proliferation and differentiation into two types of helper cells—T helper 1 (Th1) and T helper 2 (Th2). Th1 cells secrete IFN-γ to induce macrophages to break down and process the pathogens that they have engulfed; IFN-γ also stimulates cytotoxic T cells to destroy infected cells. Th2 cells secrete a number of ILs such as IL4 and IL13 that are responsible for inducing a systemic reaction against intestinal nematode infection (Finkelman et al., 2004), IL10 that inhibits the synthesis of inflammatory cytokines by Th1 cells (Fiorentino et al., 1989) and IL5 that stimulates B lymphocytes to proliferate and differentiate into immunoglobulin-producing cells (Takatsu et al., 1988).

---

are thought to be the primary cause of lupus nephritis, although much of the responsible interactions are not yet fully determined. Experimental evidence supports two different, albeit potentially complementary, models—anti-DNA antibodies may cross-react with structurally unrelated renal antigens or they may react with exposed chromatin in the kidney glomeruli (Rekvig et al., 2016).

An additional immunological hallmark of SLE is the overproduction of interleukin 10 (IL10) by non-T cells (Llorente et al., 1994). In general, IL10 (Box 22.1) is deficient in autoimmune diseases, leading to an increase in inflammatory response. In SLE, in spite of enhanced levels of this cytokine, inflammation is part of the disease's etiology. IL10 stimulates B lymphocytes to produce autoantibodies (Llorente et al., 1995). Several other immune system cell types are affected in SLE. There is a substantial increase in the relative number of a particular type of CD4+ T cells that produce IL17, a very strong pro-inflammatory interleukin (Qu et al., 2013; Talaat et al., 2015). These cells invade the kidneys (Crispin et al., 2008) and contribute to the development of lupus-associated nephritis. Other affected cells are CD8+ T cells that have reduced cytotoxicity in SLE (Filaci et al., 2001).

Interleukins and interferons perform their functions by initiating a regulatory cascade that activates a large number of genes. As expected, CD4+ T cells of SLE patients exhibit significant changes in the levels of histone acetyl transferases, deacetylases and methyl transferases, and in the concomitant levels of covalent histone modifications (Hu et al., 2008). In addition to these epigenetic modifications, a number of miRNAs are differentially expressed in peripheral mononuclear cells (immune system lymphocytes and monocytes that are precursors of macrophages and dendritic cells) from SLE patients (Dai et al., 2007). Early evidence of the role of DNA methylation in the function of the immune system was provided by the observation that inhibitors of this modification induced autoreactivity in cultured T cells (Cornacchia et al., 1988). Studies with isolated mouse CD4+ T cells supported the hypothesis that DNA methylation is involved in lupus-like disease (Quddus et al., 1993). A particularly informative experiment consisted of the induction of autoimmunity by the injection into mice of activated CD4+ T cells that had been treated *in vitro* with DNA methyl transferase inhibitors (Yung et al., 1995).

Specific examples of the effect of hypomethylation in SLE were eventually obtained with CD4+ T cells isolated from patients (Zhao et al., 2010). In normal individuals, cytotoxic CD8+ T cells exert their killing function by secreting **perforin**, a protein that inserts into the cell membrane and oligomerizes to form pores through which secreted proteases can penetrate. In the CD4+ T cells of normal individuals, the *PRF1* gene that encodes perforin is silent; in SLE patients, a region between the *PRF1* promoter and the enhancer is dem-ethylated (as is the case in normal CD8+ T cells) and CD4+ cells express perforin, presumably explaining the excessive death of monocytes and macrophages (Kaplan et al., 2004). A screen of over 200 transcription factors revealed that RFX1 (regulatory factor X 1) recruits the epigenetic co-repressors histone deacetylase 1 and DNA methyl transferase 1 to the promoters of genes encoding two CD proteins (CD11a, an integrin subunit that contributes to cell adhesion, and CD70, a cytokine that contributes to T and B cell activation and immuno-globulin synthesis). In SLE CD4+ T cells, RFX1 is down-regulated, leading to the over-expression of both CD11a and CD70 and contributing to the autoimmune response. More recent genome-wide DNA methylation studies on CD4+ T cells, B cells and monocytes isolated from SLE patients revealed a high degree of hypo-methylation of genes that are regulated by type I inter-ferons in all three cell types (Absher et al., 2013; Coit et al., 2013).

Rheumatoid arthritis (RA)

This disease affects primarily peripheral joints. It exhibits a sex bias in favor of females, although less pronounced than the bias in the occurrence of SLE. A substantial number of genes (around 50) that confer a risk for the disease have been identified (Eyre et al., 2012). As in the case of SLE and other autoimmune dis-eases, onset is triggered by a combination of genomic susceptibility and particular environmental factors. Although some serological features, such as the so-called rheumatoid factor, have been considered to characterize RA, the hallmark of the disease is the presence of autoantibodies against cyclic citrullinated peptides (Schellekens et al., 1998). The arginine found in histones can be converted by hydrolysis to the amino acid citrulline. This modification, which can occur in histones H2A, H3 and H4, is involved in tran-scriptional regulation of pluripotency, but also in the immune response to infection; it causes chromatin decondensation by displacing histone H1 and usually results in gene activation (Christophorou et al., 2014). Citrullinated proteins are considered to be new anti-gens that stimulate the formation of auto-reactive anti-bodies. These antibodies, which are present at low levels in normal individuals, can be induced at high levels by different environmental factors in individuals who are genetically at risk for RA. Anti-citrullinated

protein antibodies are found in a number of other inflammatory diseases.

In a similar manner to SLE, T cells from RA patients are hypomethylated (Richardson et al., 1990). This early observation was supported by studying the role of DNA methylation on specific genes involved in the etiology of RA, for example the *IL6* gene that encodes a cytokine present at high levels in patients (van Leeuwen et al., 1995). As in many genes, the *IL6* promoter is located within a CpG island. In mono-nuclear cells of normal individuals, the island exhibits widespread methylation; in RA patients, a single, specific CpG site is less frequently methylated and the lower level of methylation at this site is correlated with a higher level of *IL6* transcription (Nile et al., 2008). A genome-wide determination of DNA methyla-tion was performed on fibroblast-like synovial cells (Nakano et al., 2013). These cells are found in the inner layer of the synovial membrane that separates the two elements of a joint; they contribute to the pathology of RA by synthesizing cytokines that induce inflammation and proteases that attack the joint's cartilage (Bartok and Firestein, 2010). Several hundred hypomethylated and hypermethylated genes were found.

In contrast to the enhancement of transcription of the numerous hypomethylated genes involved in the inflammation phenotype of RA, a few genes exhibit hypermethylation and silencing. An example is the *DR3* (death receptor 3) gene, a member of the TNF receptor family of genes that induce programmed cell death. In RA patients, the CpG island associated with the *DR3* promoter is hypermethylated in the cells pre-sent in the synovial fluid of joints, but not in peripheral blood mononuclear cells (Takami et al., 2006). The reduced level of the DR3 protein may decrease the rate of apoptosis of synovial cells in RA patients. Some of the CD4+ regulatory T cells that accumulate in the synovial fluid lose their suppressive activity and produce the IL17 cytokine, thereby contributing substantially to the inflammatory condition (Wang et al., 2015). Furthermore, these cells exhibit high levels of anti-apoptotic regulators that allow them to increase in number (van der Geest et al., 2015).

As expected, consistent with the CpG methylation-mediated transcriptional modulation of genes in synovial fibroblasts of RA patients, promoters are enriched in the classical histone marks characteristic of gene activation or silencing (Karouzakis et al., 2014; Trenkmann et al., 2011; Wada et al., 2014). Unfortunately, the more significant questions regarding the identity of the triggers responsible for the hypomethylation of specific CpG sites or of entire CpG islands leading to gene activation have not been answered.

Several miRNAs have been associated with the pathology of RA (Smigielska-Czepiel et al., 2014; Stanczyk et al., 2011). For example, miR-34, which normally inhibits cell death by targeting a pro-apoptosis factor, is repressed in RA synovial cells (Niederer et al., 2012). Another example, but with an opposite effect, is miR-155 that is elevated in peripheral blood mononuclear cells and in fibroblast-like synovial cells by the action of TNF, which is a pro-inflammatory agent (Long et al., 2013). In these cells, miR-155 decreased the levels of a matrix metallopeptidase (MMP)—an enzyme that breaks down extra-cellular matrix; it also decreased the ability of the synovial cells to proliferate and invade other tissues.

Chapter summary

Organisms are formed by differential gene activation and the associated epigenetic chromatin modifications that occur during embryonic development and in adult life. These gene expression programs can be heritably maintained in the absence of the initiating signals and can be influenced by environmental factors during pre- and postnatal development. All systems within an organism are mediated by sets of genetic and epigenetic directives, and alterations in these directives can lead to malfunction and disease. Illustrative examples of these interactions are provided by the nervous, cardiovascular and immune systems.

A large proportion of protein-coding and non-coding RNA transcription units are differentially expressed in the brain, mostly during the prenatal developmental period. A specific function performed by the brain is learning—the acquisition of new information, its storage as short-term memory and its consolidation as long-term memory. Memory establishment involves the activation of CREB (cAMP responsive element binding protein), a transcription factor that, in turn, activates a number of genes, including several transcription factors. As expected, the histone modifications involved with gene activation are associated with the target genes of these factors. Memory consolidation involves the standard (CpG) and less common (non-CpG or CpH) DNA methylation and demethylation of specific genes. The establishment of short- and long-term memories also involves the action of a neuron-specific nucleosome remodeling BAF complex. Finally, several miRNAs and a number of lncRNAs have been shown to play a role in memory formation. In general, TAD boundaries are not affected, although neuronal activity-dependent genes are spatially relocated to transcription factories.

Epigenetic alterations have been correlated with the etiologies of most neuropathies, including developmental disorders such as Rubinstein–Taybi, Rett or fragile X syndromes, as well as neurodegenerative disorders such as Alzheimer's, Parkinson's or Huntington's diseases discussed in previous chapters. All of these pathologies present with severe cognitive deficits. Rubinstein–Taybi syndrome is caused by loss of activity of either CBP or the closely related p300 histone acetyl transferase. Rett syndrome is caused by the absence of the methyl-CpG-binding protein MeCP2. Alzheimer's disease may be caused by the presence of elevated levels of the histone deacetylase HDAC2. In fragile X syndrome, fully affected individuals carry a high number of CGG triplets in a CpG island that is methylated, with concomitant loss of histones H3 and H4 acetylation in the 5′ region.

In vertebrates, the heart develops from cells of the mesoderm through cascades of transcription factors that mediate the differentiation of the many cell types of this organ. Histone-modifying enzymes, such as the histone acetyl transferase p300, have significant effects on cardiovascular development. Distinct groups of genes, miRNAs and lncRNAs are expressed during the different stages of heart development, exhibiting the appropriate activation or repression histone marks, and ATP-dependent remodeling BAF complexes have been implicated in specific steps during cardiovascular differentiation.

A number of studies have reported cases of epigenetic misregulation associated with specific cardiopathic phenotypes. Wolf–Hirschhorn and Kabuki syndromes are caused by the absence of two different histone methyl transferases. Changes in the expression of various miRNAs have been correlated to different types of heart failure, and changes involving covalent modifications of DNA and histones, as well as alterations in miRNA levels, have been found in cardiovascular diseases.

The innate immune system consists of cells that can either ingest invading bacteria or recognize and destroy virus-infected cells, and of circulating proteins that attach themselves to the surface of the pathogens and destroy them. The adaptive immune system consists of three major types of cells: macrophages and dendritic cells, T lymphocytes and B lymphocytes. Macrophages and dendritic cells acquire a foreign antigen from an invading pathogen, process it and present it to T lymphocytes, now said to be activated. One type of T cells (cytotoxic) recognizes target cells by the presence of the antigen and destroys them. A second type of T cells (helper) associates specifically with B lymphocytes that had acquired and processed the same foreign antigen. The stimulated B cells, referred to as plasma cells, secrete large quantities of the antibodies (immunoglobulins) specific to the antigen. A subset of B cells and T cells are retained as memory cells, to be expanded in case of the organism's re-exposure to the same antigen. A third type of T cells (regulatory) targets other cells of the immune system that are self-reactive and prevent autoimmunity.

Failure to recognize antigens as one's own can lead to a variety of autoimmune diseases. Among the characteristics of systemic lupus erythematosus (SLE) is a high level of anti-DNA antibodies thought to originate from the presence of necrotic cells or cells undergoing apoptosis. Interleukins are cytokines that are produced by particular cell types to stimulate, via matching specific cell surface receptors, the proliferation or to modulate the transcription of other cells. SLE affects several cell types of the immune system and the levels of interleukins that they produce. Interleukins initiate regulatory cascades that activate large numbers of genes. SLE patients' cells exhibit significant changes in the levels of histone acetyl transferases, deacetylases and methyl transferases, and in the concomitant levels of covalent histone modifications. In addition, a number of miRNAs are differentially expressed in SLE cells. Another autoimmune disease—rheumatoid arthritis (RA)—is characterized by the presence of autoantibodies against cyclic citrullinated peptides. Arginines in histones and other proteins can be converted to citrulline, and citrullinated proteins are considered as non-self by the immune system. In addition, hundreds of genes are hypo- or hypermethylated, and the levels of a number of miRNAs are altered in cells from RA patients.

References

Absher, D. M., Li, X., Waite, L. L., Gibson, A., Roberts, K., Edberg, J., Chatham, W. W. & Kimberly, R. P. 2013. Genome-wide DNA methylation analysis of systemic lupus erythematosus reveals persistent hypomethylation of interferon genes and compositional changes to CD4+ T-cell populations. *PLoS Genet,* 9, e1003678.

Alarcon-Segovia, D., Alarcon-Riquelme, M. E.,Cardiel, M. H., Caeiro, F., Massardo, L., Villa, A. R., Pons-Estel, B. A. & Grupo Latinoamericano De Estudio Del Lupus, E. 2005. Familial aggregation of systemic lupus erythematosus, rheumatoid arthritis, and other autoimmune diseases in 1,177 lupus patients from the GLADEL cohort. *Arthritis Rheum,* 52, 1138–47.

Albensi, B. C. & Mattson, M. P. 2000. Evidence for the involvement of TNF and NF-kappaB in hippocampal synaptic plasticity. *Synapse,* 35, 151–9.

Barbosa, A. C., Kim, M. S., Ertunc, M., Adachi, M., Nelson, E. D., Mcanally, J., Richardson, J. A., Kavalali, E. T., Monteggia, L. M., Bassel-Duby, R. & Olson, E. N. 2008. MEF2C, a transcription factor that facilitates learning and memory by negative regulation of synapse numbers and function. *Proc Natl Acad Sci U S A,* 105, 9391–6.

Bartok, B. & Firestein, G. S. 2010. Fibroblast-like synoviocytes: key effector cells in rheumatoid arthritis. *Immunol Rev,* 233, 233–55.

Bell, M. V., Hirst, M. C., Nakahori, Y., Mackinnon, R. N., Roche, A., Flint, T. J., Jacobs, P. A., Tommerup, N., Tranebjaerg, L., Froster-Iskenius, U. & ET AL. 1991. Physical mapping across the fragile X: hypermethylation and clinical expression of the fragile X syndrome. *Cell,* 64, 861–6.

Bernard, D., Prasanth, K. V., Tripathi, V., Colasse, S., Nakamura, T., Xuan, Z., Zhang, M. Q., Sedel, F., Jourdren, L., Coulpier, F., Triller, A., Spector, D. L. & Bessis, A. 2010. A long nuclear-retained non-coding RNA regulates synaptogenesis by modulating gene expression. *EMBO J,* 29, 3082–93.

Bondue, A., Lapouge, G., Paulissen, C., Semeraro, C., Iacovino, M., Kyba, M. & Blanpain, C. 2008. Mesp1 acts as a master regulator of multipotent cardiovascular progenitor specification. *Cell Stem Cell,* 3, 69–84.

Buenrostro, J. D., Giresi, P. G., Zaba, L. C., Chang, H. Y. & Greenleaf, W. J. 2013. Transposition of native chromatin for fast and sensitive epigenomic profiling of open chromatin, DNA-binding proteins and nucleosome position. *Nat Methods,* 10, 1213–18.

Ceppellini, R., Polli, E. & Celada, F. 1957. A DNA-reacting factor in serum of a patient with lupus erythematosus diffusus. *Proc Soc Exp Biol Med,* 96, 572–4.

Chang, S., Mckinsey, T. A., Zhang, C. L., Richardson, J. A., Hill, J. A. & Olson, E. N. 2004. Histone deacetylases 5 and 9 govern responsiveness of the heart to a subset of stress signals and play redundant roles in heart development. *Mol Cell Biol,* 24, 8467–76.

Chistiakov, D. A., Orekhov, A. N., Sobenin, I. A. & Bobryshev, Y. V. 2014. Plasmacytoid dendritic cells: development, functions, and role in atherosclerotic inflammation. *Front Physiol,* 5, 279.

Christophorou, M. A., Castelo-Branco, G., Halley-Stott, R. P., Oliveira, C. S., Loos, R., Radzisheuskaya, A., Mowen, K. A., Bertone, P., Silva, J. C., Zernicka-Goetz, M., Nielsen, M. L., Gurdon, J. B. & Kouzarides, T. 2014. Citrullination regulates pluripotency and histone H1 binding to chromatin. *Nature,* 507, 104–8.

Coffee, B., Zhang, F., Warren, S. T. & Reines, D. 1999. Acetylated histones are associated with FMR1 in normal but not fragile X-syndrome cells. *Nat Genet,* 22, 98–101.

Coit, P., Jeffries, M., Altorok, N., Dozmorov, M. G., Koelsch, K. A., Wren, J. D., Merrill, J. T., Mccune, W. J. & Sawalha, A. H. 2013. Genome-wide DNA methylation study suggests epigenetic accessibility and transcriptional poising of interferon-regulated genes in naive CD4+ T cells from lupus patients. *J Autoimmun,* 43, 78–84.

Colak, D., Zaninovic, N., Cohen, M. S., Rosenwaks, Z., Yang, W. Y., Gerhardt, J., Disney, M. D. & Jaffrey, S. R. 2014. Promoter-bound trinucleotide repeat mRNA drives epigenetic silencing in fragile X syndrome. *Science,* 343, 1002–5.

Cornacchia, E., Golbus, J., Maybaum, J., Strahler, J., Hanash, S. & Richardson, B. 1988. Hydralazine and procainamide inhibit T cell DNA methylation and induce autoreactivity. *J Immunol,* 140, 2197–200.

Costello, I., Pimeisl, I. M., Drager, S., Bikoff, E. K., Robertson, E. J. & Arnold, S. J. 2011. The T-box transcription factor Eomesodermin acts upstream of Mesp1 to specify cardiac mesoderm during mouse gastrulation. *Nat Cell Biol,* 13, 1084–91.

Crepaldi, L., Policarpi, C., Coatti, A., Sherlock, W. T., Jongbloets, B. C., Down, T. A. & Riccio, A. 2013. Binding of TFIIIC to sine elements controls the relocation of activity-dependent neuronal genes to transcription factories. *PLoS Genet,* 9, e1003699.

Crispin, J. C., Oukka, M., Bayliss, G., Cohen, R. A., Van Beek, C. A., Stillman, I. E., Kyttaris, V. C., Juang, Y. T. & Tsokos, G. C. 2008. Expanded double negative T cells in patients with systemic lupus erythematosus produce IL-17 and infiltrate the kidneys. *J Immunol,* 181, 8761–6.

Dai, Y., Huang, Y. S., Tang, M., Lv, T. Y., Hu, C. X., Tan, Y. H., Xu, Z. M. & Yin, Y. B. 2007. Microarray analysis of micro-RNA expression in peripheral blood cells of systemic lupus erythematosus patients. *Lupus,* 16, 939–46.

Dai, Y. S. & Markham, B. E. 2001. p300 Functions as a coactivator of transcription factor GATA-4. *J Biol Chem,* 276, 37178–85.

Deapen, D., Escalante, A., Weinrib, L., Horwitz, D., Bachman, B., Roy-Burman, P., Walker, A. & Mack, T. M. 1992. A revised estimate of twin concordance in systemic lupus erythematosus. *Arthritis Rheum,* 35, 311–18.

Deng, Y. & Tsao, B. P. 2014. Advances in lupus genetics and epigenetics. *Curr Opin Rheumatol,* 26, 482–92.

Derrien, T., Johnson, R., Bussotti, G., Tanzer, A., Djebali, S., Tilgner, H., Guernec, G., Martin, D., Merkel, A., Knowles, D. G., Lagarde, J., Veeravalli, L., Ruan, X., Ruan, Y., Lassmann, T., Carninci, P., Brown, J. B., Lipovich, L., Gonzalez, J. M., Thomas, M., Davis, C. A., Shiekhattar, R., Gingeras, T. R., Hubbard, T. J., Notredame, C., Harrow, J. & Guigo, R. 2012. The GENCODE v7 catalog of human long noncoding RNAs: analysis of their gene structure, evolution, and expression. *Genome Res,* 22, 1775–89.

Eyre, S., Bowes, J., Diogo, D., Lee, A., Barton, A., Martin, P., Zhernakova, A., Stahl, E., Viatte, S., Mcallister, K., Amos, C. I., Padyukov, L., Toes, R. E., Huizinga, T. W., Wijmenga, C., Trynka, G., Franke, L., Westra, H. J., Alfredsson, L., Hu, X., Sandor, C., De Bakker, P. I., Davila, S., Khor, C. C., Heng, K. K., Andrews, R., Edkins, S., Hunt, S. E., Langford, C., Symmons, D., Biologics In Rheumatoid Arthritis, G., Genomics Study, S., Wellcome Trust Case Control, C., Concannon, P., Onengut-Gumuscu, S., Rich, S. S., Deloukas, P., Gonzalez-Gay, M. A., Rodriguez-Rodriguez, L., Arlsetig, L., Martin, J., Rantapaa-Dahlqvist, S., Plenge, R. M., Raychaudhuri, S., Klareskog, L., Gregersen, P. K. & Worthington, J. 2012. High-density genetic mapping identifies new susceptibility loci for rheumatoid arthritis. *Nat Genet,* 44, 1336–40.

Feng, Y., Zhang, F., Lokey, L. K., Chastain, J. L., Lakkis, L., Eberhart, D. & Warren, S. T. 1995. Translational suppression by trinucleotide repeat expansion at FMR1. *Science,* 268, 731–4.

Filaci, G., Bacilieri, S., Fravega, M., Monetti, M., Contini, P., Ghio, M., Setti, M., Puppo, F. & Indiveri, F. 2001. Impairment of CD8+ T suppressor cell function in patients with active systemic lupus erythematosus. *J Immunol,* 166, 6452–7.

Finkelman, F. D., Shea-Donohue, T., Morris, S. C., Gildea, L., Strait, R., Madden, K. B., Schopf, L. & Urban, J. F.,JR. 2004. Interleukin-4- and interleukin-13-mediated host protection against intestinal nematode parasites. *Immunol Rev,* 201, 139–55.

Fiorentino, D. F., Bond, M. W. & Mosmann, T. R. 1989. Two types of mouse T helper cell. IV. Th2 clones secrete a factor that inhibits cytokine production by Th1 clones. *J Exp Med,* 170, 2081–95.

Friso, S., Carvajal, C. A., Fardella, C. E. & Olivieri, O. 2015. Epigenetics and arterial hypertension: the challenge of emerging evidence. *Transl Res,* 165, 154–65.

Fu, Y. H., Kuhl, D. P., Pizzuti, A., Pieretti, M., Sutcliffe, J. S., Richards, S., Verkerk, A. J., Holden, J. J., Fenwick, R. G.,JR., Warren, S. T. & Et Al. 1991. Variation of the CGG repeat at the fragile X site results in genetic instability: resolution of the Sherman paradox. *Cell,* 67, 1047–58.

Globisch, D., Munzel, M., Muller, M., Michalakis, S., Wagner, M., Koch, S., Bruckl, T., Biel, M. & Carell, T. 2010. Tissue distribution of 5-hydroxymethylcytosine and search for active demethylation intermediates. *PLoS One,* 5, e15367.

Gomez-Velazquez, M., Badia-Careaga, C., Lechuga-Vieco, A. V., Nieto-Arellano,R., Tena, J. J., Rollan, I., Alvarez, A., Torroja, C., Caceres, E.f., Roy, A. R., Galjart, N., Delgado-Olguin, P., Sanchez-Cabo, F., Enriquez, J. A., Gomez-Skarmeta, J. L. & Manzanares, M. 2017. CTCF counter-regulates cardiomyocyte development and maturation programs in the embryonic heart. *PLoS Genet,* 13, e1006985.

Grimaldi, V., Vietri, M. T., Schiano, C., Picascia, A., De Pascale, M. R., Fiorito, C., Casamassimi, A. & Napoli, C. 2015. Epigenetic reprogramming in atherosclerosis. *Curr Atheroscler Rep,* 17, 476.

Gupta, S., Kim, S. Y., Artis, S., Molfese, D. L., Schumacher, A., Sweatt, J. D., Paylor, R. E. & Lubin, F. D. 2010. Histone methylation regulates memory formation. *J Neurosci*, 30, 3589–99.

Guzowski, J. F. 2002. Insights into immediate-early gene function in hippocampal memory consolidation using antisense oligonucleotide and fluorescent imaging approaches. *Hippocampus*, 12, 86–104.

He, B., Xing, S., Chen, C., Gao, P., Teng, L., Shan, Q., Gullicksrud, J. A., Martin, M. D., Yu, S., Harty, J. T., Badovinac, V. P., Tan, K. & Xue, H. H. 2016. CD8(+) T cells utilize highly dynamic enhancer repertoires and regulatory circuitry in response to infections. *Immunity*, 45, 1341–54.

Hernandez-Rapp, J., Smith, P. Y., Filali, M., Goupil, C., Planel, E., Magill, S. T., Goodman, R. H. & Hebert, S. S. 2015. Memory formation and retention are affected in adult miR-132/212 knockout mice. *Behav Brain Res*, 287, 15–26.

Hu, N., Qiu, X., Luo, Y., Yuan, J., Li, Y., Lei, W., Zhang, G., Zhou, Y., Su, Y. & Lu, Q. 2008. Abnormal histone modification patterns in lupus CD4+ T cells. *J Rheumatol*, 35, 804–10.

Jordan, B. A., Fernholz, B. D., Khatri, L. & Ziff, E. B. 2007. Activity-dependent AIDA-1 nuclear signaling regulates nucleolar numbers and protein synthesis in neurons. *Nat Neurosci*, 10, 427–35.

Kandel, E. R. 2012. The molecular biology of memory: cAMP, PKA, CRE, CREB-1, CREB-2, and CPEB. *Mol Brain*, 5. doi: 10.1186/1756-6606-5-14.

Kandel, E. R., Dudai, Y. & Mayford, M. R. 2014. The molecular and systems biology of memory. *Cell*, 157, 163–86.

Kang, H. J., Kawasawa, Y. I., Cheng, F., Zhu, Y., Xu, X., Li, M., Sousa, A. M., Pletikos, M., Meyer, K. A., Sedmak, G., Guennel, T., Shin, Y., Johnson, M. B., Krsnik, Z., Mayer, S., Fertuzinhos, S., Umlauf, S., Lisgo, S. N., Vortmeyer, A., Weinberger, D. R., Mane, S., Hyde, T. M., Huttner, A., Reimers, M., Kleinman, J. E. & Sestan, N. 2011. Spatio-temporal transcriptome of the human brain. *Nature*, 478, 483–9.

Kaplan, M. J., Lu, Q., Wu, A., Attwood, J. & Richardson, B. 2004. Demethylation of promoter regulatory elements contributes to perforin overexpression in CD4+ lupus T cells. *J Immunol*, 172, 3652–61.

Kappler, J. W., Roehm, N. & Marrack, P. 1987. T cell tolerance by clonal elimination in the thymus. *Cell*, 49, 273–80.

Karouzakis, E., Trenkmann, M., Gay, R. E., Michel, B. A., Gay, S. & Neidhart, M. 2014. Epigenome analysis reveals TBX5 as a novel transcription factor involved in the activation of rheumatoid arthritis synovial fibroblasts. *J Immunol*, 193, 4945–51.

Kawai, T., Takahashi, K., Sato, S., Coban, C., Kumar, H., Kato, H., Ishii, K. J., Takeuchi, O. & Akira, S. 2005. IPS-1, an adaptor triggering RIG-I- and Mda5-mediated type I interferon induction. *Nat Immunol*, 6, 981–8.

Kenneson, A., Zhang, F., Hagedorn, C. H. & Warren, S. T. 2001. Reduced FMRP and increased FMR1 transcription is proportionally associated with CGG repeat number in intermediate-length and premutation carriers. *Hum Mol Genet*, 10, 1449–54.

Kerimoglu, C., Agis-Balboa, R. C., Kranz, A., Stilling, R., Bahari-Javan, S., Benito-Garagorri, E., Halder, R., Burkhardt, S., Stewart, A. F. & Fischer, A. 2013. Histone-methyltransferase MLL2 (KMT2B) is required for memory formation in mice. *J Neurosci*, 33, 3452–64.

Kim, S., Yu, N-K., Shim, K-W., Kim, J-I., Kim, H., Han, D. H., Choi, J. E., Lee, S-W., Choi, D. I., Kim, M. W., Lee, D-S., Lee, K., Galjart, N., Lee, Y-S., Lee, J-H. & Kaang, B-K. 2018. Remote memory and cortical synaptic plasticity require neuronal CCCTC binding factor (CTCF). *J Neurosci*, DOI: 10.1523/JNEUROSCI.2738-17.2018.

Kim, T. K., Hemberg, M., Gray, J. M., Costa, A. M., Bear, D. M., Wu, J., Harmin, D. A., Laptewicz, M., Barbara-Haley, K., Kuersten, S., Markenscoff-Papadimitriou, E., Kuhl, D., Bito, H., Worley, P. F., Kreiman, G. & Greenberg, M. E. 2010. Widespread transcription at neuronal activity-regulated enhancers. *Nature*, 465, 182–7.

Kisielow, P., Bluthmann, H., Staerz, U. D., Steinmetz, M. & Von Boehmer, H. 1988. Tolerance in T-cell-receptor transgenic mice involves deletion of nonmature CD4+8+ thymocytes. *Nature*, 333, 742–6.

Klein, M. E., Lioy, D. T., Ma, L., Impey, S., Mandel, G. & Goodman, R. H. 2007. Homeostatic regulation of MeCP2 expression by a CREB-induced microRNA. *Nat Neurosci*, 10, 1513–14.

Koshibu, K., Graff, J., Beullens, M., Heitz, F. D., Berchtold, D., Russig, H., Farinelli, M., Bollen, M. & Mansuy, I. M. 2009. Protein phosphatase 1 regulates the histone code for long-term memory. *J Neurosci*, 29, 13079–89.

Kuss, A. W. & Chen, W. 2008. MicroRNAs in brain function and disease. *Curr Neurol Neurosci Rep*, 8, 190–7.

Levenson, J. M., O'riordan, K. J., Brown, K. D., Trinh, M. A., Molfese, D. L. & Sweatt, J. D. 2004. Regulation of histone acetylation during memory formation in the hippocampus. *J Biol Chem*, 279, 40545–59.

Lickert, H., Takeuchi, J. K., Von Both, I., Walls, J. R., Mcauliffe, F., Adamson, S. L., Henkelman, R. M., Wrana, J. L., Rossant, J. & Bruneau, B. G. 2004. Baf60c is essential for function of BAF chromatin remodelling complexes in heart development. *Nature*, 432, 107–12.

Lister, R., Mukamel, E. A., Nery, J. R., Urich, M., Puddifoot, C. A., Johnson, N. D., Lucero, J., Huang, Y., Dwork, A. J., Schultz, M. D., Yu, M., Tonti-Filippini, J., Heyn, H., Hu, S., Wu, J. C., Rao, A., Esteller, M., He, C., Haghighi, F. G., Sejnowski, T. J., Behrens, M. M. & Ecker, J. R. 2013. Global epigenomic reconfiguration during mammalian brain development. *Science*, 341, 1237905.

Lister, R., Pelizzola, M., Dowen, R. H., Hawkins, R. D., Hon, G., Tonti-Filippini, J., Nery, J. R., Lee, L., Ye, Z., Ngo,

Q. M., Edsall, L., Antosiewicz-Bourget, J., Stewart, R., Ruotti, V., Millar, A. H., Thomson, J. A., Ren, B. & Ecker, J. R. 2009. Human DNA methylomes at base resolution show widespread epigenomic differences. *Nature*, 462, 315–22.

Llorente, L., Richaud-Patin, Y., Fior, R., Alcocer-Varela, J., Wijdenes, J., Fourrier, B. M., Galanaud, P. & Emilie, D. 1994. *In vivo* production of interleukin-10 by non-T cells in rheumatoid arthritis, Sjogren's syndrome, and systemic lupus erythematosus. A potential mechanism of B lymphocyte hyperactivity and autoimmunity. *Arthritis Rheum*, 37, 1647–55.

Llorente, L., Zou, W., Levy, Y., Richaud-Patin, Y., Wijdenes, J., Alcocer-Varela, J., Morel-Fourrier, B., Brouet, J. C., Alarcon-Segovia, D., Galanaud, P. & Emilie, D. 1995. Role of interleukin 10 in the B lymphocyte hyperactivity and autoantibody production of human systemic lupus erythematosus. *J Exp Med*, 181, 839–44.

Long, L., Yu, P., Liu, Y., Wang, S., Li, R., Shi, J., Zhang, X., Li, Y., Sun, X., Zhou, B., Cui, L. & Li, Z. 2013. Upregulated micro-RNA-155 expression in peripheral blood mononuclear cells and fibroblast-like synoviocytes in rheumatoid arthritis. *Clin Dev Immunol*, 2013, 296139.

Lubs, H. A. 1969. A marker X chromosome. *Am J Hum Genet*, 21, 231–44

Madabhushi, R., Gao, F., Pfenning, A. R., Pan, L., Yamakawa, S., Seo, J., Rueda, R., Phan, T. X., Yamakawa, H., Pao, P. C., Stott, R. T., Gjoneska, E., Nott, A., Cho, S., Kellis, M. & Tsai, L. H. 2015. Activity-induced DNA breaks govern the expression of neuronal early-response genes. *Cell*, 161, 1592–605.

Malik, A. N., Vierbuchen, T., Hemberg, M., Rubin, A. A., Ling, E., Couch, C. H., Stroud, H., Spiegel, I., Farh, K. K., Harmin, D. A. & Greenberg, M. E. 2014. Genome-wide identification and characterization of functional neuronal activity-dependent enhancers. *Nat Neurosci*, 17, 1330–9.

Malmevik, J., Petri, R., Knauff, P., Brattas, P. L., Akerblom, M. & Jakobsson, J. 2016. Distinct cognitive effects and underlying transcriptome changes upon inhibition of individual miRNAs in hippocampal neurons. *Sci Rep*, 6, 19879.

Martin, J. P. & Bell, J. 1943. A pedigree of mental defect showing sex-linkage. *J Neurol Psychiatry*, 6, 154–7.

Martinez, G., Vidal, R. L., Mardones, P., Serrano, F. G., Ardiles, A. O., Wirth, C., Valdes, P., Thielen, P., Schneider, B. L., Kerr, B., Valdes, J. L., Palacios, A. G., Inestrosa, N. C., Glimcher, L. H. & Hetz, C. 2016. Regulation of memory formation by the transcription factor XBP1. *Cell Rep*, 14, 1382–94.

Mercer, T. R., Dinger, M. E., Sunkin, S. M., Mehler, M. F. & Mattick, J. S. 2008. Specific expression of long noncoding RNAs in the mouse brain. *Proc Natl Acad Sci U S A*, 105, 716–21.

Miescher, P. & Strassle, R. 1957. New serological methods for the detection of the L. E. factor. *Vox Sang*, 2, 283–7.

Miller, C. A., Gavin, C. F., White, J. A., Parrish, R. R., Honasoge, A., Yancey, C. R., Rivera, I. M., Rubio, M. D., Rumbaugh, G. & Sweatt, J. D. 2010. Cortical DNA methylation maintains remote memory. *Nat Neurosci*, 13, 664–6.

Miller, C. A. & Sweatt, J. D. 2007. Covalent modification of DNA regulates memory formation. *Neuron*, 53, 857–69.

Miller, F. W., Alfredsson, L., Costenbader, K. H., Kamen, D. L., Nelson, L. M., Norris, J. M. & De Roos, A. J. 2012. Epidemiology of environmental exposures and human autoimmune diseases: findings from a National Institute of Environmental Health Sciences Expert Panel Workshop. *J Autoimmun*, 39, 259–71.

Montgomery, R. L., Davis, C. A., Potthoff, M. J., Haberland, M., Fielitz, J., Qi, X., Hill, J. A., Richardson, J. A. & Olson, E. N. 2007. Histone deacetylases 1 and 2 redundantly regulate cardiac morphogenesis, growth, and contractility. *Genes Dev*, 21, 1790–802.

Murakami, M., Nakagawa, M., Olson, E. N. & Nakagawa, O. 2005. A WW domain protein TAZ is a critical coactivator for TBX5, a transcription factor implicated in Holt-Oram syndrome. *Proc Natl Acad Sci U S A*, 102, 18034–9.

Nakano, K., Whitaker, J. W., Boyle, D. L., Wang, W. & Firestein, G. S. 2013. DNA methylome signature in rheumatoid arthritis. *Ann Rheum Dis*, 72, 110–17.

Ng, S. B., Bigham, A. W., Buckingham, K. J., Hannibal, M. C., Mcmillin, M. J., Gildersleeve, H. I., Beck, A. E., Tabor, H. K., Cooper, G. M., Mefford, H. C., Lee, C., Turner, E. H., Smith, J. D., Rieder, M. J., Yoshiura, K., Matsumoto, N., Ohta, T., Nikawa, N., Nickerson, D. A., Bamshad, M. J. & Shendure, J. 2010. Exome sequencing identifies *MLL2* mutations as a cause of Kabuki syndrome. *Nat Genet*, 42, 790–3.

Niederer, F., Trenkmann, M., Ospelt, C., Karouzakis, E., Neidhart, M., Stanczyk, J., Kolling, C., Gay, R. E., Detmar, M., Gay, S., Jungel, A. & Kyburz, D. 2012. Down-regulation of microRNA-34a* in rheumatoid arthritis synovial fibroblasts promotes apoptosis resistance. *Arthritis Rheum*, 64, 1771–9.

Nile, C. J., Read, R. C., Akil, M., Duff, G. W. & Wilson, A. G. 2008. Methylation status of a single CpG site in the IL6 promoter is related to IL6 messenger RNA levels and rheumatoid arthritis. *Arthritis Rheum*, 58, 2686–93.

Nimura, K., Ura, K., Shiratori, H., Ikawa, M., Okabe, M., Schwartz, R. J. & Kaneda, Y. 2009. A histone H3 lysine 36 trimethyltransferase links Nkx2-5 to Wolf-Hirschhorn syndrome. *Nature*, 460, 287–91.

Olave, I., Wang, W., Xue, Y., Kuo, A. & Crabtree, G. R. 2002. Identification of a polymorphic, neuron-specific chromatin remodeling complex. *Genes Dev*, 16, 2509–17.

Ovchinnikova, E. S., Schmitter, D., Vegter, E. L., Ter Maaten, J. M., Valente, M. A., Liu, L. C., Van Der Harst, P., Pinto, Y. M., De Boer, R. A., Meyer, S., Teerlink, J. R., O'connor,C. M., Metra, M., Davison, B. A., Bloomfield, D. M., Cotter, G., Cleland, J. G., Mebazaa, A., Laribi, S., Givertz, M. M.,

Ponikowski, P., Van Der Meer, P., Van Veldhuisen, D. J., Voors, A. A. & Berezikov, E. 2016. Signature of circulating microRNAs in patients with acute heart failure. *Eur J Heart Fail*, 18, 414–23.

Ponjavic, J., Oliver, P. L., Lunter, G. & Ponting, C. P. 2009. Genomic and transcriptional co-localization of protein-coding and long non-coding RNA pairs in the developing brain. *PLoS Genet*, 5, e1000617.

Qu, N., Xu, M., Mizoguchi, I., Furusawa, J., Kaneko, K., Watanabe, K., Mizuguchi, J., Itoh, M., Kawakami, Y. & Yoshimoto, T. 2013. Pivotal roles of T-helper 17-related cytokines, IL-17, IL-22, and IL-23, in inflammatory diseases. *Clin Dev Immunol*, 2013, 968549.

Quddus, J., Johnson, K. J., Gavalchin, J., Amento, E. P., Chrisp, C. E., Yung, R. L. & Richardson, B. C. 1993. Treating activated CD4+ T cells with either of two distinct DNA methyltransferase inhibitors, 5-azacytidine or procainamide, is sufficient to cause a lupus-like disease in syngeneic mice. *J Clin Invest*, 92, 38–53.

Rekvig, O. P., Thiyagarajan, D., Pedersen, H. L., Horvei, K. D. & Seredkina, N. 2016. Future perspectives on pathogenesis of lupus nephritis: facts, problems, and potential causal therapy modalities. *Am J Pathol*, 186, 2772–82.

Richardson, B., Scheinbart, L., Strahler, J., Gross, L., Hanash, S. & Johnson, M. 1990. Evidence for impaired T cell DNA methylation in systemic lupus erythematosus and rheumatoid arthritis. *Arthritis Rheum*, 33, 1665–73.

Robbins, W. C., Holman, H. R., Deicher, H. & Kunkel, H. G. 1957. Complement fixation with cell nuclei and DNA in lupus erythematosus. *Proc Soc Exp Biol Med*, 96, 575–9.

Ronnblom, L. 2016. The importance of the type I interferon system in autoimmunity. *Clin Exp Rheumatol*, 34, 21–4.

Rovozzo, R., Korza, G., Baker, M. W., Li, M., Bhattacharyya, A., Barbarese, E. & Carson, J. H. 2016. CGG Repeats in the 5′ UTR of FMR1 RNA regulate translation of other RNAs localized in the same RNA granules. *PLoS One*, 11, e0168204.

Schellekens, G. A., De Jong, B. A., Van Den Hoogen, F. H., Van De Putte, L. B. & Van Venrooij, W. J. 1998. Citrulline is an essential constituent of antigenic determinants recognized by rheumatoid arthritis-specific autoantibodies. *J Clin Invest*, 101, 273–81.

Scott, H. L., Tamagnini, F., Narduzzo, K. E., Howarth, J. L., Lee, Y. B., Wong, L. F., Brown, M. W., Warburton, E. C., Bashir, Z. I. & Uney, J. B. 2012. MicroRNA-132 regulates recognition memory and synaptic plasticity in the perirhinal cortex. *Eur J Neurosci*, 36, 2941–8.

Scott-Browne, J. P., Lopez-Moyado, I. F., Trifari, S., Wong, V., Chavez, L., Rao, A. & Pereira, R. M. 2016. Dynamic changes in chromatin accessibility occur in CD8(+) T cells responding to viral infection. *Immunity*, 45, 1327–40.

Selmi, C., Leung, P. S., Sherr, D. H., Diaz, M., Nyland, J. F., Monestier, M., Rose, N. R. & Gershwin, M. E. 2012. Mechanisms of environmental influence on human autoimmunity: a National Institute of Environmental Health Sciences expert panel workshop. *J Autoimmun*, 39, 272–84.

Smigielska-Czepiel, K., Van Den Berg, A., Jellema, P., Van Der Lei, R. J., Bijzet, J., Kluiver, J., Boots, A. M., Brouwer, E. & Kroesen, B. J. 2014. Comprehensive analysis of miRNA expression in T-cell subsets of rheumatoid arthritis patients reveals defined signatures of naive and memory Tregs. *Genes Immun*, 15, 115–25.

Soibam, B., Benham, A., Kim, J., Weng, K. C., Yang, L., Xu, X., Robertson, M., Azares, A., Cooney, A. J., Schwartz, R. J. & Liu, Y. 2015. Genome-wide identification of MESP1 targets demonstrates primary regulation over mesendoderm gene activity. *Stem Cells*, 33, 3254–65.

Stanczyk, J., Ospelt, C., Karouzakis, E., Filer, A., Raza, K., Kolling, C., Gay, R., Buckley, C. D., Tak, P. P., Gay, S. & Kyburz, D. 2011. Altered expression of microRNA-203 in rheumatoid arthritis synovial fibroblasts and its role in fibroblast activation. *Arthritis Rheum*, 63, 373–81.

Stefanelli, G., Azam, A. B., Walters, B. J., Brimble, M. A., Gettens, C. P., Bouchard-Cannon, P., Cheng, H. M., Davidoff, A. M., Narkaj, K., Day, J. J., Kennedy, A. J. & Zovkic, I. B. 2018. Learning and age-related changes in genome-wide H2A.Z binding in the mouse hippocampus. *Cell Rep*, 22, 1124–31.

Suberbielle, E., Sanchez, P. E., Kravitz, A. V., Wang, X., Ho, K., Eilertson, K., Devidze, N., Kreitzer, A. C. & Mucke, L. 2013. Physiologic brain activity causes DNA double-strand breaks in neurons, with exacerbation by amyloid-beta. *Nat Neurosci*, 16, 613–21.

Surface, L. E., Thornton, S. R. & Boyer, L. A. 2010. Polycomb group proteins set the stage for early lineage commitment. *Cell Stem Cell*, 7, 288–98.

Sutcliffe, J. S., Nelson, D. L., Zhang, F., Pieretti, M., Caskey, C. T., Saxe, D. & Warren, S. T. 1992. DNA methylation represses FMR-1 transcription in fragile X syndrome. *Hum Mol Genet*, 1, 397–400.

Swank, M. W. & Sweatt, J. D. 2001. Increased histone acetyltransferase and lysine acetyltransferase activity and biphasic activation of the ERK/RSK cascade in insular cortex during novel taste learning. *J Neurosci*, 21, 3383–91.

Takami, N., Osawa, K., Miura, Y., Komai, K., Taniguchi, M., Shiraishi, M., Sato, K., Iguchi, T., Shiozawa, K., Hashiramoto, A. & Shiozawa, S. 2006. Hypermethylated promoter region of *DR3*, the death receptor 3 gene, in rheumatoid arthritis synovial cells. *Arthritis Rheum*, 54, 779–87.

Takatsu, K., Tominaga, A., Harada, N., Mita, S., Matsumoto, M., Takahashi, T., Kikuchi, Y. & Yamaguchi, N. 1988. T cell-replacing factor (TRF)/interleukin 5 (IL-5): molecular and functional properties. *Immunol Rev*, 102, 107–35.

Takeuchi, J. K. & Bruneau, B. G. 2009. Directed transdifferentiation of mouse mesoderm to heart tissue by defined factors. *Nature*, 459, 708–11.

Talaat, R. M., Mohamed, S. F., Bassyouni, I. H. & Raouf, A. A. 2015. Th1/Th2/Th17/Treg cytokine imbalance in systemic lupus erythematosus (SLE) patients: correlation with disease activity. *Cytokine*, 72, 146–53.

Trenkmann, M., Brock, M., Gay, R. E., Kolling, C., Speich, R., Michel, B. A., Gay, S. & Huber, L. C. 2011. Expression and function of EZH2 in synovial fibroblasts: epigenetic repression of the Wnt inhibitor SFRP1 in rheumatoid arthritis. *Ann Rheum Dis*, 70, 1482–8.

Tuoc, T., Dere, E., Radyushkin, K., Pham, L., Nguyen, H., Tonchev, A. B., Sun, G., Ronnenberg, A., Shi, Y., Staiger, J. F., Ehrenreich, H. & Stoykova, A. 2017. Ablation of BAF170 in developing and postnatal dentate gyrus affects neural stem cell proliferation, differentiation, and learning. *Mol Neurobiol*, 54, 4618–35.

Unterholzner, L., Keating, S. E., Baran, M., Horan, K. A., Jensen, S. B., Sharma, S., Sirois, C. M., Jin, T., Latz, E., Xiao, T. S., Fitzgerald, K. A., Paludan, S. R. & Bowie, A. G. 2010. IFI16 is an innate immune sensor for intracellular DNA. *Nat Immunol*, 11, 997–1004.

Vallin, H., Blomberg, S., Alm, G. V., Cederblad, B. & Ronnblom, L. 1999. Patients with systemic lupus erythematosus (SLE) have a circulating inducer of interferon-alpha (IFN-alpha) production acting on leucocytes resembling immature dendritic cells. *Clin Exp Immunol*, 115, 196–202.

Van Der Geest, K. S., Smigielska-Czepiel, K., Park, J. A., Abdulahad, W. H., Kim, H. W., Kroesen, B. J., Van Den Berg, A., Boots, A. M., Lee, E. B. & Brouwer, E. 2015. SF Treg cells transcribing high levels of Bcl-2 and micro-RNA-21 demonstrate limited apoptosis in RA. *Rheumatology (Oxford)*, 54, 950–8.

Van Leeuwen, M. A., Westra, J., Limburg, P. C., Van Riel, P. L. & Van Rijswijk, M. H. 1995. Interleukin-6 in relation to other proinflammatory cytokines, chemotactic activity and neutrophil activation in rheumatoid synovial fluid. *Ann Rheum Dis*, 54, 33–8.

Van Rooij, E., Sutherland, L. B., Liu, N., Williams, A. H., Mcanally, J., Gerard, R. D., Richardson, J. A. & Olson, E. N. 2006. A signature pattern of stress-responsive micro-RNAs that can evoke cardiac hypertrophy and heart failure. *Proc Natl Acad Sci U S A*, 103, 18255–60.

Varley, K. E., Gertz, J., Bowling, K. M., Parker, S. L., Reddy, T. E., Pauli-Behn, F., Cross, M. K., Williams, B. A., Stamatoyannopoulos, J. A., Crawford, G. E., Absher, D. M., Wold, B. J. & Myers, R. M. 2013. Dynamic DNA methylation across diverse human cell lines and tissues. *Genome Res*, 23, 555–67.

Verkerk, A. J., Pieretti, M., Sutcliffe, J. S., Fu, Y. H., Kuhl, D. P., Pizzuti, A., Reiner, O., Richards, S., Victoria, M. F., Zhang, F. P. & Et Al. 1991. Identification of a gene (FMR-1) containing a CGG repeat coincident with a breakpoint cluster region exhibiting length variation in fragile X syndrome. *Cell*, 65, 905–14.

Vogel-Ciernia, A., Matheos, D. P., Barrett, R. M., Kramar, E. A., Azzawi, S., Chen, Y., Magnan, C. N., Zeller, M., Sylvain, A., Haettig, J., Jia, Y., Tran, A., Dang, R., Post, R. J., Chabrier, M., Babayan, A. H., Wu, J. I., Crabtree, G. R., Baldi, P., Baram, T. Z., Lynch, G. & Wood, M. A. 2013. The neuron-specific chromatin regulatory subunit BAF53b is necessary for synaptic plasticity and memory. *Nat Neurosci*, 16, 552–61.

Wada, T. T., Araki, Y., Sato, K., Aizaki, Y., Yokota, K., Kim, Y. T., Oda, H., Kurokawa, R. & Mimura, T. 2014. Aberrant histone acetylation contributes to elevated interleukin-6 production in rheumatoid arthritis synovial fibroblasts. *Biochem Biophys Res Commun*, 444, 682–6.

Wamstad, J. A., Alexander, J. M., Truty, R. M., Shrikumar, A., Li, F., Eilertson, K. E., Ding, H., Wylie, J. N., Pico, A. R., Capra, J. A., Erwin, G., Kattman, S. J., Keller, G. M., Srivastava, D., Levine, S. S., Pollard, K. S., Holloway, A. K., Boyer, L. A. & Bruneau, B. G. 2012. Dynamic and coordinated epigenetic regulation of developmental transitions in the cardiac lineage. *Cell*, 151, 206–20.

Wang, J. C. 2002. Cellular roles of DNA topoisomerases: a molecular perspective. *Nat Rev Mol Cell Biol*, 3, 430–40.

Wang, T., Sun, X., Zhao, J., Zhang, J., Zhu, H., Li, C., Gao, N., Jia, Y., Xu, D., Huang, F. P., Li, N., Lu, L. & Li, Z. G. 2015. Regulatory T cells in rheumatoid arthritis showed increased plasticity toward Th17 but retained suppressive function in peripheral blood. *Ann Rheum Dis*, 74, 1293–301.

Weeber, E. J., Beffert, U., Jones, C., Christian, J. M., Forster, E., Sweatt, J. D. & Herz, J. 2002. Reelin and ApoE receptors cooperate to enhance hippocampal synaptic plasticity and learning. *J Biol Chem*, 277, 39944–52.

Weiler, I. J., Irwin, S. A., Klintsova, A. Y., Spencer, C. M., Brazelton, A. D., Miyashiro, K., Comery, T. A., Patel, B., Eberwine, J. & Greenough, W. T. 1997. Fragile X mental retardation protein is translated near synapses in response to neurotransmitter activation. *Proc Natl Acad Sci U S A*, 94, 5395–400.

Won, H., De La Torre-Ubieta, L., Stein, J. L., Parikshak, N. N., Huang, J., Opland, C. K., Gandal, M. J., Sutton, G. J., Hormozdiari, F., Lu, D., Lee, C., Eskin, E., Voineagu, I., Ernst, J. & Geschwind, D. H. 2016. Chromosome conformation elucidates regulatory relationships in developing human brain. *Nature*, 538, 523–7.

Xie, W., Barr, C. L., Kim, A., Yue, F., Lee, A. Y., Eubanks, J., Dempster, E. L. & Ren, B. 2012. Base-resolution analyses of sequence and parent-of-origin dependent DNA methylation in the mouse genome. *Cell*, 148, 816–31.

Yang, Y., Shu, X., Liu, D., Shang, Y., Wu, Y., Pei, L., Xu, X., Tian, Q., Zhang, J., Qian, K., Wang, Y. X., Petralia, R. S., Tu, W., Zhu, L. Q., Wang, J. Z. & Lu, Y. 2012. EPAC null mutation impairs learning and social interactions via aberrant regulation of miR-124 and Zif268 translation. *Neuron*, 73, 774–88.

Yung, R. L., Quddus, J., Chrisp, C. E., Johnson, K. J. & Richardson, B. C. 1995. Mechanism of drug-induced lupus. I. Cloned Th2 cells modified with DNA methylation inhibitors *in vitro* cause autoimmunity *in vivo. J Immunol*, 154, 3025–35.

Zaidi, S., Choi, M., Wakimoto, H., Ma, L., Jiang, J., Overton, J. D., Romano-Adesman, A., Bjornson, R. D., Breitbart, R. E., Brown, K. K., Carriero, N. J., Cheung, Y. H., Deanfield, J., Depalma, S., Fakhro, K. A., Glessner, J., Hakonarson, H., Italia, M. J., Kaltman, J. R., Kaski, J., Kim, R., Kline, J. K., Lee, T., Leipzig, J., Lopez, A., Mane, S. M., Mitchell, L. E., Newburger, J. W., Parfenov, M., Pe'er, I., Porter, G., Roberts, A. E., Sachidanandam, R., Sanders, S. J., Seiden, H. S., State, M. W., Subramanian, S., Tikhonova, I. R., Wang, W., Warburton, D., White, P. S., Williams, I. A., Zhao, H., Seidman, J. G., Brueckner, M., Chung, W. K., Gelb, B. D., Goldmuntz, E., Seidman, C. E. & Lifton, R. P. 2013. *De novo*

mutations in histone-modifying genes in congenital heart disease. *Nature*, 498, 220–3.

Zalfa, F., Giorgi, M., Primerano, B., Moro, A., Di Penta, A., Reis, S., Oostra, B. & Bagni, C. 2003. The fragile X syndrome protein FMRP associates with BC1 RNA and regulates the translation of specific mRNAs at synapses. *Cell*, 112, 317–27.

Zhao, M., Sun, Y., Gao, F., Wu, X., Tang, J., Yin, H., Luo, Y., Richardson, B. & Lu, Q. 2010. Epigenetics and SLE: RFX1 downregulation causes CD11a and CD70 overexpression by altering epigenetic modifications in lupus CD4+ T cells. *J Autoimmun*, 35, 58–69.

Zovkic, I. B., Paulukaitis, B. S., Day, J. J., Etikala, D. M. & Sweat, D. 2014. Histone H2A.Z subunit exchange controls consolidation of recent and remote memory. *Nature*, 515, 582–6.

Index

Notes

Figures, tables, and boxes are indicated by an italic *f*, *t*, or *b* following the page number.